BROOKS/COLE
CENGAGE Learning™

MATH APPS
Ronald J. Harshbarger
James J. Reynolds

VP/Editor-in-Chief: Michelle Julet

Publisher: Richard Stratton

Senior Development Editor: Laura Wheel

Development Editor: Sue Warne

Editorial Assistant: Haeree Chang

Media Editor: Heleny Wong

Content Project Manager: Alison Eigel Zade

Product Development Manager, 4LTR Press:
Steven E. Joos

Senior Art Director: Jill Ort

Rights Acquisition Specialist: Amanda Groszko

Senior Print Buyer: Diane Gibbons

Marketing Director: Mandee Eckersley

Marketing Manager: Ashley Pickering

Marketing Communications Manager: Mary
Anne Payumo

Production Service: Lachina Publishing Services

Cover Designer: Hannah Wellman

Cover Image: ©iStockphoto.com/Lefthome

Editorial Development and Composition:
Lachina Publishing Services

Library of Congress Control Number: 2010930878

ISBN-13: 978-0-8400-5822-5

ISBN-10: 0-8400-5822-5

Brooks/Cole
20 Channel Center Street
Boston, MA 02210
USA

Cengage Learning is a leading provider of customized learning solutions with office locations around the globe, including Singapore, the United Kingdom, Australia, Mexico, Brazil and Japan. Locate your local office at **international.cengage.com/region**

Cengage Learning products are represented in Canada by Nelson Education, Ltd.

For your course and learning solutions, visit **www.cengage.com.**

Purchase any of our products at your local college store or at our preferred online store **www.cengagebrain.com.**

Printed in the United States of America
1 2 3 4 5 6 7 14 13 12 11 10

Brief Contents

math apps

©iStockphoto.com/Martin McElligot

STUDY YOUR WAY

At no additional cost, you have access to online learning resources that include **tutorial videos, printable flashcards, quizzes,** and more!

Watch videos that offer step-by-step conceptual explanations and guidance for each chapter in the text.

Along with the printable flashcards and other online resources, you will have a multitude of ways to check your comprehension of key mathematical concepts.

You can find these resources at **login.cengagebrain.com.**

Contents

Shutterstock

Shutterstock

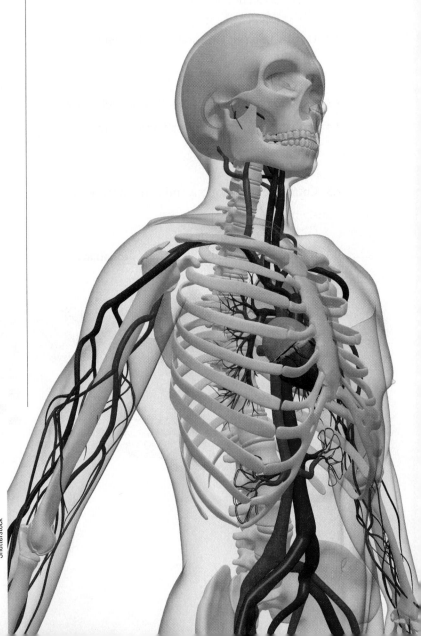

Shutterstock

Shutterstock

Linear Equations
and Functions

A wide variety of problems from business, the social sciences, and the life sciences may be solved using equations. Managers and economists use equations and their graphs to study costs, sales, national consumption, or supply and demand. Social scientists may plot demographic data or try to develop equations that predict population growth, voting behavior, or learning and retention rates. Life scientists use equations to model the flow of blood or the conduction of nerve impulses and to test theories or develop new models by using experimental data.

This chapter introduces two important applications that will be expanded and used throughout the text as increased mathematical skills permit: supply and demand as functions of price (market analysis); and total cost, total revenue, and profit as functions of the quantity produced and sold (theory of the firm).

objectives

1.1 Solutions of Linear Equations and Inequalities in One Variable

1.2 Functions

1.3 Linear Functions

1.4 Solutions of Systems of Linear Equations

1.5 Applications of Functions in Business and Economics

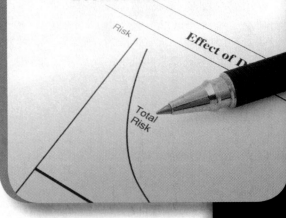

1.1 Solutions of Linear Equations and Inequalities in One Variable

Equations

An **equation** is a statement that two quantities or algebraic expressions are equal. Finding the value(s) of the variable(s) that make the equation true—that is, finding the **solutions**—is called **solving the equation**. The set of solutions of an equation is called a **solution set** of the equation. The variable in an equation is sometimes called the **unknown**.

Equations that are true for all values of the variables are called **identities.** The equation $2(x - 1) = 2x - 2$ is an example of an identity. Equations that are true only for certain values of the variables are called **conditional equations** or simply **equations.**

We can often solve a complicated linear equation by finding an equivalent equation whose solution is easily found. We use the properties of equality in the table on the next page to reduce an equation to a simple equivalent equation.

Solving Linear Equations

If an equation contains one variable and if the variable occurs to the first degree, the equation is called a **linear equation in one variable.** We can solve linear equations in one variable by using the procedure shown on the next page.

Properties of Equality

Substitution

The equation formed by substituting one expression for an equal expression is equivalent to the original equation.

Addition

The equation formed by adding the same quantity to both sides of an equation is equivalent to the original equation.

Multiplication

The equation formed by multiplying both sides of an equation by the same nonzero quantity is equivalent to the original equation.

©iStockphoto.com

Procedure

To solve a linear equation in one variable:

1. If the equation contains fractions, multiply both sides by the least common denominator (LCD) of the fractions.

2. Remove any parentheses in the equation.

3. Perform any additions or subtractions to get all terms containing the variable on one side and all other terms on the other side.

4. Divide both sides of the equation by the coefficient of the variable.

5. Check the solution by substitution in the original equation.

Solving a Linear Equation

Image Source/Getty Images

EXAMPLE 1.1 Linear Equations

➡ Solve for x: $\dfrac{3x + 1}{2} = \dfrac{x}{3} - 3$

SOLUTION

$$\frac{3x + 1}{2} = \frac{x}{3} - 3$$

$$6\left(\frac{3x + 1}{2}\right) = 6\left(\frac{x}{3} - 3\right) \qquad \text{Multiply both sides by the LCD, 6.}$$

$$3(3x + 1) = 6\left(\frac{x}{3} - 3\right) \qquad \text{Simplify the fraction on the left side.}$$

$$9x + 3 = 2x - 18 \qquad \text{Distribute to remove parentheses.}$$

$$7x = -21 \qquad \text{Add } (-2x) + (-3) \text{ to both sides.}$$

$$x = -3 \qquad \text{Divide both sides by 7.}$$

Check: $\dfrac{3(-3) + 1}{2} \overset{?}{=} \dfrac{-3}{3} - 3$ gives $-4 = -4$ ✔

EXAMPLE 1.2 Future Value of an Investment

➡ The future value of a simple interest investment is given by $S = P + Prt$, where P is the principal invested, r is the annual interest rate (as a decimal), and t is the time in years. At what simple interest rate r must $P = 1500$ dollars be invested so that the future value is $2940 after 8 years?

SOLUTION

Entering the values $S = 2940$, $P = 1500$, and $t = 8$ into $S = P + Prt$ gives

$$2940 = 1500 + 1500(r)(8)$$

©iStockphoto.com/Anna Subotina

Hence we solve the equation

$$2940 = 1500 + 12{,}000r$$
$$1440 = 12{,}000r \quad \text{Subtract 1500 from both sides.}$$
$$\frac{1440}{12{,}000} = r \quad \text{Divide both sides by 12,000.}$$
$$0.12 = r$$

Thus the interest rate is 0.12, or 12%.

Checking, we see that $2940 = 1500 + 1500(0.12)(8)$. ✔

Fractional Equations

A **fractional equation** is an equation that contains a variable in a denominator. It is solved by first multiplying both sides of the equation by the least common denominator (LCD) of the fractions in the equation. Note that the solution to any fractional equation *must* be checked in the original equation, because multiplying both sides of a fractional equation by a variable expression may result in an equation that is not equivalent to the original equation. If a solution to the fraction-free linear equation makes a denominator of the original equation equal to zero, that value cannot be a solution to the original equation. Some fractional equations have no solutions.

EXAMPLE 1.3 Solving Fractional Equations

➤ Solve for x: $\dfrac{3x}{2x+10} = 1 + \dfrac{1}{x+5}$

SOLUTION

First multiply each term on both sides by the LCD, $2x + 10$. Then simplify and solve.

$$(2x+10)\left(\frac{3x}{2x+10}\right) = (2x+10)(1) + (2x+10)\left(\frac{1}{x+5}\right)$$
$$3x = (2x+10) + 2 \quad \text{gives} \quad x = 12$$

Check:

$$\frac{3(12)}{2(12)+10} \overset{?}{=} 1 + \frac{1}{12+5} \quad \text{gives} \quad \frac{36}{34} = \frac{18}{17} \quad ✔$$

To see that it is important to check solutions to fractional equations, consider

$$\frac{2x-1}{x-3} = 4 + \frac{5}{x-3}$$

Performing the operations to solve $\dfrac{2x-1}{x-3} = 4 + \dfrac{5}{x-3}$ gives $x = 3$. Notice that $x = 3$ does not check because

the denominators equal 0 when $x = 3$. Hence the equation has no solution.

Linear Equations with Two Variables

The steps used to solve linear equations in one variable can also be used to solve linear equations in more than one variable for one of the variables in terms of the other. Solving an equation such as the one in the following example is important when using a graphing utility.

EXAMPLE 1.4 Profit

➤ Suppose that the relationship between a firm's profit P and the number x of items sold can be described by the equation

$$5x - 4P = 1200$$

(a) How many units must be produced and sold for the firm to make a profit of $150?
(b) Solve this equation for P in terms of x.
(c) Find the profit when 240 units are sold.

SOLUTION

(a) $5x - 4(150) = 1200$
$$5x - 600 = 1200$$
$$5x = 1800$$
$$x = 360 \text{ units}$$
Check: $5(360) - 4(150) = 1800 - 600 = 1200$ ✔

(b) $5x - 4P = 1200$
$$5x - 1200 = 4P$$
$$P = \frac{5x}{4} - \frac{1200}{4} = \frac{5}{4}x - 300$$

(c) $P = \dfrac{5}{4}x - 300$

$$P = \frac{5}{4}(240) - 300 = 0$$

Because $P = 0$ when $x = 240$, we know that profit is $0 when 240 units are sold, and we say that the firm **breaks even** when 240 units are sold.

Applied Problems

With an applied problem, it is frequently necessary to convert the problem from its stated form into one or more equations from which the problem's solution can be found. The guidelines on the next page may be useful in solving stated problems.

Guidelines for Solving Stated Problems

1. Begin by reading the problem carefully to determine what you are to find. Use variables to represent the quantities to be found.

2. Reread the problem and use your variables to translate given information into algebraic expressions. Often, drawing a figure is helpful.

3. Use the algebraic expressions and the problem statement to formulate an equation (or equations).

4. Solve the equation(s).

5. Check the solution in the problem, not just in your equation or equations. The answer should satisfy the conditions.

EXAMPLE 1.5 Investment Mix

Jill Bell has $90,000 to invest. She has chosen one relatively safe investment fund that has an annual yield of 10% and another, riskier one that has a 15% annual yield. How much should she invest in each fund if she would like to earn $10,000 in one year from her investments?

SOLUTION

We want to find the amount of each investment, so we begin as follows:

Let x = the amount invested at 10%,   then

$90,000 - x$ = the amount invested at 15% (because the two investments total $90,000)

If P is the amount of an investment and r is the annual rate of yield (expressed as a decimal), then the annual earnings $I = Pr$. Using this relationship, we can summarize the information about these two investments in a table.

	P	r	I
10% invest.	x	0.10	$0.10x$
15% invest.	$90,000 - x$	0.15	$0.15(90,000 - x)$
Total invest.	90,000		10,000

The column under I shows that the sum of the earnings is

$$0.10x + 0.15(90,000 - x) = 10,000$$

We solve this as follows.

$$0.10x + 13,500 - 0.15x = 10,000$$
$$-0.05x = -3500 \quad \text{or} \quad x = 70,000$$

Thus the amount invested at 10% is $70,000, and the amount invested at 15% is $90,000 - 70,000 = 20,000$.

Check: To check, we return to the problem and note that 10% of $70,000 plus 15% of $20,000 gives a yield of $7000 + $3000 = $10,000. ✔

Linear Inequalities

An **inequality** is a statement that one quantity is greater than (or less than) another quantity. Certain values of the variable will satisfy the inequality. These values form the solution set of the inequality. *Solving* an inequality means finding its solution set, and two inequalities are *equivalent* if they have the same solution set. As with equations, we find the solutions to inequalities by finding equivalent inequalities from which the solutions can be easily seen. We use the properties on the next page to reduce an inequality to a simple equivalent inequality. For some inequalities, it requires several operations to find their solution sets. In this case, the order in which the operations are performed is the same as that used in solving linear equations.

EXAMPLE 1.6 Solution of an Inequality

Solve the inequality $2(x - 4) < \dfrac{x - 3}{3}$.

SOLUTION

$$2(x - 4) < \frac{x - 3}{3}$$

$6(x - 4) < x - 3$ Clear fractions.

$6x - 24 < x - 3$ Remove parentheses.

$5x < 21$ Perform additions and subtractions.

$x < \dfrac{21}{5}$ Multiply by $\frac{1}{5}$.

We may also solve inequalities of the form $a \le b$. This means "a is less than b or a equals b."

The solution of the inequality $3x - 2 \le 7$ is $3x \le 9$, or $x \le 3$. The graph of the solution set includes the point $x = 3$ and all points $x < 3$ (see Figure 1.1).

Figure 1.1

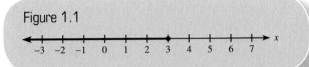

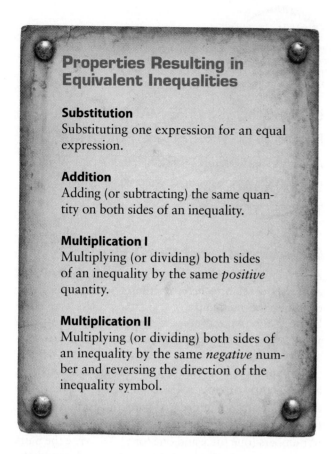

Properties Resulting in Equivalent Inequalities

Substitution
Substituting one expression for an equal expression.

Addition
Adding (or subtracting) the same quantity on both sides of an inequality.

Multiplication I
Multiplying (or dividing) both sides of an inequality by the same *positive* quantity.

Multiplication II
Multiplying (or dividing) both sides of an inequality by the same *negative* number and reversing the direction of the inequality symbol.

EXAMPLE 1.7 Normal Height for a Given Age

For boys between 4 and 16 years of age, height and age are linearly related. That relation can be expressed as

$$H = 2.31A + 31.26$$

where H is height in inches and A is age in years. To account for natural variation among individuals, normal is considered to be any measure falling within $\pm 5\%$ of the height obtained from the equation.* Write as an inequality the range of normal height for a boy who is 9 years old.

SOLUTION
The boy's height from the formula is $H = 2.31(9) + 31.26 = 52.05$ inches. For a 9-year-old boy to be considered of normal height, H would have to be within $\pm 5\%$ of 52.05 inches. That is, the boy's height H is considered normal if $H \geq 52.05 - (0.05)(52.05)$ and $H \leq 52.05 + (0.05)(52.05)$. We can express this range of normal height by the compound inequality

$$52.05 - (0.05)(52.05) \leq H \leq 52.05 + (0.05)(52.05)$$

or $\qquad 49.45 \leq H \leq 54.65$

*Adapted from data from the National Center for Health Statistics.

1.1 Exercises

In Problems 1–3, solve each equation.

1. $x + 8 = 8(x + 1)$

2. $2(x - 7) = 5(x + 3) - x$

3. $\dfrac{5x}{2} - 4 = \dfrac{2x - 7}{6}$

In Problems 4 and 5, solve each equation and check.

4. $\dfrac{2x}{2x + 5} = \dfrac{2}{3} - \dfrac{5}{4x + 10}$

5. $\dfrac{2x}{x - 1} + \dfrac{1}{3} = \dfrac{5}{6} + \dfrac{2}{x - 1}$

In Problem 6, solve for y in terms of x.

6. $3x - 4y = 15$

In Problems 7 and 8, solve each inequality and graph the solution.

7. $2(x - 1) - 3 > 4x + 1$

8. $\dfrac{-3x}{2} > 3 - x$

Applications

9. **Depreciation** A $648,000 property is depreciated for tax purposes by its owner with the straight-line depreciation method. The value of the building, y, after x months of use is given by $y = 648{,}000 - 1800x$ dollars. After how many months will the value of the building be $387,000?

10. **Credit card debt** High interest rates make it difficult for people to pay off credit card debt in a reasonable period of time. The interest I (in dollars) paid on a $10,000 debt over 3 years when the interest rate is $r\%$ can be approximated by the equation

$$\frac{I}{175.393} + 0.663 = r$$

(*Source:* Consumer Federation of America) If the credit card interest rate is 19.8%, find the amount of interest paid during the 3 years.

11. **Break-even** Burnem, Inc. manufactures blank CDs and sells them to a distributor in packs of 500 CDs. Burnem's total cost and total revenue (in dollars) for x packs of 500 CDs are given by

Total cost $= 2x + 7920$ and   Total revenue $= 20x$

How many packs of 500 CDs must Burnem sell to break even?

12. **Investment mix** A retired woman has $120,000 to invest. She has chosen one relatively safe investment fund that has an annual yield of 9% and another, riskier fund that has a 13% annual yield. How much should she invest in each fund if she would like to earn $12,000 per year from her investments?

13. **Profit** For a certain product, the revenue function is $R(x) = 40x$ and the cost function is $C(x) = 20x + 1600$. To obtain a profit, the revenue must be greater than the cost. For what values of x will there be a profit? Graph the solution.

14. **Car rental** Thrift rents a compact car for $33 per day, and General rents a similar car for $20 per day plus an initial fee of $78. For how many days would it be cheaper to rent from General? Graph the solution.

Need more practice?
Find more here: cengagebrain.com

1.2   Functions

UNDERSTANDING THE MATHEMATICAL MEANING OF THE PHRASE **FUNCTION OF** AND LEARNING TO INTERPRET AND APPLY SUCH RELATIONSHIPS ARE THE GOALS OF THIS SECTION.

Relations and Functions

An equation or inequality containing two variables expresses a **relation** between those two variables. For example, the inequality $R \geq 35x$ expresses a relation between the two variables x and R, and the equation $y = 4x - 3$ expresses a relation between the two variables x and y.

In addition to defining a relation by an equation, an inequality, or rule of correspondence, we may also define it as any set of **ordered pairs** of real numbers (a, b). For example, the solutions to $y = 4x - 3$ are pairs of numbers (one for x and one for y). We write the pairs (x, y) so that the first number is the x-value and the second is the y-value, and these ordered pairs define the relation between x and y. Some relations may be defined by a table, a graph, or an equation.

For example, the set of ordered pairs

$$\{(1, 3), (1, 6), (2, 6), (3, 9), (3, 12), (4, 12)\}$$

expresses a relation between the set of first components, $\{1, 2, 3, 4\}$, and the set of second components, $\{3, 6, 9, 12\}$. The set of first components is called the **domain** of the relation, and the set of second components is called the **range** of the relation.

Relation: A **relation** is defined by a set of ordered pairs or by a rule that determines how the ordered pairs are found. It may also be defined by a table, a graph, an equation, or an inequality.

AN EQUATION OR INEQUALITY CONTAINING TWO VARIABLES EXPRESSES A RELATION BETWEEN THOSE TWO VARIABLES.

An equation frequently expresses how the second component (the output) is obtained from the first component (the input). For example, the equation

$$y = 4x - 3$$

expresses a special relation between x and y, because each value of x that is substituted into the equation results in exactly one value of y. If each value of x put into an equation results in one value of y, we say that the equation expresses y as a **function** of x.

> **Definition of a Function:** A **function** is a relation between two sets such that to each element of the domain (input) there corresponds exactly one element of the range (output). A function may be defined by a set of ordered pairs, a table, a graph, or an equation.

When a function is defined, the variable that represents the numbers in the domain (input) is called the **independent variable** of the function, and the variable that represents the numbers in the range (output) is called the **dependent variable** (because its values depend on the values of the independent variable).

The data given in Table 1.1 (the tax brackets for U.S. income tax for single wage earners) represents the tax rate as a function of the income. The relation defined in Figure 1.2 is not a function, however, because the graph representing the Dow Jones Utilities Average shows that for each day there are at least three different values—the actual high, the actual low, and the close. This particular figure also has historical interest because it shows a break in the graph when the New York Stock Exchange closed following the terrorist attacks of 9/11/2001.

Table 1.1   **2010 Federal Income Tax**

Income	Rate
$0–$8375	10%
$8376–$34,000	15%
$34,001–$82,400	25%
$82,401–$171,850	28%
$171,851–$373,650	33%
over $373,650	35%

Source: Internal Revenue Service

Figure 1.2

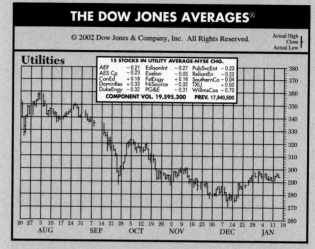

Source: The Wall Street Journal, January 17, 2002. Copyright © 2002 Dow Jones & Co. Reprinted by permission of Dow Jones & Co. via Copyright Clearance Center.

EXAMPLE 1.8 Functions

 Does $y^2 = 2x$ express y as a function of x?

SOLUTION

No, because some values of x are associated with more than one value of y. In fact, there are two y-values for each $x > 0$. For example, if $x = 8$, then $y = 4$ or $y = -4$, two different y-values for the same x-value.

Graphs of Functions

It is possible to picture geometrically the relations and functions that we have been discussing by sketching their graphs on a rectangular coordinate system.

The **graph** of an equation that defines a function (or relation) is the picture that results when we plot the points whose coordinates (x, y) satisfy the equation. To sketch the graph, we plot enough points to suggest the shape of the graph and draw a smooth curve through the points. This is called the **point-plotting method** of sketching a graph.

EXAMPLE 1.9 Graphing a Function

 Graph the function $y = 4x^2$.

SOLUTION

We choose some sample values of x and find the corresponding values of y. Placing these in a table, we have sample points to plot. When we have enough to determine the shape of the graph, we connect the points to complete the graph. The table and graph are shown in Figure 1.3(a).

We can determine whether a relation is a function by inspecting its graph. If the relation is a function, then no one input (*x*-value) has two different outputs (*y*-values). This means that no two points on the graph will have the same first coordinate (component).

Vertical-Line Test: If no vertical line exists that intersects the graph at more than one point, then the graph is that of a function.

On the graph of $y = 4x^2$ (Figure 1.3(a)), we easily see that this equation describes a function. The graph of $y^2 = 2x$ is shown in Figure 1.3(b), and we can see that the vertical-line test indicates that this is not a function. For example, a vertical line at $x = 2$ intersects the curve at $(2, 2)$ and $(2, -2)$.

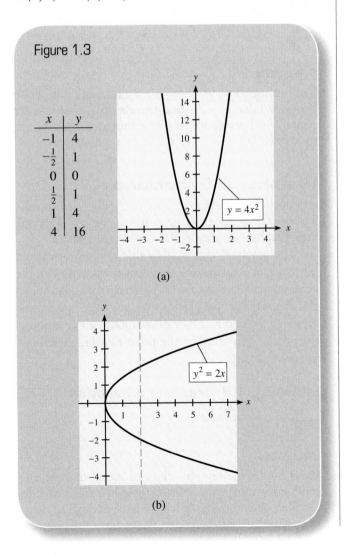

Figure 1.3

x	y
-1	4
$-\frac{1}{2}$	1
0	0
$\frac{1}{2}$	1
1	4
4	16

(a)

(b)

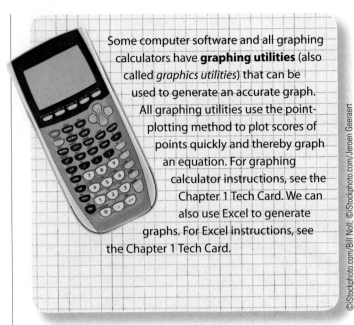

Some computer software and all graphing calculators have **graphing utilities** (also called *graphics utilities*) that can be used to generate an accurate graph. All graphing utilities use the point-plotting method to plot scores of points quickly and thereby graph an equation. For graphing calculator instructions, see the Chapter 1 Tech Card. We can also use Excel to generate graphs. For Excel instructions, see the Chapter 1 Tech Card.

Function Notation

We can use function notation to indicate that *y* is a function of *x*. The function is denoted by *f*, and we write $y = f(x)$. This is read "*y* is a function of *x*" or "*y* equals *f* of *x*." For specific values of *x*, $f(x)$ represents the values of the function (that is, outputs, or *y*-values) at those *x*-values. Thus if

$$f(x) = 3x^2 + 2x + 1$$

then

$$f(2) = 3(2)^2 + 2(2) + 1 = 17$$

and

$$f(-3) = 3(-3)^2 + 2(-3) + 1 = 22$$

Figure 1.4 shows the function notation $f(x)$ as (a) an operator on *x* and (b) a *y*-coordinate for a given *x*-value.

Letters other than *f* may also be used to denote functions. For example, $y = g(x)$ or $y = h(x)$ may be used.

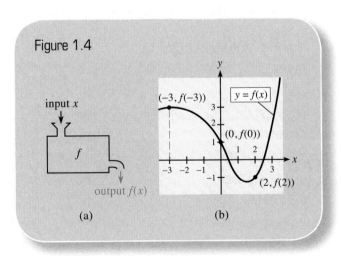

Figure 1.4

input x

f

output $f(x)$

(a)

(b)

EXAMPLE 1.10 Electronic Income Tax Returns

The relationship between the number of individual income tax returns filed electronically and the number of years after 1995 can be described by the equation

$$y = f(x) = 5.091x + 11.545$$

where y is in millions and x is the number of years after 1995. (*Source:* Internal Revenue Service)

(a) Find $f(8)$.

(b) Write a sentence that explains the meaning of the result in part (a).

SOLUTION

(a) $f(8) = 5.091(8) + 11.545 = 52.273$

(b) The statement $f(8) = 52.273$ means that in $1995 + 8 = 2003$, 52.273 million income tax returns were filed electronically.

The Tech Card also shows how to evaluate functions with a graphing calculator and with Excel.

EXAMPLE 1.11 Function Notation

Given $f(x) = x^2 - 3x + 8$, find $\dfrac{f(x + h) - f(x)}{h}$ and simplify (if $h \neq 0$).

SOLUTION

We find $f(x + h)$ by replacing each x in $f(x)$ with the expression $x + h$.

$$\frac{f(x + h) - f(x)}{h} = \frac{[(x + h)^2 - 3(x + h) + 8] - [x^2 - 3x + 8]}{h}$$

$$= \frac{[(x^2 + 2xh + h^2) - 3x - 3h + 8] - x^2 + 3x - 8}{h}$$

$$= \frac{x^2 + 2xh + h^2 - 3x - 3h + 8 - x^2 + 3x - 8}{h}$$

$$= \frac{2xh + h^2 - 3h}{h} = \frac{h(2x + h - 3)}{h} = 2x + h - 3$$

Domains and Ranges

We will limit our discussion in this text to **real functions,** which are functions whose domains and ranges contain only real numbers. If the domain and range of a function are not specified, it is assumed that the domain consists of all real inputs (*x*-values) that result in real outputs (*y*-values), making the range a subset of the real numbers.

For the types of functions we are now studying, if the domain is unspecified, it will include all real numbers except

1. values that result in a denominator of 0, and
2. values that result in an even root of a negative number.

EXAMPLE 1.12 Domain and Range

Find the domain of each of the following functions; find the range for the functions in (a) and (b).

(a) $y = 4x^2$ (b) $y = \sqrt{4 - x}$ (c) $y = 1 + \dfrac{1}{x - 2}$

SOLUTION

(a) There are no restrictions on the numbers substituted for x, so the domain consists of all real numbers. Because the square of any real number is nonnegative, $4x^2$ must be nonnegative. Thus the range is $y \geq 0$.

(b) We note the restriction that $4 - x$ cannot be negative. Thus the domain consists of only numbers less than or equal to 4. That is, the domain is the set of real numbers satisfying $x \leq 4$. Because $\sqrt{4 - x}$ is always nonnegative, the range is all $y \geq 0$. Figure 1.5(a) shows the graph of $y = \sqrt{4 - x}$.

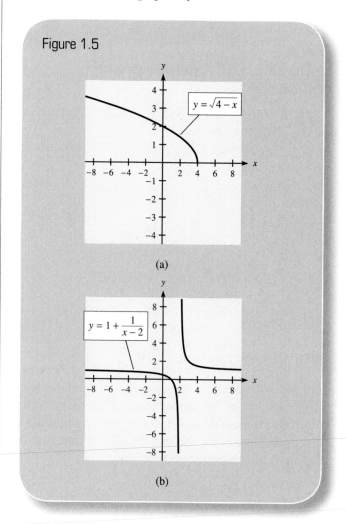

Figure 1.5

$y = \sqrt{4 - x}$

(a)

$y = 1 + \dfrac{1}{x - 2}$

(b)

(c) $y = 1 + \dfrac{1}{x-2}$ is undefined at $x = 2$ because $\dfrac{1}{0}$ is undefined. Hence, the domain consists of all real numbers except 2. Figure 1.5(b) shows the graph of $y = 1 + \dfrac{1}{x-2}$.

Operations with Functions

We can form new functions by performing algebraic operations with two or more functions. We define new functions that are the sum, difference, product, and quotient of two functions as follows.

Operations with Functions: Let f and g be functions of x, and define the following.

Sum	$(f + g)(x) = f(x) + g(x)$
Difference	$(f - g)(x) = f(x) - g(x)$
Product	$(f \cdot g)(x) = f(x) \cdot g(x)$
Quotient	$\left(\dfrac{f}{g}\right)(x) = \dfrac{f(x)}{g(x)}$ if $g(x) \neq 0$

EXAMPLE 1.13 Operations with Functions

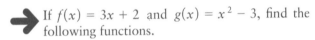 If $f(x) = 3x + 2$ and $g(x) = x^2 - 3$, find the following functions.

(a) $(f + g)(x)$ (b) $(f - g)(x)$

(c) $(f \cdot g)(x)$ (d) $\left(\dfrac{f}{g}\right)(x)$

SOLUTION

(a) $(f + g)(x) = f(x) + g(x) = (3x + 2) + (x^2 - 3)$
    $= x^2 + 3x - 1$

(b) $(f - g)(x) = f(x) - g(x) = (3x + 2) - (x^2 - 3)$
    $= -x^2 + 3x + 5$

(c) $(f \cdot g)(x) = f(x) \cdot g(x) = (3x + 2)(x^2 - 3)$
    $= 3x^3 + 2x^2 - 9x - 6$

(d) $\left(\dfrac{f}{g}\right)(x) = \dfrac{f(x)}{g(x)} = \dfrac{3x + 2}{x^2 - 3}$, if $x^2 - 3 \neq 0$

We now consider a new way to combine two functions. Just as we can substitute a number for the independent variable in a function, we can substitute a second function for the variable. This creates a new function called a **composite function.**

Composite Functions: Let f and g be functions. Then the **composite functions** g of f (denoted $g \circ f$) and f of g (denoted $f \circ g$) are defined as follows:

$$(g \circ f)(x) = g(f(x))$$
$$(f \circ g)(x) = f(g(x))$$

EXAMPLE 1.14 Composite Functions

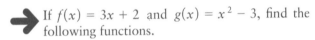 If $f(x) = 2x^3 + 1$ and $g(x) = x^2$, find the following.

(a) $(g \circ f)(x)$ (b) $(f \circ g)(x)$

SOLUTION

(a) $(g \circ f)(x) = g(f(x))$
    $= g(2x^3 + 1)$
    $= (2x^3 + 1)^2 = 4x^6 + 4x^3 + 1$

(b) $(f \circ g)(x) = f(g(x))$
    $= f(x^2)$
    $= 2(x^2)^3 + 1$
    $= 2x^6 + 1$

Figure 1.6 illustrates both composite functions found in Example 1.14.

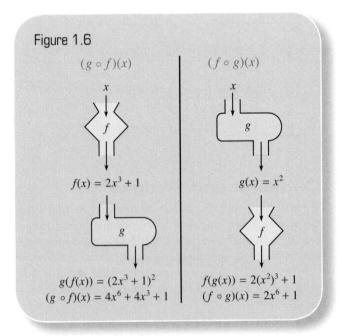

Figure 1.6

1.2 Exercises

1. Is the relation defined by the table a function? Explain why or why not and give the domain and range.

x	1	2	3	8	9
y	−4	−4	5	16	5

2. Does either of the graphs in Figure 1.7 represent y as a function of x? Explain your choices.

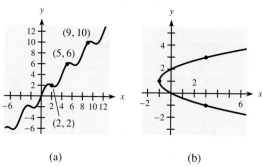

(a)   (b)

Figure 1.7

3. If $C(x) = 4x^2 - 3$, find the following.
   (a) $C(0)$
   (b) $C(-1)$
   (c) $C(-2)$
   (d) $C(-\frac{3}{2})$

4. Let $f(x) = 1 + x + x^2$ and $h \neq 0$.
   (a) Is $f(2 + 1) = f(2) + f(1)$?
   (b) Find $f(x + h)$.
   (c) Does $f(x + h) = f(x) + f(h)$?
   (d) Does $f(x + h) = f(x) + h$?
   (e) Find $\dfrac{f(x + h) - f(x)}{h}$ and simplify.

5. If $f(x) = x - 2x^2$ and $h \neq 0$, find the following and simplify.
   (a) $f(x + h)$
   (b) $\dfrac{f(x + h) - f(x)}{h}$

6. If $y = f(x)$ in Figure 1.7(a), find the following.
   (a) $f(9)$ (b) $f(5)$

7. State the domain and range of $y = \sqrt{x + 4}$.

8. Given $f(x) = 3x$ and $g(x) = x^3$, find:
   (a) $(f + g)(x)$
   (b) $(f - g)(x)$
   (c) $(f \cdot g)(x)$
   (d) $(f/g)(x)$

9. Given $f(x) = (x - 1)^3$ and $g(x) = 1 - 2x$, find:
   (a) $(f \circ g)(x)$
   (b) $(g \circ f)(x)$
   (c) $f(f(x))$
   (d) $f^2(x) = (f \cdot f)(x)$

Applications

10. **Mortgage** A couple seeking to buy a home decides that a monthly payment of $800 fits their budget. Their bank's interest rate is 7.5%. The amount they can borrow, A, is a function of the time t, in years, it will take to repay the debt. If we denote this function by $A = f(t)$, then the following table defines the function.

t	A	t	A
5	40,000	20	103,000
10	69,000	25	113,000
15	89,000	30	120,000

Source: *Comprehensive Mortgage Payment Tables*, Publication No. 492, Financial Publishing Co., Boston

   (a) Find $f(20)$ and write a sentence that explains its meaning.
   (b) Does $f(5 + 5) = f(5) + f(5)$? Explain.
   (c) If the couple is looking at a house that requires them to finance $89,000, how long must they make payments? Write this correspondence in the form $A = f(t)$.

11. **Social Security benefits funding** Social Security benefits paid to eligible beneficiaries are funded by individuals who are currently employed. The following graph, based on known data until 2005, with projections into the future, defines a function that gives the number of workers, n, supporting each retiree as a function of time t (given by calendar year). Let us denote this function by $n = f(t)$.
   (a) Find $f(1950)$ and explain its meaning.
   (b) Find $f(1990)$.
   (c) If, after the year 2050, actual data through 2050 regarding workers per Social Security beneficiary were graphed, what parts of the new graph *must* be the same as this graph and what parts *might* be the same? Explain.

(d) Find the domain and range of $n = f(t)$ if the function is defined by the graph.

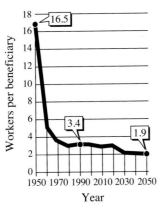

Source: Social Security Administration

12. **Cost** The total cost of producing a product is given by
$$C(x) = 300x + 0.1x^2 + 1200$$
where x represents the number of units produced. Give
(a) the total cost of producing 10 units
(b) the value of $C(100)$
(c) the meaning of $C(100)$

13. **Pollution** Suppose that the cost C (in dollars) of removing p percent of the particulate pollution from the smokestacks of an industrial plant is given by
$$C(p) = \frac{7300p}{100 - p}$$
(a) Find the domain of this function. Recall that p represents the percent pollution that is removed.
In parts (b)–(e), find the functional values and explain what each means.
(b) $C(45)$ (c) $C(90)$
(d) $C(99)$ (e) $C(99.6)$

14. **Profit** Suppose that the profit from the production and sale of x units of a product is given by
$$P(x) = 180x - \frac{x^2}{100} - 200$$
In addition, suppose that for a certain month the number of units produced on day t of the month is
$$x = q(t) = 1000 + 10t$$
(a) Find $(P \circ q)(t)$ to express the profit as a function of the day of the month.
(b) Find the number of units produced, and the profit, on the 15th day of the month.

Need more practice?
Find more here: cengagebrain.com

1.3 Linear Functions

IN THIS SECTION, WE WILL FIND THE SLOPES AND INTERCEPTS OF GRAPHS OF LINEAR FUNCTIONS AND WRITE EQUATIONS OF LINES.

Linear Function: A **linear function** is a function of the form
$$y = f(x) = ax + b$$
where a and b are constants.

Intercepts

It is frequently possible to use **intercepts** to graph a linear function. The point(s) where a graph intersects the x-axis are called the x-intercept points, and the x-coordinates of these points are the **x-intercepts**. Similarly, the points where a graph intersects the y-axis are the y-intercept points, and the y-coordinates of these points are the **y-intercepts**. We find intercepts as follows.

Intercepts
(a) To find the **y-intercept(s)** of the graph of an equation, set $x = 0$ in the equation and solve for y. Note: A function of x has at most one y-intercept.
(b) To find the **x-intercept(s)**, set $y = 0$ and solve for x.

EXAMPLE 1.15 Intercepts

Find the intercepts and graph the following.

(a) $3x + y = 9$ (b) $x = 4y$

SOLUTION

(a) We set $x = 0$ and solve for y, getting $y = 9$, so the y-intercept is 9. We set $y = 0$ and solve for x, getting $x = 3$, so the x-intercept is 3. Using the intercepts gives the graph, shown in Figure 1.8.

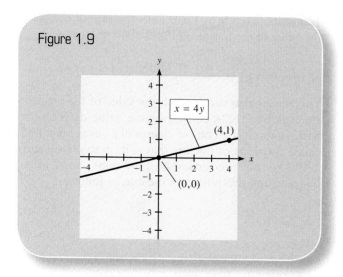

Figure 1.8

$3x + y = 9$

$(0, 9)$

$(3, 0)$

Figure 1.9

$x = 4y$

$(4, 1)$

$(0, 0)$

(b) Letting $x = 0$ gives $y = 0$, and letting $y = 0$ gives $x = 0$, so the only intercept of the graph is $(0, 0)$. A second point is needed to graph the line. If we let $y = 1$ in $x = 4y$, we get $x = 4$ and have a second point $(4, 1)$ on the graph. The graph is shown in Figure 1.9.

Note that the equation graphed in Figure 1.8 can be rewritten as

$$y = 9 - 3x \quad \text{or} \quad f(x) = 9 - 3x$$

We see in Figure 1.8 that $(3, 0)$ is the point where the function value is zero, so the *x*-intercept is 3. The *x*-coordinate of such a point is also called a **zero of the function.** Thus we see that the *x-intercepts of the graph of a function are the same as its zeros.*

EXAMPLE 1.16 Depreciation

A business property is purchased for $122,880 and depreciated over a period of 10 years. Its value y is related to the number of months of service x by the equation

$$4096x + 4y = 491,520$$

Find the *x*-intercept and the *y*-intercept and use them to sketch the graph of the equation.

SOLUTION

x-intercept: $y = 0$ gives $4096x = 491,520$
$$x = 120$$

Thus 120 is the *x*-intercept.

y-intercept: $x = 0$ gives $4y = 491,520$
$$y = 122,880$$

Thus 122,880 is the *y*-intercept. The graph is shown in Figure 1.10. Note that the units on the *x*- and *y*-axes

are different and that the *y*-intercept gives the purchase price. The *x*-intercept corresponds to the number of months that have passed before the value is 0; that is, the property is fully depreciated after 120 months, or 10 years. Note that only positive values for x and y make sense in this application, so only the Quadrant I portion of the graph is shown.

Rate of Change; Slope of a Line

Note that in Figure 1.10, as the graph moves from the *y*-intercept point $(0, 122,880)$ to the *x*-intercept point $(120, 0)$, the *y*-value on the line changes $-122,880$ units (from 122,880 to 0), whereas the *x*-value changes

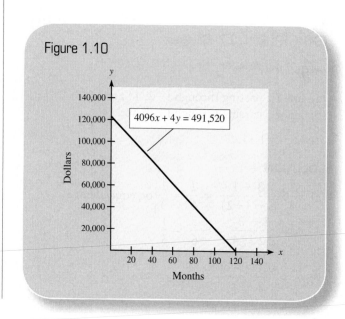

Figure 1.10

$4096x + 4y = 491,520$

Dollars

Months

120 units (from 0 to 120). Thus the **rate of change** of the value of the business property is

$$\frac{-122,880}{120} = -1024 \text{ dollars per month}$$

This means that each month the value of the property changes by -1024 dollars, or the value decreases by 1024/month. This **rate of change** of a linear function is called the **slope** of the line that is its graph (see Figure 1.10). For any nonvertical line, the slope can be found by using any two points on the line, as follows.

> **Slope of a Line:** If a nonvertical line passes through the points $P_1(x_1, y_1)$ and $P_2(x_2, y_2)$, its **slope,** denoted by m, is found by using either
>
> $$m = \frac{y_2 - y_1}{x_2 - x_1} = \frac{\Delta y}{\Delta x}$$
>
> where Δy is read "delta y" and means "change in y" and Δx means "change in x." The slope of a vertical line is undefined.

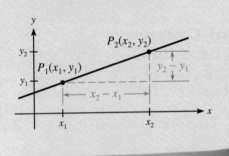

EXAMPLE 1.17 Slopes

 Find the slope of

(a) line ℓ_1, passing through $(-2, 1)$ and $(4, 3)$
(b) line ℓ_2, passing through $(3, 0)$ and $(4, -3)$

See Figure 1.11.

SOLUTION

(a) $m = \dfrac{3 - 1}{4 - (-2)} = \dfrac{2}{6} = \dfrac{1}{3}$ or, equivalently,

$m = \dfrac{1 - 3}{-2 - 4} = \dfrac{-2}{-6} = \dfrac{1}{3}$

(b) $m = \dfrac{0 - (-3)}{3 - 4} = \dfrac{3}{-1} = -3$

Figure 1.11

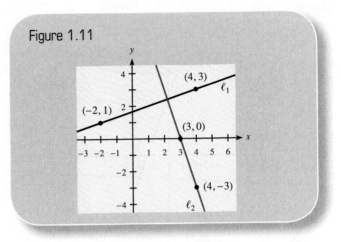

We see that the slope describes the direction of a line as follows.

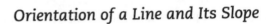

Orientation of a Line and Its Slope

1. The slope is *positive* if the line slopes upward toward the right. The function is increasing.

2. The slope is *negative* if the line slopes downward toward the right. The function is decreasing.

3. The slope of a *horizontal line* is 0, because $\Delta y = 0$. The function is constant.

4. The slope of a *vertical line* is undefined, because $\Delta x = 0$.

> **Parallel Lines:** Two distinct nonvertical lines are *parallel* if and only if their slopes are *equal.*

Note in Example 1.17 that the slope of ℓ_1, $\frac{1}{3}$, is the negative reciprocal of the slope of ℓ_2, -3. In fact, as with lines ℓ_1 and ℓ_2, any two nonvertical lines that are perpendicular have slopes that are negative reciprocals of each other.

Shutterstock

Slopes of Perpendicular Lines: A line ℓ_1 with slope m, where $m \neq 0$, is *perpendicular* to line ℓ_2 if and only if the slope of ℓ_2 is $-1/m$. (The slopes are *negative reciprocals*.)

Writing Equations of Lines

If the slope of a line is m, then the slope between a fixed point (x_1, y_1) and any other point (x, y) on the line is also m. That is,

$$m = \frac{y - y_1}{x - x_1}$$

Solving for $y - y_1$ gives the point-slope form of the equation of a line.

Point-Slope Form: The equation of the line passing through the point (x_1, y_1) and with slope m can be written in the **point-slope form**

$$y - y_1 = m(x - x_1)$$

EXAMPLE 1.18 Equations of Lines

Write the equation for the line that passes through $(1, -2)$ and has slope $\frac{2}{3}$.

SOLUTION

Here $m = \frac{2}{3}$, $x_1 = 1$, and $y_1 = -2$. An equation of the line is

$$y - (-2) = \frac{2}{3}(x - 1)$$
$$y + 2 = \frac{2}{3}x - \frac{2}{3}$$
$$y = \frac{2}{3}x - \frac{8}{3}$$

This equation also may be written in **general form** as $2x - 3y - 8 = 0$. Figure 1.12 shows the graph of this line; the point $(1, -2)$ and the slope are highlighted.

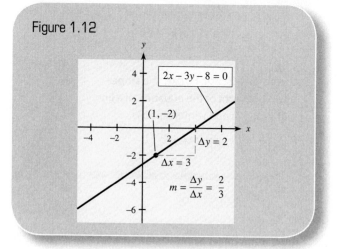

Figure 1.12

If a line passes through $(1, -2)$ and $(2, 3)$, we first find the slope.

$$m = \frac{3 - (-2)}{2 - 1} = 5$$

Using $m = 5$ and the point $(1, -2)$ (the other point could also be used) gives the equation

$$y - (-2) = 5(x - 1) \quad \text{or} \quad y = 5x - 7$$

If the slope m is undefined, we cannot use the point-slope form. If a line is vertical and through $(1, 2)$, its equation is $x = 1$. The graph of $x = 1$ is a vertical line, as shown in Figure 1.13(a) on the next page; the graph of $y = 1$ has slope 0, and its graph is a horizontal line, as shown in Figure 1.13(b).

In general, **vertical lines** have undefined slope and the equation form $x = a$, where a is the x-coordinate of each point on the line. **Horizontal lines** have $m = 0$ and the equation form $y = b$, where b is the y-coordinate of each point on the line.

The point-slope form, with the y-intercept point $(0, b)$, can be used to derive a special form for the equation of a line.

$$y - b = m(x - 0)$$
$$y = mx + b$$

Figure 1.13

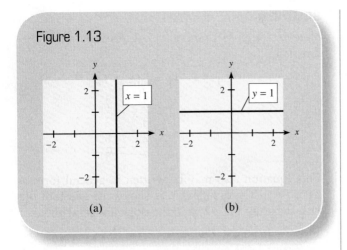

(a)   (b)

Slope-Intercept Form: The **slope-intercept form** of the equation of a line with slope m and y-intercept b is

$$y = mx + b$$

Note that if a linear equation has the form $y = mx + b$, then the coefficient of x is the slope and the constant term is the y-intercept.

EXAMPLE 1.19 Writing the Equation of a Line

 Write the equation of the line with slope $\frac{1}{2}$ and y-intercept -3.

SOLUTION

Substituting $m = \frac{1}{2}$ and $b = -3$ in the equation $y = mx + b$ gives $y = \frac{1}{2}x + (-3)$ or $y = \frac{1}{2}x - 3$.

EXAMPLE 1.20 Banks

 The number of banks in the United States for selected years from 1980 to 2005 is given by

$$y = -416.454x + 18{,}890.75$$

where x is the number of years after 1980.

(a) Find the slope and the y-intercept.
(b) What does the y-intercept tell us about the banks?
(c) Interpret the slope as a rate of change.

SOLUTION

(a) The slope is $m = -416.454$ and the y-intercept is $b = 18{,}890.75$.

(b) Because x is the number of years after 1980, $x = 0$ represents 1980 and the y-intercept tells us that there were approximately 18,891 banks in 1980.
(c) The slope is -416.454, which tells us that the number of banks was decreasing at a rate of approximately 416.5 banks per year during this period.

When a linear equation does not appear in slope-intercept form (and does not have the form $x = a$), it can be put into slope-intercept form by solving the equation for y.

EXAMPLE 1.21 Slope-Intercept Form

 Find the slope and y-intercept of the line whose equation is $x + 2y = 8$.

SOLUTION

To put the equation in slope-intercept form, we must solve it for y.

$$2y = -x + 8 \quad \text{or} \quad y = -\frac{1}{2}x + 4$$

Thus the slope is $-\frac{1}{2}$ and the y-intercept is 4.

It is also possible to graph a straight line if we know its slope and any point on the line; we simply plot the point that is given and then use the slope to plot other points.

The following summarizes the forms of equations of lines.

Forms of Linear Equations:
General form: $ax + by + c = 0$
Point-slope form: $y - y_1 = m(x - x_1)$
Slope-intercept form: $y = mx + b$
Vertical line: $x = a$
Horizontal line: $y = b$

1.3 Exercises

Find the intercepts and graph the following function.

1. $5x - 8y = 60$

In Problems 2–4, find the slope of the line passing through the given pair of points.

2. $(22, 11)$ and $(15, -17)$

3. $(3, 2)$ and $(-1, 2)$

4. If a line is horizontal, then its slope is ___?___.

For the graph in Problem 5, determine whether each line has a slope that is positive, negative, 0, or undefined.

5.

6. Find the slope and y-intercept of the line $y = \dfrac{7}{3} x - \dfrac{1}{4}$. Then graph the line.

In Problems 7 and 8, write the equation and sketch the graph of each line described.

7. Slope -2 and y-intercept $\frac{1}{2}$

8. Through $(-1, 4)$ with slope $-\frac{3}{4}$

9. Write the equation of the line passing through the points $(7, 3)$ and $(-6, 2)$.

10. Determine whether the pair of equations

$$3x + 2y = 6; \quad 2x - 3y = 6$$

represent parallel lines, perpendicular lines, or neither of these.

11. Write the equation of the line through $(-2, -7)$ that is parallel to $3x + 5y = 11$.

Applications

12. ***Depreciation*** A \$360,000 building is depreciated by its owner. The value y of the building after x months of use is $y = 360,000 - 1500x$.
    (a) Graph this function for $x \geq 0$.
    (b) How long is it until the building is completely depreciated (its value is zero)?
    (c) The point $(60, 270,000)$ lies on the graph. Explain what this means.

13. ***Earnings and gender*** According to the U.S. Bureau of the Census, the relation between the average annual earnings of males and females with various levels of educational attainment can be modeled by the function

$$F = 0.78M - 1.316$$

where M and F represent the average annual earnings (in thousands of dollars) of males and females, respectively.
    (a) Viewing F as a function of M, what is the slope of the graph of this function?
    (b) Interpret the slope as a rate of change.
    (c) When the average annual earnings for males reach \$60,000, what does the equation predict for the average annual earnings for females?

14. ***Residential electric costs*** An electric utility company determines the monthly bill for a residential customer by adding an energy charge of 8.38 cents per kilowatt-hour to its base charge of \$16.37 per month. Write an equation for the monthly charge y in terms of x, the number of kilowatt-hours used.

15. ***Earnings and race*** Data from 2005 for various age groups show that for each \$100 increase in the median weekly income for whites, the median weekly income of blacks increases by \$105. Also, for workers of ages 25 to 54, the median weekly income for whites was \$676 and for blacks was \$527. (*Source:* U.S. Department of Labor)
    (a) Let W represent the median weekly income for whites and B the median weekly income for blacks, and write the equation of the line that gives B as a linear function of W.
    (b) When the median weekly income for whites is \$850, what does the equation in part (a) predict for the median weekly income for blacks?

> **Need more practice?**
> Find more here: **cengagebrain.com**

1.4 Solutions of Systems of Linear Equations

OUR GOAL IS TO EXAMINE THREE WAYS OF FINDING SOLUTIONS FOR SYSTEMS OF LINEAR EQUATIONS: GRAPHICAL SOLUTIONS, SOLUTIONS BY SUBSTITUTION, AND SOLUTIONS BY ELIMINATION.

Graphical Solution

When we find a set of values (a, b) that satisfies both of two equations in two variables, we have found a solution to a system of two equations in two variables. We can use graphing to find the solution of a system of two linear equations by finding the point of intersection of the graphs of the two equations.

EXAMPLE 1.22 Graphical Solution of a System

 Use graphing to find the solution of the system

$$\begin{cases} 4x + 3y = 11 \\ 2x - 5y = -1 \end{cases}$$

SOLUTION

The graphs of the two equations intersect (meet) at the point (2, 1). (See Figure 1.14.) The solution of the system is $x = 2$, $y = 1$. Note that these values satisfy both equations.

Figure 1.14

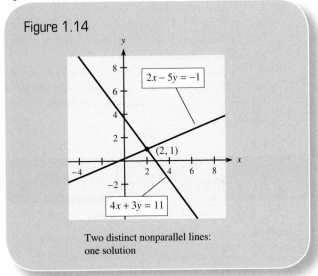

Two distinct nonparallel lines:
one solution

See the Tech Card for details on how to solve systems with a graphing calculator or Excel.

If the graphs of two equations are parallel lines, they have no point in common, and thus the system has no solution. Such a system of equations is called **inconsistent**. For example,

$$\begin{cases} 4x + 3y = 4 \\ 8x + 6y = 18 \end{cases}$$

is an **inconsistent system** (see Figure 1.15(a)).

It is also possible that two equations describe the same line. When this happens, the equations are equivalent, and values that satisfy either equation are solutions of the system. For example,

$$\begin{cases} 4x + 3y = 4 \\ 8x + 6y = 8 \end{cases}$$

is called a **dependent system** because all points that satisfy one equation also satisfy the other (see Figure 1.15(b)).

Figures 1.14, 1.15(a), and 1.15(b) represent the three possibilities that can occur when we are solving a system of two linear equations in two variables.

Figure 1.15

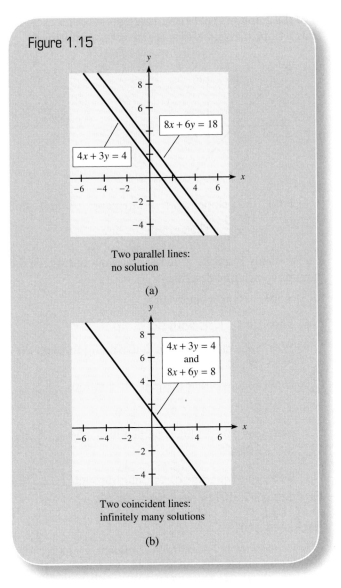

Two parallel lines:
no solution

(a)

Two coincident lines:
infinitely many solutions

(b)

Solution by Substitution

Graphical solution methods may yield only approximate solutions to some systems. Exact solutions can be found using algebraic methods, which are based on the fact that equivalent systems result when any of the following operations are performed.

Equivalent Systems

Equivalent systems result when
1. One expression is replaced by an equivalent expression.
2. Two equations are interchanged.
3. A multiple of one equation is added to another equation.
4. An equation is multiplied by a nonzero constant.

The **substitution method** is based on operation (1).

Substitution Method for Solving Systems

To solve a system of two equations in two variables by substitution:

1. Solve one of the equations for one of the variables in terms of the other.

2. Substitute this expression into the other equation to give one equation in one unknown.

3. Solve this linear equation for the unknown.

4. Substitute this solution into the equation in Step 1 or into one of the original equations to solve for the other variable.

5. Check the solution by substituting for x and y in both original equations.

EXAMPLE 1.23 Solution by Substitution

 Solve the system

$$\begin{cases} 4x + 5y = 18 & (1) \\ 3x - 9y = -12 & (2) \end{cases}$$

SOLUTION

$$x = \frac{9y - 12}{3} = 3y - 4 \quad \text{Solve for } x \text{ in equation (2).}$$

$$4(3y - 4) + 5y = 18 \quad \text{Substitute for } x \text{ in equation (1).}$$

$$12y - 16 + 5y = 18 \quad \text{Solve for } y.$$

$$17y = 34$$

$$y = 2$$

$$x = 3(2) - 4 \quad \text{Use } y = 2 \text{ to find } x.$$

$$x = 2$$

$$4(2) + 5(2) = 18 \text{ and } 3(2) - 9(2) = -12 \; ✔ \; \text{Check.}$$

Thus the solution is $x = 2$, $y = 2$. This means that when the two equations are graphed simultaneously, their point of intersection is (2, 2). See Figure 1.16.

Figure 1.16

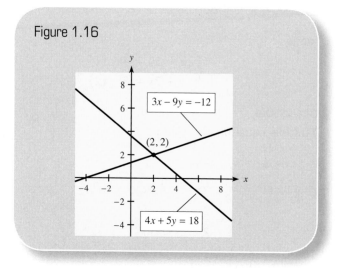

Solution by Elimination

We can also eliminate one of the variables in a system by the **elimination method,** which uses addition or subtraction of equations.

Procedure

To solve a system of two equations in two variables by the elimination method:

1. If necessary, multiply one or both equations by a nonzero number that will make the coefficients of one of the variables identical, except perhaps for signs.

2. Add or subtract the equations to eliminate one of the variables.

3. Solve for the variable in the resulting equation.

4. Substitute the solution into one of the original equations and solve for the other variable.

5. Check the solutions in both original equations.

Elimination Method for Solving Systems

EXAMPLE 1.24

 Solve the system

$$\begin{cases} 2x - 5y = 4 & (1) \\ x + 2y = 3 & (2) \end{cases}$$

SOLUTION

Multiply equation (2) by -2.

$$\begin{array}{r} 2x - 5y = 4 \\ -2x - 4y = -6 \end{array}$$

Adding gives $0x - 9y = -2$

$$y = \frac{2}{9}$$

$$2x - 5\left(\frac{2}{9}\right) = 4$$

$$2x = 4 + \frac{10}{9} = \frac{36}{9} + \frac{10}{9}$$

$$2x = \frac{46}{9} \quad \text{so} \quad x = \frac{23}{9}$$

$$2\left(\frac{23}{9}\right) - 5\left(\frac{2}{9}\right) = 4 ✔$$

$$\frac{23}{9} + 2\left(\frac{2}{9}\right) = 3 ✔$$

Note: When solving algebraically, it is possible to reach an identity (such as $0 = 0$), which indicates a system has infinitely many solutions, or an impossibility (such as $0 = 10$), which indicates no solution.

EXAMPLE 1.25 Investment Mix

 A person has $200,000 invested, part at 9% and part at 8%. If the total yearly income from the two investments is $17,200, how much is invested at 9% and how much at 8%?

SOLUTION

If x represents the amount invested at 9% and y represents the amount invested at 8%, then $x + y$ is the total investment,

$$x + y = 200,000 \quad (1)$$

and $0.09x + 0.08y$ is the total income earned.

$$0.09x + 0.08y = 17,200 \quad (2)$$

We solve these equations as follows:

$$\begin{array}{rl} -8x - 8y = -1,600,000 & (3) \quad \text{Multiply equation (1) by } -8. \\ \underline{9x + 8y = 1,720,000} & (4) \quad \text{Multiply equation (2) by 100.} \\ x = 120,000 & \text{Add (3) and (4).} \end{array}$$

We find y by using $x = 120,000$ in equation (1).

$$120,000 + y = 200,000 \quad \text{gives} \quad y = 80,000$$

Thus $120,000 is invested at 9%, and $80,000 is invested at 8%.

As a check, we note that equation (1) is satisfied and

$$\begin{aligned} 0.09(120,000) + 0.08(80,000) &= 10,800 + 6400 \\ &= 17,200 ✔ \end{aligned}$$

Three Equations in Three Variables

If a, b, c, and d represent constants, then

$$ax + by + cz = d$$

is a first-degree (linear) equation in three variables. When equations of this form are graphed in a three-dimensional coordinate system, their graphs are planes. Two different planes may intersect in a line (like two walls) or may not intersect at all (like a floor and ceiling). Three different planes may intersect in a single point (as when two walls meet the ceiling), may intersect in a line (as in a paddle wheel), or may not have a common intersection. (See Figures 1.17, 1.18, and 1.19.) Thus three linear equations in three variables may have a unique solution, infinitely many solutions, or no solution. In this section, we will discuss only systems of three linear equations in three variables that have unique solutions.

We can solve three equations in three variables using a systematic procedure called the **left-to-right elimination method**.

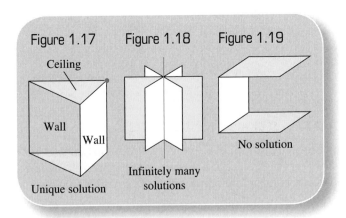

Figure 1.17   Figure 1.18   Figure 1.19

Ceiling

Wall

Wall

Unique solution

Infinitely many solutions

No solution

Procedure

To solve a system of three equations in three variables by the left-to-right elimination method:

1. If necessary, interchange two equations or use multiplication to make the coefficient of the first variable in equation (1) a factor of the other first variable coefficients.

2. Add multiples of the first equation to each of the following equations so that the coefficients of the first variable in the second and third equations become zero.

3. Add a multiple of the second equation to the third equation so that the coefficient of the second variable in the third equation becomes zero.

4. Solve the third equation and *back substitute* from the bottom to find the remaining variables.

Left-to-Right Elimination Method

EXAMPLE 1.26

➡ Solve: $\begin{cases} 2x + 4y + 5z = 4 \\ x - 2y - 3z = 5 \\ x + 3y + 4z = 1 \end{cases}$

SOLUTION

Interchange the first two equations:

$$\begin{aligned} x - 2y - 3z &= 5 \quad (1) \\ 2x + 4y + 5z &= 4 \quad (2) \\ x + 3y + 4z &= 1 \quad (3) \end{aligned}$$

Add $(-2) \times$ equation (1) to equation (2) and add $(-1) \times$ equation (1) to equation (3):

$$\begin{aligned} x - 2y - 3z &= 5 \quad (1) \\ 0x + 8y + 11z &= -6 \quad (2) \\ 0x + 5y + 7z &= -4 \quad (3) \end{aligned}$$

Add $\left(-\frac{5}{8}\right) \times$ equation (2) to equation (3):

$$\begin{aligned} x - 2y - 3z &= 5 \quad (1) \\ 8y + 11z &= -6 \quad (2) \\ 0y + \frac{1}{8}z &= -\frac{2}{8} \quad (3) \end{aligned}$$

$z = -2$ from equation (3)

$y = \frac{1}{8}(-6 - 11z) = 2$ from equation (2)

$x = 5 + 2y + 3z = 3$ from equation (1)

so $x = 3, y = 2, z = -2$

....

1.4 Exercises

The graphs of two equations are shown in Problems 1 and 2. Decide whether the system of equations in each problem has one solution, no solution, or an infinite number of solutions. If the system has one solution, estimate it.

1.

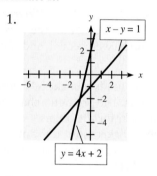

2.

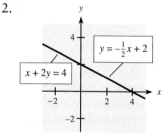

3. Solve the following system by using graphical methods.

$$\begin{cases} 4x - 2y = 4 \\ x - 2y = -2 \end{cases}$$

In Problems 4 and 5, solve the systems of equations by substitution.

4. $\begin{cases} 3x - 2y = 6 \\ 4y = 8 \end{cases}$

5. $\begin{cases} 2x - y = 2 \\ 3x + 4y = 6 \end{cases}$

In Problems 6 and 7, solve each system by elimination or by any convenient method.

6. $\begin{cases} 3x + 4y = 1 \\ 2x - 3y = 12 \end{cases}$

7. $\begin{cases} -4x + 3y = -5 \\ 3x - 2y = 4 \end{cases}$

8. Use the left-to-right elimination method to solve the system

$$\begin{cases} x + 4y - 2z = 9 \\ x + 5y + 2z = -2 \\ x + 4y - 28z = 22 \end{cases}$$

Applications

9. **Personal expenditures** Since 1995, the total personal expenditures (in billions of dollars) in the United States for food, $f(x)$, and for housing, $h(x)$, can be described by

$$f(x) = 40.74x + 742.65 \quad \text{and} \quad h(x) = 47.93x + 725$$

where x is the number of years past 1995. (*Source:* U.S. Department of Commerce) Find the year in which these expenditures were equal and the amount spent on each.

10. **Pricing** A concert promoter needs to make $42,000 from the sale of 1800 tickets. The promoter charges $20 for some tickets and $30 for the others.
    (a) If there are x of the $20 tickets sold and y of the $30 tickets sold, write an equation that states that the sum of the tickets sold is 1800.
    (b) How much money is received from the sale of x tickets for 20 dollars each?
    (c) How much money is received from the sale of y tickets for 30 dollars each?
    (d) Write an equation that states that the total amount received from the sale is 42,000 dollars.
    (e) Solve the equations simultaneously to find how many tickets of each type must be sold to yield the $42,000.

11. **Investment yields** One safe investment pays 10% per year, and a more risky investment pays 18% per year. A woman who has $145,600 to invest would like to have an income of $20,000 per year from her investments. How much should she invest at each rate?

12. **Medications** A nurse has two solutions that contain different concentrations of a certain medication. One is a 20% concentration and the other is a 5% concentration. How many cubic centimeters of each should he mix to obtain 10 cc of a 15.5% solution?

Application Problem 13 requires a system of equations in three variables.

13. **Nutrition** Each ounce of substance A supplies 5% of the nutrition a patient needs. Substance B supplies 15% of the required nutrition per ounce, and substance C supplies 12% of the required nutrition per ounce. If digestive restrictions require that substances A and C be given in equal amounts, and the amount of substance B be one-fifth of either of these other amounts, find the number of ounces of each substance that should be in the meal to provide 100% of the required nutrition.

Need more practice? Find more here: cengagebrain.com

1.5 Applications of Functions in Business and Economics

IN THIS SECTION WE FORM TOTAL COST, TOTAL REVENUE, AND PROFIT AS FUNCTIONS OF THE QUANTITY *X* THAT IS PRODUCED AND SOLD AND WHICH IS CALLED THE **THEORY OF THE FIRM.** WE WILL ALSO DISCUSS **MARKET ANALYSIS,** IN WHICH SUPPLY AND DEMAND ARE FOUND AS FUNCTIONS OF PRICE, AND MARKET EQUILIBRIUM IS FOUND.

Total Cost, Total Revenue, and Profit

The **profit** a firm makes on its product is the difference between the amount it receives from sales (its revenue) and its cost. If x units are produced and sold, we can write

$$P(x) = R(x) - C(x)$$

where
$P(x)$ = profit from sale of x units
$R(x)$ = total revenue from sale of x units
$C(x)$ = total cost of production and sale of x units

In general, we note the following about these functions. **Revenue** is is found by using the equation

$$\text{Revenue} = (\text{price per unit})(\text{number of units})$$

Cost is composed of two parts: fixed costs and variable costs.

$$\text{Cost} = \text{variable costs} + \text{fixed costs}$$

Additionally,

Revenue: 0 units produce 0 revenue; $R(0) = 0$.

Cost: 0 units' costs equal fixed costs; $C(0) = FC$.

Profit: 0 units yield a loss equal to fixed costs; $P(0) = -FC$.

Marginals

The slope of a profit function represents the rate of change in profit with respect to the number of units produced and sold. This is called the **marginal profit** ($\overline{MP}$) for the product. Similarly, the **marginal cost** ($\overline{MC}$) for this product is the slope of the cost function, and the **marginal revenue** ($\overline{MR}$) is the slope of the revenue function.

EXAMPLE 1.27 Marginal Cost

Suppose that the cost (in dollars) for a product is $C = 21.75x + 4890$. What is the marginal cost for this product, and what does it mean?

SOLUTION

The equation has the form $C = mx + b$, so the slope is 21.75. Thus the marginal cost is $\overline{MC} = 21.75$ dollars per unit.

Because the marginal cost is the slope of the cost line, production of each additional unit will cost $21.75 more, at any level of production.

Break-Even Analysis

We can solve the equations for total revenue and total cost simultaneously to find the **break-even point** where cost and revenue are equal. If we use x to represent the quantity produced and y to represent the dollar value of revenue *and* cost, the point where the total revenue line crosses the total cost line is the break-even point.

EXAMPLE 1.28 Break-Even Point

A manufacturer sells a product for $10 per unit. The manufacturer's fixed costs are $1200 per month, and the variable costs are $2.50 per unit. How many units must the manufacturer produce each month to break even?

SOLUTION

The total revenue for x units of the product is $R = R(x) = 10x$. The total cost for x units is $C = C(x) = 2.50x + 1200$. We find the break-even point by solving the two equations simultaneously ($R = C$ at the break-even point). By substitution,

$$10x = 2.50x + 1200$$
$$7.5x = 1200$$
$$x = 160$$

Thus the manufacturer will break even if 160 units are produced per month. (See Figure 1.20.)

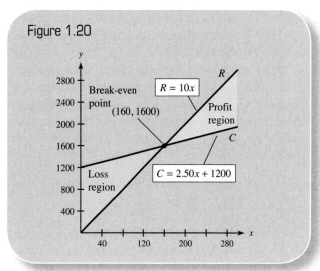

Figure 1.20

The profit function for the previous example is given by

$$P(x) = R(x) - C(x) = 7.50x - 1200$$

We can find the point where the profit is zero (the break-even point) by setting $P(x) = 0$ and solving for x, getting

$$x = 160$$

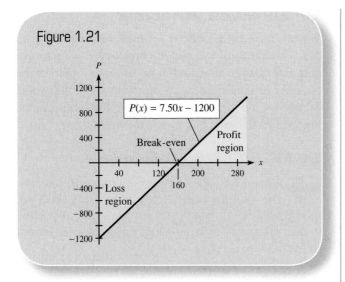

Figure 1.21

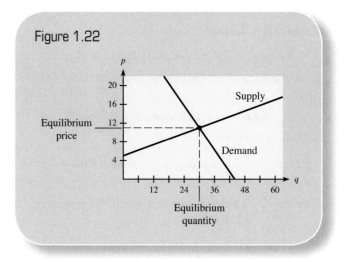

Figure 1.22

Note that this is the same break-even point that we found by solving the total revenue and total cost equations simultaneously (see Figure 1.21).

Supply, Demand, and Market Equilibrium

Demand by consumers for a commodity is related to the price of the commodity. The **law of demand** states that the quantity demanded will increase as price decreases and that the quantity demanded will decrease as price increases. The **law of supply** states that the quantity supplied for sale will increase as the price of a product increases. When graphing supply and demand, price is placed on the vertical axis.

If the supply and demand curves for a commodity are graphed on the same coordinate system, with the same units, market equilibrium occurs at the point where the curves intersect. The price at that point is the **equilibrium price,** and the quantity at that point is the **equilibrium quantity.**

For the supply and demand functions $p = \frac{1}{5}q + 5$ and $p = \frac{-11}{14}q + \frac{242}{7}$, respectively, shown in Figure 1.22, we see that the curves intersect at the point (30, 11). This means that when the price is $11, consumers are willing to purchase the same number of units (30) that producers are willing to supply.

EXAMPLE 1.29 Market Equilibrium

Suppose the supply and demand functions for a clothes dryer are $p = 2q + 170$ and $p = -5q + 450$, respectively. Find the equilibrium point for the market.

SOLUTION
Because the prices are equal at market equilibrium, we have

$$-5q + 450 = 2q + 170$$
$$280 = 7q$$
$$q = 40$$
$$p = 250$$

The equilibrium point is (40, 250). See Figure 1.23 for the graphs of these functions.

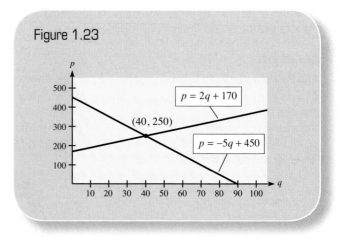

Figure 1.23

Supply, Demand, and Taxation

Let's examine how supply and demand are affected when the supplier's product is taxed. Suppose a supplier is taxed $K per unit sold, and the tax is passed on to the consumer by adding $K to the selling price of the product. If the original supply function $p = f(q)$ gives the supply price per unit, then passing the tax on gives a new supply function, $p = f(q) + K$. Because the value of the product is not changed by the tax, the demand function is unchanged. Figure 1.24 shows the effect that this has on market equilibrium.

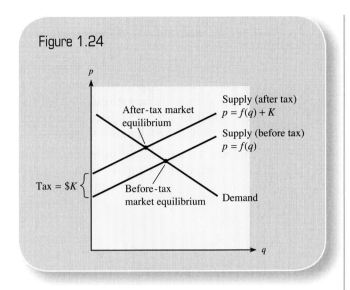

Figure 1.24

EXAMPLE 1.30 Taxation

➡ Suppose the supply and demand functions for dryers are given as follows.

$$\text{Supply:} \quad p = 2q + 170$$
$$\text{Demand:} \quad p = -5q + 450$$

The equilibrium point was $q = 40$, $p = \$250$. If the wholesaler is taxed \$14 per unit sold, what is the new equilibrium point?

SOLUTION

The \$14 tax per unit is passed on by the wholesaler, so the new supply function is

$$p = 2q + 170 + 14$$

and the demand function is unchanged. Thus we solve the system

$$\begin{cases} p = 2q + 184 \\ p = -5q + 450 \end{cases}$$

$$2q + 184 = -5q + 450$$
$$7q = 266$$
$$q = 38$$
$$p = 2(38) + 184 = 260$$

The new equilibrium point is $q = 38$, $p = \$260$.
    Checking, we see that

$$260 = 2(38) + 184 \checkmark \quad \text{and} \quad 260 = -5(38) + 450 \checkmark$$

1.5 Exercises

Total Cost, Total Revenue, and Profit

1. Suppose a radio manufacturer has the total cost function $C(x) = 43x + 1850$ and the total revenue function $R(x) = 80x$.
    (a) What is the equation of the profit function for this commodity?
    (b) What is the profit on 30 units? Interpret your result.
    (c) How many radios must be sold to avoid losing money?

2. A linear cost function is $C(x) = 5x + 250$.
    (a) What are the slope and the C-intercept?
    (b) What is the marginal cost, and what does it mean?
    (c) How are your answers to parts (a) and (b) related?
    (d) What is the cost of producing *one more* item if 50 are currently being produced? What is it if 100 are currently being produced?

3. A linear revenue function is $R = 27x$.
    (a) What is the slope?
    (b) What is the marginal revenue, and what does it mean?
    (c) What is the revenue received from selling *one more* item if 50 are currently being sold? If 100 are being sold?

4. Let $C(x) = 5x + 250$ and $R(x) = 27x$.
    (a) Write the profit function $P(x)$.
    (b) What is the slope of the profit function?
    (c) What is the marginal profit?
    (d) Interpret the marginal profit.

5. Extreme Protection, Inc. manufactures helmets for skiing and snowboarding. The fixed costs for one model of helmet are \$6600 per month. Materials and labor for each helmet of this model are \$35, and the company sells this helmet to dealers for \$60 each.
    (a) For this helmet, write the function for monthly total costs.
    (b) Write the function for total revenue.
    (c) Write the function for profit.
    (d) Find $C(200)$, $R(200)$, and $P(200)$ and interpret each answer.
    (e) Find $C(300)$, $R(300)$, and $P(300)$ and interpret each answer.
    (f) Find the marginal profit and write a sentence that explains its meaning.

Break-Even Analysis

6. The figure shows graphs of the total cost function and the total revenue function for a commodity.

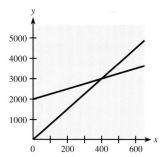

   (a) Label each function correctly.
   (b) Determine the fixed costs.
   (c) Locate the break-even point and determine the number of units sold to break even.
   (d) Estimate the marginal cost and marginal revenue.

7. A manufacturer sells belts for $12 per unit. The fixed costs are $1600 per month, and the variable costs are $8 per unit.
   (a) Write the equations of the revenue and cost functions.
   (b) Find the break-even point.

8. Electronic equipment manufacturer Dynamo Electric, Inc. makes several types of surge protectors. Their base model surge protector has monthly fixed costs of $1045. This particular model wholesales for $10 each and costs $4.50 per unit to manufacture.
   (a) Write the function for Dynamo's monthly total costs.
   (b) Write the function for Dynamo's monthly total revenue.
   (c) Write the function for Dynamo's monthly profit.
   (d) Find the number of this type of surge protector that Dynamo must produce and sell each month to break even.

In Problem 9, *some* of the graphs of total revenue (R), total cost (C), variable cost (VC), fixed cost (FC), and profit (P) are shown as functions of the number of units, x.
(a) Correctly label the graphs shown.
(b) Carefully sketch and label the graphs of the other functions. Explain your method.

9. $

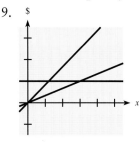

Supply, Demand, and Market Equilibrium

10. If the demand for a pair of shoes is given by $2p + 5q = 200$ and the supply function for it is $p - 2q = 10$, compare the quantity demanded and the quantity supplied when the price is $60. Will there be a surplus or shortfall at this price?

Complete Problems 11 and 12 by using the accompanying figure, which shows a supply function and a demand function.

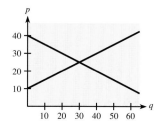

11. (a) Label each function as "demand" or "supply."
    (b) Label the equilibrium point and determine the price and quantity at which market equilibrium occurs.

12. (a) If the price is $20, what quantity is supplied?
    (b) If the price is $20, what quantity is demanded?
    (c) Is there a surplus or a shortage when the price is $20? How many units is this surplus or shortage?

13. Find the market equilibrium point for the following demand and supply functions.

$$\text{Demand:} \qquad 2p = -q + 56$$
$$\text{Supply:} \quad 3p - q = 34$$

14. A group of retailers will buy 80 televisions from a wholesaler if the price is $350 and 120 if the price is $300. The wholesaler is willing to supply 60 if the price is $280 and 140 if the price is $370. Assuming the resulting supply and demand functions are linear, find the equilibrium point for the market.

Supply, Demand, and Taxation

Use the figure to answer Problem 15.

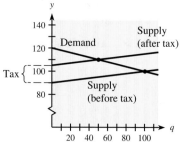

15. (a) What is the amount of the tax?
    (b) What are the original equilibrium price and quantity?

(c) What are the new equilibrium price and quantity?

(d) Does the supplier suffer from the tax even though it is passed on?

16. Suppose that in a certain market the demand function for a product is given by $60p + q = 2100$ and the supply function is given by $120p - q = 540$. Then a tax of \$0.50 per item is levied on the supplier, who passes it on to the consumer as a price increase. Find the equilibrium price and quantity after the tax is levied.

Need more practice?
Find more here: cengagebrain.com

Chapter Exercises

Solve the equations in Problems 1–4.

1. $2x - 8 = 3x + 5$

2. $2x + \dfrac{1}{2} = \dfrac{x}{2} + \dfrac{1}{3}$

3. $\dfrac{6}{3x - 5} = \dfrac{6}{2x + 3}$

4. $\dfrac{2x + 5}{x + 7} = \dfrac{1}{3} + \dfrac{x - 11}{2x + 14}$

5. Solve for y: $3(y - 2) = -2(x + 5)$

6. Solve $3x - 9 \le 4(3 - x)$ and graph the solution.

7. Solve $5x + 1 \ge \frac{2}{3}(x - 6)$ and graph the solution.

8. If $p = 3q^3$, is p a function of q?

9. If $y^2 = 9x$, is y a function of x?

10. What are the domain and range of the function $y = \sqrt{9 - x}$?

11. If $f(x) = x^2 + 4x + 5$, find the following.

(a) $f(-3)$ (b) $f(4)$ (c) $f\left(\dfrac{1}{2}\right)$

12. If $g(x) = x^2 + 1/x$, find the following.

(a) $g(-1)$ (b) $g\left(\dfrac{1}{2}\right)$ (c) $g(0.1)$

13. If $f(x) = 9x - x^2$, find $\dfrac{f(x + h) - f(x)}{h}$ and simplify.

14. Does the graph in Figure 1.25 represent y as a function of x?

15. Does the graph in Figure 1.26 represent y as a function of x?

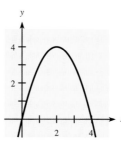

Figure 1.25

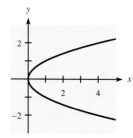

Figure 1.26

16. For the function f graphed in Figure 1.25, what is $f(2)$?

17. The following table defines y as a function of x, denoted $y = f(x)$.

x	-2	-1	0	1	3	4
y	8	2	-3	4	2	7

Use the table to complete the following.
(a) Identify the domain and range of $y = f(x)$.
(b) Find $f(4)$.
(c) Find all x-values for which $f(x) = 2$.
(d) Graph $y = f(x)$.
(e) Does the table define x as a function of y? Explain.

18. If $f(x) = 3x + 5$ and $g(x) = x^2$, find
(a) $(f + g)(x)$ (b) $(f/g)(x)$
(c) $f(g(x))$ (d) $(f \circ f)(x)$

In Problems 19 and 20, find the intercepts and graph.

19. $5x + 2y = 10$

20. $x = -2$

21. Find the slope of the line that passes through the points $(2, -1)$ and $(-1, -4)$.

22. Find the slope and y-intercept of the line $2x + 5y = 10$.

In Problems 23–29, write the equation of each line described.

23. Slope 4 and y-intercept 2

24. Slope $-\frac{1}{2}$ and y-intercept 3

25. Through $(-2, 1)$ with slope $\frac{2}{5}$

26. Through $(-2, 7)$ and $(6, -4)$

27. Through $(-1, 8)$ and $(-1, -1)$

28. Through $(1, 6)$ and parallel to $y = 4x - 6$

29. Through $(-1, 2)$ and perpendicular to
$3x + 4y = 12$

In Problems 30–35, solve each system of equations.

30. $\begin{cases} 2x + y = 19 \\ x - 2y = 12 \end{cases}$

31. $\begin{cases} 3x + 2y = 5 \\ 2x - 3y = 12 \end{cases}$

32. $\begin{cases} 6x + 3y = 1 \\ y = -2x + 1 \end{cases}$

33. $\begin{cases} 4x - 3y = 253 \\ 13x + 2y = -12 \end{cases}$

34. $\begin{cases} x + 2y + 3z = 5 \\ y + 11z = 21 \\ 5y + 9z = 13 \end{cases}$

35. $\begin{cases} x + y - z = 12 \\ 2y - 3z = -7 \\ 3x + 3y - 7z = 0 \end{cases}$

Applications

36. **Course grades** In a certain course, grades are based on three tests worth 100 points each, three quizzes worth 50 points each, and a final exam worth 200 points. A student has test grades of 91, 82, and 88, and quiz grades of 50, 42, and 42. What is the lowest percent the student can get on the final and still earn an A (90% or more of the total points) in the course?

37. **Cost analysis** The owner of a small construction business needs a new truck. He can buy a diesel truck for $38,000 and it will cost him $0.24 per mile to operate. He can buy a gas engine truck for $35,600 and it will cost him $0.30 per mile to operate. Find the number of miles he must drive before the costs are equal. If he normally keeps a truck for 5 years, which is the better buy?

38. **Heart disease risk** The Multiple Risk Factor Intervention Trial (MRFIT) used data from 356,222 men aged 35 to 57 to investigate the relationship between serum cholesterol and coronary heart disease (CHD) risk. Figure 1.27 shows the graph of the relationship of CHD risk and cholesterol, where a risk of 1 is assigned to 200 mg/dl of serum

cholesterol and where the CHD risk is 4 times as high when serum cholesterol is 300 mg/dl.

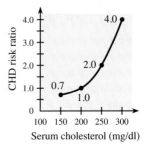

Figure 1.27

(a) Does this graph indicate that the CHD risk is a function of the serum cholesterol?
(b) Is the relationship a linear function?
(c) If CHD risk is a function f of serum cholesterol, what is $f(300)$?

39. **Mortgage loans** When a couple purchases a home, one of the first questions they face deals with the relationship between the amount borrowed and the monthly payment. In particular, if a bank offers 25-year loans at 7% interest, then the data in the following table would apply.

Amount Borrowed	Monthly Payment
$40,000	$282.72
50,000	353.39
60,000	424.07
70,000	494.75
80,000	565.44
90,000	636.11
100,000	706.78

Source: Comprehensive Mortgage Payment Tables, Publication No. 492, Financial Publishing Co., Boston

Assume that the monthly payment P is a function of the amount borrowed A (in thousands) and is denoted by $P = f(A)$, and answer the following.
(a) Find $f(80)$.
(b) Write a sentence that explains the meaning of $f(70) = 494.75$.

40. **Profit** Suppose that the profit from the production and sale of x units of a product is given by

$$P(x) = 330x - 0.05x^2 - 5000$$

In addition, suppose that for a certain month, the number of units produced on day t of the month is

$$x = q(t) = 100 + 10t$$

(a) Find $(P \circ q)(t)$ to express the profit as a function of the day of the month.
(b) Find the number of units produced, and the profit, on the 15th day of the month.

41. ***Distance to a thunderstorm*** The distance d (in miles) to a thunderstorm is given by

$$d = \frac{t}{4.8}$$

where t is the number of seconds that elapse between seeing the lightning and hearing the thunder.
(a) Graph this function for $0 \le t \le 20$.
(b) The point $(9.6, 2)$ satisfies the equation. Explain its meaning.

42. ***Profit*** A company charting its profits notices that the relationship between the number of units sold, x, and the profit, P, is linear.
(a) If 200 units sold results in \$3100 profit and 250 units sold results in \$6000 profit, write the profit function for this company.
(b) Interpret the slope from part (a) as a rate of change.

43. ***Health care costs*** The average annual cost per consumer for health care can be modeled by

$$A = 71.72x + 1401.36$$

where x is the number of years from 1990. (*Source:* www.infoplease.com)
(a) Is A a linear function of x?
(b) Find the slope and A-intercept of this function.
(c) Write a sentence that interprets the A-intercept.
(d) Write a sentence that interprets the slope of this function as a rate of change.

44. ***Temperature*** Write the equation of the linear relationship between temperature in Celsius (C) and Fahrenheit (F) if water freezes at $0°C$ and $32°F$ and boils at $100°C$ and $212°F$.

45. ***Investment mix*** A retired couple have \$150,000 to invest and want to earn \$15,000 per year in interest. The safer investment yields 9.5%, but they can supplement their earnings by investing some of their money at 11%. How much should they invest at each rate to earn \$15,000 per year?

46. ***Botany*** A botanist has a 20% solution and a 70% solution of an insecticide. How much of each must be used to make 4.0 liters of a 35% solution?

47. ***Supply and demand*** A certain product has supply and demand functions $p = 4q + 5$ and $p = -2q + 81$, respectively.
(a) If the price is \$53, how many units are supplied and how many are demanded?
(b) Does this give a shortfall or a surplus?
(c) Is the price likely to increase from \$53 or decrease from it?

48. ***Market analysis*** Of the equations $p + 6q = 420$ and $p = 6q + 60$, one is the supply function for a product and one is the demand function for that product.
(a) Graph these equations on the same set of axes.
(b) Label the supply function and the demand function.
(c) Find the market equilibrium point.

49. ***Cost, revenue, and profit*** The total cost and total revenue for a certain product are given by the following:

$$C(x) = 38.80x + 4500$$
$$R(x) = 61.30x$$

(a) Find the marginal cost.
(b) Find the marginal revenue.
(c) Find the marginal profit.
(d) Find the number of units required to break even.

50. ***Cost, revenue, and profit*** A certain commodity has the following costs for a period.

Fixed costs:	\$1500
Variable costs:	\$22 per unit

The commodity is sold for \$52 per unit.
(a) What is the total cost function?
(b) What is the total revenue function?
(c) What is the profit function?
(d) What is the marginal cost?
(e) What is the marginal revenue?
(f) What is the marginal profit?
(g) What is the break-even point?

51. ***Market analysis*** The supply function and the demand function for a product are linear and are determined by the tables that follow. Find the quantity and price that will give market equilibrium.

Supply Function		**Demand Function**	
Price	Quantity	Price	Quantity
100	200	200	200
200	400	100	400
300	600	0	600

52. ***Market analysis*** Suppose that for a certain product the supply and demand functions prior to any taxation are

$$\text{Supply:} \quad p = \frac{q}{10} + 8$$
$$\text{Demand:} \quad 10p + q = 1500$$

If a tax of \$2 per item is levied on the supplier and is passed on to the consumer as a price increase, find the market equilibrium after the tax is levied.

Need more practice?
Find more here: cengagebrain.com

chapter 2

Quadratic and
Other Special Functions

In this chapter we will discuss quadratic functions and their applications, and we will also discuss other types of functions, including identity, constant, power, absolute value, piecewise defined, and reciprocal functions. Graphs of polynomial and rational functions will also be introduced; they will be studied in detail in Chapter 10.

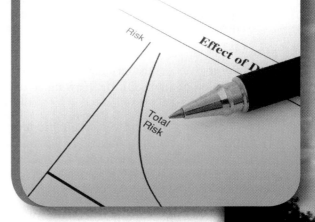

2.1 Quadratic Equations

IN THIS SECTION, WE LEARN HOW TO SOLVE EQUATIONS OF THIS TYPE, USING FACTORING METHODS AND USING THE QUADRATIC FORMULA.

Factoring Methods

A **quadratic equation** in one variable is an equation that can be written in the *general form*

$$ax^2 + bx + c = 0 \quad (a \neq 0)$$

where *a*, *b*, and *c* represent constants. For example, the equations

$$3x^2 + 4x + 1 = 0 \quad \text{and} \quad 2x^2 + 1 = x^2 - x$$

are quadratic equations; the first of these is in general form, and the second can easily be rewritten in general form.

When we solve quadratic equations, we will be interested only in real number solutions and will consider two methods of solution: factoring and the quadratic formula. We will discuss solving quadratic equations by factoring first.

Solution by factoring is based on the Zero Product property of the real numbers.

EXAMPLE 2.1 Solving Quadratic Equations

Solve: (a) $6x^2 + 3x = 4x + 2$ (b) $6x^2 = 9x$

SOLUTION

(a) $6x^2 + 3x = 4x + 2$

$6x^2 - x - 2 = 0$ Form for factoring

$(3x - 2)(2x + 1) = 0$

$3x - 2 = 0$ or $2x + 1 = 0$ Set factors equal to zero.

$$x = \frac{2}{3} \qquad\qquad x = -\frac{1}{2}$$

(b) $6x^2 = 9x$

$6x^2 - 9x = 0$

$3x(2x - 3) = 0$

$3x = 0$ or $2x - 3 = 0$

$x = 0$ or $2x = 3$

$$x = \frac{3}{2}$$

Thus the solutions are $x = 0$ and $x = \frac{3}{2}$.

Shutterstock

> **Zero Product Property:** For real numbers a and b, $ab = 0$ if and only if $a = 0$ or $b = 0$ or both. Hence, to solve by factoring, we must first write the equation with zero on one side.

Note that in Example 2.1(b) it is tempting to divide both sides of the equation by x, but this is incorrect because it results in the loss of the solution $x = 0$. Never divide both sides of an equation by an expression containing the variable.

EXAMPLE 2.2 Solving by Factoring

 Solve: $\dfrac{x + 1}{3x + 6} = \dfrac{3}{x} + \dfrac{2x + 6}{x(3x + 6)}$

SOLUTION

The LCD of all fractions is $x(3x + 6)$. Multiplying both sides of the equation by this LCD gives a quadratic equation that is equivalent to the original equation for $x \neq 0$ and $x \neq -2$. (The original equation is undefined for these values.)

$$\frac{x + 1}{3x + 6} = \frac{3}{x} + \frac{2x + 6}{x(3x + 6)}$$

$$x(3x + 6)\frac{x + 1}{3x + 6} = x(3x + 6)\left(\frac{3}{x} + \frac{2x + 6}{x(3x + 6)}\right) \begin{smallmatrix}\text{Multiply}\\\text{both}\\\text{sides by}\\x(3x+6).\end{smallmatrix}$$

$$x(x + 1) = 3(3x + 6) + (2x + 6)$$

$$x^2 + x = 9x + 18 + 2x + 6$$

$$x^2 - 10x - 24 = 0$$

$$(x - 12)(x + 2) = 0$$

$$x - 12 = 0 \quad \text{or} \quad x + 2 = 0$$

$$x = 12 \quad \text{or} \quad x = -2$$

Checking $x = 12$ and $x = -2$ in the original equation, we see that $x = -2$ does not check because it makes the denominator equal to zero. Hence the only solution is $x = 12$.

EXAMPLE 2.3 Falling Object

A tennis ball is thrown into a swimming pool from the top of a tall hotel. The height of the ball from the pool is modeled by $D(t) = -16t^2 - 4t + 300$ feet, where t is the time, in seconds, after the ball was thrown. How long after the ball was thrown was it 144 feet above the pool?

SOLUTION

To find the number of seconds until the ball is 144 feet above the pool, we solve the equation

$$144 = -16t^2 - 4t + 300$$

for t. The solution follows.

$$144 = -16t^2 - 4t + 300$$

$$0 = -16t^2 - 4t + 156$$

$$0 = 4t^2 + t - 39$$

$$0 = (t - 3)(4t + 13)$$

$$t - 3 = 0 \quad \text{or} \quad 4t + 13 = 0$$

$$t = 3 \quad \text{or} \quad t = -13/4$$

This indicates that the ball will be 144 feet above the pool 3 seconds after it is thrown. The negative value for t has no meaning in this application.

The Quadratic Formula

In general, we can solve quadratic equations of the form $x^2 = C$ (no x-term) by taking the square root of both sides.

Square Root Property: The solution of $x^2 = C$ is

$$x = \pm\sqrt{C}$$

EXAMPLE 2.4 Square Root Method

Solve the following equations.
(a) $4x^2 = 5$ (b) $(3x - 4)^2 = 9$

SOLUTION

We can use the square root property for both parts.
(a) $4x^2 = 5$ is equivalent to $x^2 = \frac{5}{4}$. Thus

$$x = \pm\sqrt{\frac{5}{4}} = \pm\frac{\sqrt{5}}{\sqrt{4}} = \pm\frac{\sqrt{5}}{2}$$

(b) $(3x - 4)^2 = 9$ is equivalent to $3x - 4 = \pm\sqrt{9}$. Thus

$$3x - 4 = 3 \qquad\qquad 3x - 4 = -3$$
$$3x = 7 \qquad\qquad 3x = 1$$
$$x = \frac{7}{3} \qquad\qquad x = \frac{1}{3}$$

The solution of the general quadratic equation $ax^2 + bx + c = 0$, where $a \neq 0$ is called the **quadratic formula.**

Quadratic Formula:

If $ax^2 + bx + c = 0$, where $a \neq 0$ then
$$x = \frac{-b \pm \sqrt{b^2 - 4ac}}{2a}$$
See the Enrichment material on Course-Mate for MATH APPS at cengagebrain.com for a proof.

EXAMPLE 2.5 Quadratic Formula

Use the quadratic formula to solve $2x^2 - 3x - 6 = 0$ for x.

SOLUTION

The equation is already in general form, with $a = 2$, $b = -3$, and $c = -6$. Hence

$$x = \frac{-b \pm \sqrt{b^2 - 4ac}}{2a}$$
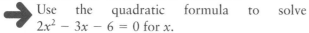
$$= \frac{-(-3) \pm \sqrt{(-3)^2 - 4(2)(-6)}}{2(2)}$$
$$= \frac{3 \pm \sqrt{57}}{4}$$

Thus the solutions are

$$x = \frac{3 + \sqrt{57}}{4} \quad \text{and} \quad x = \frac{3 - \sqrt{57}}{4}$$

EXAMPLE 2.6 Nonreal Solution

Using the quadratic formula, find the (real) solutions to $x^2 = x - 1$.

SOLUTION

We must rewrite the equation in general form before we can determine the values of a, b, and c.

$$x^2 = x - 1$$
$$x^2 - x + 1 = 0 \qquad \text{Note: } a = 1, b = -1, \text{ and } c = 1.$$
$$x = \frac{-(-1) \pm \sqrt{(-1)^2 - 4(1)(1)}}{2(1)}$$
$$x = \frac{1 \pm \sqrt{-3}}{2}$$

Because $\sqrt{-3}$ is not a real number, the values of x are not real. Hence there are no real solutions to the given equation.

Graphing Utilities
GRAPHICAL SOLUTIONS OF EQUATIONS

We can use graphing utilities to find the x-intercepts of graphs of functions. Because an x-intercept is an x-value that makes the function equal to 0, the x-intercept is also a zero of the function. Thus if an equation is written in the form $0 = f(x)$, the x-intercept of $y = f(x)$ is a solution of the equation. This method of solving an equation is called the **x-intercept method.**

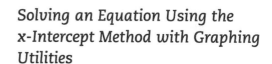

Solving an Equation Using the x-Intercept Method with Graphing Utilities

1. Rewrite the equation to be solved with 0 (and nothing else) on one side of the equation.

2. Enter the nonzero side of the equation found in the previous step in the equation editor of your graphing utility, and graph the equation in an appropriate viewing window. Be certain that you can see the graph cross the horizontal axis.

3. Find or approximate the x-intercept by inspection with TRACE or by using the graphical solver that is often called ZERO or ROOT. The x-intercept is the value of x that makes the equation equal to zero, so it is the solution to the equation. The value of x found by this method is often displayed as a decimal approximation of the exact solution rather than as the exact solution.

4. Verify the solution in the original equation.

We can use graphing utilities to determine (or approximate) the solutions of quadratic equations. See Solving Equations with a Graphing Calculator on the Tech Card.

Spreadsheet Note

A spreadsheet such as Excel can also be used to solve a quadratic equation. See Solving Equations with Excel on the Tech Card.

....

2.1 Exercises

In Problems 1 and 2, solve each equation by factoring.

1. $9 - 4x^2 = 0$

2. $4t^2 - 4t + 1 = 0$

In Problems 3 and 4, solve each equation by using the quadratic formula. Give real answers (a) exactly and (b) rounded to two decimal places.

3. $x^2 - 4x - 4 = 0$

4. $2w^2 + w + 1 = 0$

In Problems 5 and 6, find the exact real solutions to each equation, if they exist.

5. $5x^2 = 80$

6. $(x + 4)^2 = 25$

In Problems 7–9, use any method to find the exact real solutions, if they exist.

7. $x^2 + 5x = 21 + x$

8. $\dfrac{w^2}{8} - \dfrac{w}{2} - 4 = 0$

9. $(x - 1)(x + 5) = 7$

In Problems 10 and 11, multiply both sides of the equation by the LCD, and solve the resulting quadratic equation.

10. $x + \dfrac{8}{x} = 9$

11. $\dfrac{x}{x - 1} = 2x + \dfrac{1}{x - 1}$

Applications

12. **Profit** If the profit from the sale of x units of a product is $P = 90x - 200 - x^2$, what level(s) of production will yield a profit of $1200?

13. **Profit** Suppose the profit from the sale of x units of a product is

$$P = 6400x - 18x^2 - 400$$

(a) What level(s) of production will yield a profit of $61,800?

(b) Can a profit of more than $61,800 be made?

14. **Flight of a ball** If a ball is thrown upward at 96 feet per second from the top of a building that is 100 feet high, the height of the ball can be modeled by

$$S = 100 + 96t - 16t^2 \text{ feet}$$

where t is the number of seconds after the ball is thrown. How long after it is thrown is the height 100 feet?

15. **Wind and pollution** The amount of airborne particulate pollution p from a power plant depends on the wind speed s, among other things, with the relationship between p and s approximated by

$$p = 25 - 0.01s^2$$

(a) Find the value(s) of s that will make $p = 0$.

(b) What value of s makes sense in the context of this application? What does $p = 0$ mean in this application?

16. **Corvette acceleration** The time t, in seconds, that it takes a 2008 Corvette to accelerate to x miles per hour can be described by

$$t = 0.001(0.732x^2 + 15.417x + 607.738)$$

(*Source:* Motor Trend). How fast is the Corvette going after 8.99 seconds? Give your answer to the nearest tenth.

17. **Marijuana use** For the years from 1995 to 2006, the percentage p of high school seniors who have tried marijuana can be considered as a function of time t according to

$$p = f(t) = -0.22t^2 + 4.49t + 26.3$$

where t is the number of years past 1990. (*Source:* National Institute on Drug Abuse) In what year after 1990 will the percentage predicted by the function fall to 18.6% if this function remains valid?

....

Need more practice?
Find more here: cengagebrain.com

2.2 Quadratic Functions: Parabolas

IN THIS SECTION WE WILL DESCRIBE WAYS TO FIND THE MAXIMUM POINT OR MINIMUM POINT FOR A QUADRATIC FUNCTION.

Parabolas

The general equation of a quadratic function has the form

$$y = f(x) = ax^2 + bx + c$$

where a, b, and c are real numbers and $a \neq 0$.

The graph of a quadratic function,

$$y = ax^2 + bx + c \quad (a \neq 0).$$

has a distinctive shape called a **parabola.**

The basic function $y = x^2$ and a variation of it, $y = -\frac{1}{2}x^2$, are parabolas whose graphs are shown in Figure 2.1.

Vertex of a Parabola

As these examples illustrate, the graph of $y = ax^2$ is a parabola that opens upward if $a > 0$ and downward if $a < 0$. The **vertex,** where the parabola turns, is a **minimum point** if $a > 0$ and a **maximum point** if $a < 0$. The vertical line through the vertex of a parabola is called the **axis of symmetry** because one half of the graph is a reflection of the other half through this line.

The graph of $y = (x - 2)^2 - 1$ is the graph of $y = x^2$ shifted to a new location that is 2 units to the right and 1 unit down; its vertex is shifted from $(0, 0)$ to $(2, -1)$ and its axis of symmetry is shifted 2 units to the right. (See Figure 2.2.)

If we know the location of the vertex and the direction in which the parabola opens, we need very few other points to make a good sketch.

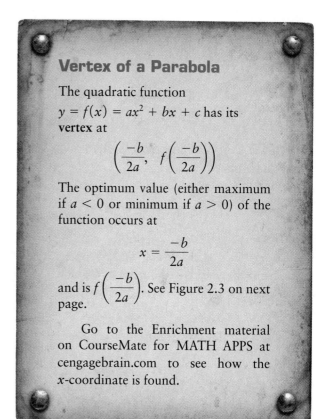

Vertex of a Parabola

The quadratic function
$$y = f(x) = ax^2 + bx + c \text{ has its}$$
vertex at

$$\left(\frac{-b}{2a}, \ f\left(\frac{-b}{2a}\right) \right)$$

The optimum value (either maximum if $a < 0$ or minimum if $a > 0$) of the function occurs at

$$x = \frac{-b}{2a}$$

and is $f\left(\dfrac{-b}{2a}\right)$. See Figure 2.3 on next page.

Go to the Enrichment material on CourseMate for MATH APPS at cengagebrain.com to see how the x-coordinate is found.

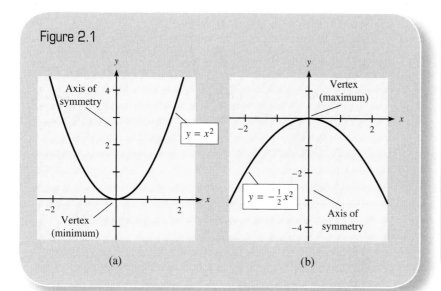

Figure 2.1

(a)

(b)

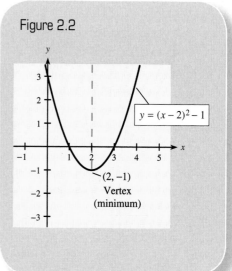

Figure 2.2

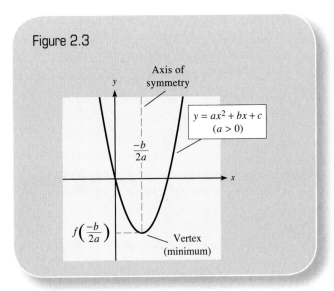

Figure 2.3

EXAMPLE 2.7 Vertex and Graph of a Parabola

 Find the vertex and sketch the graph of

$$f(x) = 2x^2 - 4x + 4$$

SOLUTION

Because $a = 2 > 0$, the graph of $f(x)$ opens upward and the vertex is the minimum point. We can calculate its coordinates as follows:

$$x = \frac{-b}{2a} = \frac{-(-4)}{2(2)} = 1$$

$$y = f(1) = 2$$

Thus the vertex is $(1, 2)$. Using x-values on either side of the vertex to plot additional points enables us to sketch the graph accurately. (See Figure 2.4.)

We can also use the coordinates of the vertex to find maximum or minimum values without a graph.

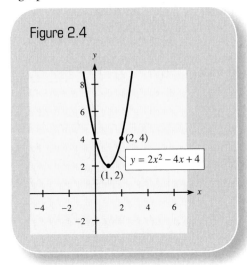

Figure 2.4

EXAMPLE 2.8 Profit Maximization

 For the profit function

$$P(x) = -0.1x^2 + 300x - 1200$$

find the number of units that give maximum profit and find the maximum profit.

SOLUTION

$P(x)$ is a quadratic function with $a < 0$. Thus the graph of $y = P(x)$ is a parabola that opens downward, so the vertex is a maximum point. The coordinates of the vertex are

$$x = \frac{-b}{2a} = \frac{-300}{2(-0.1)} = 1500$$

$$P = P(1500) = -0.1(1500)^2 + 300(1500) - 1200 = 223,800$$

Therefore, the maximum profit is $223,800 when 1500 units are sold.

The information that is useful in graphing quadratic functions is summarized as follows.

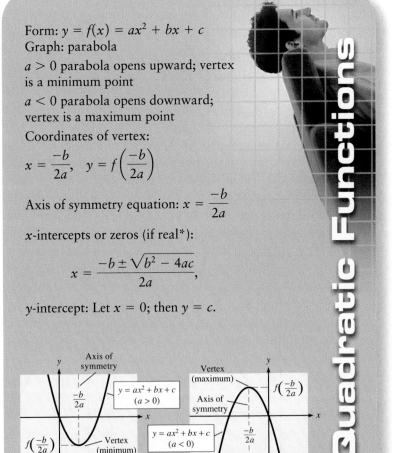

Form: $y = f(x) = ax^2 + bx + c$
Graph: parabola

$a > 0$ parabola opens upward; vertex is a minimum point

$a < 0$ parabola opens downward; vertex is a maximum point

Coordinates of vertex:

$$x = \frac{-b}{2a}, \quad y = f\left(\frac{-b}{2a}\right)$$

Axis of symmetry equation: $x = \frac{-b}{2a}$

x-intercepts or zeros (if real*):

$$x = \frac{-b \pm \sqrt{b^2 - 4ac}}{2a},$$

y-intercept: Let $x = 0$; then $y = c$.

Quadratic Functions

*If the zeros are not real, the graph does not cross the x-axis.

EXAMPLE 2.9 Graph of a Quadratic Function

➡️ For the function $y = 4x - x^2$, determine whether its vertex is a maximum point or a minimum point and find the coordinates of this point; find the zeros, if any exist; and sketch the graph.

SOLUTION

The proper form is $y = -x^2 + 4x + 0$, so $a = -1$. Thus the parabola opens downward, and the vertex is the highest (maximum) point.

The vertex occurs at $x = \dfrac{-b}{2a} = \dfrac{-4}{2(-1)} = 2$.

The y-coordinate of the vertex is
$$f(2) = -(2)^2 + 4(2) = 4.$$
The zeros of the function are solutions to
$$-x^2 + 4x = 0$$
$$x(-x + 4) = 0$$
$$\text{or } x = 0 \text{ and } x = 4$$

The graph of the function can be found by drawing a parabola with these three points (see Figure 2.5(a)), or by using a graphing utility or a spreadsheet (see Figure 2.5(b)). Also see the Tech Card for additional examples.

EXAMPLE 2.10 Maximizing Revenue

➡️ Ace Cruises offers a sunset cruise to a group of 50 people for a price of $30 per person, but it reduces the price per person by $0.50 for each additional person above the 50.

(a) Does reducing the price per person to get more people in the group give the cruise company more revenue?

(b) How many people will provide maximum revenue for the cruise company?

Shutterstock

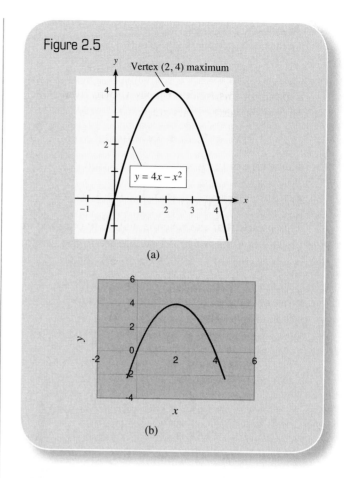

Figure 2.5

Vertex (2, 4) maximum

$y = 4x - x^2$

(a)

(b)

SOLUTION

(a) The revenue to the company if 50 people are in the group and each pays $30 is $50(\$30) = \$1500$. The table shows the revenue for the addition of people to the group. The table shows that as the group size begins to increase past 50 people, the revenue also increases. However, it also shows that increasing the group size too much (to 70 people) reduces revenue.

Increase in Group Size	Number of People	Decrease in Price	New Price ($)	Revenue ($)
0	50	0	30	$50(30) = 1500$
1	51	$0.50(1)$	29.50	$51(29.50) = 1504.50$
2	52	$0.50(2) = 1$	29	$52(29) = 1508$
3	53	$0.50(3) = 1.50$	28.50	$53(28.50) = 1510.50$
⋮	⋮	⋮	⋮	⋮
20	70	$0.50(20) = 10$	20	$70(20) = 1400$
⋮	⋮	⋮	⋮	⋮
x	$50 + x$	$0.50(x)$	$30 - 0.50x$	$(50 + x)(30 - 0.50x)$

(b) The last entry in the table shows the revenue for an increase of x people in the group.

$$R(x) = (50 + x)(30 - 0.50x)$$

Expanding this function gives a form from which we can find the vertex of its graph.

$$R(x) = 1500 + 5x - 0.50x^2$$

The vertex of the graph of this function is

$$x = \frac{-b}{2a} = \frac{-5}{2(-0.50)} = 5, \; R(5) = 1512.50$$

This means that the revenue will be maximized at $1512.50 when $50 + 5 = 55$ people are in the group, with each paying $30 - 0.50(5) = 27.50$ dollars.

Average Rate of Change: The average rate of change of $f(x)$ with respect to x over the interval from $x = a$ to $x = b$ (where $a < b$) is calculated as

$$\text{Average rate of change} = \frac{\text{change in } f(x)}{\text{corresponding change in } x\text{-values}}$$

$$= \frac{f(b) - f(a)}{b - a}$$

The average rate of change is also the slope of the segment (or secant line) joining the points $(a, f(a))$ and $(b, f(b))$.

$$m = \frac{f(b) - f(a)}{b - a}$$

EXAMPLE 2.11 Average Rate of Change of Revenue

The revenue for a sunset cruise is defined by the quadratic function

$$R(x) = 1500 + 5x - 0.50x^2$$

where x is the increase in the group size beyond 50 people. What is the average rate of change of revenue if the group increases from 50 to 55 people?

SOLUTION

The average rate of change of revenue from 50 to 55 people is the average rate of change of the function from $x = 0$ to $x = 5$.

$$\frac{R(5) - R(0)}{5 - 0} = \frac{1512.50 - 1500}{5} = 2.50 \text{ dollars per person}$$

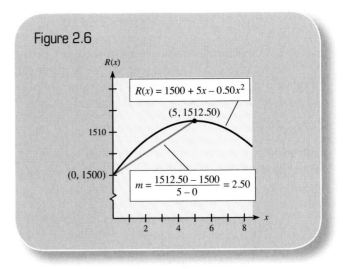

Figure 2.6

$R(x)$

$R(x) = 1500 + 5x - 0.50x^2$

(5, 1512.50)

1510

(0, 1500)

$$m = \frac{1512.50 - 1500}{5 - 0} = 2.50$$

This average rate of change is also the slope of the secant line connecting the points $(0, 1500)$ and $(5, 1512.50)$ on the graph of the function (see Figure 2.6).

2.2 Exercises

In Problems 1–3, (a) find the vertex of the graph of the equation, (b) determine if the vertex is a maximum or minimum point, (c) determine what value of x gives the optimal value of the function, and (d) determine the optimal (maximum or minimum) value of the function.

1. $y = \frac{1}{2}x^2 + x$

2. $y = 8 + 2x - x^2$

3. $f(x) = 6x - x^2$

In Problems 4–6, determine whether each function's vertex is a maximum point or a minimum point and find the coordinates of this point. Find the zeros, if any exist, and sketch the graph of the function.

4. $y = x - \frac{1}{4}x^2$

5. $y = x^2 + 4x + 4$

6. $\frac{1}{2}x^2 + x - y - 3 = 0$

In Problems 7 and 8, find the average rate of change of the function between the given values of x.

7. $y = -5x - x^2$ between $x = -1$ and $x = 1$.

8. $y = 8 + 3x + 0.5x^2$ between $x = 2$ and $x = 4$.

Applications

9. **Profit** The daily profit from the sale of a product is given by $P = 16x - 0.1x^2 - 100$ dollars.
   (a) What level of production maximizes profit?
   (b) What is the maximum possible profit?

10. **Crop yield** The yield in bushels from a grove of orange trees is given by $Y = x(800 - x)$, where x is the number of orange trees per acre. How many trees will maximize the yield?

11. **Projectiles** Two projectiles are shot into the air from the same location. The paths of the projectiles are parabolas and are given by
    (a) $y = -0.0013x^2 + x + 10$ and
    (b) $y = \dfrac{-x^2}{81} + \dfrac{4}{3}x + 10$

    where x is the horizontal distance and y is the vertical distance, both in feet. Determine which projectile goes higher by locating the vertex of each parabola.

12. **Cost** The figure shows the graph of a total cost function, with x equal to the number of units produced.
    (a) Is the average rate of change of cost greater from $x = a$ to $x = b$ or from $x = b$ to $x = c$? Explain.
    (b) Would the number of units d need to satisfy $d < b$ or $d > b$ for the average rate of change of cost from $x = a$ to $x = d$ to be greater than that from $x = a$ to $x = b$? Explain.

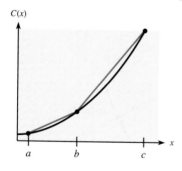

13. **Apartment rental** The owner of an apartment building can rent all 50 apartments if she charges $600 per month, but she rents one fewer apartment for each $20 increase in monthly rent.
    (a) Construct a table that gives the revenue generated if she charges $600, $620, and $640.
    (b) Does her revenue from the rental of the apartments increase or decrease as she increases the rent from $600 to $640?
    (c) Write an equation that gives the revenue from rental of the apartments if she makes x increases of $20 in the rent.
    (d) Find the rent she should charge to maximize her revenue.

14. **Pension resources** The Pension Benefit Guaranty Corporation is the agency that insures pensions. The figure shows one study's projection for the agency's total resources, initially rising (from taking over the assets of failing plans) but then falling (as more workers retire and payouts increase).
    (a) What kind of function might be used to model the agency's total resources?
    (b) If a function of the form $f(x) = ax^2 + bx + c$ were used to model these total resources, would $f(x)$ have $a > 0$ or $a < 0$? Explain.
    (c) If the model from part (b) used x as the number of years past 2004, explain why the model would have $b > 0$ and $c > 0$.

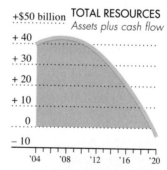

Source: The New York Times, September 14, 2004. Copyright .
© 2004 The New York Times Co. Reprinted by permission.

Need more practice?
Find more here: cengagebrain.com

2.3 Business Applications of Quadratic Functions

IN THIS SECTION, WE GRAPH QUADRATIC SUPPLY AND DEMAND FUNCTIONS AND FIND MARKET EQUILIBRIUM BY SOLVING SUPPLY AND DEMAND FUNCTIONS SIMULTANEOUSLY USING QUADRATIC METHODS. WE WILL ALSO DISCUSS QUADRATIC REVENUE, COST, AND PROFIT FUNCTIONS, INCLUDING BREAK-EVEN POINTS AND PROFIT MAXIMIZATION.

Supply, Demand, and Market Equilibrium

The first-quadrant parts of parabolas or other quadratic equations are frequently used to represent supply and demand functions. When quadratic equations are used to represent supply or demand curves, we can solve their equations simultaneously to find the market equilibrium.

EXAMPLE 2.12 Market Equilibrium

If the supply function for a commodity is given by $p = q^2 + 100$ and the demand function is given by $p = -20q + 2500$, find the point of market equilibrium.

In 2010, harsh weather threatened Florida citrus orchards, and growers sprayed their trees with water to prevent widespread damage to the fruit (supply). What effect would damaged crops have on the market equilibrium price for, say, orange juice, assuming constant demand?

Red Huber/AP Images

SOLUTION

At market equilibrium, both equations will have the same p-value. Thus substituting $q^2 + 100$ for p in $p = -20q + 2500$ yields

$$q^2 + 100 = -20q + 2500$$

$$q^2 + 20q - 2400 = 0$$

$$(q - 40)(q + 60) = 0$$

$$q = 40 \quad \text{or} \quad q = -60$$

Because a negative quantity has no meaning, the equilibrium point occurs when 40 units are sold, at (40, 1700). The graphs of the functions are shown (in the first quadrant only) in Figure 2.7.

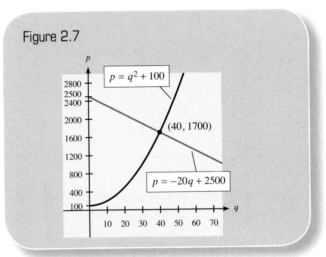

©iStockphoto.com/Jeroen Geeraert

Graphing Utilities: Graphing utilities also can be used to sketch these graphs. Check out the Tech Card to see how.

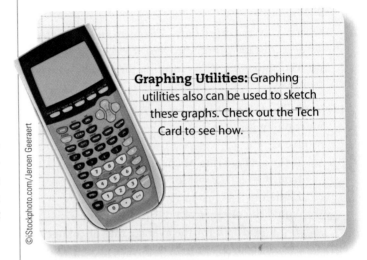

Figure 2.7

Break-Even Points and Maximization

When we know the functions for $C(x)$, total cost, and $R(x)$, total revenue, we can find the break-even point by finding the quantity x that makes $C(x) = R(x)$.

EXAMPLE 2.13 Break-Even Point

Suppose that the total cost per week of producing a high-tech product is given by $C = 3600 + 100x + 2x^2$. Suppose further that the weekly demand function for this product is $p = 500 - 2x$. Find the number of units that will give the break-even point for the product.

SOLUTION

The total cost function is $C(x) = 3600 + 100x + 2x^2$, and the total revenue function is $R(x) = px = (500 - 2x)x = 500x - 2x^2$.

Setting $C(x) = R(x)$ and solving for x gives

$$3600 + 100x + 2x^2 = 500x - 2x^2$$

$$4x^2 - 400x + 3600 = 0$$

$$x^2 - 100x + 900 = 0$$

$$(x - 90)(x - 10) = 0$$

$$x = 90 \quad \text{or} \quad x = 10$$

Does this mean the firm will break even at 10 units and at 90 units? Yes. Figure 2.8 shows the graphs of $C(x)$ and $R(x)$. From the graph we can observe that the firm makes a profit after $x = 10$ *until* $x = 90$, because $R(x) > C(x)$ in that interval. At $x = 90$, the profit is 0, and the firm loses money if it produces more than 90 units per week.

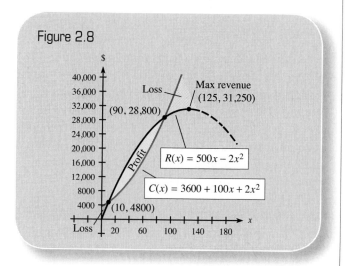

Figure 2.8

EXAMPLE 2.14 Profit Maximization

For the total cost function $C(x) = 3600 + 100x + 2x^2$ and the total revenue function $R(x) = 500x - 2x^2$ (from Example 2.13), find the number of units that maximizes profit and find the maximum profit.

SOLUTION

Using Profit = revenue − cost, we can determine the profit function:

$$P(x) = (500x - 2x^2) - (3600 + 100x + 2x^2)$$
$$= -3600 + 400x - 4x^2$$

This profit function is a parabola that opens downward, so the vertex will be the maximum point.

$$\text{Vertex: } x = \frac{-b}{2a} = \frac{-400}{2(-4)} = \frac{-400}{-8} = 50$$

Furthermore, when $x = 50$, we have

$$P(50) = -3600 + 400(50) - 4(50)^2 = 6400 \text{ (dollars)}$$

Thus, when 50 items are produced and sold, a maximum profit of \$6400 is made (see Figure 2.9).

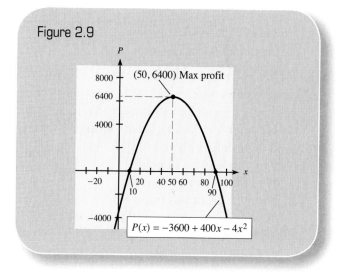

Figure 2.9

2.3 Exercises

Supply, Demand, and Market Equilibrium

1. If the supply function for a commodity is $p = q^2 + 8q + 16$ and the demand function is $p = -3q^2 + 6q + 436$, find the equilibrium quantity and equilibrium price.

2. If the demand function for a commodity is given by the equation $p^2 + 4q = 1600$ and the supply function is given by the equation $300 - p^2 + 2q = 0$, find the equilibrium quantity and equilibrium price.

3. The supply function for a product is $2p - q - 10 = 0$, while the demand function for the same product is $(p + 10)(q + 30) = 7200$. Find the market equilibrium point.

Break-Even Points and Maximization

4. The total costs for a company are given by

$$C(x) = 2000 + 40x + x^2$$

and the total revenues are given by

$$R(x) = 130x$$

Find the break-even points.

5. If a company has total costs $C(x) = 15,000 + 35x + 0.1x^2$ and total revenues given by $R(x) = 385x - 0.9x^2$, find the break-even points.

6. Find the maximum revenue for the revenue function $R(x) = 385x - 0.9x^2$.

7. The profit function for a certain commodity is $P(x) = 110x - x^2 - 1000$. Find the level of production that yields maximum profit, and find the maximum profit.

8. Suppose a company has fixed costs of \$28,000 and variable costs of $\frac{2}{5}x + 222$ dollars per unit, where x is the total number of units produced. Suppose further that the selling price of its product is $1250 - \frac{3}{5}x$ dollars per unit.
   (a) Find the break-even points.
   (b) Find the maximum revenue.
   (c) Form the profit function from the cost and revenue functions and find maximum profit.
   (d) What price will maximize the profit?

....

Need more practice?
Find more here: cengagebrain.com

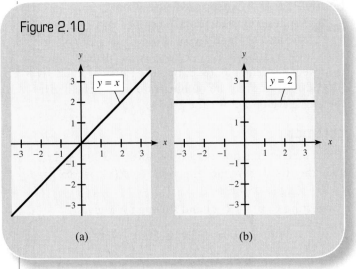

Figure 2.10

(a)   (b)

2.4 Special Functions and Their Graphs

IN THIS SECTION, WE DISCUSS POLYNOMIAL, RATIONAL, AND OTHER SPECIAL FUNCTIONS.

Basic Functions

The special linear function

$$y = f(x) = x$$

is called the **identity function** (Figure 2.10(a)), and a linear function defined by

$$y = f(x) = C \qquad C \text{ a constant}$$

is called a **constant function.** Figure 2.10(b) shows the graph of the constant function $y = f(x) = 2$. (Note that the slope of the graph of any constant function is 0.)

The functions of the form $y = ax^b$, where $b > 0$, are called **power functions.** Examples of power functions include $y = x^2$, $y = x^3$, $y = \sqrt{x} = x^{1/2}$, and $y = \sqrt[3]{x} = x^{1/3}$. (See Figure 2.11(a)–(d).) The functions $y = \sqrt{x}$ and $y = \sqrt[3]{x}$ are also called **root functions,** and $y = x^2$ and $y = x^3$ are basic **polynomial functions.**

The general shape for the power function $y = ax^b$, where $b > 0$, depends on the value of b. Figure 2.12 shows the first-quadrant portions of typical graphs of $y = x^b$ for different values of b. Note how the direction in which the graph bends differs for $b > 1$ and for $0 < b < 1$. Getting accurate graphs of these functions requires plotting a number of points by hand or with a graphing utility. Our goal at this stage is to recognize the basic shapes of certain functions.

Polynomial and Rational Functions

A **polynomial function of degree n** has the form

$$y = a_n x^n + a_{n-1} x^{n-1} + \ldots + a_1 x + a_0$$

where $a_n \neq 0$ and n is an integer, $n \geq 0$.

Some characteristics of the graphs of polynomial functions of degrees 2, 3, and 4 are summarized in Table 2.1. Using this information and point plotting or using a graphing utility yields the graphs of these functions.

Figure 2.11

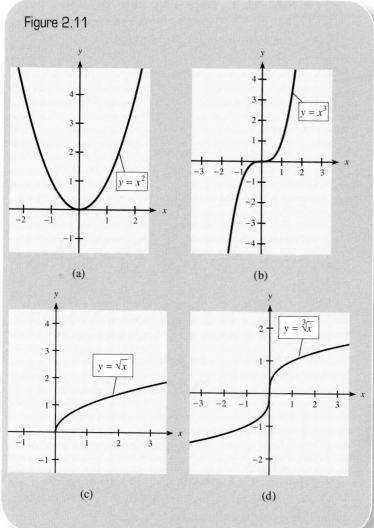

(a)

(b)

(c)

(d)

Figure 2.12

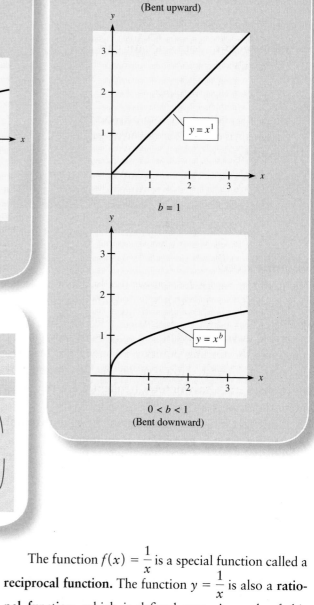

$b > 1$
(Bent upward)

$b = 1$

$0 < b < 1$
(Bent downward)

Table 2.1 **Graphs of Some Polynomials**

	Degree 2	Degree 3	Degree 4
Turning points	1	0 or 2	1 or 3
x-intercepts	0, 1, or 2	1, 2, or 3	0, 1, 2, 3, or 4
Possible shapes	⋃ or ⋂ (a) (b)	(a) (b) (c) (d)	(a) (b) (c) (d) (e) (f)

Step-by-step instructions for graphing polynomial functions with a graphing calculator and with Excel are shown on the Tech Card.

The function $f(x) = \dfrac{1}{x}$ is a special function called a **reciprocal function**. The function $y = \dfrac{1}{x}$ is also a **rational function**, which is defined next. A graph of this function is shown in Figure 2.13 on the following page.

Figure 2.13

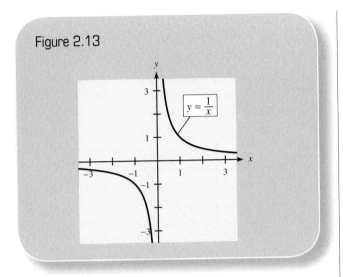

Rational Function: A rational function is a function of the form

$$y = \frac{f(x)}{g(x)} \text{ with } g(x) \neq 0$$

where $f(x)$ and $g(x)$ are both polynomials. Its domain is the set of all real numbers for which $g(x) \neq 0$.

Asymptotes

The graph in Figure 2.13 shows that $|y|$ gets very large as x approaches 0, and that y is undefined at $x = 0$. In this case, we call the line $x = 0$ (that is, the y-axis) a **vertical asymptote.** On the graphs of polynomial functions, the turning points are usually the features of greatest interest. However, on graphs of rational functions, vertical asymptotes frequently are the most interesting features.

Rational functions sometimes have **horizontal asymptotes** as well as vertical asymptotes. Whenever the values of y approach some finite number b as $|x|$ becomes very large, we say that there is a horizontal asymptote at $y = b$.

Note that the graph of $y = \dfrac{1}{x}$ appears to get close to the x-axis as $|x|$ becomes large.

Testing values of x for which $|x|$ is large, we see that y is close to 0. Thus, we say that the line $y = 0$ (or the x-axis) is a horizontal asymptote for the graph of $y = \dfrac{1}{x}$. The graph in the following example has both a vertical and a horizontal asymptote.

EXAMPLE 2.15 Rational Functions

(a) Use values of x from -5 to 5 to develop a table of function values for the graph of

$$y = \frac{12x + 8}{3x - 9}$$

(b) Sketch the graph.

SOLUTION

(a) Because $3x - 9 = 0$ when $x = 3$, it follows that $x = 3$ is not in the domain of this function.

x	-5	-4	-3	-2	-1	0	1
y	2.17	1.90	1.56	1.07	0.33	-0.89	-3.33

2	2.5	2.9	3.1	3.5	4	5
-10.7	-25.33	-142.67	150.67	33.33	18.67	11.33

The values in the table indicate that the graph is approaching the vertical asymptote at $x = 3$.

To see if the function has a horizontal asymptote, we calculate y as $|x|$ becomes larger.

x	-10000	-1000	1000	10,000
y	3.999	3.99	4.01	4.001

This table indicates that the graph is approaching $y = 4$ as $|x|$ increases, so we have a horizontal asymptote at $y = 4$.

(b) Using the information about vertical and horizontal asymptotes and plotting these points gives the graph in Figure 2.14.

Figure 2.14

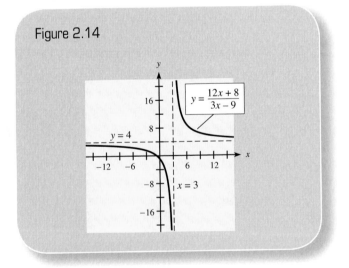

Vertical Asymptote

The graph of the rational function $y = \dfrac{f(x)}{g(x)}$ has a **vertical asymptote** at $x = c$ if $g(c) = 0$ and $f(c) \neq 0$.

Horizontal Asymptote

Consider the rational function

$$y = \frac{f(x)}{g(x)} = \frac{a_n x^n + \ldots + a_1 x + a_0}{b_m x^m + \ldots + b_1 x + b_0}.$$

1. If $n < m$ (that is, if the degree of the numerator is less than that of the denominator), a **horizontal asymptote** occurs at $y = 0$ (the x-axis).

2. If $n = m$ (that is, if the degree of the numerator equals that of the denominator), a horizontal asymptote occurs at $y = \dfrac{a_n}{b_m}$ (the ratio of the leading coefficients).

3. If $n > m$ (that is, if the degree of the numerator is greater than that of the denominator), there is no horizontal asymptote.

Recall that the graph of the function $y = \dfrac{12x + 8}{3x - 9}$ (shown in Figure 2.14) has a vertical asymptote at $x = 3$, and that $x = 3$ makes the denominator 0 but does not make the numerator 0. Observe also that this graph has a horizontal asymptote at $y = 4$, and that the degree of the numerator is the same as that of the denominator with the ratio of the leading coefficients equal to $12/3 = 4$.

See the Tech Card for detailed instructions on graphing rational functions with a graphing calculator and on graphing discontinuous functions with Excel.

A rational function may or may not have turning points. For example, the graph in Figure 2.15 appears to have a minimum near $x = 30$. By plotting points or using MINIMUM on a graphing utility, we can verify that the minimum point occurs at $x = 30$ and $\overline{C} = 63$. Thus the minimum average cost is \$63 per unit when 30 units are produced.

Piecewise Defined Functions

Another special function comes from the definition of $|x|$. The **absolute value function** can be written as

$$f(x) = |x| \quad \text{or} \quad f(x) = \begin{cases} x & \text{if } x \geq 0 \\ -x & \text{if } x < 0 \end{cases}$$

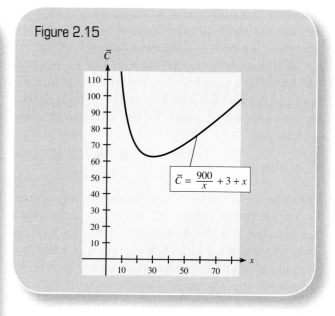

Figure 2.15

$$\overline{C} = \frac{900}{x} + 3 + x$$

Note that restrictions on the domain of the absolute value function specify different formulas for different parts of the domain. To graph $f(x) = |x|$, we graph the portion of the line $y = x$ for $x \geq 0$ (see Figure 2.16(a)) and the portion of the line $y = -x$ for $x < 0$ (see Figure 2.16(b)). When we put these pieces on the same graph

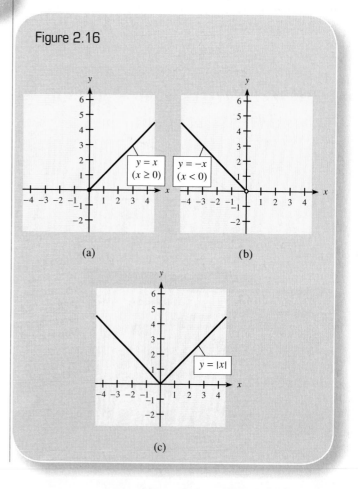

Figure 2.16

(a) $y = x$ $(x \geq 0)$

(b) $y = -x$ $(x < 0)$

(c) $y = |x|$

(Figure 2.16(c)), they give us the graph of $f(x) = |x|$. Because the absolute value function is defined by two equations, we say it is a **piecewise defined function**.

It is possible for the selling price S of a product to be defined as a piecewise function of the cost C of the product. For example, the selling price might be defined by two different equations on two different intervals, as follows:

$$S = f(C) = \begin{cases} 3C & \text{if } 0 \le C \le 20 \\ 1.5C + 30 & \text{if } C > 20 \end{cases}$$

When we write the equations in this way, the value of S depends on the value of C, so C is the independent variable and S is the dependent variable. Each of the two pieces of the graph of this function is a line and is easily graphed. It remains only to graph each in the proper interval. The graph is shown in Figure 2.17. See the Tech Card for instructions on graphing piecewise defined functions with a graphing calculator.

EXAMPLE 2.16 Residential Electrical Costs

➡ The 2010 monthly charge in dollars for x kilowatt-hours of electricity used by a residential customer of Excelsior Electric Membership Corporation during the months of November through June is given by the function

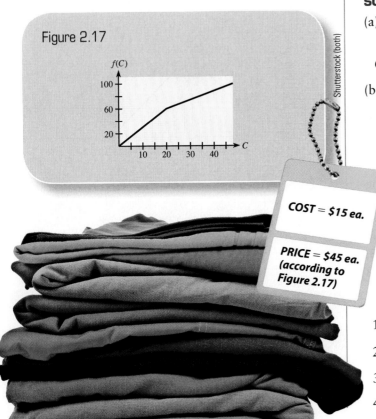

Figure 2.17

What if you don't use any kilowatt-hours? According to the first formula line, you'd still pay $10.

$$C(x) = \begin{cases} 10 + 0.094x & \text{if } 0 \le x \le 100 \\ 19.40 + 0.075(x - 100) & \text{if } 100 < x \le 500 \\ 49.40 + 0.05(x - 500) & \text{if } x > 500 \end{cases}$$

(a) What is the monthly charge if 1100 kilowatt-hours of electricity are consumed in a month?

(b) What is the monthly charge if 450 kilowatt-hours are consumed in a month?

SOLUTION

(a) We need to find $C(1100)$. Because $1100 > 500$, we use it in the third formula line.

$$C(1100) = 49.40 + 0.05(1100 - 500) = \$79.40$$

(b) We evaluate $C(450)$ by using the second formula line for $C(x)$.

$$C(450) = 19.40 + 0.075(450 - 100) = \$45.65$$

COST = $15 ea.

PRICE = $45 ea. (according to Figure 2.17)

2.4 Exercises

In Problems 1–5, match each of the functions with one of the graphs labeled (a)–(e) shown following these functions. Recognizing special features of certain types of functions and plotting points for the functions will be helpful.

1. $f(x) = -3$

2. $y = \sqrt[3]{x}$

3. $y = (x + 4)^3 + 1$

4. $y = |x|$

5. $f(x) = \begin{cases} x^2 & \text{if } x \le 2 \\ 4 & \text{if } x > 2 \end{cases}$

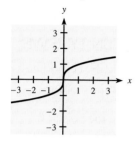

(a)

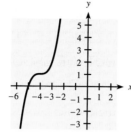

(b)

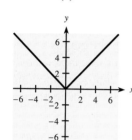

(c)

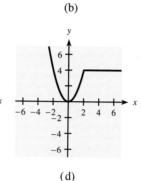

(d)

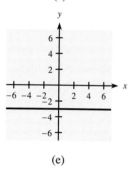

(e)

6. Decide whether each function whose graph is shown is the graph of a cubic (third-degree) or quartic (fourth-degree) function.

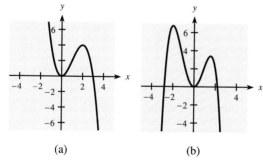

(a)   (b)

In Problems 7–9, match each equation with the correct graph from among those labeled (a)–(c) by recognizing shapes and features of polynomial and rational functions.

7. $y = x^3 - x$

8. $y = 16x^2 - x^4$

9. $y = \dfrac{x - 3}{x + 1}$

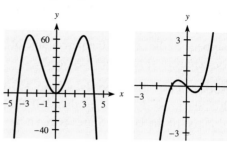

(a)   (b)

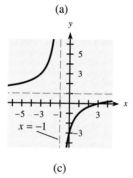

(c)

In Problems 10 and 11, graph the function.

10. $y = (x + 1)(x - 3)(x - 1)$

11. $y = \dfrac{1 - 2x}{x}$

12. If $k(x) = \begin{cases} 2 & \text{if } x < 0 \\ x + 4 & \text{if } 0 \le x < 1, \\ 1 - x & \text{if } x \ge 1 \end{cases}$

find the following.
(a) $k(-5)$ (b) $k(0)$ (c) $k(1)$ (d) $k(-0.001)$

Applications

13. **Postal restrictions** If a box with a square cross section is to be sent by the postal service, there are restrictions on its size such that its volume is given by $V = x^2(108 - 4x)$, where x is the length of each side of the cross section (in inches).
    (a) If $V = V(x)$, find $V(10)$ and $V(20)$.
    (b) What restrictions must be placed on x (the domain) so that the problem makes physical sense?

14. **Pollution** Suppose that the cost C (in dollars) of removing p percent of the particulate pollution from the smokestacks of an industrial plant is given by
$$C(p) = \frac{7300p}{100 - p}$$
    (a) Is $C(p)$ discontinuous at any p-value? If so, what value?

(b) What is the domain of $C(p)$ as given by the equation?

(c) What is the domain of $C(p)$ in the context of the application?

(d) What happens to the cost as the percentage of pollution removed approaches 100%?

15. **Gross domestic product** By using data from 1950 through 2005, the gross domestic product (GDP, in billions of dollars) of the United States is given by

$$y = \begin{cases} 90.742x + 210.29 & \text{if } 10 \leq x \leq 60 \\ 66.786x^2 - 7820.9x + 238,570 & \text{if } 60 < x \leq 65 \end{cases}$$

where x is the number of years past 1940. (*Source:* Bureau of Economic Analysis)

(a) Graph the function for $10 \leq x \leq 65$.

(b) What does the model estimate the GDP to be in 1980?

(c) What does the model estimate the GDP to be in 2005?

....

Need more practice?
Find more here: cengagebrain.com

2.5 Modeling Data with Graphing Utilities (optional)

CREATING AN EQUATION, OR **MATHEMATI-CAL MODEL,** FOR A SET OF DATA IS A COMPLEX PROCESS THAT INVOLVES CAREFUL ANALYSIS OF THE DATA IN ACCORDANCE WITH A NUMBER OF GUIDELINES. OUR APPROACH TO CREATING A MODEL IS MUCH SIMPLER AND RELIES ON A KNOWLEDGE OF THE APPEARANCE OF LINEAR, QUADRATIC, POWER, CUBIC, AND QUARTIC FUNCTIONS AND THE CAPABILITIES OF COMPUTERS AND GRAPHING CALCULATORS. WE SHOULD NOTE THAT THE EQUATIONS THE COMPUTERS AND CALCULATORS PROVIDE ARE BASED ON SOPHISTICATED

FORMULAS THAT ARE DERIVED USING CALCULUS OF SEVERAL VARIABLES. THESE FORMULAS PROVIDE THE BEST FIT FOR THE DATA USING THE FUNCTION TYPE WE CHOOSE, BUT WE SHOULD KEEP THE FOLLOWING LIMITATIONS IN MIND.

1. The computer/calculator will give the best fit for whatever type of function we choose (even if the selected function is a bad fit for the points), so we must choose a function type carefully. To choose a function type, compare the graph of the given data points (called a **scatter plot**) with the graphs of the functions discussed earlier in this chapter. The model will not fit the data if we choose a function type whose graph does not match the shape of the plotted data.

2. Some sets of data have no pattern, so they cannot be modeled by functions. Other data sets cannot be modeled by the functions we have studied.

3. Modeling provides a formula that relates the data, but the relationship is usually not perfect. Also, even though a model may perfectly fit a data set, it may not be a good predictor for values outside the data set.

EXAMPLE 2.17 Curve Fitting

Graphs of three sets of data points are shown in Figure 2.18(a–c). Determine what type of function is your choice as the best-fitting curve for each scatter plot. If it is a polynomial function, state the degree.

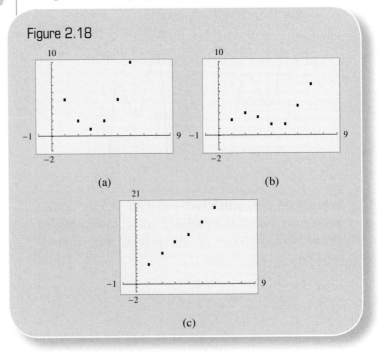

Figure 2.18

(a)

(b)

(c)

SOLUTION

(a) It appears that a parabola will fit the data points in Figure 2.18(a), so the data points can be modeled by a second-degree (quadratic) function.

(b) The two "turns" in the scatter plot in Figure 2.18(b) suggest that it could be modeled by a cubic function.

(c) The data points in Figure 2.18(c) appear to lie (approximately) along a line, so the data can be modeled by a linear function.

After the type of function that gives the best fit for the data is chosen, a graphing utility or spreadsheet can be used to develop the best-fitting equation for the function chosen. The following steps are used to find an equation that models data.

Modeling Data

1. Enter the data and use a graphing utility or spreadsheet to plot the data points and create the scatter plot. (The Tech Card shows how to create a scatter plot.)

2. Visually determine what type of function (including degree, if it is a polynomial function) would have a graph that best fits the data.

3. Use a graphing utility or spreadsheet to determine the equation of the chosen type that gives the best fit for the data. (The Tech Card shows how to create an equation that models the data.)

4. To see how well the equation models the data, graph the equation and the data points on the same set of axes. If the graph of the equation does not fit the data points well, another type of function may model the data better.

5. After the model for a data set has been found, it can be rounded for reporting purposes. However, use the unrounded model in graphing and in calculations, unless otherwise instructed. Numerical answers found using a model should be rounded in a way that agrees with the context of the problem and with no more accuracy than the original output data.

If it is not obvious what model will best fit a given set of data, several models can be developed and compared graphically with the data.

EXAMPLE 2.18 Expected Life Span

The expected life span of people in the United States depends on their year of birth, as shown in the table.

Year	Life Span (years)	Year	Life Span (years)	Year	Life Span (years)	Year	Life Span (years)
1920	54.1	1960	69.7	1985	74.7	2005	77.9
1930	59.7	1970	70.8	1990	75.4	2007	77.9
1940	62.9	1975	72.6	1995	75.8		
1950	68.2	1980	73.7	2000	77.0		

Source: National Center for Health Statistics

(a) Create linear and quadratic models that give life span as a function of birth year with $x = 0$ representing 1900 and, by visual inspection, decide which model gives the better fit.

(b) Use both models to estimate the life span of a person born in the year 2000.

(c) Which model's prediction for the life span in 2010 seems better?

SOLUTION

(a) The scatter plot for the data is shown in Figure 2.19(a). It appears that a linear function could be used to model the data. The linear equation that is the best fit for the data is

$$y = 0.251x + 52.68$$

The graph in Figure 2.19(b) shows how well the line fits the data points. The quadratic function that is the best fit for the data is

$$y = -0.00213x^2 + 0.530x + 45.2$$

Its graph is shown in Figure 2.19(c). The quadratic model appears to fit the data points better than the linear model. A table can be used to compare the models (see Table 2.2 on the next page, with y_1 giving values from the linear model and y_2 giving values from the quadratic model).

(b) We can estimate the life span of people born in the year 2000 with either model by evaluating the functions at $x = 100$. The x-values in Table 2.2 represent the years 1985, 1990, 2000, and 2010. From the linear model the expected life span in 2000 is 77.8, and from the quadratic model the expected life span is 76.9.

Figure 2.19

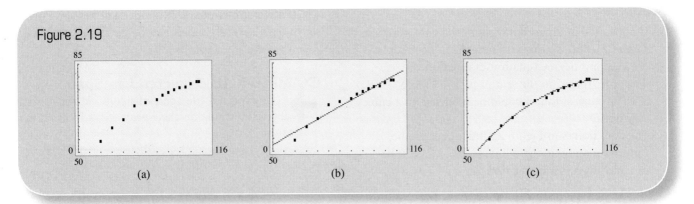

(a)   (b)   (c)

(c) Looking at Table 2.2, we see that the linear model may be giving optimistic values in 2010, when $x = 110$, so the quadratic model may be better in the years up to 2010.

Table 2.2

X	Y₁	Y₂
85	74.015	74.861
90	75.27	75.647
100	77.78	76.9
110	80.29	77.727

X =

EXAMPLE 2.19 Corvette Acceleration

The table shows the times that it takes a 2005 Corvette to reach speeds from 0 miles per hour to 100 miles per hour, in increments of 10 miles per hour after 30 miles per hour.

Time (sec)	Speed (mph)
1.7	30
2.4	40
3.3	50
4.1	60
5.3	70
6.5	80
7.9	90
9.5	100

(a) Use a power function to model the data, and graph the points and the function to see how well the function fits the points.

(b) What does the model indicate the speed is 5 seconds after the car starts to move?

(c) Use the model to determine the number of seconds until the Corvette reaches 110 miles per hour.

SOLUTION

(a) Using x as the time and y as the speed, the scatter plot is shown in Figure 2.20(a). An equation found using the power function $y = ax^b$ (graphed in Figure 2.20(b)) is

$$y = 21.551x^{0.697}$$

Figure 2.20

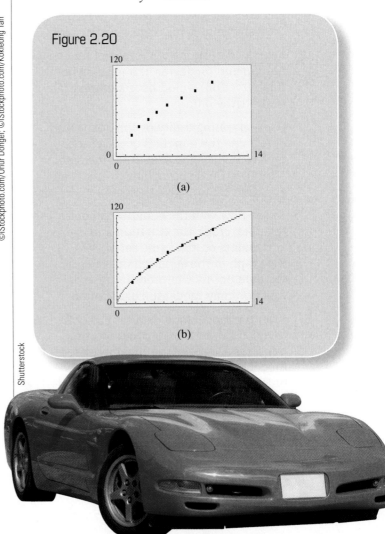

(a)

(b)

(b) Using $x = 5$ in this equation gives 66 miles per hour as the speed of the Corvette.

(c) Tracing along the graph of $y_1 = 21.551x^{0.697}$ until the y-value is 110 gives 10.4, so the time required to reach 110 miles per hour is 10.4 seconds.

Spreadsheet Note

The *Excel Chart Wizard* can be used to create scatter plots of data and to find the best-fitting linear or polynomial function for a set of data. The Tech Card has detailed instructions for creating scatter plots and for modeling sets of data with functions.

2.5 Exercises

In Problems 1–3, determine whether the scatter plot should be modeled by a linear, power, quadratic, cubic, or quartic function.

1.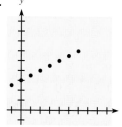

2.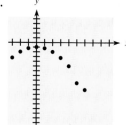

3.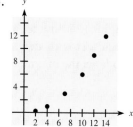

In Problems 4 and 5, find the equation of the function of the specified type that is the best fit for the given data. Plot the data and the equation.

4. Linear

x	y
-2	-7
-1	-5
0	-3
1	-1
2	1
3	3
4	5

5. Power

x	y
1	2
2	2.8284
3	3.4641
4	4
5	4.4721
6	4.899

In Problem 6, (a) plot the given points, (b) determine what type of function best models the data, and (c) find the equation that is the best fit for the data.

6.

x	y
-1	2
0	0
1	2
2	3.1748
3	4.1602
4	5.0397
5	5.848

Applications

7. **Earnings and gender** The table shows the 2007 average earnings of year-round full-time workers by gender and educational attainment.

(a) Let x represent earnings for men and y represent earnings for women, and create a scatter plot of these data.

(b) Find a linear model that expresses women's annual earnings as a function of men's.

(c) Find the slope of the linear model in part (b) and write a sentence that interprets it.

Educational Attainment	Average Annual Earnings (thousands)	
	Men	Women
Less than 9th grade	$21.659	$17.023
Some high school	26.277	19.162
High school graduate	35.725	26.029
Some college	41.875	30.816
Associate degree	44.404	33.481
Bachelor's degree	57.220	41.681
Master's degree	71.530	51.316
Doctorate	82.401	68.875

Source: U.S. Bureau of the Census

8. *National health care* The table shows the national expenditures for health care in the United States for selected years, with projections to 2015.
   (a) Use a scatter plot with x as the number of years past 1950 and y as the total expenditures for health care (in billions) to identify what type (or types) of function(s) would make a good model for these data.
   (b) Find a power model and a quadratic model for the data.
   (c) Which model from part (b) more accurately estimates the 2010 expenditures for national health care?
   (d) Use the better model from part (c) to estimate the 2020 expenditures for national health care.

Year	National Expenditures for Health Care (billions)
1960	$28
1970	75
1980	255
1990	717
1995	1020
2000	1359
2005	2016
2010	2879
2015	4032

Source: U.S. Centers for Medicare and Medicaid Services

Need more practice?
Find more here: cengagebrain.com

Chapter Exercises

In Problems 1–7, find the real solutions to each quadratic equation.

1. $3x^2 + 10x = 5x$

2. $x^2 + 5x + 6 = 0$

3. $11 - 10x - 2x^2 = 0$

4. $(x - 1)(x + 3) = -8$

5. $4x^2 = 3$

6. $8x^2 + 8x = 1 - 8x^2$

7. $7 = 2.07x - 0.02x^2$

In Problems 8 and 9, solve each equation by using a graphing utility to find the zeros of the function. Solve the equation algebraically to check your results.

8. $4z^2 + 25 = 0$ 9. $z(z + 6) = 27$

10. Approximate the real solutions to
$$23.1 - 14.1x - 0.002x^2 = 0$$
to two decimal places.

In Problems 11–14, find the vertex of each function and determine if it is a maximum or minimum point, find the zeros if they exist, and sketch the graph.

11. $y = \frac{1}{2}x^2 + 2x$

12. $y = 6 + x - x^2$

13. $y = x^2 - 4x + 5$

14. $y = 12x - 9 - 4x^2$

In Problems 15–16, use a graphing utility to graph each function. Use the vertex and zeros to determine an appropriate range. Be sure to label the maximum or minimum point.

15. $y = -10 + 7x - x^2$

16. $y = 20x - 0.1x^2$

17. Find the average rate of change of $f(x) = 100x - x^2$ from $x = 30$ to $x = 50$.

In Problems 18–19, a graph is given. Use the graph to
(a) locate the vertex,
(b) determine the zeros, and

(c) match the graph with one of the equations A, B, C, or D.

A. $y = 7x - \frac{1}{2}x^2$

B. $y = \frac{1}{2}x^2 - x - 4$

C. $y = 8 - 2x - x^2$

D. $y = 49 - x^2$

18.

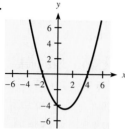

19.

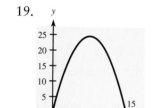

20. Sketch a graph of each of the following basic functions.

(a) $f(x) = x^2$ (b) $f(x) = 1/x$ (c) $f(x) = x^{1/4}$

21. If $f(x) = \begin{cases} x & \text{if } x \le 1 \\ 3x - 2 & \text{if } x > 1 \end{cases}$, find the following.

(a) $f(-2)$ (b) $f(0)$

(c) $f(1)$ (d) $f(2)$

22. Graph: $f(x) = \begin{cases} x & \text{if } x \le 1 \\ 3x - 2 & \text{if } x > 1 \end{cases}$

23. Graph: $y = x^3 - 9x$. Use a graphing utility and find any turning points.

In Problems 24 and 25, graph each function. Find and identify any asymptotes.

24. $y = \dfrac{1}{x - 2}$

25. $y = \dfrac{2x - 1}{x + 3}$

26. **Modeling** Consider the data given in the table.

(a) Make a scatter plot.

(b) Fit a linear function to the data and comment on the fit.

(c) Try other function types and find one that fits better than a linear function.

x	0	4	8	12	16	20	24
y	153	151	147	140	128	115	102

27. **Modeling** Consider the data given in the table.

(a) Make a scatter plot.

(b) Fit a linear function to the data and comment on the fit.

(c) Try other function types and find one that fits better than a linear function.

x	3	5	10	15	20	25	30
y	35	45	60	70	80	87	95

Applications

28. **Profit** The profit for a product is given by $P(x) = 82x - 0.10x^2 - 1600$, where x is the number of units produced and sold. Break-even points will occur at values of x where $P(x) = 0$. How many units will give a break-even point for the product?

29. **Market equilibrium** The supply function for a product is given by $p = q^2 + 300$, and the demand is given by $p + q = 410$. Find the equilibrium quantity and price.

30. **Break-even points** If total costs for a product are given by $C(x) = 1760 + 8x + 0.6x^2$ and total revenues are given by $R(x) = 100x - 0.4x^2$, find the break-even quantities.

31. **Maximum profit** Given $C(x) = 360 + 10x + 0.2x^2$ and $R(x) = 50x - 0.2x^2$, find the level of production that gives maximum profit and find the maximum profit.

32. **Break-even and profit maximization** A certain company has fixed costs of $15,000 for its product and variable costs given by $140 + 0.04x$ dollars per unit, where x is the total number of units. The selling price of the product is given by $300 + 0.06x$ dollars per unit.

(a) Formulate the functions for total cost and total revenue.

(b) Find the break-even quantities.

(c) Find the level of sales that maximizes revenue.

(d) Find the profit function and find the level of production and sales that maximizes profit.

(e) Find the profit (or loss) at the production levels found in parts (c) and (d).

33. **Photosynthesis** The amount y of photosynthesis that takes place in a certain plant depends on the intensity x of the light present, according to

$$y = 120x^2 - 20x^3 \qquad \text{for } x \geq 0$$

(a) Graph this function with a graphing utility. (Use y-min $= -100$ and y-max $= 700$.)
(b) The model is valid only when $f(x) \geq 0$ (that is, on or above the x-axis). For what x-values is this true?

34. **Cost-benefit** Suppose the cost C, in dollars, of eliminating p percent of the pollution from the emissions of a factory is given by

$$C(p) = \frac{4800p}{100 - p}$$

(a) What type of function is this?
(b) Given that p represents the percentage of pollution removed, what is the domain of $C(p)$?
(c) Find $C(0)$ and interpret its meaning.
(d) Find the cost of removing 99% of the pollution.

35. **Municipal water costs** The Borough Municipal Authority of Beaver, Pennsylvania, used the following function to determine charges for water.

$$C(x) = \begin{cases} 2.557x & 0 \leq x \leq 100 \\ 255.70 + 2.04(x - 100) & 100 < x \leq 1000 \\ 2091.07 + 1.689(x - 1000) & x > 1000 \end{cases}$$

where $C(x)$ is the cost in dollars for x thousand gallons of water.

(a) Find the monthly charge for 12,000 gallons of water.
(b) Find the monthly charge for 825,000 gallons of water.

36. **Modeling (Subaru WRX)** The table shows the times that it takes a Subaru WRX to accelerate from 0 miles per hour to speeds of 30 miles per hour, 40 miles per hour, etc., up to 90 miles per hour, in increments of 10 miles per hour.

Time (sec)	Speed (mph)	Time (sec)	Speed (mph)
1.6	30	7.8	70
2.7	40	10.2	80
4.0	50	12.9	90
5.6	60		

Source: Motor Trend

(a) Represent the times by x and the speeds by y, and model the function that is the best fit for the points.
(b) Graph the points and the function to see how well the function fits the points.
(c) What does the model indicate the speed is 5 seconds after the car starts to move?
(d) According to the model, in how many seconds will the car reach 79.3 miles per hour?

Need more practice?
Find more here: **cengagebrain.com**

LEARN YOUR WAY

With **MATH APPS**, you have a multitude of study aids at your fingertips.

The Student Solutions Manual contains worked-out solutions to every odd-numbered exercise, to further reinforce your understanding of mathematical concepts in **MATH APPS**.

In addition to the Student Solutions Manual, Cengage Learning's **CourseMate** offers exercises and questions that correspond to every section and chapter in the text for extra practice.

For access to these study aids sign in at **login.cengagebrain.com.**

Matrices

A wide variety of application problems can be solved using matrices. A matrix (plural: matrices) is a rectangular array of numbers. In addition to storing data in a matrix and making comparisons of data, we can analyze data and make business decisions by defining the operations of addition, subtraction, scalar multiplication, and matrix multiplication. Matrices can be used to solve systems of linear equations, as we will see in this chapter. Matrices and matrix operations are the basis for computer spreadsheets such as Excel, which are used extensively in education, research, and business. Matrices are also useful in linear programming, which is discussed in Chapter 4.

objectives

3.1 Operations with Matrices

3.2 Multiplication of Matrices

3.3 Gauss-Jordan Elimination: Solving Systems of Equations

3.4 Inverse of a Square Matrix

3.1 Operations with Matrices

MATRICES CAN BE USED TO STORE DATA AND TO PERFORM OPERATIONS WITH THE DATA. WE WILL DISCUSS HOW TO FIND THE DIFFERENCE OF TWO MATRICES, HOW TO FIND THE SUM OF TWO MATRICES, AND HOW TO MULTIPLY A MATRIX BY A REAL NUMBER.

A business may collect and store or analyze various types of data as a regular part of its record-keeping procedures. The data may be presented in tabular form. For example, a building contractor who builds four different styles of houses may catalog the number of units of certain materials needed to build each style in a table like Table 3.1.

In the following rectangular array, the rows correspond to the types of building materials, and the columns correspond to the types of houses. The rows of a matrix are numbered from the top to the bottom, and the columns are numbered from left to right.

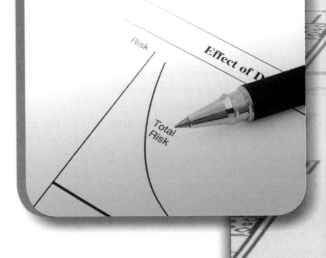

mathematics

TABLE 3.1

Required Units	House Style			
	Ranch	Colonial	Split Level	Cape Cod
Wood	20	25	22	18
Siding	27	40	31	25
Roofing	16	16	19	16

$$A = \begin{array}{c} \text{Column 1} \quad \text{Column 2} \quad \text{Column 3} \quad \text{Column 4} \\ \begin{bmatrix} 20 & 25 & 22 & 18 \\ 27 & 40 & 31 & 25 \\ 16 & 16 & 19 & 16 \end{bmatrix} \begin{array}{l} \text{Row 1} \\ \text{Row 2} \\ \text{Row 3} \end{array} \end{array}$$

Shutterstock

Matrices are classified in terms of the numbers of rows and columns they have. Matrix A just presented has three rows and four columns, so we say this is a 3×4 (read "three by four") matrix.

The matrix

$$A = \begin{bmatrix} a_{11} & a_{12} & a_{13} & \cdots & a_{1n} \\ a_{21} & a_{22} & a_{23} & \cdots & a_{2n} \\ & & \vdots & & \vdots \\ a_{m1} & a_{m2} & a_{m3} & \cdots & a_{mn} \end{bmatrix}$$

has m rows and n columns, so it is an $m \times n$ matrix. When we designate A as an $m \times n$ matrix, we are indicating the size of the matrix. Two matrices are said to have the same **order** (be the same size) if they have the same number of rows and the same number of columns.

The numbers in a matrix are called its **entries** or **elements**. Note that the subscripts on an entry in the preceding matrix A correspond respectively to the row and column in which the entry is located. Thus a_{23} represents the entry in the second row and the third column, and we refer to it as the "two-three entry." In matrix B that follows, the entry denoted by b_{23} is 1.

Some matrices take special names because of their size. If the number of rows equals the number of columns, we say the matrix is a **square matrix.** Matrix B is a 3×3 square matrix.

$$B = \begin{bmatrix} 4 & 3 & 0 \\ 0 & 0 & 1 \\ -4 & 0 & 0 \end{bmatrix}$$

EXAMPLE 3.1 Population and Labor Force

Figure 3.1 gives the U.S. population and labor force (in millions) for the years 1992 and 2002, and projections for 2012.

A partial table of values from Figure 3.1 is shown next. Complete the table that summarizes these data, and use it to create a 2×3 matrix describing the information.

	1992	2002	2012
Population, 16 Plus	193		
Labor Force		145	

SOLUTION

Reading the data from Figure 3.1 and completing the table gives

	1992	2002	2012
Population, 16 Plus	193	218	242
Labor Force	128	145	162

Shutterstock

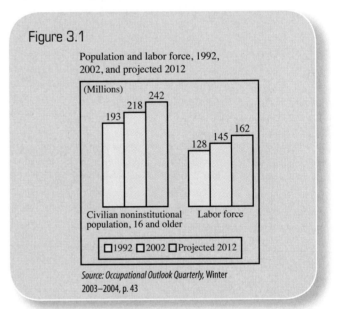

Figure 3.1

Population and labor force, 1992, 2002, and projected 2012

Source: *Occupational Outlook Quarterly,* Winter 2003–2004, p. 43

Writing these data in a matrix gives

$$\begin{bmatrix} 193 & 218 & 242 \\ 128 & 145 & 162 \end{bmatrix}$$

A matrix with one row, such as [9 5] or [3 2 1 6], is called a **row matrix,** and a matrix with one column, such as

$$\begin{bmatrix} 1 \\ 3 \\ 5 \end{bmatrix}$$

is called a **column matrix.** Row and column matrices are also called **vectors.**

Any matrix in which every entry is zero is called a **zero matrix;** examples include

$$\begin{bmatrix} 0 & 0 \\ 0 & 0 \end{bmatrix}, \begin{bmatrix} 0 & 0 \\ 0 & 0 \\ 0 & 0 \end{bmatrix}, \text{ and } \begin{bmatrix} 0 & 0 & 0 & 0 \\ 0 & 0 & 0 & 0 \\ 0 & 0 & 0 & 0 \end{bmatrix}$$

We define two matrices to be **equal** if they are of the same order and if each entry in one equals the corresponding entry in the other. When the columns and rows of matrix A are interchanged to create a matrix B, and vice versa, we say that A and B are **transposes** of each other and write $A^T = B$ and $B^T = A$. We will see valuable uses for transposes of matrices in Chapter 4, "Inequalities and Linear Programming."

EXAMPLE 3.2 Matrices

(a) Which element of

$$A = \begin{bmatrix} 1 & 0 & 3 \\ 3 & 4 & 2 \\ 7 & 8 & 3 \end{bmatrix} \quad B = \begin{bmatrix} 1 & 0 & 3 \\ 3 & 4 & 2 \\ 7 & 8 & 4 \end{bmatrix}$$

is represented by a_{32}?
(b) Is A a square matrix?
(c) Find the transpose of matrix A.
(d) Does $A = B$?

SOLUTION

(a) a_{32} represents the element in row 3 and column 2 of matrix A—that is, $a_{32} = 8$.
(b) Yes, it is a 3×3 (square) matrix.
(c) $A^T = \begin{bmatrix} 1 & 3 & 7 \\ 0 & 4 & 8 \\ 3 & 2 & 3 \end{bmatrix}$
(d) No, $a_{33} \neq b_{33}$.

Addition and Subtraction of Matrices

If two matrices have the same number of rows and columns, we can add the matrices by adding their corresponding entries.

EXAMPLE 3.3 Matrix Sums

Find the sum of A and B if

$$A = \begin{bmatrix} 1 & 2 & 3 \\ 4 & -1 & -2 \end{bmatrix} \quad \text{and} \quad B = \begin{bmatrix} -1 & 2 & -3 \\ -2 & 0 & 1 \end{bmatrix}$$

SOLUTION

$A + B =$
$$\begin{bmatrix} 1 + (-1) & 2 + 2 & 3 + (-3) \\ 4 + (-2) & -1 + 0 & -2 + 1 \end{bmatrix} = \begin{bmatrix} 0 & 4 & 0 \\ 2 & -1 & -1 \end{bmatrix}$$

We can define the difference $A - B$ (when A and B have the same order) by subtracting corresponding elements.

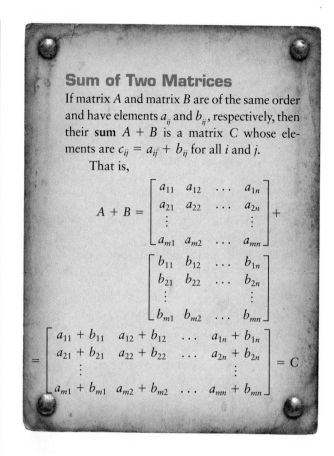

Sum of Two Matrices

If matrix A and matrix B are of the same order and have elements a_{ij} and b_{ij}, respectively, then their **sum** $A + B$ is a matrix C whose elements are $c_{ij} = a_{ij} + b_{ij}$ for all i and j.

That is,

$$A + B = \begin{bmatrix} a_{11} & a_{12} & \cdots & a_{1n} \\ a_{21} & a_{22} & \cdots & a_{2n} \\ \vdots & & & \vdots \\ a_{m1} & a_{m2} & \cdots & a_{mn} \end{bmatrix} +$$

$$\begin{bmatrix} b_{11} & b_{12} & \cdots & b_{1n} \\ b_{21} & b_{22} & \cdots & b_{2n} \\ \vdots & & & \vdots \\ b_{m1} & b_{m2} & \cdots & b_{mn} \end{bmatrix}$$

$$= \begin{bmatrix} a_{11} + b_{11} & a_{12} + b_{12} & \cdots & a_{1n} + b_{1n} \\ a_{21} + b_{21} & a_{22} + b_{22} & \cdots & a_{2n} + b_{2n} \\ \vdots & & & \vdots \\ a_{m1} + b_{m1} & a_{m2} + b_{m2} & \cdots & a_{mn} + b_{mn} \end{bmatrix} = C$$

EXAMPLE 3.4 Matrix Differences

For matrices A and B in Example 3.3, find $A - B$.

SOLUTION

$A - B$ can be found by subtracting corresponding elements.

$A - B =$
$$\begin{bmatrix} 1 - (-1) & 2 - 2 & 3 - (-3) \\ 4 - (-2) & -1 - 0 & -2 - 1 \end{bmatrix} = \begin{bmatrix} 2 & 0 & 6 \\ 6 & -1 & -3 \end{bmatrix}$$

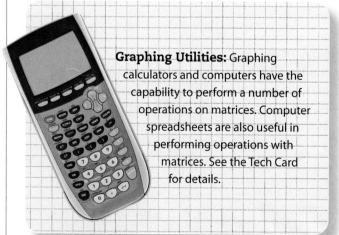

Graphing Utilities: Graphing calculators and computers have the capability to perform a number of operations on matrices. Computer spreadsheets are also useful in performing operations with matrices. See the Tech Card for details.

Scalar Multiplication

Consider the matrix

$$A = \begin{bmatrix} 3 & 2 \\ 1 & 1 \\ 2 & 0 \end{bmatrix}$$

Because $2A$ is $A + A$, we see that

$$2A = \begin{bmatrix} 3 & 2 \\ 1 & 1 \\ 2 & 0 \end{bmatrix} + \begin{bmatrix} 3 & 2 \\ 1 & 1 \\ 2 & 0 \end{bmatrix} = \begin{bmatrix} 6 & 4 \\ 2 & 2 \\ 4 & 0 \end{bmatrix}$$

Note that $2A$ could have been found by multiplying each entry of A by 2.

We can define **scalar multiplication** as follows.

Scalar Multiplication: Multiplying a matrix by a real number (called a scalar) results in a matrix in which each entry of the original matrix is multiplied by the real number. Thus, if

$$A = \begin{bmatrix} a_{11} & a_{12} \\ a_{21} & a_{22} \end{bmatrix}, \quad \text{then} \quad cA = \begin{bmatrix} ca_{11} & ca_{12} \\ ca_{21} & ca_{22} \end{bmatrix}$$

EXAMPLE 3.5 Supply Charges

Suppose the purchase prices and delivery costs (per unit) for wood, siding, and roofing used in construction are given by Table 3.2. Then the table of unit costs may be represented by the matrix C.

$$C = \begin{bmatrix} 6 & 4 & 2 \\ 1 & 1 & 0.5 \end{bmatrix}$$

If the supplier announces a 10% increase on both purchase and delivery of these items, find the new unit cost matrix.

SOLUTION

A 10% increase means that the new unit costs are the former costs plus 0.10 times the former cost. That is, the new costs are 1.10 times the former, so the new unit cost matrix is given by

$$1.10C = 1.10 \begin{bmatrix} 6 & 4 & 2 \\ 1 & 1 & 0.5 \end{bmatrix}$$

$$= \begin{bmatrix} 6.60 & 4.40 & 2.20 \\ 1.10 & 1.10 & 0.55 \end{bmatrix}$$

Table 3.2

	Wood	Siding	Roofing
Purchase	6	4	2
Delivery	1	1	0.5

3.1 Exercises

In Problems 1–8, use the following matrices.

$$A = \begin{bmatrix} 1 & 0 & -2 \\ 3 & 2 & 1 \\ 4 & 0 & 3 \end{bmatrix} \qquad B = \begin{bmatrix} 1 & 1 & 3 & 0 \\ 4 & 2 & 1 & 1 \\ 3 & 2 & 0 & 1 \end{bmatrix}$$

$$C = \begin{bmatrix} 5 & 3 \\ 1 & 2 \end{bmatrix} \quad D = \begin{bmatrix} 4 & 2 \\ 3 & 5 \end{bmatrix} \quad E = \begin{bmatrix} 1 & 0 & 4 \\ 5 & 1 & 0 \end{bmatrix}$$

$$F = \begin{bmatrix} 1 & 2 & 3 \\ -1 & 0 & 1 \\ 2 & -3 & -4 \end{bmatrix} \qquad G = \begin{bmatrix} 2 & 1 & 1 \\ 1 & 0 & 4 \\ 5 & 1 & 0 \end{bmatrix}$$

$$Z = \begin{bmatrix} 0 & 0 & 0 \\ 0 & 0 & 0 \\ 0 & 0 & 0 \end{bmatrix}$$

1. Which of the matrices A, B, C, D, E, F, G, and Z are square?

2. What is the element a_{23}?

3. Write the transpose of matrix A.

4. What is the sum of matrix A and its negative?

In Problems 5–8, perform the operations, if possible.

5. $C + D$

6. $A - F$

7. $3B$

8. $4C + 2D$

9. Find x, y, z, and w if $\begin{bmatrix} x & 1 & 0 \\ 0 & y & z \\ w & 2 & 1 \end{bmatrix} = \begin{bmatrix} 3 & 1 & 0 \\ 0 & 2 & 3 \\ 4 & 2 & 1 \end{bmatrix}$.

10. Solve for x, y, and z if

$$3 \begin{bmatrix} x & y \\ y & z \end{bmatrix} + 2 \begin{bmatrix} 2x & -y \\ 3y & -4z \end{bmatrix} = \begin{bmatrix} 14 & 4-y \\ 18 & 15 \end{bmatrix}$$

Applications

11. *Endangered species* The tables give the numbers of some species of threatened and endangered wildlife

in the United States and in foreign countries in 2007.

United States

	Mammals	Birds	Reptiles	Amphibians	Fishes
Endangered	69	75	13	13	74
Threatened	12	14	24	10	65

Foreign

	Mammals	Birds	Reptiles	Amphibians	Fishes
Endangered	256	176	65	8	11
Threatened	20	6	16	1	1

Source: U.S. Fish and Wildlife Service

(a) Write a matrix A that contains the number of each of these species in the United States in 2007 and a matrix B that contains the number of each of these species outside the United States in 2007.

(b) Find a matrix with the total number of these species. Assume that U.S. and foreign species are different.

(c) Find the matrix $B - A$. What do the negative entries in matrix $B - A$ mean?

12. **Sales** Let matrix A represent the sales (in thousands of dollars) for the Walbash Company in 2010 in various cities, and let matrix B represent the sales (in thousands of dollars) for the same company in 2011 in the same cities.

$$A = \begin{bmatrix} \overset{\text{Chicago}}{450} & \overset{\text{Atlanta}}{280} & \overset{\text{Memphis}}{850} \\ 400 & 350 & 150 \end{bmatrix} \begin{matrix} \text{Wholesale} \\ \text{Retail} \end{matrix}$$

$$B = \begin{bmatrix} \overset{\text{Chicago}}{375} & \overset{\text{Atlanta}}{300} & \overset{\text{Memphis}}{710} \\ 410 & 300 & 200 \end{bmatrix} \begin{matrix} \text{Wholesale} \\ \text{Retail} \end{matrix}$$

(a) Write the matrix that represents the total sales by type and city for both years.

(b) Write the matrix that represents the change in sales by type and city from 2010 to 2011.

13. **International trade** The tables show the value (in billions of dollars) of U.S. international trade in goods and services for the years 2004–2006.

	Exports			Imports	
	Goods	Services		Goods	Services
2004	807.5	349.7	2004	1477.1	292.2
2005	894.6	388.4	2005	1681.8	315.7
2006	1023.1	422.6	2006	1861.3	342.8

Source: Bureau of Economic Analysis

Let E be a 3×2 matrix representing the export data, and let M be a similar matrix for the import data. Find each of the following matrices and indicate what information each provides.

(a) $E - M$ (b) $E + M$ (c) $\dfrac{1}{12}E$

3.2 Multiplication of Matrices

IN THIS SECTION WE WILL DISCUSS FINDING THE PRODUCT OF TWO MATRICES.

Product of Two Matrices

Given an $m \times n$ matrix A and an $n \times p$ matrix B, the **matrix product** AB is an $m \times p$ matrix C, with the ij entry of C given by the formula

$$c_{ij} = a_{i1}b_{1j} + a_{i2}b_{2j} + \ldots + a_{in}b_{nj}$$

which is illustrated in Figure 3.2 (and developed in the Enrichment material on CourseMate for MATH APPS at cengagebrain.com).

Figure 3.2

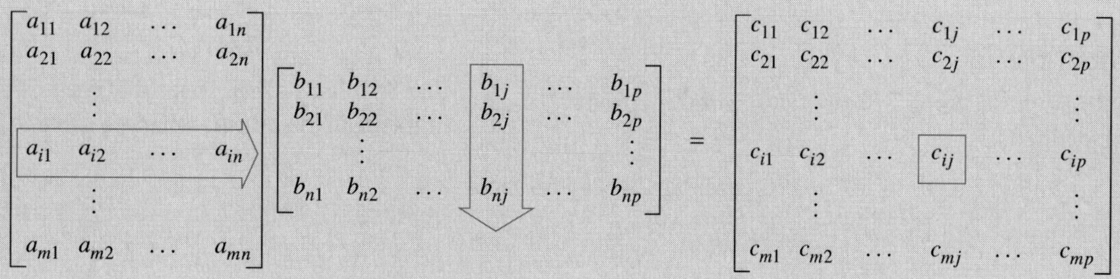

$$\begin{bmatrix} a_{11} & a_{12} & \cdots & a_{1n} \\ a_{21} & a_{22} & \cdots & a_{2n} \\ & & \vdots & \\ a_{i1} & a_{i2} & \cdots & a_{in} \\ & & \vdots & \\ a_{m1} & a_{m2} & \cdots & a_{mn} \end{bmatrix} \begin{bmatrix} b_{11} & b_{12} & \cdots & b_{1j} & \cdots & b_{1p} \\ b_{21} & b_{22} & \cdots & b_{2j} & \cdots & b_{2p} \\ & \vdots & & & & \vdots \\ b_{n1} & b_{n2} & \cdots & b_{nj} & \cdots & b_{np} \end{bmatrix} = \begin{bmatrix} c_{11} & c_{12} & \cdots & c_{1j} & \cdots & c_{1p} \\ c_{21} & c_{22} & \cdots & c_{2j} & \cdots & c_{2p} \\ & & \vdots & & & \vdots \\ c_{i1} & c_{i2} & \cdots & c_{ij} & \cdots & c_{ip} \\ & & \vdots & & & \vdots \\ c_{m1} & c_{m2} & \cdots & c_{mj} & \cdots & c_{mp} \end{bmatrix}$$

Shutterstock

EXAMPLE 3.6 Matrix Product

Find AB if

$$A = \begin{bmatrix} 3 & 4 \\ 2 & 5 \\ 6 & 10 \end{bmatrix} \quad \text{and} \quad B = \begin{bmatrix} a & b & c & d \\ e & f & g & h \end{bmatrix}$$

SOLUTION

A is a 3×2 matrix and B is a 2×4 matrix, so the number of columns of A equals the number of rows of B. Thus we can find the product AB, which is a 3×4 matrix, as follows:

$$AB = \begin{bmatrix} 3 & 4 \\ 2 & 5 \\ 6 & 10 \end{bmatrix} \begin{bmatrix} a & b & c & d \\ e & f & g & h \end{bmatrix}$$

$$= \begin{bmatrix} 3a + 4e & 3b + 4f & 3c + 4g & 3d + 4h \\ 2a + 5e & 2b + 5f & 2c + 5g & 2d + 5h \\ 6a + 10e & 6b + 10f & 6c + 10g & 6d + 10h \end{bmatrix}$$

This example shows that if $AB = C$, element c_{32} is found by multiplying each entry of A's *third* row by the corresponding entry of B's *second* column and then adding these products.

EXAMPLE 3.7 Matrix Multiplication

Find AB and BA if

$$A = \begin{bmatrix} 3 & 2 \\ 1 & 0 \end{bmatrix} \quad \text{and} \quad B = \begin{bmatrix} 1 & 2 \\ 3 & 1 \end{bmatrix}$$

SOLUTION

Both A and B are 2×2 matrices, so both AB and BA are defined.

$$AB = \begin{bmatrix} 3 & 2 \\ 1 & 0 \end{bmatrix}\begin{bmatrix} 1 & 2 \\ 3 & 1 \end{bmatrix} = \begin{bmatrix} 3{\cdot}1 + 2{\cdot}3 & 3{\cdot}2 + 2{\cdot}1 \\ 1{\cdot}1 + 0{\cdot}3 & 1{\cdot}2 + 0{\cdot}1 \end{bmatrix} = \begin{bmatrix} 9 & 8 \\ 1 & 2 \end{bmatrix}$$

$$BA = \begin{bmatrix} 1 & 2 \\ 3 & 1 \end{bmatrix}\begin{bmatrix} 3 & 2 \\ 1 & 0 \end{bmatrix} = \begin{bmatrix} 1{\cdot}3 + 2{\cdot}1 & 1{\cdot}2 + 2{\cdot}0 \\ 3{\cdot}3 + 1{\cdot}1 & 3{\cdot}2 + 1{\cdot}0 \end{bmatrix} = \begin{bmatrix} 5 & 2 \\ 10 & 6 \end{bmatrix}$$

Note that for these matrices, the two products AB and BA are matrices of the same size, but they are not equal. That is, $BA \neq AB$. Thus we say that *matrix multiplication is not commutative.*

An $n \times n$ (square) matrix (where n is any natural number) that has 1s down its diagonal and 0s everywhere else is called an **identity matrix**. The matrix

$$I = \begin{bmatrix} 1 & 0 & 0 \\ 0 & 1 & 0 \\ 0 & 0 & 1 \end{bmatrix}$$

is a 3×3 identity matrix. The matrix I is called an identity matrix because for any 3×3 matrix A, $AI = IA = A$. That is, if I is multiplied by a 3×3 matrix A, the product matrix is A. Note that when one of two square matrices being multiplied is an identity matrix, the product is *commutative.*

EXAMPLE 3.8 Identity Matrix

(a) Write the 2×2 identity matrix.

(b) Given $A = \begin{bmatrix} 4 & -7 \\ 13 & 2 \end{bmatrix}$, show that $AI = IA = A$.

SOLUTION

(a) We denote the 2×2 identity matrix by

$$I = \begin{bmatrix} 1 & 0 \\ 0 & 1 \end{bmatrix}$$

(b)

$$AI = \begin{bmatrix} 4 & -7 \\ 13 & 2 \end{bmatrix}\begin{bmatrix} 1 & 0 \\ 0 & 1 \end{bmatrix} = \begin{bmatrix} 4+0 & 0-7 \\ 13+0 & 0+2 \end{bmatrix} = \begin{bmatrix} 4 & -7 \\ 13 & 2 \end{bmatrix}$$

$$IA = \begin{bmatrix} 1 & 0 \\ 0 & 1 \end{bmatrix}\begin{bmatrix} 4 & -7 \\ 13 & 2 \end{bmatrix} = \begin{bmatrix} 4 & -7 \\ 13 & 2 \end{bmatrix}$$

Therefore, $AI = IA = A$.

3.2 Exercises

1. Multiply the matrices.

(a) $\begin{bmatrix} 1 & 2 & 3 \end{bmatrix}\begin{bmatrix} 4 \\ 5 \\ 6 \end{bmatrix}$ (b) $\begin{bmatrix} 1 & 2 \end{bmatrix}\begin{bmatrix} 3 & 5 \\ 4 & 6 \end{bmatrix}$

In Problems 2–6, use matrices A–F. Perform the indicated operations, when possible.

$$A = \begin{bmatrix} 1 & 0 & 2 \\ 3 & 2 & 1 \\ 4 & 0 & 3 \end{bmatrix} \qquad B = \begin{bmatrix} 1 & 1 & 3 & 0 \\ 4 & 2 & 1 & 1 \\ 3 & 2 & 0 & 1 \end{bmatrix}$$

$$C = \begin{bmatrix} 5 & 3 \\ 1 & 2 \end{bmatrix} \qquad D = \begin{bmatrix} 4 & 2 \\ 3 & 5 \end{bmatrix}$$

$$E = \begin{bmatrix} 1 & 0 & 4 \\ 5 & 1 & 0 \end{bmatrix} \qquad F = \begin{bmatrix} 1 & 0 & -1 & 3 \\ 2 & -1 & 3 & -4 \end{bmatrix}$$

2. CD

3. DE

4. AB

5. BA

6. A^2

In Problems 7 and 8, use the following matrices. Perform the indicated operations.

$$A = \begin{bmatrix} 2 & 5 & 4 \\ 1 & 4 & 3 \\ 1 & -3 & -2 \end{bmatrix} \qquad B = \begin{bmatrix} -1 & 2 & 1 \\ -5 & 8 & 2 \\ -7 & 11 & -3 \end{bmatrix}$$

$$I = \begin{bmatrix} 1 & 0 & 0 \\ 0 & 1 & 0 \\ 0 & 0 & 1 \end{bmatrix} \qquad C = \begin{bmatrix} 3 & 0 & 4 \\ 1 & 7 & -1 \\ 3 & 0 & 4 \end{bmatrix}$$

$$D = \begin{bmatrix} 4 & 4 & -8 \\ -1 & -1 & 2 \\ -3 & -3 & 6 \end{bmatrix} \qquad F = \begin{bmatrix} 0 & 1 & 2 \\ 0 & 0 & -4 \\ 0 & 0 & 0 \end{bmatrix}$$

$$Z = \begin{bmatrix} 0 & 0 & 0 \\ 0 & 0 & 0 \\ 0 & 0 & 0 \end{bmatrix}$$

7. AI

8. ZCI

9. Substitute the given values of x, y, and z into the matrix equation and use matrix multiplication to see whether the values are the solution of the equation.

$$\begin{bmatrix} 1 & 2 & 1 \\ 3 & 4 & -2 \\ 2 & 0 & -1 \end{bmatrix} \begin{bmatrix} x \\ y \\ z \end{bmatrix} = \begin{bmatrix} 2 \\ -2 \\ 2 \end{bmatrix} \quad x = 2, y = -1, z = 2$$

Applications

10. **Car pricing** A car dealer can buy midsize cars for 12% under the list price, and he can buy luxury cars for 15% under the list price. The table gives the list prices for two midsize and two luxury cars.

Midsize	36,000	42,000
Luxury	50,000	56,000

Write these data in a matrix and multiply it by the matrix

$$\begin{bmatrix} 0.88 & 0 \\ 0 & 0.85 \end{bmatrix}$$

What does each entry in this product matrix represent?

11. **Oil refineries** When gasoline is refined from crude oil, each gallon of regular, midgrade, and premium uses a different blend of two different grades of crude oil according to the matrix.

	Gasoline Blends		
Crude Grade	Regular	Midgrade	Premium
Black	$\frac{3}{4}$	$\frac{2}{5}$	$\frac{1}{4}$
Gold	$\frac{1}{4}$	$\frac{3}{5}$	$\frac{3}{4}$

In addition, the demand (in thousands of gallons) from a Houston refinery is 22 regular, 12 midgrade, and 8 premium, and the demand from a Gulfport refinery is 30 regular, 20 midgrade, and 11 premium.

(a) Make a 2 × 3 matrix B that contains the blending information and a 3 × 2 matrix D for the demands at each refinery. Then use a matrix product to find each refinery's need for each type of crude oil.

(b) If the cost of black crude is \$3.17 per gallon and for gold crude is \$3.32 per gallon, then the matrix

$$P = \begin{bmatrix} 3.17 & 3.32 \end{bmatrix}$$

represents these crude prices. Write a matrix product involving B, D, and P that could be used to find the total cost for crude oil at each refinery. Then find those total costs.

Need more practice?
Find more here: cengagebrain.com

3.3 Gauss-Jordan Elimination: Solving Systems of Equations

THERE ARE SEVERAL WAYS TO SOLVE SYSTEMS OF EQUATIONS. IN THIS SECTION, WE'LL EXPLORE A TECHNIQUE CALLED GAUSS-JORDAN ELIMINATION.

Systems with Unique Solutions

In solving systems with the left-to-right elimination method we've already discussed, we operated on the coefficients of the variables x, y, and z and on the constants. If we keep the coefficients of the variables x, y, and z in distinctive columns, we do not need to write the equations. In solving a system of linear equations with matrices, we first write the coefficients and constants from the system (on the left) in the **augmented matrix** (on the right).

$$\begin{cases} x + 2y + 3z = 6 \\ x \quad\;\; - z = 0 \\ x - y - z = -4 \end{cases} \qquad \begin{bmatrix} 1 & 2 & 3 & 6 \\ 1 & 0 & -1 & 0 \\ 1 & -1 & -1 & -4 \end{bmatrix}$$

In the augmented matrix, the numbers on the left side of the solid line form the **coefficient matrix,** with each column containing the coefficients of a variable (0 represents any missing variable). The column on the right side of the line (called the *augment*) contains the constants. Each row of the matrix gives the corresponding coefficients of an equation.

We can use matrices to solve systems of linear equations by performing the same operations on the rows of a matrix to reduce it as we do on equations in a linear system. The three different operations we can use to reduce the matrix are called **elementary row operations** and are similar to the operations with equations that result in equivalent systems. These operations are as follows:

1. Interchange two rows.

2. Add a multiple of one row to another row.

3. Multiply a row by a nonzero constant.

When a new matrix results from one or more of these elementary row operations being performed on a matrix, the new matrix is said to be **equivalent** to the original because these matrices represent equivalent systems of equations. Thus, if it can be shown that the augmented matrix

$$\begin{bmatrix} 1 & 2 & 3 & 6 \\ 1 & 0 & -1 & 0 \\ 1 & -1 & -1 & -4 \end{bmatrix} \text{ is equivalent to } \begin{bmatrix} 1 & 0 & 0 & -0.5 \\ 0 & 1 & 0 & 4 \\ 0 & 0 & 1 & -0.5 \end{bmatrix},$$

then the systems corresponding to these matrices have the same solution. That is,

$$\begin{cases} x + 2y + 3z = 6 \\ x \quad\;\; - z = 0 \\ x - y - z = -4 \end{cases} \text{ is equivalent to } \begin{cases} x = -0.5 \\ y = 4 \\ z = -0.5 \end{cases}$$

Thus, if a system of linear equations has a unique solution, we can solve the system by reducing the associated matrix to one whose coefficient matrix is the identity matrix and then "reading" the values of the variables that give the solution.

The process that we use to solve a system of equations with matrices (called the **elimination method** or **Gauss-Jordan elimination method**; see the following page) is a systematic procedure that uses row operations to attempt to reduce the coefficient matrix to an identity matrix.

EXAMPLE 3.9 Using the Gauss-Jordan Elimination Method

 Solve the system

$$\begin{aligned} x_1 + x_2 + x_3 + 2x_4 &= 6 \\ x_1 + 2x_2 + \quad\quad x_4 &= -2 \\ x_1 + x_2 + 3x_3 - 2x_4 &= 12 \\ x_1 + x_2 - 4x_3 + 5x_4 &= -16 \end{aligned}$$

SOLUTION

To solve this system, we must reduce the augmented matrix.

$$\begin{bmatrix} 1 & 1 & 1 & 2 & 6 \\ 1 & 2 & 0 & 1 & -2 \\ 1 & 1 & 3 & -2 & 12 \\ 1 & 1 & -4 & 5 & -16 \end{bmatrix}$$

The first step in the Gauss-Jordan Process (see next page) is to get a 1 in the row 1, column 1 position, and zeros below it. The entry in row 1, column 1 is 1. To get zeros in the first column, add -1 times row 1 to row 2, and put the result in row 2; add -1 times row 1 to row 3 and put the result in row 3; and add -1 times row 1 to row 4 and put the result in row 4.

$$\begin{matrix} (-1)R_1 + R_2 \to R_2 \\ (-1)R_1 + R_3 \to R_3 \\ (-1)R_1 + R_4 \to R_4 \end{matrix} \begin{bmatrix} 1 & 1 & 1 & 2 & 6 \\ 0 & 1 & -1 & -1 & -8 \\ 0 & 0 & 2 & -4 & 6 \\ 0 & 0 & -5 & 3 & -22 \end{bmatrix}$$

To solve $\begin{cases} 2x + 5y + 4z = 4 \\ x + 4y + 3z = 1 \\ x - 3y - 2z = 5 \end{cases}$ we reduce the matrix $\begin{bmatrix} 2 & 5 & 4 & | & 4 \\ 1 & 4 & 3 & | & 1 \\ 1 & -3 & -2 & | & 5 \end{bmatrix}$.

Gauss-Jordan Elimination Method

Goal	Row Operation	Equivalent Matrix
1. Get a 1 in row 1, column 1.	Interchange row 1 and row 2.	$\begin{bmatrix} 1 & 4 & 3 & | & 1 \\ 2 & 5 & 4 & | & 4 \\ 1 & -3 & -2 & | & 5 \end{bmatrix}$
2. Add multiples of row 1 *only* to the other rows to get zeros in other entries of column 1.	Add -2 times row 1 to row 2; put the result in row 2. Add -1 times row 1 to row 3; put the result in row 3.	$\begin{bmatrix} 1 & 4 & 3 & | & 1 \\ 0 & -3 & -2 & | & 2 \\ 0 & -7 & -5 & | & 4 \end{bmatrix}$
3. Use rows below row 1 to get a 1 in row 2, column 2.	Multiply row 2 by $-\frac{1}{3}$; put the result in row 2.	$\begin{bmatrix} 1 & 4 & 3 & | & 1 \\ 0 & 1 & \frac{2}{3} & | & -\frac{2}{3} \\ 0 & -7 & -5 & | & 4 \end{bmatrix}$
4. Add multiples of row 2 *only* to the other rows to get zeros as the other entries in column 2.	Add -4 times row 2 to row 1; put the result in row 1. Add 7 times row 2 to row 3; put the result in row 3.	$\begin{bmatrix} 1 & 0 & \frac{1}{3} & | & \frac{11}{3} \\ 0 & 1 & \frac{2}{3} & | & -\frac{2}{3} \\ 0 & 0 & -\frac{1}{3} & | & -\frac{2}{3} \end{bmatrix}$
5. Use rows below row 2 to get a 1 in row 3, column 3.	Multiply row 3 by -3; put the result in row 3.	$\begin{bmatrix} 1 & 0 & \frac{1}{3} & | & \frac{11}{3} \\ 0 & 1 & \frac{2}{3} & | & -\frac{2}{3} \\ 0 & 0 & 1 & | & 2 \end{bmatrix}$
6. Add multiples of row 3 *only* to the other rows to get zeros as the other entries in column 3.	Add $-\frac{1}{3}$ times row 3 to row 1; put the result in row 1. Add $-\frac{2}{3}$ times row 3 to row 2; put the result in row 2.	$\begin{bmatrix} 1 & 0 & 0 & | & 3 \\ 0 & 1 & 0 & | & -2 \\ 0 & 0 & 1 & | & 2 \end{bmatrix}$
7. Repeat the process until it cannot be continued.	All rows have been used. The matrix is in reduced form.	Thus $x = 3$, $y = -2$, and $z = 2$.

Continuing Example 3.9 (from the previous page), we note that the entry in row 2, column 2 is 1. To get zeros in the second column, add -1 times row 2 to row 1, and put the result in row 1.

$$\xrightarrow{(-1)R_2 + R_1 \to R_1} \begin{bmatrix} 1 & 0 & 2 & 3 & | & 14 \\ 0 & 1 & -1 & -1 & | & -8 \\ 0 & 0 & 2 & -4 & | & 6 \\ 0 & 0 & -5 & 3 & | & -22 \end{bmatrix}$$

The entry in row 3, column 3 is 2. To get a 1 in this position, multiply row 3 by $\frac{1}{2}$.

$$\xrightarrow{\frac{1}{2}R_3 \to R_3} \begin{bmatrix} 1 & 0 & 2 & 3 & | & 14 \\ 0 & 1 & -1 & -1 & | & -8 \\ 0 & 0 & 1 & -2 & | & 3 \\ 0 & 0 & -5 & 3 & | & -22 \end{bmatrix}$$

Using row 3, add -2 times row 3 to row 1 and put the result in row 1; add row 3 to row 2 and put the result in row 2; and add 5 times row 3 to row 4 and put the result in row 4.

$$\begin{matrix}(-2)R_3 + R_1 \to R_1 \\ R_3 + R_2 \to R_2 \\ \hline 5R_3 + R_4 \to R_4\end{matrix} \longrightarrow \left[\begin{array}{ccc c|c} 1 & 0 & 0 & 7 & 8 \\ 0 & 1 & 0 & -3 & -5 \\ 0 & 0 & 1 & -2 & 3 \\ 0 & 0 & 0 & -7 & -7 \end{array}\right]$$

Multiplying $-\frac{1}{7}$ times row 4 and putting the result in row 4 gives a 1 in row 4, column 4.

$$\xrightarrow{\frac{1}{7}R_4 \to R_4} \left[\begin{array}{ccc c|c} 1 & 0 & 0 & 7 & 8 \\ 0 & 1 & 0 & -3 & -5 \\ 0 & 0 & 1 & -2 & 3 \\ 0 & 0 & 0 & 1 & 1 \end{array}\right]$$

Adding appropriate multiples of row 4 to the other rows gives

$$\begin{matrix}(-7)R_4 + R_1 \to R_1 \\ 3R_4 + R_2 \to R_2 \\ \hline 2R_4 + R_3 \to R_3\end{matrix} \longrightarrow \left[\begin{array}{cccc|c} 1 & 0 & 0 & 0 & 1 \\ 0 & 1 & 0 & 0 & -2 \\ 0 & 0 & 1 & 0 & 5 \\ 0 & 0 & 0 & 1 & 1 \end{array}\right]$$

This corresponds to the system

$$\begin{aligned} x_1 & & & = 1 \\ & x_2 & & = -2 \\ & & x_3 & = 5 \\ & & & x_4 = 1 \end{aligned}$$

Thus the solution is $x_1 = 1$, $x_2 = -2$, $x_3 = 5$, and $x_4 = 1$.

When a system of linear equations has a unique solution, the coefficient part of the reduced augmented matrix will be an identity matrix. Note that for this to occur, the coefficient matrix must be square; that is, the number of equations must equal the number of variables. The Gauss-Jordan elimination method may be used with systems of any size.

Systems with Nonunique Solutions

All the systems considered so far had unique solutions, but it is also possible for a system of linear equations to have an infinite number of solutions or no solution at all. Although coefficient matrices for systems with an infinite number of solutions or no solution will not reduce to identity matrices, row operations can be used to obtain a reduced form from which the solutions, if they exist, can be determined.

A matrix is said to be in **reduced form** when it is in the following form:

1. The first nonzero element in each row is 1.

2. Every column containing a first nonzero element for some row has zeros everywhere else.

3. The first nonzero element of each row is to the right of the first nonzero element of every row above it.

4. All rows containing zeros are grouped together below the rows containing nonzero entries.

The following matrices are in reduced form because they satisfy these conditions.

$$\left[\begin{array}{cc|c} 1 & 0 & 4 \\ 0 & 1 & 2 \\ 0 & 0 & 0 \end{array}\right] \quad \left[\begin{array}{ccc|c} 1 & 0 & 0 & 0 \\ 0 & 1 & 0 & 3 \\ 0 & 0 & 1 & 1 \\ 0 & 0 & 0 & 0 \end{array}\right] \quad \left[\begin{array}{ccc|c} 1 & 4 & 0 & 1 \\ 0 & 0 & 1 & 2 \\ 0 & 0 & 0 & 0 \end{array}\right]$$

The following matrices are *not* in reduced form.

$$\left[\begin{array}{cc|c} 1 & ① & 0 \\ 0 & 1 & 3 \\ 0 & 0 & 0 \end{array}\right] \quad \left[\begin{array}{cc|c} 1 & 0 & 1 \\ 0 & ② & 1 \\ 0 & 0 & 0 \end{array}\right] \quad \left[\begin{array}{cc|c} 0 & 1 & 0 \\ 1 & 0 & 0 \\ 0 & 0 & 1 \end{array}\right]$$

We can solve a system of linear equations by using row operations on the augmented matrix until the coefficient matrix is transformed to an equivalent matrix in reduced form.

EXAMPLE 3.10 Matrix Solution of a System

Solve the system

$$\begin{cases} x_1 + x_2 - x_3 + x_4 = 3 \\ x_2 + x_3 + x_4 = 1 \\ x_1 - 2x_3 + x_4 = 6 \\ 2x_1 - x_2 - 5x_3 - 3x_4 = -5 \end{cases}$$

SOLUTION

To solve this system we must reduce the augmented matrix

$$\left[\begin{array}{cccc|c} 1 & 1 & -1 & 1 & 3 \\ 0 & 1 & 1 & 1 & 1 \\ 1 & 0 & -2 & 1 & 6 \\ 2 & -1 & -5 & -3 & -5 \end{array}\right]$$

Attempting to reduce the coefficient matrix to an identity matrix requires the following.

$$\begin{array}{c}(-1)R_1 + R_3 \to R_3 \\ \hline (-2)R_1 + R_4 \to R_4\end{array} \xrightarrow{} \left[\begin{array}{cccc|c} 1 & 1 & -1 & 1 & 3 \\ 0 & 1 & 1 & 1 & 1 \\ 0 & -1 & -1 & 0 & 3 \\ 0 & -3 & -3 & -5 & -11 \end{array}\right]$$

$$\begin{array}{c}(-1)R_2 + R_1 \to R_1 \\ R_2 + R_3 \to R_3 \\ \hline 3R_2 + R_4 \to R_4\end{array} \xrightarrow{} \left[\begin{array}{cccc|c} 1 & 0 & -2 & 0 & 2 \\ 0 & 1 & 1 & 1 & 1 \\ 0 & 0 & 0 & 1 & 4 \\ 0 & 0 & 0 & -2 & -8 \end{array}\right]$$

The entry in row 3, column 3 cannot be made 1 using rows *below* row 2. Moving to column 4, the entry in row 3, column 4 is a 1. Using row 3 gives

$$\begin{array}{c}(-1)R_3 + R_2 \to R_2 \\ \hline 2R_3 + R_4 \to R_4\end{array} \xrightarrow{} \left[\begin{array}{cccc|c} 1 & 0 & -2 & 0 & 2 \\ 0 & 1 & 1 & 0 & -3 \\ 0 & 0 & 0 & 1 & 4 \\ 0 & 0 & 0 & 0 & 0 \end{array}\right]$$

The augmented matrix is now reduced; this matrix corresponds to the system

$$\begin{cases} x_1 - 2x_3 = 2 \\ x_2 + x_3 = -3 \\ \qquad\quad x_4 = 4 \\ \qquad\quad 0 = 0 \end{cases}$$

If we solve each of the equations for the leading variable (the variable corresponding to the first 1 in each row of the reduced form of a matrix) and let any nonleading variables equal any real number (x_3 in this case), we obtain the **general solution** of the system.

$$x_1 = 2 + 2x_3$$

$$x_2 = -3 - x_3$$

$$x_4 = 4$$

$$x_3 = \text{any real number}$$

The general solution gives the values of x_1 and x_2 dependent on the value of x_3, so we can get many different solutions of the system by specifying different values of x_3. For example, if $x_3 = 1$, then $x_1 = 4$, $x_2 = -4$, $x_3 = 1$, and $x_4 = 4$ is a solution of the system; if we let $x_3 = -2$, then $x_1 = -2$, $x_2 = -1$, $x_3 = -2$, and $x_4 = 4$ is another solution.

The system

$$\begin{cases} x + 2y - z = 3 \\ 3x + y = 4 \\ 2x - y + z = 2 \end{cases}$$

results in the following reduced matrix.

$$\left[\begin{array}{ccc|c} 1 & 0 & \frac{1}{5} & 1 \\ 0 & 1 & -\frac{3}{5} & 1 \\ 0 & 0 & 0 & 1 \end{array}\right]$$

The system of equations corresponding to this reduced matrix is

$$\begin{cases} x + \frac{1}{5}z = 1 \\ y - \frac{3}{5}z = 1 \\ \qquad\quad 0 = 1 \end{cases}$$

Because this system has $0 = 1$ as an equation (an impossibility), there is no solution. Table 3.3 summarizes our discussion to this point.

Table 3.3

Reduced Form of Augmented Matrix for *n* Equations in *n* Variables	Solution to System
1. Coefficient array is an identity matrix.	Unique solution (see Example 3.9).
2. Coefficient array is *not* an identity matrix with either: (a) A row of 0s in the coefficient array with a nonzero entry in the augment. (b) Or otherwise.	(a) No solution (see preceding discussion). (b) Infinitely many solutions. Solve for lead variables in terms of nonlead variables; nonlead variables can equal any real number (see Example 3.10).

Nonsquare Systems

Finally, we consider systems with fewer equations than variables. Because the coefficient matrix for such a system is not square, it cannot reduce to an identity matrix. Hence these systems of equations cannot have a unique solution but may have no solution or infinitely many solutions. Nevertheless, they are solved in the same manner as those considered so far but with the following two possible results.

1. If the reduced augmented matrix contains a row of 0s in the coefficient matrix with a nonzero number in the augment, the system has no solution.

2. Otherwise, the system has infinitely many solutions.

EXAMPLE 3.11 Investment

A trust account manager has $500,000 to be invested in three different accounts. The accounts pay annual interest rates of 8%, 10%, and 14%, and the goal is to earn $49,000 a year. To accomplish this, assume that x dollars is invested at 8%, y dollars at 10%, and z dollars at 14%. Find how much should be invested in each account to satisfy the conditions.

SOLUTION

The sum of the three investments is $500,000, so we have the equation

$$x + y + z = 500,000$$

The interest earned from the 8% investment is $0.08x$, the amount earned from the 10% investment is $0.10y$, and the amount earned from the 14% investment is $0.14z$, so the total amount earned from the investments is given by the equation

$$0.08x + 0.10y + 0.14z = 49,000$$

These two equations represent all the given information, so we attempt to answer the question by solving a system of *two* equations in *three* variables. (Because there are fewer equations than variables, we cannot find a unique solution to the system.) The system is

$$\begin{cases} x + y + z = 500,000 \\ 0.08x + 0.10y + 0.14z = 49,000 \end{cases}$$

We solve this system using matrices, as follows:

$$\begin{bmatrix} 1 & 1 & 1 & 500,000 \\ 0.08 & 0.10 & 0.14 & 49000 \end{bmatrix}$$

$$\xrightarrow{-0.08R_1 + R_2 \rightarrow R_2} \begin{bmatrix} 1 & 1 & 1 & 500,000 \\ 0 & 0.02 & 0.06 & 9,000 \end{bmatrix}$$

$$\xrightarrow{50R_2 \rightarrow R_2} \begin{bmatrix} 1 & 1 & 1 & 500,000 \\ 0 & 1 & 3 & 450,000 \end{bmatrix}$$

$$\xrightarrow{-R_2 + R_1 \rightarrow R_1} \begin{bmatrix} 1 & 0 & -2 & 50,000 \\ 0 & 1 & 3 & 450,000 \end{bmatrix}$$

This matrix represents the equivalent system

$$\begin{cases} x - 2z = 50,000 \\ y + 3z = 450,000 \end{cases}$$

so

$$\begin{aligned} x &= 50,000 + 2z \\ y &= 450,000 - 3z \\ z &= \text{any real number} \end{aligned}$$

Because a negative amount cannot be invested, x, y, and z must all be nonnegative. Note that $x \geq 0$ when $z \geq 0$ and that $y \geq 0$ when $450,000 - 3z \geq 0$, or when $z \leq 150,000$. Thus the amounts invested are

$$\$z \text{ at } 14\%, \text{ where } 0 \leq z \leq 150,000$$

$$\$x \text{ at } 8\%, \text{ where } x = 50,000 + 2z$$

$$\$y \text{ at } 10\%, \text{ where } y = 450,000 - 3z$$

There are many possible investment plans. One example would be with $z = \$100,000$, $x = \$250,000$, and $y = \$150,000$. Observe that the values for x and y depend on z, and once z is chosen, x will be between $50,000 and $350,000, and y will be between $0 and $450,000.

··

3.3 Exercises

1. Write the augmented matrix associated with the system of linear equations.

$$\begin{cases} x - 3y + 4z = 2 \\ 2x + 2z = 1 \\ x + 2y + z = 1 \end{cases}$$

2. Solve the system whose reduced augmented matrix

is $\begin{bmatrix} 1 & 0 & 0 & 2 \\ 0 & 1 & 0 & \frac{1}{2} \\ 0 & 0 & 1 & -5 \end{bmatrix}$.

In Problems 3 and 4, use row operations on augmented matrices to solve the given systems of linear equations.

3. $\begin{cases} x + y - z = 0 \\ x + 2y + 3z = -5 \\ 2x - y - 13z = 17 \end{cases}$

4. $\begin{cases} 2x - 6y - 12z = 6 \\ 3x - 10y - 20z = 5 \\ 2x - 17z = -4 \end{cases}$

5. Solve the following system (if a solution exists) by using its given reduced matrix.

$$\begin{cases} x + 2y + 3z = 1 \\ 2x - y = 3 \\ x + 2y + 3z = 2 \end{cases} \quad \begin{bmatrix} 1 & 0 & \frac{3}{5} & | & 0 \\ 0 & 1 & \frac{6}{5} & | & 0 \\ 0 & 0 & 0 & | & 1 \end{bmatrix}$$

In Problems 6–9, the systems of equations may have unique solutions, an infinite number of solutions, or no solution. Use matrices to find the general solution of each system, if a solution exists.

6. $\begin{cases} 3x + 2y + z = 0 \\ x + y + 2z = 2 \\ 2x + y - z = -1 \end{cases}$

7. $\begin{cases} x - 3y + 3z = 7 \\ x + 2y - z = -2 \\ 3x - 2y + 4z = 5 \end{cases}$

8. $\begin{cases} 2x - 5y + z = -9 \\ x + 4y - 6z = 2 \\ 3x - 4y - 2z = -10 \end{cases}$

9. $\begin{cases} 3x - 2y + 5z = 14 \\ 2x - 3y + 4z = 8 \end{cases}$

Applications

10. **Nutrition** A preschool has Campbell's Chunky Beef soup, which contains 2.5 grams of fat and 15 milligrams of cholesterol per serving (cup), and Campbell's Chunky Sirloin Burger soup, which contains 7 grams of fat and 15 milligrams of cholesterol per serving. By combining the soups, it is possible to get 10 servings of soup that will have 61 grams of fat and 150 milligrams of cholesterol. How many cups of each soup should be used?

11. **Investment** A man has $235,000 invested in three properties. One earns 12%, one 10%, and one 8%.

His annual income from the properties is $22,500 and the amount invested at 8% is twice that invested at 12%.
(a) How much is invested in each property?
(b) What is the annual income from each property?

12. **Car rental patterns** A car rental agency in a major city has a total of 2200 cars that it rents from three locations: Metropolis Airport, downtown, and at the smaller City Airport. Some weekly rental and return patterns are shown in the table.

	Rented from		
Returned to	MP	DT	CA
Metropolis Airport (MP)	90%	10%	10%
Downtown (DT)	5%	80%	5%

At the beginning of a week, how many cars should be at each location so that same number of cars will be there at the end of the week (and hence at the start of the next week)?

13. **Nutrition** A botanist can purchase plant food of four different types, I, II, III, and IV. Each food comes in the same size bag, and the table summarizes the number of grams of each of three nutrients that each bag contains.

	Food (grams)			
	I	II	III	IV
Nutrient A	5	5	10	5
Nutrient B	10	5	30	10
Nutrient C	5	15	10	25

The botanist wants to use a food that has these nutrients in a different proportion and determines that he will need a total of 10,000 grams of A, 20,000 grams of B, and 20,000 grams of C. Find the number of bags of each type of food that should be ordered.

14. **Investment** An investment club has set a goal of earning 15% on the money they invest in stocks. The members are considering purchasing three possible stocks, with their cost per share (in dollars) and their projected growth per share (in dollars) summarized in the table.

	Stocks		
	Computer (C)	Utility (U)	Retail (R)
Cost/share	30	44	26
Growth/share	6.00	6.00	2.40

(a) If they have \$392,000 to invest, how many shares of each stock should they buy to meet their goal?

(b) If they buy 1000 shares of retail stock, how many shares of the other stocks do they buy? What if they buy 2000 shares of retail stock?

(c) What is the minimum number of shares of computer stock they will buy, and what is the number of shares of the other stocks in this case?

(d) What is the maximum number of shares of computer stock purchased, and what is the number of shares of the other stocks in this case?

Need more practice?
Find more here: cengagebrain.com

3.4 Inverse of a Square Matrix

IF THE PRODUCT OF *A* AND *B* IS THE IDENTITY MATRIX, *I*, WE SAY THAT *B* IS THE INVERSE OF *A* (AND *A* IS THE INVERSE OF *B*). THE MATRIX *B* IS CALLED THE **INVERSE MATRIX** OF *A*, DENOTED A^{-1}.

Inverse Matrices: Two square matrices, *A* and *B*, are called inverses of each other if
$$AB = I \quad \text{and} \quad BA = I$$
In this case, $B = A^{-1}$ and $A = B^{-1}$.

EXAMPLE 3.12 Inverse Matrices

➡ Is *B* the inverse of *A* if $A = \begin{bmatrix} 1 & 2 \\ 1 & 1 \end{bmatrix}$ and $B = \begin{bmatrix} -1 & 2 \\ 1 & -1 \end{bmatrix}$?

SOLUTION

$$AB = \begin{bmatrix} 1 & 2 \\ 1 & 1 \end{bmatrix} \cdot \begin{bmatrix} -1 & 2 \\ 1 & -1 \end{bmatrix} = \begin{bmatrix} 1 & 0 \\ 0 & 1 \end{bmatrix} = I$$

$$BA = \begin{bmatrix} -1 & 2 \\ 1 & -1 \end{bmatrix} \cdot \begin{bmatrix} 1 & 2 \\ 1 & 1 \end{bmatrix} = \begin{bmatrix} 1 & 0 \\ 0 & 1 \end{bmatrix} = I$$

Thus *B* is the inverse of *A*, or $B = A^{-1}$. We also say *A* is the inverse of *B*, or $A = B^{-1}$. Thus $(A^{-1})^{-1} = A$.

We have seen how to use elementary row operations on augmented matrices to solve systems of linear equations. We can also find the inverse of a matrix by using elementary row operations.

If the inverse exists for a square matrix *A*, we find A^{-1} as follows.

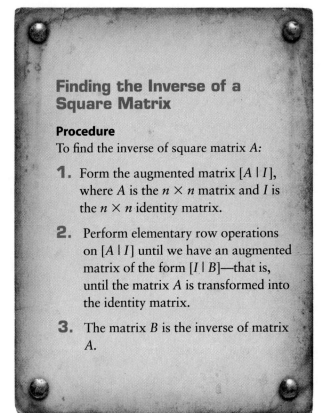

Finding the Inverse of a Square Matrix

Procedure
To find the inverse of square matrix *A*:

1. Form the augmented matrix [*A* | *I*], where *A* is the $n \times n$ matrix and *I* is the $n \times n$ identity matrix.

2. Perform elementary row operations on [*A* | *I*] until we have an augmented matrix of the form [*I* | *B*]—that is, until the matrix *A* is transformed into the identity matrix.

3. The matrix *B* is the inverse of matrix *A*.

Note that if *A* has no inverse, the reduction process on [*A*|*I*] will yield a row of zeros in the left half.

EXAMPLE 3.13 Finding an Inverse

➡ Find the inverse of matrix *A*.

$$A = \begin{bmatrix} 2 & 5 & 4 \\ 1 & 4 & 3 \\ 1 & -3 & -2 \end{bmatrix}$$

SOLUTION

To find the inverse of matrix A, we reduce A in $[A|I]$ (the matrix A augmented with the identity matrix I).

$$\begin{bmatrix} 2 & 5 & 4 & | & 1 & 0 & 0 \\ 1 & 4 & 3 & | & 0 & 1 & 0 \\ 1 & -3 & -2 & | & 0 & 0 & 1 \end{bmatrix}$$

When the left side becomes an identity matrix, the inverse is in the augment.

$$\xrightarrow{\text{Switch } R_1 \text{ and } R_2} \begin{bmatrix} 1 & 4 & 3 & | & 0 & 1 & 0 \\ 2 & 5 & 4 & | & 1 & 0 & 0 \\ 1 & -3 & -2 & | & 0 & 0 & 1 \end{bmatrix}$$

$$\xrightarrow[-R_1 + R_3 \to R_3]{-2R_1 + R_2 \to R_2} \begin{bmatrix} 1 & 4 & 3 & | & 0 & 1 & 0 \\ 0 & -3 & -2 & | & 1 & -2 & 0 \\ 0 & -7 & -5 & | & 0 & -1 & 1 \end{bmatrix}$$

$$\xrightarrow{-\frac{1}{3}R_2 \to R_2} \begin{bmatrix} 1 & 4 & 3 & | & 0 & 1 & 0 \\ 0 & 1 & \frac{2}{3} & | & -\frac{1}{3} & \frac{2}{3} & 0 \\ 0 & -7 & -5 & | & 0 & -1 & 1 \end{bmatrix}$$

$$\xrightarrow[7R_2 + R_3 \to R_3]{-4R_2 + R_1 \to R_1} \begin{bmatrix} 1 & 0 & \frac{1}{3} & | & \frac{4}{3} & -\frac{5}{3} & 0 \\ 0 & 1 & \frac{2}{3} & | & -\frac{1}{3} & \frac{2}{3} & 0 \\ 0 & 0 & -\frac{1}{3} & | & -\frac{7}{3} & \frac{11}{3} & 1 \end{bmatrix}$$

$$\xrightarrow[\substack{2R_3 + R_2 \to R_2 \\ -3R_3 \to R_3}]{R_3 + R_1 \to R_1} \begin{bmatrix} 1 & 0 & 0 & | & -1 & 2 & 1 \\ 0 & 1 & 0 & | & -5 & 8 & 2 \\ 0 & 0 & 1 & | & 7 & -11 & -3 \end{bmatrix}$$

The inverse we seek is

$$A^{-1} = \begin{bmatrix} -1 & 2 & 1 \\ -5 & 8 & 2 \\ 7 & -11 & -3 \end{bmatrix}$$

The technique of reducing $[A|I]$ can be used to find A^{-1} or to find that A^{-1} doesn't exist for any square matrix A. However, a special formula can be used to find the inverse of a 2×2 matrix, if the inverse exists.

Inverse of a 2 × 2 Matrix:

If $A = \begin{bmatrix} a & b \\ c & d \end{bmatrix}$, then $A^{-1} = \dfrac{1}{ad - bc}\begin{bmatrix} d & -b \\ -c & a \end{bmatrix}$, provided $ad - bc \neq 0$. If $ad - bc = 0$, then A^{-1} does not exist.

This result can be verified by direct calculation or by reducing $[A|I]$.

EXAMPLE 3.14 Inverse of a 2 × 2 Matrix

Find the inverse, if it exists, for each of the following.

(a) $A = \begin{bmatrix} 3 & 7 \\ 2 & 5 \end{bmatrix}$ (b) $B = \begin{bmatrix} 2 & -1 \\ -4 & 2 \end{bmatrix}$

SOLUTION

(a) For A, $ad - bc = (3)(5) - (2)(7) = 1 \neq 0$, so A^{-1} exists.

$$A^{-1} = \frac{1}{1}\begin{bmatrix} 5 & -7 \\ -2 & 3 \end{bmatrix} = \begin{bmatrix} 5 & -7 \\ -2 & 3 \end{bmatrix}$$

(b) For B, $ad - bc = (2)(2) - (-4)(-1) = 4 - 4 = 0$, so B^{-1} does not exist.

Matrix Equations

This system

$$\begin{cases} 2x + 5y + 4z = 4 \\ x + 4y + 3z = 1 \\ x - 3y - 2z = 5 \end{cases}$$

can be written as either of the following matrix equations.

$$\begin{bmatrix} 2x + 5y + 4z \\ x + 4y + 3z \\ x - 3y - 2z \end{bmatrix} = \begin{bmatrix} 4 \\ 1 \\ 5 \end{bmatrix} \text{ or } \begin{bmatrix} 2 & 5 & 4 \\ 1 & 4 & 3 \\ 1 & -3 & -2 \end{bmatrix}\begin{bmatrix} x \\ y \\ z \end{bmatrix} = \begin{bmatrix} 4 \\ 1 \\ 5 \end{bmatrix}$$

We can see that this second form of a matrix equation is equivalent to the first if we carry out the multiplication. With the system written in this second form, we could also solve the system by multiplying both sides of the equation by the inverse of the coefficient matrix.

EXAMPLE 3.15 Solution with Inverse Matrices

Use an inverse matrix to solve the system

$$\begin{cases} 2x + 5y + 4z = 4 \\ x + 4y + 3z = 1 \\ x - 3y - 2z = 5 \end{cases}$$

SOLUTION

To solve the system, we multiply both sides of the associated matrix equation, presented earlier, by the inverse of the coefficient matrix, which we found in Example 3.13.

$$\begin{bmatrix} -1 & 2 & 1 \\ -5 & 8 & 2 \\ 7 & -11 & -3 \end{bmatrix}\begin{bmatrix} 2 & 5 & 4 \\ 1 & 4 & 3 \\ 1 & -3 & -2 \end{bmatrix}\begin{bmatrix} x \\ y \\ z \end{bmatrix} = \begin{bmatrix} -1 & 2 & 1 \\ -5 & 8 & 2 \\ 7 & -11 & -3 \end{bmatrix}\begin{bmatrix} 4 \\ 1 \\ 5 \end{bmatrix}$$

Note that we must be careful to multiply both sides *from the left* because matrix multiplication is not commutative. If we carry out the multiplications, we obtain

$$\begin{bmatrix} 1 & 0 & 0 \\ 0 & 1 & 0 \\ 0 & 0 & 1 \end{bmatrix}\begin{bmatrix} x \\ y \\ z \end{bmatrix} = \begin{bmatrix} 3 \\ -2 \\ 2 \end{bmatrix}, \text{ or } \begin{bmatrix} x \\ y \\ z \end{bmatrix} = \begin{bmatrix} 3 \\ -2 \\ 2 \end{bmatrix}$$

which yields $x = 3$, $y = -2$, $z = 2$.

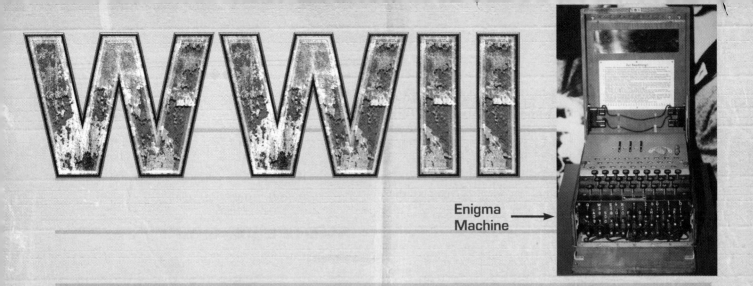

Enigma
Machine →

During World War II, the military commanders sent instructions that used simple substitution codes. These codes were easily broken, so they were further coded by using a coding matrix and matrix multiplication. The resulting coded messages, when received, could be unscrambled with a decoding matrix that was the inverse matrix of the coding matrix. The following code

a	b	c	d	e	f	g	h	i	j	k	l
1	2	3	4	5	6	7	8	9	10	11	12

m	n	o	p	q	r	s	t	u	v	w	x
13	14	15	16	17	18	19	20	21	22	23	24

y	z	blank
25	26	27

and the encoding matrix

$$A = \begin{bmatrix} 3 & 5 \\ 4 & 6 \end{bmatrix}$$

can be used to convert numbers to a coded message (in pairs of numbers). To decode a message encoded by this matrix A, we find the inverse of matrix A, which is denoted A^{-1}, and multiply A^{-1} by the coded message.

To see how this works, let's use the inverse of A to decode the coded message 96, 118, 65, 84, 131, 168, 55, 72.

We first find the inverse of matrix A.

$$A^{-1} = \frac{1}{18-20}\begin{bmatrix} 6 & -5 \\ -4 & 3 \end{bmatrix} = -\frac{1}{2}\begin{bmatrix} 6 & -5 \\ -4 & 3 \end{bmatrix} = \begin{bmatrix} -3 & 2.5 \\ 2 & -1.5 \end{bmatrix}$$

Then we can write the coded message numbers (in pairs) as columns of a matrix B and then find the product $A^{-1}B$.

$$A^{-1}B = \begin{bmatrix} -3 & 2.5 \\ 2 & -1.5 \end{bmatrix}\begin{bmatrix} 96 & 65 & 131 & 55 \\ 118 & 84 & 168 & 72 \end{bmatrix} = \begin{bmatrix} 7 & 15 & 27 & 15 \\ 15 & 4 & 10 & 2 \end{bmatrix}$$

Reading the numbers (down the respective columns) gives the result 7, 15, 15, 4, 27, 10, 15, 2, which is the message "good job."

DECODER

Just as we wrote the system of three equations as a matrix equation of the form $AX = B$, this can be done in general. If A is an $n \times n$ matrix and if B and X are $n \times 1$ matrices, then

$$AX = B$$

is a **matrix equation.**

If the inverse of a matrix A exists, then we can use that inverse to solve the matrix equation for the matrix X. The general solution method follows.

$$AX = B$$

Multiplying both sides of the equation (from the left) by A^{-1} gives

$$A^{-1}(AX) = A^{-1}B$$

$$(A^{-1}A)X = A^{-1}B$$

$$IX = A^{-1}B$$

$$X = A^{-1}B$$

Thus inverse matrices can be used to solve systems of equations. Unfortunately, this method will work only if the solution to the system is unique. In fact, a system $AX = B$ has a unique solution if and only if A^{-1} exists. If the inverse of the coefficient matrix exists, the preceding solution method can be used to solve the system.

Determinants

Recall that the inverse of $A = \begin{bmatrix} a & b \\ c & d \end{bmatrix}$ can be found with the formula

$$A^{-1} = \frac{1}{ad - bc}\begin{bmatrix} d & -b \\ -c & a \end{bmatrix}$$

if $ad - bc \neq 0$

The value $ad - bc$ used in finding the inverse of this 2×2 matrix is used so frequently that it is given a special name, the **determinant** of A (denoted *det A* or $|A|$).

Determinant of a 2 × 2 Matrix:

$$\det\begin{bmatrix} a & b \\ c & d \end{bmatrix} = \begin{vmatrix} a & b \\ c & d \end{vmatrix} = ad - bc$$

EXAMPLE 3.16 Determinant of a 2 × 2 Matrix

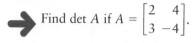 Find det A if $A = \begin{bmatrix} 2 & 4 \\ 3 & -4 \end{bmatrix}$.

SOLUTION

$$\det A = (2)(-4) - (3)(4) = -8 - 12 = -20$$

There are formulas for finding the determinants of square matrices of orders larger than 2×2, but for the matrices used in this text, calculators and computers make it easy to find determinants. We see from its definition that the inverse of a 2×2 matrix A does not exist if and only if det $A = 0$. In general, the inverse of an $n \times n$ matrix B does not exist if det $B = 0$. This also means that if the coefficient matrix of a system of linear equations has a determinant equal to 0, the system does not have a unique solution.

3.4 Exercises

1. If $A = \begin{bmatrix} 1 & 2 & 1 \\ 0 & 0 & 3 \\ 1 & 0 & 1 \end{bmatrix}$ and $B = \begin{bmatrix} 0 & -\frac{1}{3} & 1 \\ \frac{1}{2} & 0 & -\frac{1}{2} \\ 0 & \frac{1}{3} & 0 \end{bmatrix}$, does $B = A^{-1}$?

2. Find the inverse matrix of $\begin{bmatrix} 4 & 7 \\ 1 & 2 \end{bmatrix}$.

3. Find the inverse of matrix A and check it by calculating AA^{-1}.

   $$A = \begin{bmatrix} 3 & 0 & 0 \\ 0 & 3 & 0 \\ 0 & 0 & 3 \end{bmatrix}$$

4. Find the inverse matrix of $\begin{bmatrix} 3 & 1 & 2 \\ 1 & 2 & 3 \\ 1 & 1 & 1 \end{bmatrix}$

5. Use inverse matrices to solve the system of linear equations.

   $$\begin{cases} x + 2y = 4 \\ 3x + 4y = 10 \end{cases}$$

Applications

In Problems 6–9, set up each system of equations and then solve it by using inverse matrices.

6. *Medication* Medication A is given every 4 hours and medication B is given twice per day, and the

ratio of the dosage of A to the dosage of B is always 5 to 8.

(a) For patient I, the total intake of the two medications is 50.6 milligrams per day. Find the dosage of each administration of each medication for patient I.

(b) For patient II, the total intake of the two medications is 92 milligrams per day. Find the dosage of each administration of each medication for patient II.

7. **Investment** One safe investment pays 10% per year, and a more risky investment pays 18% per year. A woman has $145,600 to invest and would like to have an income of $20,000 per year from her investments. How much should she invest at each rate?

8. **Manufacturing** A manufacturer of table saws has three models—Deluxe, Premium, and Ultimate—which must be painted, assembled, and packaged for shipping. The table gives the number of hours required for each of these operations for each type of table saw.

(a) If the manufacturer has 96 hours available per day for painting, 156 hours for assembly, and 37 hours for packaging, how many of each type of saw can be produced each day?

(b) If 8 more hours of painting time become available, find the new production strategy, and tell how it is related to the inverse matrix used in part (a).

	Deluxe	Premium	Ultimate
Painting	1.6	2	2.4
Assembly	2	3	4
Packaging	0.5	0.5	1

9. **Investment** A trust account manager has $1,000,000 to be invested in three different accounts. The accounts pay 6%, 8%, and 10%, and the goal is to earn $86,000 with the amount invested at 10% equal to the sum of the other two investments. To accomplish this, assume that x dollars are invested at 8%, y dollars at 10%, and z dollars at 6%. Find how much should be invested in each account to satisfy the conditions.

....

Need more practice?
Find more here: cengagebrain.com

Chapter Exercises

In Problems 1–7, use the following matrices as needed to perform the indicated operations.

$$A = \begin{bmatrix} 4 & 4 & 2 & -5 \\ 6 & 3 & -1 & 0 \\ 0 & 0 & -3 & 5 \end{bmatrix} \quad B = \begin{bmatrix} 2 & -5 & -11 & 8 \\ 4 & 0 & 0 & 4 \\ -2 & -2 & 1 & 9 \end{bmatrix}$$

$$C = \begin{bmatrix} 4 & -2 \\ 5 & 0 \\ 6 & 0 \\ 1 & 3 \end{bmatrix} \quad D = \begin{bmatrix} 3 & 5 \\ 1 & 2 \end{bmatrix} \quad E = \begin{bmatrix} 1 & 1 \\ 1 & 1 \\ 4 & 6 \\ 0 & 5 \end{bmatrix}$$

$$F = \begin{bmatrix} -1 & 6 \\ 4 & 11 \end{bmatrix} \quad G = \begin{bmatrix} 2 & -5 \\ -1 & 3 \end{bmatrix} \quad I = \begin{bmatrix} 1 & 0 \\ 0 & 1 \end{bmatrix}$$

1. $A + B$

2. $C - E$

3. $3C$

4. $3A - 5B$

5. AC

6. FI

7. DG^T

In Problems 8–10, the reduced matrix for a system of equations is given. (a) Identify the type of solution for the system (unique solution, no solution, or infinitely many solutions). Explain your reasoning. (b) For each system that has a solution, find it. If the system has infinitely many solutions, find two different specific solutions.

8. $\begin{bmatrix} 1 & 0 & -2 & | & 6 \\ 0 & 1 & 3 & | & 7 \\ 0 & 0 & 0 & | & 0 \end{bmatrix}$

9. $\begin{bmatrix} 1 & 0 & -4 & | & 0 \\ 0 & 1 & 3 & | & 0 \\ 0 & 0 & 0 & | & 1 \end{bmatrix}$

10. $\begin{bmatrix} 1 & 0 & 0 & | & 0 \\ 0 & 1 & 0 & | & -10 \\ 0 & 0 & 1 & | & 14 \end{bmatrix}$

In Problems 11–15, solve each system using matrices.

11. $\begin{cases} x - 2y = 4 \\ -3x + 10y = 24 \end{cases}$

12. $\begin{cases} x + y + z = 4 \\ 3x + 4y - z = -1 \\ 2x - y + 3z = 3 \end{cases}$

13. $\begin{cases} x + y - 2z = 5 \\ 3x + 2y + 5z = 10 \\ -2x - 3y + 15z = 2 \end{cases}$

14. $\begin{cases} x - 3y + z = 4 \\ 2x - 5y - z = 6 \end{cases}$

15. $\begin{cases} x_1 + x_2 + x_3 + x_4 = 3 \\ x_1 - 2x_2 - x_3 - 4x_4 = -5 \\ x_1 - x_3 + x_4 = 0 \\ x_2 + x_3 + x_4 = 2 \end{cases}$

16. Are D and G inverse matrices if

$$D = \begin{bmatrix} 3 & 5 \\ 1 & 2 \end{bmatrix} \quad \text{and} \quad G = \begin{bmatrix} 2 & -5 \\ -1 & 3 \end{bmatrix}?$$

In Problems 17 and 18, find the inverse of each matrix.

17. $\begin{bmatrix} 7 & -1 \\ -10 & 2 \end{bmatrix}$

18. $\begin{bmatrix} 1 & 0 & 2 \\ 3 & 4 & -1 \\ 1 & 1 & 0 \end{bmatrix}$

In Problems 19 and 20, solve each system of equations by using inverse matrices.

19. $\begin{cases} x + 2z = 5 \\ 3x + 4y - z = 2 \\ x + y = -3 \end{cases}$ (See Problem 18.)

20. $\begin{cases} x + 3y + z = 0 \\ x + 4y + 3z = 2 \\ 2x - y - 11z = -12 \end{cases}$

21. Does the following matrix have an inverse?

$$\begin{bmatrix} 1 & 2 & 4 \\ 2 & 0 & -4 \\ 1 & 0 & -2 \end{bmatrix}$$

In Problems 22 and 23, (a) find the determinant of the matrix and (b) use it to decide whether the matrix has an inverse.

22. $\begin{bmatrix} 4 & 4 \\ -2 & 2 \end{bmatrix}$

23. $\begin{bmatrix} 1 & 2 & 3 \\ 4 & -1 & 8 \\ 6 & 3 & 14 \end{bmatrix}$

Applications

In Problems 24 and 25, the Burr Cabinet Company manufactures bookcases and filing cabinets at two plants, A and B. Matrix M gives the production for the two plants during June, and matrix N gives the production for July.

$$M = \begin{bmatrix} 150 & 80 \\ 280 & 300 \end{bmatrix} \begin{matrix} \text{Bookcases} \\ \text{Filing cabinets} \end{matrix}$$

with column headers A   B

$$N = \begin{bmatrix} 100 & 60 \\ 200 & 400 \end{bmatrix} \quad S = \begin{bmatrix} 120 & 80 \\ 180 & 300 \end{bmatrix} \quad P = \begin{bmatrix} 1000 & 800 \\ 600 & 1200 \end{bmatrix}$$

with column headers A   B for each.

24. **Production** Write the matrix that represents total production at the two plants for the two months.

25. **Production** If the company sells its bookcases to wholesalers for $100 and its filing cabinets for $120, for which month was the value of production higher (a) at plant A? (b) at plant B?

In Problems 26 and 27, a small church choir is made up of men and women who wear choir robes in the sizes shown in matrix A. Matrix B contains the prices (in dollars) of new robes and hoods according to size.

$$A = \begin{bmatrix} 1 & 14 \\ 12 & 10 \\ 8 & 3 \end{bmatrix} \begin{matrix} \text{Small} \\ \text{Medium} \\ \text{Large} \end{matrix}$$

with column headers Men   Women

$$B = \begin{bmatrix} 25 & 40 & 45 \\ 10 & 10 & 10 \end{bmatrix} \begin{matrix} \text{Robes} \\ \text{Hoods} \end{matrix}$$

with column headers S   M   L

26. **Cost** Find the product BA, and label the rows and columns to show what each entry represents.

27. **Cost** To find a matrix that gives the cost of new robes and the cost of new hoods, find

$$BA\begin{bmatrix} 1 \\ 1 \end{bmatrix}$$

28. **Manufacturing** Two departments of a firm, A and B, need different amounts of the same products. The table gives the amounts of the products needed by the two departments.

	Steel	Plastic	Wood
Department A	30	20	10
Department B	20	10	20

These three products are supplied by two suppliers, Ace and Kink, with the unit prices given in the table.

	Ace	Kink
Steel	300	280
Plastic	150	100
Wood	150	200

(a) Use matrix multiplication to find how much these two orders will cost at the two suppliers. The result should be a 2×2 matrix.

(b) From which supplier should each department make its purchase?

29. **Investment** A woman has $50,000 to invest. She has decided to invest all of it by purchasing some shares of stock in each of three companies: a fast-food chain that sells for $50 per share and has an expected growth of 11.5% per year, a software company that sells for $20 per share and has an expected growth of 15% per year, and a pharmaceutical company that sells for $80 per share and has an expected growth of 10% per year. She plans to buy twice as many shares of stock in the fast-food chain as in the pharmaceutical company. If her goal is 12% growth per year, how many shares of each stock should she buy?

30. **Nutrition** A biologist is growing three different types of slugs (types A, B, and C) in the same laboratory environment. Each day, the slugs are given a nutrient mixture that contains three different ingredients (I, II, and III). Each type A slug requires 1 unit of I, 3 units of II, and 1 unit of III per day. Each type B slug requires 1 unit of I, 4 units of II, and 2 units of III per day. Each type C slug

requires 2 units of I, 10 units of II, and 6 units of III per day.

(a) If the daily mixture contains 2000 units of I, 8000 units of II, and 4000 units of III, find the number of slugs of each type that can be supported.

(b) Is it possible to support 500 type A slugs? If so, how many of the other types are there?

(c) What is the maximum number of type A slugs possible? How many of the other types are there in this case?

31. **Transportation** An airline company has three types of aircraft that carry three types of cargo. The payload of each type is summarized in the table.

Units Carried	Plane Type		
	Passenger	Transport	Jumbo
First-class mail	100	100	100
Passengers	150	20	350
Air freight	20	65	35

(a) Suppose that on a given day the airline must move 1100 units of first-class mail, 460 units of air freight, and 1930 passengers. How many aircraft of each type should be scheduled? Use inverse matrices.

(b) How should the schedule from part (a) be adjusted to accommodate 730 more passengers?

(c) What column of the inverse matrix used in part (a) can be used to answer part (b)?

Need more practice?
Find more here: cengagebrain.com

TURN TO A TRUSTED RESOURCE

With the Student Solutions Manual, now available online, you will have all the learning resources you need for your course in one convenient place, at no additional charge!

Solutions to all odd-numbered exercises, organized by chapter, are easily found online.

Simply visit **login.cengagebrain.com** to access this trusted resource.

Inequalities and
Linear Programming

Most companies seek to maximize profits subject to the limitations imposed by product demand and available resources (such as raw materials and labor) or to minimize production costs subject to the need to fill customer orders. If the relationships among the various resources, production requirements, costs, and profits are all linear, then these activities may be planned (or programmed) in the best possible (optimal) way by using **linear programming.** Because linear programming provides the best possible solution to problems involving allocation of limited resources among various activities, its impact has been tremendous.

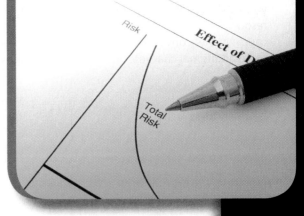

4.1 Linear Inequalities in Two Variables

BEFORE WE LOOK AT SYSTEMS OF INEQUALITIES, WE WILL DISCUSS SOLUTIONS OF ONE INEQUALITY IN TWO VARIABLES, SUCH AS $Y < X$. THE SOLUTIONS OF THIS INEQUALITY ARE THE ORDERED PAIRS (X, Y) THAT SATISFY THE INEQUALITY. THUS $(1, 0)$, $(3, 2)$, $(0, -1)$, AND $(-2, -5)$ ARE SOLUTIONS OF $Y < X$, BUT $(3, 7)$, $(-4, -3)$, AND $(2, 2)$ ARE NOT.

One Linear Inequality in Two Variables

The graph of $y < x$ consists of all points in which the y-coordinate is less than the x-coordinate. The graph of the region $y < x$ can be found by graphing the line $y = x$ (as a dashed line, because the given inequality does not include $y = x$). This line separates the xy-plane into two **half-planes,** $y < x$ and $y > x$. We can determine which half-plane is the solution region by selecting as a **test point** any point not on the line; let's choose (2, 0). Because the coordinates of this test point satisfy the inequality $y < x$, the half-plane containing this point is the solution region for $y < x$. (See Figure 4.1.) If the coordinates of the test point do not satisfy the inequality, then the other half-plane is the solution region.

EXAMPLE 4.1 Graphing Inequalities

Graph the inequality $4x - 2y \leq 6$.

SOLUTION

First we graph the line $4x - 2y = 6$, or (equivalently) $y = 2x - 3$, as a solid line, because points lying on the line satisfy the given inequality. Next we pick a

Shutterstock

Figure 4.1

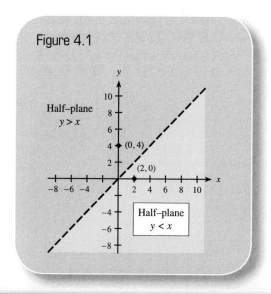

math apps

test point that is not on the line. If we use $(0, 0)$, we see that its coordinates satisfy $4x - 2y \leq 6$ (that is, $y \geq 2x - 3$). Hence the solution region is the line $y = 2x - 3$ and the half-plane that contains the test point $(0, 0)$. See Figure 4.2.

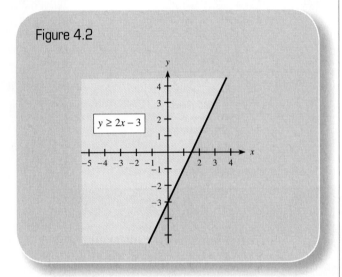

Figure 4.2

$y \geq 2x - 3$

The Tech Card shows how to use a graphing calculator to obtain the shaded solution region shown in Figure 4.2.

Systems of Linear Inequalities

If we have two or more inequalities in two variables, we can find the solutions that satisfy all the inequalities. We call the inequalities a **system of inequalities,** and the solution of the system can be found by finding the intersection of the solution sets of all the inequalities.

The solution set of the system of inequalities can be found by graphing the inequalities on the same set of axes and noting their points of intersection.

EXAMPLE 4.2 Graphical Solution of a System of Inequalities

Graph the solution of the system

$$\begin{cases} 3x - 2y \geq 4 \\ x + y - 3 > 0 \end{cases}$$

SOLUTION

Begin by graphing the equations $3x - 2y = 4$ and $x + y = 3$ (from $x + y - 3 = 0$) by the intercept method: find y when $x = 0$ and find x when $y = 0$.

$3x - 2y = 4$		$x + y = 3$	
x	y	x	y
0	-2	0	3
$\dfrac{4}{3}$	0	3	0

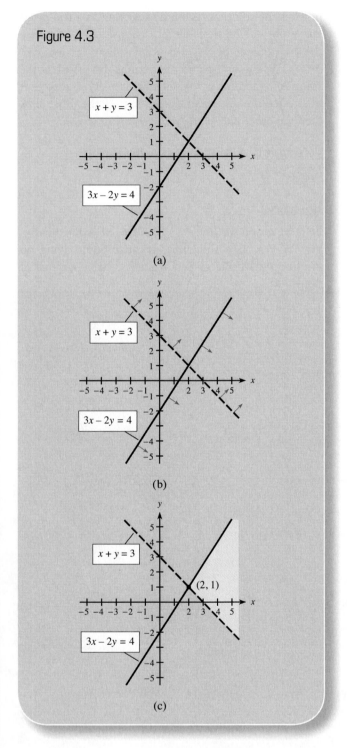

Figure 4.3

$x + y = 3$

$3x - 2y = 4$

(a)

$x + y = 3$

$3x - 2y = 4$

(b)

$x + y = 3$

$(2, 1)$

$3x - 2y = 4$

(c)

We graph $3x - 2y = 4$ as a solid line and $x + y = 3$ as a dashed line (see Figure 4.3(a)). We use any point not on either line as a test point. Let's use $(0, 0)$; the coordinates of $(0, 0)$ do not satisfy either $3x - 2y \geq 4$ or $x + y - 3 > 0$. Thus the solution region for each individual inequality is the half-plane that does not contain the point $(0, 0)$. Figure 4.3(b) indicates the half-plane solution for each inequality with arrows pointing from the line into the desired half-plane (away from the test point). The points that satisfy both of these inequalities lie in the intersection of the two individual solution regions, shown in Figure 4.3(c). This **solution region** is the graph of the solution of this system of inequalities.

The point (2, 1) in Figure 4.3(c), where the two regions form a "corner," is found by solving the equations $3x - 2y = 4$ and $x + y = 3$ simultaneously.

Many applications restrict the variables to be nonnegative (such as $x \geq 0$ and $y \geq 0$ in the next example). The effect of this restriction is to limit the solution to Quadrant I and the axes bounding Quadrant I.

EXAMPLE 4.3 Land Management

→ A farm co-op has 6000 acres available to plant with corn and soybeans. Each acre of corn requires 9 gallons of fertilizer/herbicide and 3/4 hour of labor to harvest. Each acre of soybeans requires 3 gallons of fertilizer/herbicide and 1 hour of labor to harvest. The co-op has available at most 40,500 gallons of fertilizer/herbicide and at most 5250 hours of labor for harvesting. The number of acres of each crop is limited (constrained) by the available resources: land, fertilizer/herbicide, and labor for harvesting. Write the system of inequalities that describes the constraints and graph the solution region for the system.

SOLUTION

If x represents the number of acres of corn and y represents the number of acres of soybeans, then the total acres planted is limited by $x + y \leq 6000$. The limitation on fertilizer/herbicide is given by $9x + 3y \leq 40,500$, and the labor constraint is given by $\frac{3}{4}x + y \leq 5250$. Because the number of acres planted in each crop must be nonnegative, the system of inequalities that describes the constraints is

$$
\begin{aligned}
x + \ y &\leq 6000 & (1) \\
9x + 3y &\leq 40{,}500 & (2) \\
\frac{3}{4}x + \ y &\leq 5250 & (3) \\
x &\geq 0, \ y \geq 0
\end{aligned}
$$

We can use the intercept method to graph the lines associated with this system of constraints.

The solution region is shaded in Figure 4.4 on the next page, with three of the corners at (0, 0), (4500, 0), and (0, 5250). The corner (3750, 2250) is found by solving the equations $x + y = 6000$ and $9x + 3y = 40,500$ simultaneously.

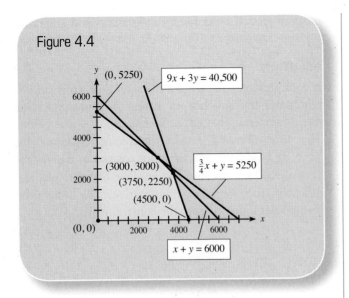

Figure 4.4

The corner (3000, 3000) is found similarly by solving simultaneously with $x + y = 6000$ and $\frac{3}{4}x + y = 5250$.

Any point in the shaded region of Figure 4.4 represents a possible number of acres of corn and of soybeans that the co-op could plant, treat with fertilizer/herbicide, and harvest.

Many computer programs save considerable time and energy in determining regions that satisfy a system of inequalities. Graphing calculators can also be used in the graphical solution of a system of inequalities. See the Tech Card for details.

4.1 Exercises

In Problems 1 and 2, graph each inequality.

1. $y \leq 2x - 1$ 2. $\dfrac{x}{2} + \dfrac{y}{4} < 1$

In Problems 3 and 4, the graph of the boundary equations for each system of inequalities is shown with that system. Locate the solution region and find the corners.

3. $\begin{cases} x + 4y \leq 60 \\ 4x + 2y \leq 100 \\ x \geq 0 \\ y \geq 0 \end{cases}$

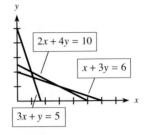

4. $\begin{cases} x + 3y \geq 6 \\ 2x + 4y \geq 10 \\ 3x + y \geq 5 \\ x \geq 0, \quad y \geq 0 \end{cases}$

In Problems 5–7, graph the solution of each system of inequalities.

5. $\begin{cases} 2x + y < 3 \\ x - 2y \geq -1 \end{cases}$

6. $\begin{cases} x + 2y \leq 48 \\ x + y \leq 30 \\ 2x + y \leq 50 \\ x \geq 0 \quad y \geq 0 \end{cases}$

7. $\begin{cases} x + 2y \geq 19 \\ 3x + 2y \geq 29 \\ x \geq 0, \quad y \geq 0 \end{cases}$

Applications

8. **Management** The Wellbuilt Company produces two types of wood chippers, economy and deluxe. The deluxe model requires 3 hours to assemble and 1/2 hour to paint, and the economy model requires 2 hours to assemble and 1 hour to paint. The maximum number of assembly hours available is 24 per day, and the maximum number of painting hours available is 8 per day.
   (a) Write the system of inequalities that describes the constraints on the number of each type of wood chipper produced. Begin by identifying what x and y represent.
   (b) Graph the solution of the system of inequalities and find the corners of the solution region.

9. **Advertising** Apex Motors manufactures luxury cars and sport-utility vehicles. The most likely customers are high-income men and women, and company managers want to initiate an advertising campaign targeting these groups. They plan to run 1-minute spots on business/investment programs, where they can reach 7 million women and 4 million men from their target groups. They also plan 1-minute

spots during sporting events, where they can reach 2 million women and 12 million men from their target groups. Apex feels that the ads must reach at least 30 million women and at least 28 million men who are prospective customers.

(a) Write the inequalities that describe the constraints on the number of each type of 1-minute spots needed to reach these target groups.

(b) Graph the region determined by these constraint inequalities.

10. *Manufacturing* A sausage company makes two different kinds of hot dogs, regular and all beef. Each pound of all-beef hot dogs requires 0.75 lb of beef and 0.2 lb of spices, and each pound of regular hot dogs requires 0.18 lb of beef, 0.3 lb of pork, and 0.2 lb of spices. Suppliers can deliver at most 1020 lb of beef, at most 600 lb of pork, and at least 500 lb of spices.

(a) Write the inequalities that describe how many pounds of each type of hot dog can be produced.

(b) Graph the region determined by these constraint inequalities.

> **Need more practice?**
> Find more here: **cengagebrain.com**

4.2 Linear Programming: Graphical Methods

MANY PRACTICAL PROBLEMS IN BUSINESS AND ECONOMICS SEEK TO OPTIMIZE SOME FUNCTION (SUCH AS COST OR PROFIT) SUBJECT TO COMPLEX RELATIONSHIPS AMONG CAPITAL, RAW MATERIALS, LABOR, AND SO FORTH.

In these problems, the optimum value of the function often can be found by using a mathematical technique called **linear programming.** Linear programming can be used to solve problems such as this if the limits on the variables (called **constraints**) can be expressed as linear inequalities and if the function that is to be maximized or minimized (called the **objective function**) is a linear function.

Solving Graphically

Linear programming is widely used by businesses for problems that involve many variables (sometimes more than 100). In this section we begin our study of this important technique by considering problems involving two variables. With two variables we can use graphical methods to help solve the problem. The constraints form a system of linear inequalities in two variables that we can solve by graphing. The solution of the system of constraint inequalities determines a region, any point of which may yield the optimal (maximum or minimum) value for the objective function.* Hence any point in the region determined by the constraints is called a **feasible solution,** and the region itself is called the **feasible region.**

Solutions of Linear Programming Problems

1. When the feasible region for a linear programming problem is closed and bounded (i.e., is entirely enclosed by, and includes, the lines associated with the constraints), the objective function has a maximum value and a minimum value.

2. When the feasible region is not closed and bounded, the objective function may have a maximum only, a minimum only, or no solution.

3. If a linear programming problem has a solution, then the optimal (maximum or minimum) value of an objective function occurs at a corner of the feasible region determined by the constraints.

4. If the objective function has its optimal value at two corners, then it also has that optimal value at any point on the line segment (boundary) connecting those two corners.

---

*The region determined by the constraints must be *convex* for the optimal to exist. A convex region is one such that for any two points in the region, the segment joining those points lies entirely within the region. We restrict our discussion to convex regions.

In a linear programming problem, we use the feasible region to find a solution that maximizes (or minimizes) the objective function, shown in the Solutions of Linear Programming Problems box on page 85.

Thus, for a closed and bounded region, we can find the maximum or minimum value of the objective function by evaluating the function at each of the corners of the feasible region formed by the solution of the constraint inequalities. If the feasible region is not closed and bounded, we must check to make sure the objective function has an optimal value.

The steps involved in solving a linear programming problem are as follows.

Linear Programming (Graphical Method) Procedure

To find the optimal value of a function subject to constraints:

1. Write the objective function and constraint inequalities from the problem.

2. Graph the solution of the constraint system.
   (a) If the feasible region is closed and bounded, proceed to Step 3.
   (b) If the region is not closed and bounded, check whether an optimal value exists. If not, state this. If so, proceed to Step 3.

3. Find the corners of the resulting feasible region. This may require simultaneous solution of two or more pairs of boundary equations.

4. Evaluate the objective function at each corner of the feasible region determined by the constraints.

5. If two corners give the optimal value of the objective function, then all points on the boundary line segment joining these two corners also optimize the function.

EXAMPLE 4.4 Linear Programming

➡ Find the maximum and minimum values of $C = 2x + 3y$ subject to the constraints

$$\begin{cases} x + 2y \le 10 \\ 2x + y \le 14 \\ x \ge 0, \ y \ge 0 \end{cases}$$

SOLUTION
Objective function: $C = 2x + 3y$
Constraints: $x + 2y \le 10$
$2x + y \le 14$
$x \ge 0, y \ge 0$

The constraint region is closed and bounded. See the figure below.

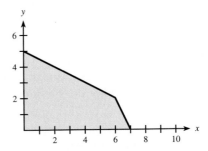

Corners are $(0, 0)$, $(0, 5)$, $(6, 2)$, $(7, 0)$.
At $(0, 0)$, $C = 2x + 3y = 2(0) + 3(0) = 0$
At $(0, 5)$, $C = 2x + 3y = 2(0) + 3(5) = 15$
At $(6, 2)$, $C = 2x + 3y = 2(6) + 3(2) = 18$
At $(7, 0)$, $C = 2x + 3y = 2(7) + 3(0) = 14$

The function is maximized at $x = 6, y = 2$. The maximum value is $C = 18$.
The function is minimized at $x = 0, y = 0$.
The minimum value is $C = 0$.

EXAMPLE 4.5 Maximizing Revenue

➡ Chairco manufactures two types of chairs, standard and plush. Standard chairs require 2 hours to construct and finish, and plush chairs require 3 hours to construct and finish. Upholstering takes 1 hour for standard chairs and 3 hours for plush chairs. There are 240 hours per day available for construction and finishing, and 150 hours per day are available for upholstering. If the revenue for standard chairs is $89 and for plush chairs is $133.50, how many of each type should be produced each day to maximize revenue?

STANDARD   PLUSH

Shutterstock (both)

SOLUTION

Let x be the number of standard chairs produced each day, and let y be the number of plush chairs produced. Then the daily revenue function is given by $R = 89x + 133.5y$. There are constraints for construction and finishing (no more than 240 hours/ day) and for upholstering (no more than 150 hours/ day). Thus we have the following.

Construction/finishing constraint: $2x + 3y \leq 240$
Upholstering constraint: $x + 3y \leq 150$

Because all quantities must be nonnegative, we also have the constraints $x \geq 0$ and $y \geq 0$.

Thus we seek to solve the following problem.

Maximize $R = 89x + 133.5y$ subject to

$$\begin{cases} 2x + 3y \leq 240 \\ x + 3y \leq 150 \\ x \geq 0, \ y \geq 0 \end{cases}$$

The feasible set is the closed and bounded region shaded in Figure 4.5. The corners of the feasible region are $(0, 0)$, $(120, 0)$, $(0, 50)$, and $(90, 20)$. All of these are obvious except $(90, 20)$, which can be found by solving $2x + 3y = 240$ and $x + 3y = 150$ simultaneously. Testing the objective function at the corners gives the following.

At $(0, 0)$, $R = 89x + 133.5y = 89(0) + 133.5(0) = 0$
At $(120, 0)$, $R = 89(120) + 133.5(0) = 10,680$ ←
At $(0, 50)$, $R = 89(0) + 133.5(50) = 6675$
At $(90, 20)$, $R = 89(90) + 133.5(20) = 10,680$ ←

Maximum at two corners

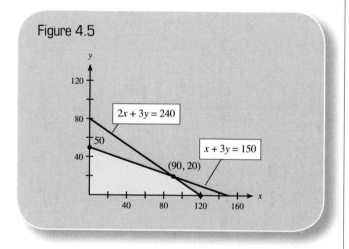

Figure 4.5

Thus the maximum revenue of $10,680 occurs at either the point $(120, 0)$ or the point $(90, 20)$. This means that the revenue function will be maximized not only at these two corner points but also at any point on the segment joining them. Since the number of each type of chair must be an integer, Chairco has maximum revenue of $10,680 at any pair of integer values along the segment joining $(120, 0)$ and $(90, 20)$. For example, the

point $(105, 10)$ is on this segment, and the revenue at this point is also $10,680:

$$89x + 133.5y = 89(105) + 133.5(10) = 10,680$$

The examples so far have involved closed and bounded regions. Similar procedures apply for an unbounded region, but optimal solutions are no longer guaranteed.

EXAMPLE 4.6 Minimizing Production Costs

Two chemical plants, one at Macon and one at Jonesboro, produce three types of fertilizer: low phosphorus (LP), medium phosphorus (MP), and high phosphorus (HP). At each plant, the fertilizer is produced in a single production run, so the three types are produced in fixed proportions. The Macon plant produces 1 ton of LP, 2 tons of MP, and 3 tons of HP in a single operation, and it charges $600 for what is produced in one operation, whereas one operation of the Jonesboro plant produces 1 ton of LP, 5 tons of MP, and 1 ton of HP, and it charges $1000 for what it produces in one operation. If a customer needs 100 tons of LP, 260 tons of MP, and 180 tons of HP, how many production runs should be ordered from each plant to minimize costs?

SOLUTION

If x represents the number of operations requested from the Macon plant and y represents the number of operations requested from the Jonesboro plant, then we seek to minimize cost

$$C = 600x + 1000y$$

The following table summarizes production capabilities and requirements.

	Macon Plant	Jonesboro Plant	Requirements
Units of LP	1	1	100
Units of MP	2	5	260
Units of HP	3	1	180

Using the number of operations requested and the fact that requirements must be met or exceeded, we can formulate the following constraints.

$$\begin{cases} x + y \geq 100 \\ 2x + 5y \geq 260 \\ 3x + y \geq 180 \\ x \geq 0, \ y \geq 0 \end{cases}$$

Figure 4.6

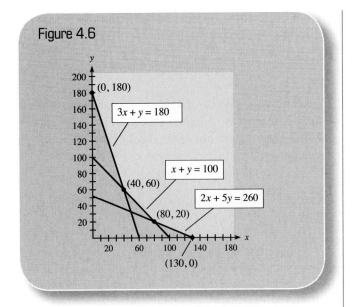

Graphing this system gives the feasible set shown in Figure 4.6. The objective function has a minimum even though the feasible set is not closed and bounded. The corners are (0, 180), (40, 60), (80, 20), and (130, 0), where (40, 60) is obtained by solving $x + y = 100$ and $3x + y = 180$ simultaneously, and where (80, 20) is obtained by solving $x + y = 100$ and $2x + 5y = 260$ simultaneously.

Evaluating $C = 600x + 1000y$ at each corner, we obtain

At (0, 180), $C = 600(0) + 1000(180) = 180,000$
At (40, 60), $C = 600(40) + 1000(60) = 84,000$
At (80, 20), $C = 600(80) + 1000(20) = 68,000$
At (130, 0), $C = 600(130) + 1000(0) = 78,000$

Thus, by placing orders requiring 80 production runs from the Macon plant and 20 production runs from the Jonesboro plant, the customer's needs will be satisfied at a minimum cost of $68,000.

See the Tech Card for details on solving linear programming problems with a graphing calculator.

....

4.2 Exercises

In Problems 1 and 2, use the given feasible region determined by the constraint inequalities to find the maximum and minimum of the given objective function (if they exist).

1. $C = 9x + 10y$

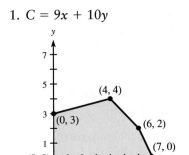

2. $C = 2x + 7y$

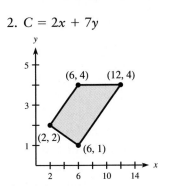

In Problems 3 and 4, the graph of the feasible region is shown. Find the corners of each feasible region, and then find the maximum and minimum of the given objective function (if they exist).

3. $f = 3x + 2y$

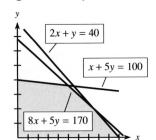

4. $g = 2x + 3y$

In Problems 5–7, solve each linear programming problem. Restrict $x \geq 0$ and $y \geq 0$.

5. Maximize $f = 2x + 6y$ subject to

$$x + y \leq 7$$
$$2x + y \leq 12$$
$$x + 3y \leq 15$$

6. Minimize $g = 7x + 6y$ subject to

$$5x + 2y \geq 16$$
$$3x + 7y \geq 27$$

7. Minimize $g = 40x + 25y$ subject to

$$x + y \geq 100$$
$$-x + y \leq 20$$
$$-2x + 3y \geq 30$$

Applications

8. **Production scheduling** Newjet, Inc. manufactures inkjet printers and laser printers. The company has the capacity to make 70 printers per day, and it has 120 hours of labor per day available. It takes 1 hour to make an inkjet printer and 3 hours to make a laser printer. The profits are $40 per inkjet printer and $60 per laser printer. Find the number of each type of printer that should be made to give maximum profit, and find the maximum profit.

9. **Manufacturing** The Janie Gioffre Drapery Company makes three types of draperies at two different locations. At location I, it can make 10 pairs of deluxe drapes, 20 pairs of better drapes, and 13 pairs of standard drapes per day. At location II, it can make 20 pairs of deluxe, 50 pairs of better, and 6 pairs of standard per day. The company has orders for 2000 pairs of deluxe drapes, 4200 pairs of better drapes, and 1200 pairs of standard drapes. If the daily costs are $500 per day at location I and $800 per day at location II, how many days should Janie schedule at each location in order to fill the orders at minimum cost? Find the minimum cost.

10. **Manufacturing** A sausage company makes two different kinds of hot dogs, regular and all beef. Each pound of all-beef hot dogs requires 0.75 lb of beef and 0.2 lb of spices, and each pound of regular hot dogs requires 0.18 lb of beef, 0.3 lb of pork, and 0.2 lb of spices. Suppliers can deliver at most 1020 lb of beef, at most 600 lb of pork, and at least 500 lb of spices. If the profit is $0.60 on each pound of all-beef hot dogs and $0.40 on each pound of

regular hot dogs, how many pounds of each should be produced to obtain maximum profit? What is the maximum profit?

11. **Shipping costs** TV Circuit has 30 large-screen televisions in a warehouse in Erie and 60 large-screen televisions in a warehouse in Pittsburgh. Thirty-five are needed in a store in Blairsville, and 40 are needed in a store in Youngstown. It costs $18 to ship from Pittsburgh to Blairsville and $22 to ship from Pittsburgh to Youngstown, whereas it costs $20 to ship from Erie to Blairsville and $25 to ship from Erie to Youngstown. How many televisions should be shipped from each warehouse to each store to minimize the shipping cost? *Hint:* If the number shipped from Pittsburgh to Blairsville is represented by x, then the number shipped from Erie to Blairsville is represented by $35 - x$.

....

Need more practice?
Find more here: cengagebrain.com

4.3 The Simplex Method: Maximization

THE GRAPHICAL METHOD FOR SOLVING LINEAR PROGRAMMING PROBLEMS IS PRACTICAL ONLY WHEN THERE ARE TWO VARIABLES. IF THERE ARE MORE THAN TWO VARIABLES, WE COULD STILL FIND THE CORNERS OF THE CONVEX REGION BY SIMULTANEOUSLY SOLVING THE EQUATIONS OF THE BOUNDARIES. THE OBJECTIVE FUNCTION COULD THEN BE EVALUATED AT THESE CORNERS. HOWEVER, AS THE NUMBER OF VARIABLES AND THE NUMBER OF CONSTRAINTS INCREASES, IT BECOMES INCREASINGLY DIFFICULT TO DISCOVER THE CORNERS AND EXTREMELY

TIME-CONSUMING TO EVALUATE THE FUNCTION. FURTHERMORE, SOME PROBLEMS HAVE NO SOLUTION, AND THIS CANNOT BE DETERMINED BY USING THE METHOD OF SOLVING THE EQUATIONS SIMULTANEOUSLY.

The method discussed in this section is called the **simplex method** for solving linear programming problems. This method gives a systematic way of moving from one feasible corner of the convex region to another in such a way that the value of the objective function increases until an optimal value is reached or it is discovered that no solution exists.

The Simplex Method

The simplex method was developed by George Dantzig in 1947. One of the earliest applications of the method was to the scheduling problem that arose in connection with the Berlin airlift, begun in 1948.

In discussing the simplex method, we initially restrict ourselves to **standard maximization problems**, which satisfy the following conditions.

1. The objective function is to be maximized.

2. All variables are nonnegative.

3. The constraints are of the form
$$a_1x_1 + a_2x_2 + \ldots + a_nx_n \leq b$$
where $b > 0$.

These conditions may seem to restrict the types of problems unduly, but in applied situations where the objective function is to be maximized, the constraints often satisfy conditions 2 and 3.

Tasks and Procedure

The simplex method involves a series of decisions and operations using matrices. It involves three major tasks.

Simplex Method Tasks

Task A: Setting up the matrix for the simplex method.
Task B: Determining necessary operations and implementing those operations to reach a solution.
Task C: Reading the solution from the simplex matrix.

The procedure that combines these three tasks to maximize a function is given below.

Linear Programming (Simplex Method)
PROCEDURE

To use the simplex method to solve linear programming problems:

1. Use a different *slack variable* to write each constraint inequality as an equation, with positive constants on the right side.

2. Set up the *simplex matrix*. It contains the system of constraint equations with the objective function in the last row (with all variables on the left and coefficient 1 for f).

3. Find the *pivot entry*:
   (a) The *pivot column* has the most negative number in the last row. (If a tie occurs, use either column.)
   (b) The positive coefficient in the pivot column that gives the smallest nonnegative quotient when divided into the constant in the constraint rows of the augment is the *pivot entry*. If there is a tie, either coefficient may be chosen. If there are no positive coefficients in the pivot column, no solution exists. Stop.

4. Use row operations with only the row containing the pivot entry to make the pivot entry a 1 and all other entries in the pivot column zeros. This process is called **pivoting**. This makes the variable of the pivot column a *basic variable* (corresponding to a column that contains a *single* entry of 1 with all other entries being 0).

5. The numerical entries in the last row are the indicators of what to do next.
   (a) If there is a negative indicator, return to Step 3.
   (b) If all indicators are positive or zero, an optimum value has been obtained for the objective function.
   (c) In a complete solution, the variables that have positive entries in the last row are nonbasic variables.

6. After setting the nonbasic variables equal to 0, read the values of the basic variables and the objective function from the rows of the matrix.

EXAMPLE 4.7 Linear Programming (Simplex Method)

➡ Maximize $f = 2x + 3y$ subject to

$$x + 2y \le 10$$
$$2x + y \le 14$$

SOLUTION

$x + 2y \le 10$ becomes $x + 2y + s_1 = 10$.
$2x + y \le 14$ becomes $2x + y + s_2 = 14$.
Maximize f in $-2x - 3y + f = 0$.

$$
\begin{array}{ccccc}
x & y & s_1 & s_2 & f \\
\end{array}
$$
$$
\left[
\begin{array}{ccccc|c}
1 & 2 & 1 & 0 & 0 & 10 \\
2 & 1 & 0 & 1 & 0 & 14 \\
-2 & -3 & 0 & 0 & 1 & 0 \\
\end{array}
\right]
$$

(a)
$$
\left[
\begin{array}{c|c|ccc|c}
1 & 2 & 1 & 0 & 0 & 10 \\
2 & 1 & 0 & 1 & 0 & 14 \\
-2 & -3 & 0 & 0 & 1 & 0 \\
\end{array}
\right]
$$

↑—— Most negative entry; pivot column

(b) Row 1 quotient: $\dfrac{10}{2} = 5$

Row 2 quotient: $\dfrac{14}{1} = 14$

Therefore, pivot entry is in row 1.

$$
\left[
\begin{array}{cccccc}
1 & ② & 1 & 0 & 0 & 10 \\
2 & 1 & 0 & 1 & 0 & 14 \\
-2 & -3 & 0 & 0 & 1 & 0 \\
\end{array}
\right]
$$
Pivot entry is 2, circled.

Row operations:
Multiply row 1 by $\frac{1}{2}$.
Add -1 times new row 1 to row 2.
Add 3 times new row 1 to row 3.

Result:
$$
\left[
\begin{array}{ccccc|c}
\frac{1}{2} & 1 & \frac{1}{2} & 0 & 0 & 5 \\
\frac{3}{2} & 0 & -\frac{1}{2} & 1 & 0 & 9 \\
-\frac{1}{2} & 0 & \frac{3}{2} & 0 & 1 & 15 \\
\end{array}
\right]
$$

Indicators

$-\frac{1}{2}$ is negative, so we identify a new pivot column and reduce again.

$$
\left[
\begin{array}{ccccc|c}
\frac{1}{2} & 1 & \frac{1}{2} & 0 & 0 & 5 \\
③\!\!\!\!\frac{3}{2} & 0 & -\frac{1}{2} & 1 & 0 & 9 \\
-\frac{1}{2} & 0 & \frac{3}{2} & 0 & 1 & 15 \\
\end{array}
\right]
$$
$5 \div \frac{1}{2} = 10$
$9 \div \frac{3}{2} = 6^*$

↑ Pivot column   *6 is smaller quotient.

The new pivot entry is $\frac{3}{2}$ (circled). Now reduce, using the new pivot row, to obtain:

$$
\left[
\begin{array}{ccccc|c}
0 & 1 & \frac{2}{3} & -\frac{1}{3} & 0 & 2 \\
1 & 0 & -\frac{1}{3} & \frac{2}{3} & 0 & 6 \\
0 & 0 & \frac{4}{3} & \frac{1}{3} & 1 & 18 \\
\end{array}
\right]
$$

Indicators all 0 or positive, so the solution is complete.
Nonbasic variables: s_1 and s_2. Basic variables: x and y.

f is maximized at 18 when $y = 2$, $x = 6$, $s_1 = 0$, and $s_2 = 0$ (This same linear programming problem was solved graphically in Example 4.4 of Section 4.2.)

EXAMPLE 4.8 Manufacturing

➡ The Solar Technology Company manufactures three different types of hand calculators and classifies them as scientific, business, and graphing according to their calculating capabilities. The three types have production requirements given by the following table.

	Scientific	Business	Graphing
Electronic circuit components	5	7	10
Assembly time (hours)	1	3	4
Cases	1	1	1

The firm has a monthly limit of 90,000 circuit components, 30,000 hours of labor, and 9000 cases. If the profit is \$6 for each scientific, \$13 for each business, and \$20 for each graphing calculator, how many of each should be produced to yield the maximum profit? What is the maximum profit?

SOLUTION

Let x_1 be the number of scientific calculators produced, x_2 the number of business calculators produced, and x_3 the number of graphing calculators produced. Then the problem is to maximize the profit $f = 6x_1 + 13x_2 + 20x_3$ subject to

Inequalities	Equations with Slack Variables
$5x_1 + 7x_2 + 10x_3 \le 90{,}000$	$5x_1 + 7x_2 + 10x_3 + s_1 = 90{,}000$
$x_1 + 3x_2 + 4x_3 \le 30{,}000$	$x_1 + 3x_2 + 4x_3 + s_2 = 30{,}000$
$x_1 + x_2 + x_3 \le 9{,}000$	$x_1 + x_2 + x_3 + s_3 = 9{,}000$

The slack variables can be interpreted as follows: s_1 is the number of unused circuit components, s_2 is the number of unused hours of labor, and s_3 is the number of unused calculator cases. The simplex matrix, with the first pivot entry circled, is as follows.

Slack variables

$$
\begin{array}{ccccccc|c}
x_1 & x_2 & x_3 & s_1 & s_2 & s_3 & f & \\
5 & 7 & \boxed{10} & 1 & 0 & 0 & 0 & 90{,}000 \\
1 & 3 & 4 & 0 & 1 & 0 & 0 & 30{,}000 \\
\hline
1 & 1 & 1 & 0 & 0 & 1 & 0 & 9000 \\
-6 & -13 & -20 & 0 & 0 & 0 & 1 & 0
\end{array}
$$

90,000/10 = 9000
30,000/4 = 7500*
9000/1 = 9000

↑
Most negative   *Smallest quotient

We now use row operations to change the pivot entry to 1 and create zeros elsewhere in the pivot column. The row operations are

1. Multiply row 2 by $\frac{1}{4}$ to convert the pivot entry to 1.

2. Using the new row 2,
   (a) Add -10 times new row 2 to row 1.
   (b) Add -1 times new row 2 to row 3.
   (c) Add 20 times new row 2 to row 4.
The result is the second simplex matrix.

Slack variables

$$
\begin{array}{ccccccc|c}
x_1 & x_2 & x_3 & s_1 & s_2 & s_3 & f & \\
\frac{5}{2} & -\frac{1}{2} & 0 & 1 & -\frac{5}{2} & 0 & 0 & 15{,}000 \\
\frac{1}{4} & \frac{3}{4} & 1 & 0 & \frac{1}{4} & 0 & 0 & 7500 \\
\boxed{\frac{3}{4}} & \frac{1}{4} & 0 & 0 & -\frac{1}{4} & 1 & 0 & 1500 \\
\hline
-1 & 2 & 0 & 0 & 5 & 0 & 1 & 150{,}000
\end{array}
$$

15,000 ÷ $\frac{5}{2}$ =   6000
7500 ÷ $\frac{1}{4}$ = 30,000
1500 ÷ $\frac{3}{4}$ =   2000*

↑
Most negative   *Smallest quotient

We check the last row indicators in the matrix above. Because there is an entry that is negative, we repeat the simplex process. As shown above the pivot entry is circled, and we apply row operations similar to those used previously. What are the row operations this time? The resulting matrix is

$$
\begin{array}{ccccccc|c}
x_1 & x_2 & x_3 & s_1 & s_2 & s_3 & f & \\
0 & -\frac{4}{3} & 0 & 1 & -\frac{5}{3} & -\frac{10}{3} & 0 & 10{,}000 \\
0 & \frac{2}{3} & 1 & 0 & \frac{1}{3} & -\frac{1}{3} & 0 & 7000 \\
1 & \frac{1}{3} & 0 & 0 & -\frac{1}{3} & \frac{4}{3} & 0 & 2000 \\
\hline
0 & \frac{7}{3} & 0 & 0 & \frac{14}{3} & \frac{4}{3} & 1 & 152{,}000
\end{array}
$$

Checking the last row, we see that all the entries in this row are 0 or positive, so the solution is complete. From the final simplex matrix we see that the nonbasic variables are x_2, s_2, and s_3, and with these equal to zero we determine all the solution values:

$$x_2 = 0, \; s_2 = 0, \; s_3 = 0 \quad \text{and}$$
$$x_1 = 2000, \; x_3 = 7000, \; s_1 = 10{,}000, \; f = 152{,}000$$

Thus the numbers of calculators that should be produced are

$$x_1 = 2000 \text{ scientific calculators}$$
$$x_2 = 0 \text{ business calculators}$$
$$x_3 = 7000 \text{ graphing calculators}$$

in order to obtain a maximum profit of \$152,000 for the month. Note also that the slack variables tell us that with this optimal solution there were $s_1 = 10{,}000$ unused circuit components, $s_2 = 0$ unused labor hours, and $s_3 = 0$ unused calculator cases.

As our examples show, the simplex method is quite complicated, even when the number of variables is small. For this reason, computer software packages such as Excel are often used to solve linear programming problems. See the Tech Card for details on using Excel to solve these problems.

Nonunique Solutions

Multiple Solutions

As we've seen with graphical methods, a linear programming problem can have multiple solutions when the optimal value for the objective function occurs at two adjacent corners of the feasible region. Consider the following problem.

$$\text{Maximize } f = 2x + y \text{ subject to}$$

$$x + \frac{1}{2}y \le 16$$
$$x + \quad y \le 24$$
$$x \ge 0$$
$$y \ge 0$$

From Figure 4.7, we can see that $f = 32$ at the two corners $(8, 16)$ and $(16, 0)$. This means that $f = 32$ at any point on the segment joining those corners. Thus, this problem has infinitely many solutions. Let's examine how we can discover these multiple solutions with the simplex method.

The simplex matrix for this problem (with the slack variables introduced) is

$$
\begin{array}{ccccc|c}
x & y & s_1 & s_2 & f & \\
\boxed{1} & \frac{1}{2} & 1 & 0 & 0 & 16 \\
1 & 1 & 0 & 1 & 0 & 24 \\
\hline
-2 & -1 & 0 & 0 & 1 & 0
\end{array}
$$

The most negative value in the last row is -2, so the x-column is the pivot column. The smallest quotient

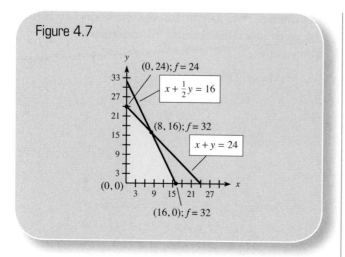

Figure 4.7

occurs in row 1; the pivot entry (circled) is 1. Making x a basic variable gives the following transformed simplex matrix.

$$\begin{array}{cccccc} & x & y & s_1 & s_2 & f \\ \left[\begin{array}{ccccc|c} 1 & \frac{1}{2} & 1 & 0 & 0 & 16 \\ 0 & \frac{1}{2} & -1 & 1 & 0 & 8 \\ \hline 0 & 0 & 2 & 0 & 1 & 32 \end{array}\right] \end{array}$$

This simplex matrix has no negative indicators, so the optimal value of f has been found ($f = 32$ when $x = 16$, $y = 0$). Note that one of the nonbasic variables (y) has a zero indicator. When this occurs, we can use the column with the 0 indicator as our pivot column and often obtain a new solution that has the same value for the objective function. In this final tableau, if we used the y-column as the pivot column, then we could find the second solution (at $x = 8$, $y = 16$) that gives the same optimal value of f. You will be asked to do this in Problem 8 of the exercises for this section.

Multiple Solutions:
When the simplex matrix for the optimal value of f has a nonbasic variable with a zero indicator in its column, there may be multiple solutions giving the same optimal value for f. We can discover whether another solution exists by using the column of that nonbasic variable as the pivot column.

Note that with Excel, multiple solutions may be found by using different initial values for the variables when the problem is entered into Excel. See the Tech Card.

No Solution

It is also possible for a linear programming problem to have an unbounded solution (and thus no maximum value for f).

EXAMPLE 4.9 Simplex Method When No Maximum Exists

➡ Maximize $f = 2x_1 + x_2 + 3x_3$ subject to

$$x_1 - 3x_2 + x_3 \leq 3$$
$$x_1 - 6x_2 + 2x_3 \leq 6$$

if a maximum exists.

SOLUTION

The simplex matrix for this problem (after introduction of slack variables) is

$$\begin{array}{cccccc} x_1 & x_2 & x_3 & s_1 & s_2 & f \\ \left[\begin{array}{ccc|ccc|c} 1 & -3 & \boxed{1} & 1 & 0 & 0 & 3 \\ 1 & -6 & 2 & 0 & 1 & 0 & 6 \\ \hline -2 & -1 & -3 & 0 & 0 & 1 & 0 \end{array}\right] \end{array}$$

We see that the x_3 column has the most negative number, so we use that column as the pivot column.

We see that the smallest quotient is 3. Both coefficients give this quotient, so we can use either coefficient as the pivot entry. Using the element in row 1, we can make x_3 basic. The new simplex matrix is

$$\begin{array}{cccccc} x_1 & x_2 & x_3 & s_1 & s_2 & f \\ \left[\begin{array}{ccccc|c} 1 & -3 & 1 & 1 & 0 & 0 & 3 \\ -1 & 0 & 0 & -2 & 1 & 0 & 0 \\ \hline 1 & -10 & 0 & 3 & 0 & 1 & 9 \end{array}\right] \end{array}$$

The most negative value in the last row of this matrix is -10 so the pivot column is the x_2-column. But there are no positive coefficients in the x_2-column; what does this mean? Even if we could pivot using the x_2-column, the variables x_1 and s_1 would remain equal to 0. And the relationships among the remaining variables x_2, x_3, and s_2 would be the same.

$$-3x_2 + x_3 = 3 \quad \text{or} \quad x_3 = 3 + 3x_2$$
$$s_2 = 0$$

As x_2 is increased, x_3 also increases because all variables are nonnegative. This means that no matter how much we increase x_2, all other variables remain nonnegative. Furthermore, as we increase x_2 (and hence x_3), the value of f also increases. That is, there is no maximum value for f.

This example indicates that when a linear programming problem has an unbounded solution (and thus no maximum value for f), this is identified by the following condition in the simplex method.

No Solution:
If, after the pivot column has been found, there are no positive coefficients in that column, no maximum solution exists.

4.3 Exercises

In Problem 1, set up the simplex matrix used to solve each linear programming problem. Assume all variables are nonnegative.

1. Maximize $f = 2x + 5y + 2z$ subject to

$$2x + 7y + 9z \leq 100$$
$$6x + 5y + z \leq 145$$
$$x + 2y + 7z \leq 90$$

In Problems 2–4, a simplex matrix for a standard maximization problem is given.

  (a) Write the values of *all* the variables (use x_1, x_2, x_3, ... and s_1, s_2, s_3, ...) and of the objective function f.

  (b) Indicate whether or not the solution from part (a) is complete (optimal).

  (c) If the solution is not complete, find the next pivot or indicate that no solution exists. When a new pivot can be found, state *all* row operations with that pivot (that is, row operations that make that pivot equal to 1, and then make other entries in the pivot column equal to 0). Do not perform the row operations.

2. $\begin{bmatrix} 10 & 27 & 1 & 0 & 0 & 0 & 200 \\ 4 & 51 & 0 & 1 & 0 & 0 & 400 \\ 15 & 27 & 0 & 0 & 1 & 0 & 350 \\ -8 & -7 & 0 & 0 & 0 & 1 & 0 \end{bmatrix}$

3. $\begin{bmatrix} 2 & 0 & 1 & -\frac{3}{4} & 0 & 14 \\ 3 & 1 & 0 & \frac{1}{3} & 0 & 45 \\ -6 & 0 & 0 & 3 & 1 & 75 \end{bmatrix}$

4. $\begin{bmatrix} 1 & 4 & 0 & 0 & \frac{3}{4} & 4 & 0 & 24 \\ 0 & -2 & 0 & 1 & -\frac{5}{8} & -2 & 0 & 16 \\ 0 & 3 & 1 & 0 & 2 & 6 & 0 & 21 \\ 0 & 4 & 0 & 0 & 2 & \frac{1}{2} & 1 & 780 \end{bmatrix}$

In Problems 5–7, use the simplex method to maximize the given functions. Assume all variables are nonnegative.

5. Maximize $f = 3x + 10y$ subject to

$$14x + 7y \leq 35$$
$$5x + 5y \leq 50$$

6. Maximize $f = 2x + 3y$ subject to

$$x + 2y \leq 10$$
$$x + y \leq 7$$

7. Maximize $f = 7x + 10y + 4z$ subject to

$$3x + 5y + 4z \leq 30$$
$$3x + 2y \leq 4$$
$$x + 2y \leq 8$$

In Problems 8 and 9, each problem involves a linear programming problem that has nonunique solutions.

8. The simplex matrix shown indicates that an optimal solution has been found but that a second solution is possible. Find the second solution.

$$\begin{bmatrix} 1 & \frac{1}{2} & 1 & 0 & 0 & 16 \\ 0 & \frac{1}{2} & -1 & 1 & 0 & 8 \\ 0 & 0 & 2 & 0 & 1 & 32 \end{bmatrix}$$

9. Use the simplex method to maximize $f = 3x + 2y$ subject to

$$x - 10y \leq 10$$
$$-x + y \leq 40$$

If there is no solution, indicate this; if multiple solutions exist, find two of them.

Applications

10. **Manufacturing** Newjet, Inc. manufactures two types of printers, an inkjet printer and a laser printer. The company can make a total of 60 printers per day, and it has 120 labor-hours per day available. It takes 1 labor-hour to make an inkjet printer and 3 labor-hours to make a laser printer. The profits are $40 per inkjet printer and $60 per laser printer.

  (a) Write the simplex matrix to maximize the daily profit.

  (b) Find the maximum profit and the number of each type of printer that will give the maximum profit.

11. **Production scheduling** At one of its factories, a jeans manufacturer makes two styles: #891 and #917. Each pair of style 891 takes 10 minutes to cut out and 20 minutes to assemble and finish.

Each pair of style 917 takes 10 minutes to cut out and 30 minutes to assemble and finish. The plant has enough workers to provide at most 7500 minutes per day for cutting and at most 19,500 minutes per day for assembly and finishing. The profit on each pair of style 891 is $6.00, and the profit on each pair of style 917 is $7.50. How many pairs of each style should be produced per day to obtain maximum profit? Find the maximum daily profit.

12. **Production scheduling** Happy Valley Ice Cream Company is planning its production for next week. Demand for Happy Valley's premium and light ice creams continues to outpace the company's production capacities. Two resources used in ice cream production are in short supply for next week. The mixing machines will be available for only 140 hours, and only 28,000 gallons of high-grade milk will be available. One hundred gallons of premium ice cream requires 0.3 hour of mixing and 90 gallons of milk. One hundred gallons of light ice cream requires 0.5 hour of mixing and 70 gallons of milk. If Happy Valley earns a profit of $100 per hundred gallons on both of its ice creams, how many hundreds of gallons of premium and of light ice cream should Happy Valley produce next week to maximize profit? How much profit will result?

13. **Manufacturing** At one of its factories, a manufacturer of television sets makes one or more of four models of HD sets: a 20-inch LCD, a 42-inch LCD, a 42-inch plasma, and a 50-inch plasma. The assembly and testing time requirements for each model are shown in the table, together with the maximum amounts of time available per week for assembly and testing. In addition to these constraints, the supplier of cabinet units indicated that it would supply no more than 200 units per week and that of these, no more than 40 could be for the 20-inch LCD model.

	20-Inch LCD	42-Inch LCD	42-Inch Plasma	50-Inch Plasma	Total Available
Assembly (time hours)	7	10	12	15	2000
Test time (hours)	2	2	4	5	500
Profit (dollars)	46	60	75	100	

Use the profit for each television shown in the table to find the number of sets of each type that should be produced to obtain the maximum profit for the week. Find the maximum profit.

....

> **Need more practice?**
> Find more here: cengagebrain.com

4.4 The Simplex Method: Duality and Minimization

WE HAVE SOLVED BOTH MAXIMIZATION AND MINIMIZATION PROBLEMS USING GRAPHICAL METHODS, BUT THE SIMPLEX METHOD, AS DISCUSSED IN THE PREVIOUS SECTION, APPLIES ONLY TO MAXIMIZATION PROBLEMS. NOW LET US TURN OUR ATTENTION TO EXTENDING THE USE OF THE SIMPLEX METHOD TO SOLVE MINIMIZATION PROBLEMS. SUCH PROBLEMS MIGHT ARISE WHEN A COMPANY SEEKS TO MINIMIZE ITS PRODUCTION COSTS YET FILL CUSTOMERS' ORDERS OR PURCHASE ITEMS NECESSARY FOR PRODUCTION.

Dual Problems

As with maximization problems, we shall consider only **standard minimization problems,** in which the constraints satisfy the following conditions.

1. All variables are nonnegative.

2. The constraints are of the form
$$a_1y_1 + a_2y_2 + \ldots + a_ny_n \geq b$$
where b is positive.

In this section, we will show how standard minimization problems can be solved using the simplex method.

EXAMPLE 4.10 A Minimization Problem and Its Dual

→ Minimize $g = 11y_1 + 7y_2$ subject to

$$y_1 + 2y_2 \geq 10$$
$$3y_1 + y_2 \geq 15$$

SOLUTION

Because the simplex method specifically seeks to increase the objective function, it does not apply to this minimization problem. Rather than solving this problem, let us solve a different but related problem called the **dual problem** or **dual.**

In forming the dual problem, we write a matrix A that has a form similar to the simplex matrix, but without slack variables and with positive coefficients for the variables in the last row (and with the function to be minimized in the augment).

The matrix for the dual problem is formed by interchanging the rows and columns of matrix A. Matrix B is the result of this procedure, and recall that B is the **transpose** of A.

$$A = \begin{bmatrix} 1 & 2 & | & 10 \\ 3 & 1 & | & 15 \\ 11 & 7 & | & g \end{bmatrix} \quad B = \begin{bmatrix} 1 & 3 & | & 11 \\ 2 & 1 & | & 7 \\ 10 & 15 & | & g \end{bmatrix}$$

Note that row 1 of A becomes column 1 of B, row 2 of A becomes column 2 of B, and row 3 of A becomes column 3 of B.

In the same way that we formed matrix A, we can now "convert back" from matrix B to the maximization problem that is the dual of the original minimization problem. We state this dual problem, using different letters to emphasize that this is a different problem from the original.

Maximize $f = 10x_1 + 15x_2$ subject to

$$x_1 + 3x_2 \leq 11$$
$$2x_1 + x_2 \leq 7$$

The simplex method does apply to this maximization problem, with the following simplex matrix.

$$\begin{array}{ccccc} x_1 & x_2 & s_1 & s_2 & f \\ \begin{bmatrix} 1 & 3 & 1 & 0 & 0 & | & 11 \\ 2 & 1 & 0 & 1 & 0 & | & 7 \\ -10 & -15 & 0 & 0 & 1 & | & 0 \end{bmatrix} \end{array}$$

The solutions of this minimization problem and its dual are shown in the following box, with the solution of the original minimization problem shown beside it.

Maximization Problem— Simplex Method

1.
$$\begin{array}{ccccc} x_1 & x_2 & s_1 & s_2 & f \\ \begin{bmatrix} 1 & ③ & 1 & 0 & 0 & | & 11 \\ 2 & 1 & 0 & 1 & 0 & | & 7 \\ -10 & -15 & 0 & 0 & 1 & | & 0 \end{bmatrix} \end{array}$$

2.
$$\begin{bmatrix} \frac{1}{3} & 1 & \frac{1}{3} & 0 & 0 & | & \frac{11}{3} \\ ⑤⁄₃ & 0 & -\frac{1}{3} & 1 & 0 & | & \frac{10}{3} \\ -5 & 0 & 5 & 0 & 1 & | & 55 \end{bmatrix}$$

3.
$$\begin{bmatrix} 0 & 1 & \frac{2}{5} & -\frac{1}{5} & 0 & | & 3 \\ 1 & 0 & -\frac{1}{5} & \frac{3}{5} & 0 & | & 2 \\ 0 & 0 & 4 & 3 & 1 & | & 65 \end{bmatrix}$$

Maximum $f = 65$ occurs at $x_1 = 2, x_2 = 3$.

Minimization Problem— Graphical Method

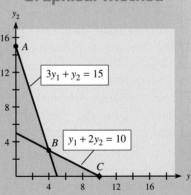

Corners:

$A = (0, 15)$
$B = (4, 3)$ } obtained by solving simultaneously
$C = (10, 0)$

$g = 11y_1 + 7y_2;$

At A, $g = 105$

At B, $g = 65$

At C, $g = 110$

Minimum $g = 65$ occurs at $y_1 = 4, y_2 = 3$.

Comparison of a Minimization Problem and Its Dual

Duality and Solving

From this example we see that the given minimization problem and its dual maximization problem are very closely related. Furthermore, it appears that when problems exhibit this relationship, the simplex method might be used to solve them both. The question is whether the simplex method will always solve them both. The answer is yes. This fact was proved by John von Neumann, and it is summarized by the Principle of Duality.

Principle of Duality:

1. When a standard minimization problem and its dual have a solution, the maximum value of the function to be maximized is the same value as the minimum value of the function to be minimized.
2. When the simplex method is used to solve the maximization problem, the values of the variables that solve the corresponding minimization problem are the last entries in the columns corresponding to the slack variables.

EXAMPLE 4.11 Purchasing

A beef producer is considering two different types of feed. Each feed contains some or all of the necessary ingredients for fattening beef. Brand 1 feed costs 20 cents per pound, and brand 2 feed costs 30 cents per pound. How much of each brand should the producer buy in order to satisfy the nutritional requirements for ingredients A and B at minimum cost? Table 4.1 contains all the relevant data about nutrition and cost of each brand and the minimum requirements per unit of beef.

Table 4.1

	Brand 1	Brand 2	Minimum Requirement
Ingredient A	3 units/lb	5 units/lb	40 units
Ingredient B	4 units/lb	3 units/lb	46 units
Cost per pound	20¢	30¢	

SOLUTION

Let y_1 be the number of pounds of brand 1, and let y_2 be the number of pounds of brand 2. Then we can formulate the problem as follows:

Minimizing cost subject to

$$3y_1 + 5y_2 \geq 40$$
$$4y_1 + 3y_2 \geq 46$$

The original linear programming problem is usually called the **primal problem**. If we write this in an augmented matrix, without the slack variables, then we can use the transpose to find the dual problem.

Primal
$$\begin{bmatrix} 3 & 5 & | & 40 \\ 4 & 3 & | & 46 \\ \hline 20 & 30 & | & C \end{bmatrix}$$

Dual (with function renamed)
$$\begin{bmatrix} 3 & 4 & | & 20 \\ 5 & 3 & | & 30 \\ \hline 40 & 46 & | & f \end{bmatrix}$$

Thus the dual maximization problem is

Maximize $f = 40x_1 + 46x_2$ subject to

$$3x_1 + 4x_2 \leq 20$$
$$5x_1 + 3x_2 \leq 30$$

The simplex matrix for this maximization problem, with slack variables included, is

1. $$\begin{bmatrix} 3 & ④ & 1 & 0 & 0 & | & 20 \\ 5 & 3 & 0 & 1 & 0 & | & 30 \\ \hline -40 & -46 & 0 & 0 & 1 & | & 0 \end{bmatrix}$$

Solving the problem gives

2. $$\begin{bmatrix} \frac{3}{4} & 1 & \frac{1}{4} & 0 & 0 & | & 5 \\ ⑪⁄₄ & 0 & -\frac{3}{4} & 1 & 0 & | & 15 \\ \hline -\frac{11}{2} & 0 & \frac{23}{2} & 0 & 1 & | & 230 \end{bmatrix}$$

3. $$\begin{bmatrix} 0 & 1 & \frac{5}{11} & -\frac{3}{11} & 0 & | & \frac{10}{11} \\ 1 & 0 & -\frac{3}{11} & \frac{4}{11} & 0 & | & \frac{60}{11} \\ \hline 0 & 0 & 10 & 2 & 1 & | & 260 \end{bmatrix}$$

The solution to this problem is

$$y_1 = 10 \text{ lb of brand 1}$$
$$y_2 = 2 \text{ lb of brand 2}$$

Minimum cost = 260 cents or \$2.60 per unit of beef

When we use Excel with Solver, minimization of an objective function is handled exactly the same as maximization, except **min** is checked and the inequality signs are $\geq$. See the Tech Card.

....

4.4 Exercises

1. (a) Form the matrix associated with the given minimization problem and find its transpose.
   (b) Write the dual maximization problem. Be sure to rename the variables.

   Minimize $g = 4y_1 + 5y_2$ subject to

   $$5y_1 + 2y_2 \geq 16$$
   $$y_1 + 2y_2 \geq 8$$

2. Suppose a primal minimization problem and its dual maximization problem were solved by using the simplex method on the dual problem, and the final simplex method is given.
   (a) Find the solution of the minimization problem. Use y_1, y_2, y_3 as the variables and g as the function.
   (b) Find the solution of the maximization problem. Use x_1, x_2, x_3 as the variables and f as the function.

$$
\left[
\begin{array}{ccccccc|c}
1 & -\frac{2}{5} & 0 & 3 & -\frac{3}{5} & 0 & 0 & 15 \\
0 & -\frac{3}{5} & 0 & -2 & -\frac{11}{5} & 1 & 0 & 13 \\
0 & -\frac{4}{5} & 1 & -1 & -\frac{2}{5} & 0 & 0 & 29 \\
\hline
0 & 2 & 0 & 7 & 4 & 0 & 1 & 452
\end{array}
\right]
$$

3. Write the dual maximization problem, and then solve both the primal and dual problems with the simplex method.

   Minimize $g = 2y_1 + 10y_2$ **subject to**

   $$2y_1 + y_2 \geq 11$$
   $$y_1 + 3y_2 \geq 11$$
   $$y_1 + 4y_2 \geq 16$$

In Problems 4 and 5, use the simplex method.

4. Minimize $g = 8x + 7y + 12z$ subject to

   $$x + y + z \geq 3$$
   $$y + 2z \geq 2$$
   $$x \geq 2$$

5. Minimize $g = 12y_1 + 48y_2 + 8y_3$ subject to

   $$y_1 + 3y_2 \qquad\ \geq 1$$
   $$4y_1 + 6y_2 + y_3 \geq 3$$
   $$4y_2 + y_3 \geq 1$$

Applications

6. **Production scheduling** CDF Appliances has assembly plants in Atlanta and Fort Worth where it produces a variety of kitchen appliances, including a 12-cup coffee maker and a cappuccino machine. In each hour at the Atlanta plant, 160 of the 12-cup models and 200 of the cappuccino machines can be assembled, and the hourly cost is \$700. In each hour at the Fort Worth plant, 800 of the 12-cup models and 200 of the cappuccino machines can be assembled, and the hourly cost is \$2100. CDF Appliances expects orders each week for at least 64,000 of the 12-cup models and at least 40,000 of the cappuccino machines. How many hours per week should each plant be operated in order to provide inventory for the orders at minimum cost? Find the minimum cost.

7. **Manufacturing** The Video Star Company makes two different models of DVD players, which are assembled on two different assembly lines. Line 1 can assemble 30 units of the Star model and 40 units of the Prostar model per hour, and line 2 can assemble 150 units of the Star model and 40 units of the Prostar model per hour. The company needs to produce at least 270 units of the Star model and 200 units of the Prostar model to fill an order. If it costs \$200 per hour to run line 1 and \$400 per hour to run line 2, how many hours should each line be run to fill the order at the minimum cost? What is the minimum cost?

8. **Production** A small company produces two products, I and II, at three facilities, A, B, and C. It has orders for 2000 of product I and 1200 of product II. The production capacities and costs per week to operate the three facilities are summarized in the following table.

	A	B	C
Product I	200	200	400
Product II	100	200	100
Cost/week	\$1000	\$3000	\$4000

How many weeks should each facility operate to fill the company's orders at a minimum cost, and what is the minimum cost?

4.5 The Simplex Method with Mixed Constraints

IN THE PREVIOUS TWO SECTIONS, WE USED THE SIMPLEX METHOD TO SOLVE TWO DIFFERENT TYPES OF STANDARD LINEAR PROGRAMMING PROBLEMS. THESE TYPES ARE SUMMARIZED AS FOLLOWS.

	Maximization Problems	Minimization Problems
Function	Maximized	Minimized
Variables	Nonnegative	Nonnegative
Constraints	$a_1x_1 + a_2x_2 + \cdots + a_nx_n \leq b\,(b \geq 0)$	$c_1y_1 + c_2y_2 + \cdots + c_ny_n \geq d\,(d \geq 0)$
Simplex method	Applied directly	Applied to dual

In this section we will show how to apply the simplex method to mixed constraint linear programming problems that do not exactly fit either of these types. The term *mixed constraint* is used because some inequalities describing the constraints contain "≤" and some contain "≥" signs.

Mixed Constraints and Maximization

We begin by considering maximization problems with constraints that have a form different from those noted in the preceding table. This can happen in one of the following ways.

1. If a constraint has the form
$a_1x_1 + a_2x_2 + \cdots + a_nx_n \leq b$, where $b < 0$, such as with $x - 2y \leq -8$

2. If a constraint has the form
$a_1x_1 + a_2x_2 + \cdots + a_nx_n \geq b$, where $b \geq 0$, such as with $2x - 3y \geq 6$

Note that in the latter case, multiplying both sides of the inequality by (-1) changes it as follows:

$$a_1x_1 + a_2x_2 + \cdots + a_nx_n \geq b$$
$$\text{becomes} \quad -a_1x_1 - a_2x_2 - \cdots - a_nx_n \leq -b$$

That is,

$$2x - 3y \geq 6 \quad \text{becomes} \quad -2x + 3y \leq -6$$

Thus we see that the two possibilities for mixed constraints are actually one: constraints that have the form $a_1x_1 + a_2x_2 + \cdots + a_nx_n \leq b$, where b is any constant. We call these **less than or equal to constraints** and denote them as **≤ constraints**.

Let's consider an example and see how we might proceed with a problem of this type.

EXAMPLE 4.12 Maximizing with Mixed Constraints

➡ Maximize $f = x + 2y$ subject to
$$x + y \leq 13$$
$$2x - 3y \geq 6$$
$$x \geq 0, y \geq 0$$

SOLUTION

We begin by expressing all constraints as "≤" constraints. In this case, we multiply $2x - 3y \geq 6$ by (-1) to obtain $-2x + 3y \leq -6$. We can now use slack variables to write the inequalities as equations and form the simplex matrix.

$$
\begin{aligned}
x + y + s_1 &= 13 \\
-2x + 3y + s_2 &= -6 \qquad \text{gives} \\
-x - 2y + f &= 0
\end{aligned}
$$

$$
\begin{array}{ccccc}
x & y & s_1 & s_2 & f \\
\end{array}
$$
$$
\left[
\begin{array}{ccccc|c}
1 & 1 & 1 & 0 & 0 & 13 \\
-2 & 3 & 0 & 1 & 0 & -6 \\
\hline
-1 & -2 & 0 & 0 & 1 & 0
\end{array}
\right]
$$

The -6 in the upper portion of the last column is a problem. If we compute values for the variables x, y, s_1, and s_2 that are associated with this matrix, we get $s_2 = -6$. This violates the condition of the simplex method that all variables be nonnegative. In order to use the simplex method, we must change the sign of any negative entry that appears in the upper portion of the last column (above the line separating the objective function from the constraints). If the problem has a solution, there will always be another negative entry in the same row as this negative entry, but in a different column. We choose this column as the pivot column, because pivoting with it will give a positive entry in the last column.

$$
\begin{array}{ccccc}
x & y & s_1 & s_2 & f \\
\end{array}
$$
$$
\left[
\begin{array}{ccccc|c}
1 & 1 & 1 & 0 & 0 & 13 \\
-2 & 3 & 0 & 1 & 0 & -6 \\
\hline
-1 & -2 & 0 & 0 & 1 & 0
\end{array}
\right]
$$
← Negative in last column

↑
Pivot column (negative entry in the same row as the negative entry in the last column)

Once we have identified the pivot column, we choose the pivot entry by forming all quotients and choosing the entry that gives the smallest positive quotient. Often this will mean pivoting by a negative number, which is the case in this problem.

$$
\begin{array}{ccccc}
x & y & s_1 & s_2 & f \\
\end{array}
$$

$$
\left[\begin{array}{ccccc|c}
1 & 1 & 1 & 0 & 0 & 13 \\
\boxed{-2} & 3 & 0 & 1 & 0 & -6 \\
\hline
-1 & -2 & 0 & 0 & 1 & 0
\end{array}\right]
\begin{array}{l}
13/1 = 13 \\
(-6)/(-2) = 3^*
\end{array}
$$

↑ Pivot entry circled   *Smallest positive quotient

Pivoting with this entry gives the following matrix.

$$
\begin{array}{ccccc}
x & y & s_1 & s_2 & f \\
\end{array}
$$

$$
\left[\begin{array}{ccccc|c}
0 & \tfrac{5}{2} & 1 & \tfrac{1}{2} & 0 & 10 \\
1 & -\tfrac{3}{2} & 0 & -\tfrac{1}{2} & 0 & 3 \\
\hline
0 & -\tfrac{7}{2} & 0 & -\tfrac{1}{2} & 1 & 3
\end{array}\right]
\Big\} \text{No negatives}
$$

↑ Most negative

This new matrix has no negatives in the upper portion of the last column, so we proceed with the simplex method as we have used it previously. The new pivot column is found from the most negative entry in the last row (indicated), and the pivot is found from the smallest positive quotient (circled). Using this pivot completes the solution.

$$
\begin{array}{ccccc}
x & y & s_1 & s_2 & f \\
\end{array}
$$

$$
\left[\begin{array}{ccccc|c}
0 & 1 & \tfrac{2}{5} & \tfrac{1}{5} & 0 & 4 \\
1 & 0 & \tfrac{3}{5} & -\tfrac{1}{5} & 0 & 9 \\
\hline
0 & 0 & \tfrac{7}{5} & \tfrac{1}{5} & 1 & 17
\end{array}\right]
$$

We see that when $x = 9$ and $y = 4$, then the maximum value of $f = x + 2y$ is 17.

Mixed Constraints and Minimization

If mixed constraints occur in a minimization problem, the simplex method does not apply to the dual problem. However, the same techniques used in Example 4.12 can be modified slightly (to maximize $-f$) and applied to minimization problems with mixed constraints as shown in the box above.

EXAMPLE 4.13 Production Scheduling for Maximum Profit

The Laser Company manufactures two models of stereo systems, the Star and the Allstar, at two plants, located in Asheville and in Cleveland. The maximum daily output at the Asheville plant is 900,

Simplex Method for Mixed Constraints

1. If the problem is to minimize f, then maximize $-f$.

2. (a) Make all constraints "≤" constraints by multiplying both sides of any "≥" constraints by (-1).
   (b) Use slack variables and form the simplex matrix.

3. In the simplex matrix, scan the upper portion of the last column for any negative entries.
   (a) If there are no negative entries, apply the simplex method.
   (b) If there are negative entries, go to Step 4.

4. When there is a negative value in the upper portion of the last column, proceed as follows:
   (a) Select any negative entry in the same row and use its column as the pivot column.
   (b) In the pivot column, compute all quotients for the entries above the line and determine the pivot from the smallest positive quotient.
   (c) After completing the pivot operations, return to Step 3.

and the profits there are \$200 per unit of the Allstar and \$100 per unit of the Star. The maximum daily output at the Cleveland plant is 800, and the profits there are \$210 per unit of the Allstar and \$80 per unit of the Star. In addition, restrictions at the Asheville plant mean that the number of units of the Star model cannot exceed 100 more than the number of units of the Allstar model produced. If the company gets a rush order for 800 units of the Allstar model and 600 units of the Star model, how many units of each model should be produced at each location to fill the order and obtain the maximum profit?

SOLUTION

The following table identifies our variables x and y and relates other important facts in the problem. In particular, we see that only variables x and y are required.

	Asheville Plant	Total Needed	Cleveland Plant
Allstar	x produced (profit = \$200 each)	800	$800 - x$ produced (profit = \$210 each)
Star	y produced (profit = \$100 each)	600	$600 - y$ produced (profit = \$80 each)
Capacity	900	—	800

We wish to maximize profit, and from the table we see that total profit is given by

$$P = 200x + 100y + 210(800 - x) + 80(600 - y)$$

or

$$P = 216{,}000 - 10x + 20y$$

We can read capacity constraints from our table.

$$x + y \leq 900 \quad \text{(Asheville)}$$

$$(800 - x) + (600 - y) \leq 800 \quad \text{or} \quad x + y \geq 600 \quad \text{(Cleve.)}$$

An additional Asheville plant constraint from the statement of the problem is

$$y \leq x + 100 \quad \text{or} \quad -x + y \leq 100$$

Thus our problem is to maximize

$$P = 216{,}000 - 10x + 20y$$

subject to

$$x + y \leq 900$$
$$x + y \geq 600$$
$$-x + y \leq 100$$
$$x \geq 0, y \geq 0$$

We must express $x + y \geq 600$ as a " $\leq$ " constraint; multiplying both sides by (-1) gives $-x - y \leq -600$. The simplex matrix is

$$
\begin{array}{ccccccc}
x & y & s_1 & s_2 & s_3 & P & \\
\end{array}
$$
$$
\left[
\begin{array}{cccccc|c}
1 & 1 & 1 & 0 & 0 & 0 & 900 \\
\boxed{-1} & -1 & 0 & 1 & 0 & 0 & -600 \\
-1 & 1 & 0 & 0 & 1 & 0 & 100 \\
\hline
10 & -20 & 0 & 0 & 0 & 1 & 216{,}000
\end{array}
\right] \leftarrow \text{Negative}
$$
$$\uparrow \text{Pivot column}$$

The negative in the upper portion of the last column is indicated. In row 2 we find negatives in both column 1 and column 2, so either of these may be our pivot column. Our choice is indicated, and the pivot entry is circled. The matrix that results from the pivot operations follows.

$$
\begin{array}{ccccccc}
x & y & s_1 & s_2 & s_3 & P & \\
\end{array}
$$
$$
\left[
\begin{array}{cccccc|c}
0 & 0 & 1 & 1 & 0 & 0 & 300 \\
1 & 1 & 0 & -1 & 0 & 0 & 600 \\
0 & \boxed{2} & 0 & -1 & 1 & 0 & 700 \\
\hline
0 & -30 & 0 & 10 & 0 & 1 & 210{,}000
\end{array}
\right]
$$
$$\uparrow \text{Pivot column}$$

No entry in the upper portion of the last column is negative, so we can proceed with the simplex method. From the indicators we see that column 2 is the pivot column (indicated previously), and the pivot entry is circled. The simplex matrix that results from pivoting follows.

ASHVILLE, NC
Capacity: 900 units
Profit: \$200/Allstar
$100/Star

CLEVELAND, OH
Capacity: 800 units
Profit: \$210/Allstar
$80/Star

$$\begin{array}{cccccc} x & y & s_1 & s_2 & s_3 & P \\ \left[\begin{array}{cccccc|c} 0 & 0 & 1 & \boxed{1} & 0 & 0 & 300 \\ 1 & 0 & 0 & -\frac{1}{2} & -\frac{1}{2} & 0 & 250 \\ 0 & 1 & 0 & -\frac{1}{2} & \frac{1}{2} & 0 & 350 \\ \hline 0 & 0 & 0 & -5 & 15 & 1 & 220{,}500 \end{array}\right] \end{array}$$

$$\underset{\text{Pivot column}}{\uparrow}$$

The new pivot column is indicated, and the pivot entry is circled. The pivot operation yields the following matrix.

$$\begin{array}{cccccc} x & y & s_1 & s_2 & s_3 & P \\ \left[\begin{array}{cccccc|c} 0 & 0 & 1 & 1 & 0 & 0 & 300 \\ 1 & 0 & \frac{1}{2} & 0 & -\frac{1}{2} & 0 & 400 \\ 0 & 1 & \frac{1}{2} & 0 & \frac{1}{2} & 0 & 500 \\ \hline 0 & 0 & 5 & 0 & 15 & 1 & 222{,}000 \end{array}\right] \end{array}$$

This matrix shows that the solution is complete. We see that $x = 400$ (so $800 - x = 400$), $y = 500$ (so $600 - y = 100$), and $P = \$222{,}000$.

Thus the company should operate the Asheville plant at capacity and produce 400 units of the Allstar model and 500 units of the Star model. The remainder of the order, 400 units of the Allstar model and 100 units of the Star model, should be produced at the Cleveland plant. This production schedule gives maximum profit $P = \$222{,}000$.

When using Excel to solve a linear programming problem with mixed constraints, it is not necessary to convert the form of the inequalities. They can simply be entered in their "mixed" form, and the objective function can be maximized or minimized with Solver. See the Tech Card for details.

....

4.5 Exercises

1. Express $3x - y \geq 5$ as a "$\leq$" constraint.

2. (a) State the given problem in a form from which the simplex matrix can be formed (that is, as a maximization problem with "$\leq$" constraints).
   (b) Form the simplex matrix, and circle the first pivot entry.
   Minimize $g = 3x + 8y$ subject to
   $$4x - 5y \leq 50$$
   $$x + y \leq 80$$
   $$-x + 2y \geq 4$$
   $$x \geq 0, \ y \geq 0$$

3. From the given final simplex matrix for a minimization problem, find the solution to the minimization problem.

$$\begin{array}{ccccccc} x & y & z & s_1 & s_2 & s_3 & -f \\ \left[\begin{array}{ccccccc|c} 0 & 1 & 0 & \frac{7}{3} & \frac{1}{3} & -\frac{2}{3} & 0 & 8 \\ 0 & 0 & 1 & \frac{2}{3} & -\frac{1}{3} & \frac{8}{3} & 0 & 12 \\ 1 & 0 & 0 & -\frac{4}{3} & \frac{2}{3} & \frac{4}{3} & 0 & 6 \\ \hline 0 & 0 & 0 & 4 & \frac{4}{3} & 2 & 1 & -120 \end{array}\right] \end{array}$$

In Problems 4 and 5, use the simplex method to find the optimal solution. Assume all variables are nonnegative.

4. Maximize $f = 4x + y$ subject to
   $$5x + 2y \leq 84$$
   $$-3x + 2y \geq 4$$

5. Minimize $f = 2x + 3y$ subject to
   $$x \geq 5$$
   $$y \leq 13$$
   $$-x + y \geq 2$$

In Problems 6 and 7, use the simplex method, Excel, or another technology. Assume all variables are nonnegative.

6. Minimize $f = x + 2y + 3z$ subject to
   $$x + z \leq 20$$
   $$x + y \geq 30$$
   $$y + z \leq 20$$

7. Maximize $f = 2x - y + 4z$ subject to
   $$x + y + z \leq 8$$
   $$x - y + z \geq 4$$
   $$x + y - z \geq 2$$

Applications

8. **Production** A sausage company makes two different kinds of hot dogs, regular and all beef. Each pound of all-beef hot dogs requires 0.75 lb of beef and 0.2 lb of spices, and each pound of regular hot dogs requires 0.18 lb of beef, 0.3 lb of pork, and 0.2 lb of spices. Suppliers can deliver at most 1020 lb of beef, at most 600 lb of pork, and at least 500 lb of spices. If the profit is $0.60 on each pound of all-beef hot dogs and $0.40 on each pound of regular hot dogs, how many pounds of each should be produced to obtain maximum profit? What is the maximum profit?

9. **Water purification** Nolan Industries manufactures water filters/purifiers that attach to a kitchen faucet. Each purifier consists of a housing unit that attaches to the faucet and a 60-day filter (sold separately) that is inserted into the housing. Past records indicate that on average the number of filters produced per week should be at least 400. It takes 20 minutes to make and assemble each filter and 40 minutes for each housing. The manufacturing facility has at most 20,000 minutes per week for making and assembling these units, but because of certain parts supply constraints, the number of housing units per week can be at most 400. If manufacturing costs (for material and labor) are $6.60 for each filter and $8.35 for each housing unit, how many of each should be produced to minimize weekly costs? Find the minimum cost.

10. **Manufacturing** A company manufactures commercial heating system components and domestic furnaces at its factories in Monaca, Pennsylvania, and Hamburg, New York. At the Monaca plant, no more than 1000 units per day can be produced, and the number of commercial components cannot exceed 100 more than half the number of domestic furnaces. At the Hamburg plant, no more than 850 units per day can be produced. The profit on each commercial component is $400 at the Monaca plant and $390 at the Hamburg plant. The profit on each domestic furnace is $200 at the Monaca plant and $215 at the Hamburg plant. If there is a rush order for 500 commercial components and 750 domestic furnaces, how many of each should be produced at each plant in order to maximize profits? Find the maximum profit.

11. **Water purification** Three water purification facilities can handle at most 10 million gallons in a certain time period. Plant I leaves 20% of certain impurities and costs $20,000 per million gallons. Plant II leaves 15% of these impurities and costs $30,000 per million gallons. Plant III leaves 10% impurities and costs $40,000 per million gallons. The desired level of impurities in the water from all three plants is at most 15%. If Plant I and Plant III combined must handle at least 6 million gallons, find the number of gallons each plant should handle so as to achieve the desired level of purity at minimum cost. Find the minimum cost.

··

Need more practice?
Find more here: cengagebrain.com

Chapter Exercises

In Problems 1–4, graph the solution set of each inequality or system of inequalities.

1. $2x + 3y \leq 12$ 　　　 2. $4x + 5y > 100$

3. $\begin{cases} x + 2y \leq 20 \\ 3x + 10y \leq 80 \\ x \geq 0, y \geq 0 \end{cases}$ 　 4. $\begin{cases} 3x + y \geq 4 \\ x + y \geq 2 \\ -x + y \leq 4 \\ x \leq 5 \end{cases}$

In Problems 5–8, a function and the graph of a feasible region are given. In each case, find both the maximum and minimum values of the function, if they exist, and the point at which each occurs.

5. $f = -x + 3y$

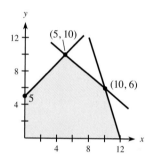

6. $f = 6x + 4y$

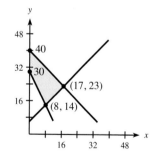

7. $f = 7x - 6y$

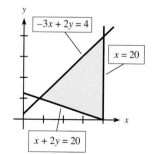

8. $f = 9x + 10y$

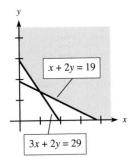

In Problems 9–14, solve each linear programming problem using graphical methods. Restrict $x \geq 0$ and $y \geq 0$.

9. Maximize $f = 5x + 6y$ subject to

$$x + 3y \leq 24$$
$$4x + 3y \leq 42$$
$$2x + y \leq 20$$

10. Maximize $f = x + 4y$ subject to

$$7x + 3y \leq 105$$
$$2x + 5y \leq 59$$
$$x + 7y \leq 70$$

11. Minimize $g = 5x + 3y$ subject to

$$3x + y \geq 12$$
$$x + y \geq 6$$
$$x + 6y \geq 11$$

12. Minimize $g = x + 5y$ subject to

$$8x + y \geq 85$$
$$x + y \geq 50$$
$$x + 4y \geq 80$$
$$x + 10y \geq 104$$

13. Maximize $f = 5x + 2y$ subject to

$$y \leq 20$$
$$2x + y \leq 32$$
$$-x + 2y \geq 4$$

14. Minimize $f = x + 4y$ subject to

$$y \leq 30$$
$$3x + 2y \geq 75$$
$$-3x + 5y \geq 30$$

In Problems 15–18, use the simplex method to solve each linear programming problem. Assume all variables are nonnegative.

15. Maximize $f = 7x + 12y$ subject to the conditions in Problem 10.

16. Maximize $f = 3x + 4y$ subject to

$$x + 4y \leq 160$$
$$x + 2y \leq 100$$
$$4x + 3y \leq 300$$

17. Maximize $f = 3x + 8y$ subject to the conditions in Problem 16.

18. Maximize $f = 3x + 2y$ subject to

$$x + 2y \leq 48$$
$$x + y \leq 30$$
$$2x + y \leq 50$$
$$x + 10y \leq 200$$

In Problems 19 and 20, each problem has a nonunique solution. If there is no solution, indicate this; if there are multiple solutions, find two different solutions. Use the simplex method, with $x \geq 0, y \geq 0$.

19. Maximize $f = 4x + 4y$ subject to

$$x + 5y \leq 500$$
$$x + 2y \leq 230$$
$$x + y \leq 160$$

20. Maximize $f = 2x + 5y$ subject to

$$-4x + y \leq 40$$
$$x - 7y \leq 70$$

In Problems 21–24, form the dual and use the simplex method to solve the minimization problem. Assume all variables are nonnegative.

21. Minimize $g = 7y_1 + 6y_2$ subject to

$$5y_1 + 2y_2 \geq 16$$
$$3y_1 + 7y_2 \geq 27$$

22. Minimize $g = 3y_1 + 4y_2$ subject to

$$3y_1 + y_2 \geq 8$$
$$y_1 + y_2 \geq 6$$
$$2y_1 + 5y_2 \geq 18$$

23. Minimize $g = 2y_1 + y_2$ subject to the conditions in Problem 22.

24. Minimize $g = 12y_1 + 11y_2$ subject to

$$y_1 + y_2 \geq 100$$
$$2y_1 + y_2 \geq 140$$
$$6y_1 + 5y_2 \geq 580$$

In Problems 25 and 26, each problem involves mixed constraints. Solve each problem with the simplex method. Assume all variables are nonnegative.

25. Maximize $f = 3x + 5y$ subject to

$$x + y \geq 19$$
$$-x + y \geq 1$$
$$-x + 10y \leq 190$$

26. Maximize $f = 4x + 6y$ subject to

$$2x + 5y \leq 37$$
$$5x - y \leq 34$$
$$-x + 2y \geq 4$$

In Problems 27–31, use the simplex method. Assume all variables are nonnegative.

27. Maximize $f = 39x + 5y + 30z$ subject to

$$x + z \leq 7$$
$$3x + 5y \leq 30$$
$$3x + y \leq 18$$

28. Minimize $g = 12y_1 + 5y_2 + 2y_3$ subject to

$$y_1 + 2y_2 + y_3 \geq 60$$
$$12y_1 + 4y_2 + 3y_3 \geq 120$$
$$2y_1 + 3y_2 + y_3 \geq 80$$

29. Minimize $f = 4x + 3y$ subject to

$$-x + y \geq 1$$
$$x + y \leq 45$$
$$10x + y \geq 45$$

30. Maximize $f = 88x_1 + 86x_2 + 100x_3 + 100x_4$ subject to

$$3x_1 + 2x_2 + 2x_3 + 5x_4 \leq 200$$
$$2x_1 + 2x_2 + 4x_3 + 5x_4 \leq 100$$
$$x_1 + x_2 + x_3 + x_4 \leq 200$$
$$x_1 \qquad\qquad \leq 40$$

31. Maximize $f = 8x_1 + 10x_2 + 12x_3 + 14x_4$ subject to

$$6x_1 + 3x_2 + 2x_3 + x_4 \geq 350$$
$$3x_1 + 2x_2 + 5x_3 + 6x_4 \leq 300$$
$$8x_1 + 3x_2 + 2x_3 + x_4 \leq 400$$
$$x_1 + x_2 + x_3 + x_4 \leq 100$$

Applications

32. **Manufacturing** A company manufactures backyard swing sets of two different sizes. The larger set requires 5 hours of labor to complete, the smaller set requires 2 hours, and 700 hours of labor are available each week. The packaging department can package at most 185 swing sets per week. If the profit is $100 on each larger set and $50 on each smaller set, how many of each should be produced to yield the maximum profit? What is the maximum profit? Use graphical methods.

33. **Production** A company produces two different grades of steel, A and B, at two different factories, 1 and 2. The table summarizes the production capabilities of the factories, the cost per day, and the number of units of each grade of steel that is required to fill orders.

	Factory 1	Factory 2	Required
Grade A steel	1 unit	2 units	80 units
Grade B steel	3 units	2 units	140 units
Cost per day	$5000	$6000	

How many days should each factory operate in order to fill the orders at minimum cost? What is the minimum cost? Use graphical methods.

In Problems 34–42, use the simplex method to solve each problem.

34. **Production** A small industry produces two items, I and II. It operates at capacity and makes a profit of $6 on each item I and $4 on each item II. The table gives the hours required to produce each item and the hours available per day.

	Item I	Item II	Hours Available
Assembly	2 hours	1 hour	100
Packaging and inspection	1 hour	1 hour	60

Find the number of items that should be produced each day to maximize profits, and find the maximum daily profit.

35. **Production** Pinocchio Crafts makes two types of wooden crafts: Jacob's ladders and locomotive engines. The manufacture of these crafts requires both carpentry and finishing. Each Jacob's ladder requires 1 hour of finishing and 1/2 hour of carpentry. Each locomotive engine requires 1 hour of finishing and 1 hour of carpentry. Pinocchio Crafts can obtain all the necessary raw materials, but only 120 finishing hours and 75 carpentry hours per week are available. Also, demand for Jacob's ladders is limited to at most 100 per week. If Pinocchio Crafts makes a profit of $3 on each Jacob's ladder and $5 on each locomotive engine, how many of each should it produce each week to maximize profits? What is the maximum profit?

36. **Profit** At its Jacksonville factory, Nolmaur Electronics manufactures 4 models of TV sets: LCD models in 27-in., 32-in., and 42-in. sizes and a 42-in. plasma model. The manufacturing and testing hours required for each model and available at the factory each week, as well as each model's profit, are shown in the table.

	27-in. LCD	32-in. LCD	42-in. LCD	42-in. Plasma	Available Hours
Manufacturing (hr)	8	10	12	15	1870
Testing (hr)	2	4	4	4	530
Profit	$80	$120	$160	$200	

In addition, the supplier of the amplifier units can provide at most 200 units per week with at most 100 of these for both types of 42-in. models. The weekly demand for the 32-in. sets is at most 120. Nolmaur wants to determine the number of each type of set that should be produced each week to obtain maximum profit.
(a) Carefully identify the variables for Nolmaur's linear programming problem.
(b) Carefully state Nolmaur's linear programming problem.
(c) Solve this linear programming problem to determine Nolmaur's manufacturing plan and maximum profit.

37. **Nutrition** A nutritionist wants to find the least expensive combination of two foods that meet minimum daily vitamin requirements, which are 5 units of A and 30 units of B. Each ounce of food I provides 2 units of A and 1 unit of B, and each ounce of food II provides 10 units of A and 10 units of B. If food I costs 30 cents per ounce and food II costs 20 cents per ounce, find the number of ounces of each food that will provide the required vitamins and minimize the cost.

38. **Nutrition** A laboratory wishes to purchase two different feeds, A and B, for its animals. The table summarizes the nutritional contents of the feeds, the required amounts of each ingredient, and the cost of each type of feed.

	Feed A	Feed B	Required
Carbohydrates	1 unit/lb	4 units/lb	40 units
Protein	2 units/lb	1 unit/lb	80 units
Cost	14¢/lb	16¢/lb	

How many pounds of each type of feed should the laboratory buy in order to satisfy its needs at minimum cost?

39. **Production** A company makes three products, I, II, and III, at three different factories. At factory A, it can make 10 units of each product per day. At factory B, it can make 20 units of II and 20 units of III per day. At factory C, it can make 20 units of I, 20 units of II, and 10 units of III per day. The company has orders for 200 units of I, 500 units of II, and 300 units of III. If the daily costs are $200 at A, $300 at B, and $500 at C, find the number of days that each factory should operate in order to fill the company's orders at minimum cost. Find the minimum cost.

40. **Profit** A company makes pancake mix and cake mix. Each pound of pancake mix uses 0.6 lb of flour and 0.1 lb of shortening. Each pound of cake mix uses 0.4 lb of flour, 0.1 lb of shortening, and 0.4 lb of sugar. Suppliers can deliver at most 6000 lb of flour, at least 500 lb of shortening, and at most 1200 lb of sugar. If the profit per pound is $0.35 for pancake mix and $0.25 for cake mix, how many pounds of each mix should be made to earn maximum profit? What is the maximum profit?

41. **Manufacturing** A company manufactures desks and computer tables at plants in Texas and Louisiana. At the Texas plant, production costs are $12 for each desk and $20 for each computer table, and the plant can produce at most 120 units per day. At the Louisiana plant, costs are $14 for each desk and $19 per computer table, and the plant can produce at most 150 units per day. The company gets a rush order for 130 desks and 130 computer tables at a time when the Texas plant is further limited by the fact that the number of computer tables it produces must be at least 10 more than the number of desks. How should production be scheduled at each location in order to fill the order at minimum cost? What is the minimum cost?

42. **Cost** Armstrong Industries makes two different grades of steel at its plants in Midland and Donora. Weekly demand is at least 500 tons for grade 1 steel and at least 450 tons for grade 2. Because of differences in the equipment and the labor force, production times and costs per ton for each grade of steel differ slightly at each location, as shown in the table.

	Grade 1		Grade 2		
	Min/Ton	Cost/Ton	Min/Ton	Cost/Ton	Furnace Time/Week (in Minutes)
Midland	40	$100	42	$120	19,460
Donora	44	$110	45	$90	21,380

Find the number of tons of each grade of steel that should be made each week at each plant so that Armstrong meets demand at minimum cost. Find the minimum cost.

Need more practice?
Find more here: cengagebrain.com

chapter 5

Exponential and
Logarithmic Functions

In this chapter we study exponential and logarithmic functions, which provide models for many applications that at first seem remote and unrelated. In our study of these functions, we will examine their descriptions, their properties, their graphs, and the special inverse relationship between these two functions. We will see how exponential and logarithmic functions are applied to some of the concerns of social scientists, business managers, and life scientists.

objectives

5.1 Exponential Functions

5.2 Logarithmic Functions and Their Properties

5.3 Exponential Equations and Applications

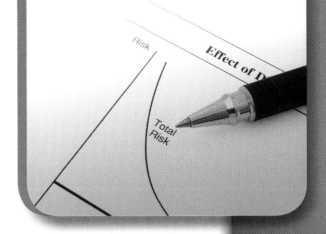

5.1   Exponential Functions

SUPPOSE A CULTURE OF BACTERIA HAS THE CHARACTERISTIC THAT EACH MINUTE, EVERY MICROORGANISM SPLITS INTO TWO NEW ORGANISMS. WE CAN DESCRIBE THE NUMBER OF BACTERIA IN THE CULTURE AS A FUNCTION OF TIME. THAT IS, IF WE BEGIN THE CULTURE WITH ONE MICROORGANISM, WE KNOW THAT AFTER 1 MINUTE WE WILL HAVE TWO ORGANISMS, AFTER 2 MINUTES, FOUR, AND SO ON. TABLE 5.1 GIVES A FEW OF THE VALUES THAT DESCRIBE THIS GROWTH. IF X REPRESENTS THE NUMBER OF MINUTES THAT HAVE PASSED AND Y REPRESENTS THE NUMBER OF ORGANISMS, THE POINTS (X, Y) LIE ON THE GRAPH OF THE FUNCTION WITH EQUATION

$$Y = 2^X$$

Table 5.1

Minutes Passed	Number of Organisms
0	1
1	2
2	4
3	8
4	16

Exponential Functions: If a is a real number with $a > 0$ and $a \neq 1$, then the function

$$f(x) = a^x$$

is an exponential function with base a.

The equation $y = 2^x$ is an example of a special group of functions called **exponential functions.** In this section we will evaluate and graph exponential functions, and we will model data with exponential functions, defined above.

Shutterstock

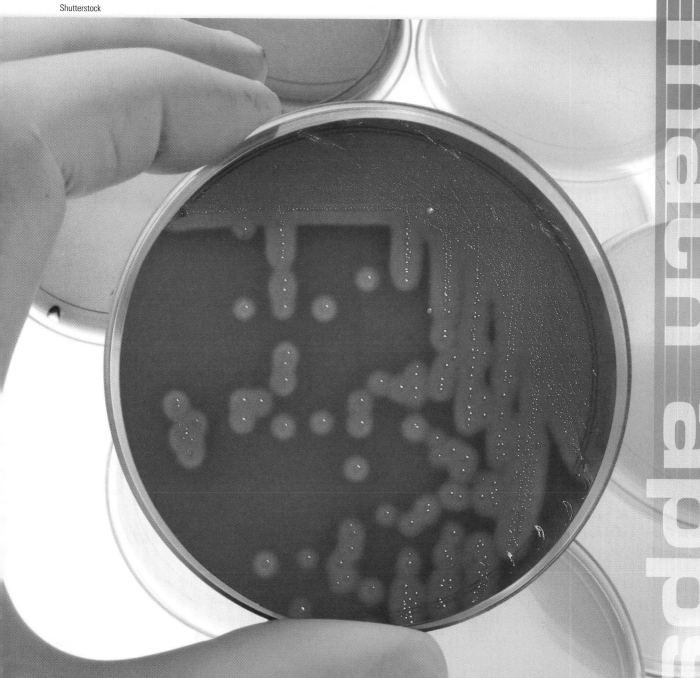

A table of some values satisfying $y = 2^x$ and the graph of this function are given in Figure 5.1. This function is said to model the growth of the number of organisms in the previous discussion, even though some points on the graph do not correspond to a time and a number of organisms. For example, time x could not be negative, and the number of organisms y could not be fractional.

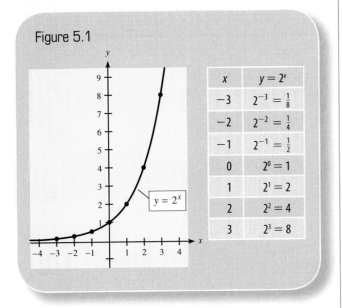

Figure 5.1

x	$y = 2^x$
−3	$2^{-3} = \frac{1}{8}$
−2	$2^{-2} = \frac{1}{4}$
−1	$2^{-1} = \frac{1}{2}$
0	$2^0 = 1$
1	$2^1 = 2$
2	$2^2 = 4$
3	$2^3 = 8$

We will assume that if we graphed $y = 2^x$ for any real values of x, those points would lie on the curve in Figure 5.1. Thus, in general, we can graph an exponential function by plotting easily calculated points, such as those in the table in Figure 5.1, and drawing a smooth curve through the points.

EXAMPLE 5.1 Graph of an Exponential Function

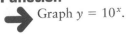 Graph $y = 10^x$.

SOLUTION

A table of values and the graph are given in Figure 5.2.

See the Tech Card for details on graphing exponentials with a graphing calculator and with Excel.

Note that the graphs of $y = 2^x$ and $y = 10^x$ are similar. The graphs of $y = 2^x$ and $y = 10^x$ clearly approach, but do not touch, the negative x-axis. That is, the negative x-axis is an asymptote for these functions.

In fact, the shapes of the graphs of functions of the form $y = f(x) = a^x$, with $a > 1$, are similar to those in Figures 5.1 and 5.2. Exponential functions of this type, and more generally of the form $f(x) = C(a^x)$, where $C > 0$ and $a > 1$, are called **exponential growth func-**

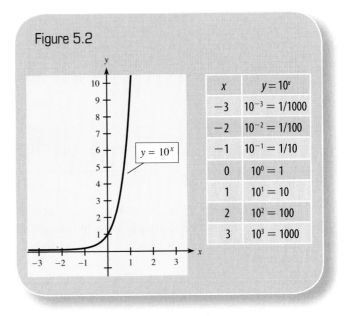

Figure 5.2

x	$y = 10^x$
−3	$10^{-3} = 1/1000$
−2	$10^{-2} = 1/100$
−1	$10^{-1} = 1/10$
0	$10^0 = 1$
1	$10^1 = 10$
2	$10^2 = 100$
3	$10^3 = 1000$

tions because they are used to model growth in diverse applications. Their graphs have the basic shape shown in the accompanying box.

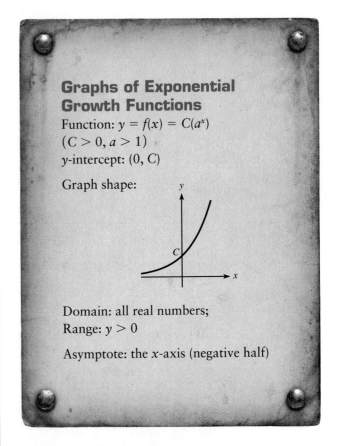

Graphs of Exponential Growth Functions

Function: $y = f(x) = C(a^x)$
$(C > 0, a > 1)$
y-intercept: $(0, C)$

Graph shape:

Domain: all real numbers;
Range: $y > 0$

Asymptote: the x-axis (negative half)

One exponential growth model concerns money invested at compound interest. Often we seek to evaluate rather than graph these functions.

EXAMPLE 5.2 Future Value of an Investment

If $10,000 is invested at 6%, compounded monthly, then the future value of the investment S after x years is given by

$$S = 10,000(1.005)^{12x}$$

Find the future value of the investment after (a) 5 years and (b) 30 years.

SOLUTION

These future values can be found with a calculator.

(a) $S =$ $10,000(1.005)^{12(5)}$
     $=$ $10,000(1.005)^{60}$
     $=$ \$13,488.50 (to the nearest cent)

(b) $S =$ $10,000(1.005)^{12(30)}$
     $=$ $10,000(1.005)^{360}$
     $=$ \$60,225.75 (to the nearest cent)

Note that the amount after 30 years is significantly more than the amount after 5 years, a result consistent with exponential growth models.

A special function that occurs frequently in economics and biology is $y = e^x$, where e is a fixed irrational number (approximately $2.71828\ldots$).

Because $e > 1$, the graph of $y = e^x$ will have the same basic shape as other growth exponentials. We can calculate the y-coordinate for points on the graph of this function with a calculator. A table of some values (with y-values rounded to two decimal places) and the graph are shown in Figure 5.3.

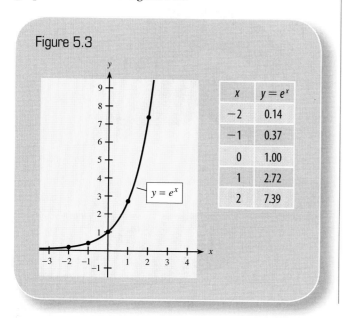

Figure 5.3

x	$y = e^x$
−2	0.14
−1	0.37
0	1.00
1	2.72
2	7.39

Exponentials whose bases are between 0 and 1, such as $y = \left(\frac{1}{2}\right)^x$, have graphs different from those of the exponentials just discussed. Using the properties of exponents, we have

$$y = \left(\frac{1}{2}\right)^x = (2^{-1})^x = 2^{-x}$$

By using this technique, all exponentials of the form $y = b^x$, where $0 < b < 1$, can be rewritten in the form $y = a^{-x}$, where $a = \frac{1}{b} > 1$. Thus, graphs of equations of the form $y = a^{-x}$, where $a > 1$, and of the form $y = b^x$, where $0 < b < 1$, will have the same shape.

EXAMPLE 5.3 Graphing an Exponential Function

Graph $y = 2^{-x}$.

SOLUTION

A table of values and the graph are given in Figure 5.4.

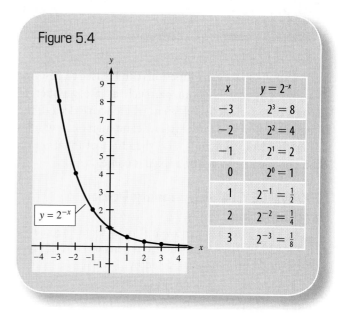

Figure 5.4

x	$y = 2^{-x}$
−3	$2^3 = 8$
−2	$2^2 = 4$
−1	$2^1 = 2$
0	$2^0 = 1$
1	$2^{-1} = \frac{1}{2}$
2	$2^{-2} = \frac{1}{4}$
3	$2^{-3} = \frac{1}{8}$

Additional examples of functions of this type would yield graphs with the same shape. **Exponential decay functions** have the form $y = a^{-x}$, where $a > 1$, or, more generally, the form $y = C(a^{-x})$, where $C > 0$ and $a > 1$. They model decay for various phenomena, and their graphs have the characteristics and shape shown in the accompanying box on the next page.

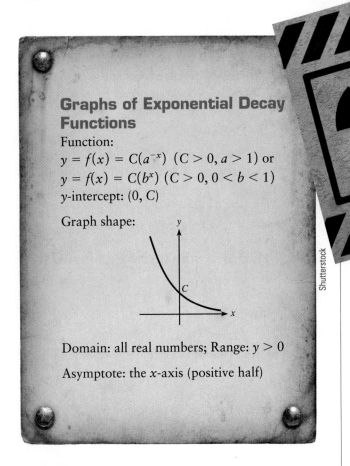

Graphs of Exponential Decay Functions

Function:
$$y = f(x) = C(a^{-x}) \ (C > 0, a > 1) \text{ or}$$
$$y = f(x) = C(b^x) \ (C > 0, 0 < b < 1)$$

y-intercept: $(0, C)$

Graph shape:

Domain: all real numbers; Range: $y > 0$

Asymptote: the x-axis (positive half)

Exponential decay functions can be used to describe various physical phenomena, such as the number of atoms of a radioactive element (radioactive decay), the lingering effect of an advertising campaign on new sales after the campaign ends, and the effect of inflation on the purchasing power of a fixed income.

EXAMPLE 5.4 Purchasing Power and Inflation

The purchasing power P of a fixed income of $30,000 per year (such as a pension) after t years of 4% inflation can be modeled by
$$P = 30,000e^{-0.04t}$$
Find the purchasing power after (a) 5 years and (b) 20 years.

SOLUTION

We can use a calculator to answer both parts.

(a) $P = 30,000e^{-0.04(5)} = 30,000e^{-0.2}$
$= \$24,562$ (to the nearest dollar)

(b) $P = 30,000e^{-0.04(20)} = 30,000e^{-0.8}$
$= \$13,480$ (to the nearest dollar)

Note that the impact of inflation over time significantly erodes purchasing power and provides some insight into the plight of elderly people who live on fixed incomes.

Modeling with Exponential Functions

Many types of data can be modeled using an exponential growth function. Figure 5.5(a) shows a graph of amounts of carbon in the atmosphere due to emissions from the burning of fossil fuels. With curve-fitting tools available with some computer software or on graphing calculators, we can develop an equation that models, or approximates, these data (see the Tech Card for instructions for modeling with a graphing calculator or with Excel). The model and its graph are shown in Figure 5.5(b).

The model in Figure 5.5(b) is an exponential growth function. As we have noted, functions of this type have the general form $y = C(a^x)$ with $a > 1$, and exponential decay functions have the general form $y = C(a^{-x})$ with $a > 1$ [or $y = C(b^x)$ with $0 < b < 1$]. Note that when technology is applied to model exponentials (either growth or decay), the model is written in an equivalent form $y = a*b^x$.

Shutterstock

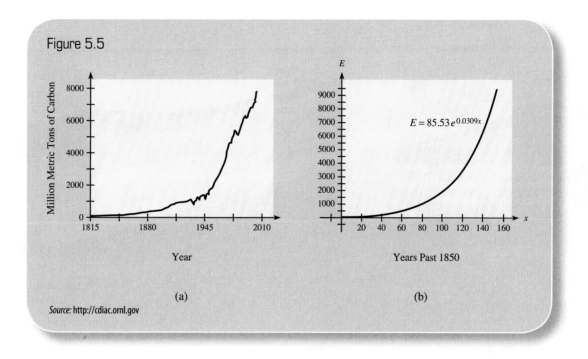

Figure 5.5

(a)

(b)

Source: http://cdiac.ornl.gov

5.1 Exercises

1. Use a calculator to evaluate each expression.
   (a) $3^{1/3}$ (b) e^2

In Problems 2–5, graph each function.

2. $y = 4^x$

3. $y = \left(\dfrac{4}{5}\right)^x$

4. $y = 5^{-x}$

5. $y = 2e^x$

Applications

6. *Compound interest* If $1000 is invested for x years at 8%, compounded quarterly, the future value that will result is
$$S = 1000(1.02)^{4x}$$
What amount will result in 8 years?

7. *Compound interest* We will show in the next chapter that if P is invested for n years at 10% compounded continuously, the future value of the investment is given by
$$S = Pe^{0.1n}$$
Use $P = 1000$ and graph this function for $0 \le n \le 20$.

8. *Product reliability* A statistical study shows that the fraction of television sets of a certain brand that are still in service after x years is given by $f(x) = e^{-0.15x}$. Graph this equation for $0 \le x \le 10$. Write a sentence that interprets the graph.

> **Need more practice?**
> Find more here: cengagebrain.com

5.2 Logarithmic Functions and Their Properties

BEFORE THE DEVELOPMENT AND EASY AVAILABILITY OF CALCULATORS AND COMPUTERS, CERTAIN ARITHMETIC COMPUTATIONS, SUCH AS $(1.37)^{13}$ AND $\sqrt[16]{3.09}$, WERE DIFFICULT TO PERFORM. THE COMPUTATIONS COULD BE PERFORMED RELATIVELY EASILY USING **LOGARITHMS**. THE USE OF LOGARITHMS AS A COMPUTING TECHNIQUE HAS ALL BUT DISAPPEARED TODAY, BUT THE STUDY OF **LOGARITHMIC FUNCTIONS** IS STILL VERY IMPORTANT BECAUSE OF THE MANY APPLICATIONS OF THESE FUNCTIONS.

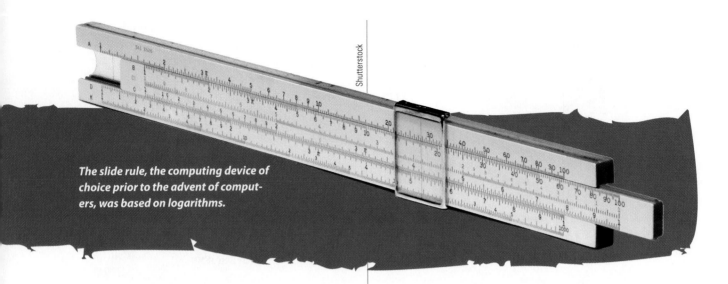

The slide rule, the computing device of choice prior to the advent of computers, was based on logarithms.

Shutterstock

Logarithmic Functions and Graphs

Let's again consider the culture of bacteria described at the beginning of the previous section. If we know that the culture is begun with one microorganism and that each minute every microorganism present splits into two new ones, then we can find the number of minutes it takes until there are 1024 organisms by solving

$$1024 = 2^y$$

The solution of this equation may be written in the form

$$y = \log_2 1024$$

which is read "y equals the logarithm of 1024 to the base 2."

In general, we may express the equation $x = a^y \ (a > 0, a \neq 1)$ in the form $y = f(x)$ by defining a logarithmic function.

Logarithmic Function: For $a > 0$ and $a \neq 1$, the logarithmic function

$$y = \log_a x \quad \text{(logarithmic form)}$$

has domain $x > 0$, base a, and is defined by

$$a^y = x \quad \text{(exponential form)}$$

From the definition, we know that $y = \log_a x$ means $x = a^y$. This means that $\log_3 81 = 4$ because $3^4 = 81$. In this case the logarithm, 4, is the exponent to which we have to raise the base 3 to obtain 81. In general, if $y = \log_a x$, then y is the exponent to which the base a must be raised to obtain x.

The a is called the **base** in both $\log_a x = y$ and $a^y = x$, and y is the *logarithm* in $\log_a x = y$ and the *exponent* in $a^y = x$. Thus **a logarithm is an exponent.**

Table 5.2 shows some logarithmic equations and their equivalent exponential forms.

Table 5.2

Logarithmic Form	Exponential Form
$\log_{10} 100 = 2$	$10^2 = 100$
$\log_{10} 0.1 = -1$	$10^{-1} = 0.1$
$\log_2 x = y$	$2^y = x$
$\log_a 1 = 0 \ (a > 0)$	$a^0 = 1$
$\log_a a = 1 \ (a > 0)$	$a^1 = a$

EXAMPLE 5.5 Logarithms and Exponential Forms

➡ (a) Write $64 = 4^3$ in logarithmic form.
(b) Write $\log_4 \left(\frac{1}{64}\right) = -3$ in exponential form.
(c) If $4 = \log_2 x$, find x.
(d) Evaluate $\log_2 8$.

SOLUTION

(a) In $64 = 4^3$, the base is 4 and the exponent (or logarithm) is 3. Thus $64 = 4^3$ is equivalent to $3 = \log_4 64$.
(b) In $\log_4 \left(\frac{1}{64}\right) = -3$, the base is 4 and the logarithm (or exponent) is -3. Thus $\log_4 \left(\frac{1}{64}\right) = -3$ is equivalent to $4^{-3} = \frac{1}{64}$.
(c) If $4 = \log_2 x$, then $2^4 = x$ and $x = 16$.
(d) If $y = \log_2 8$, then $8 = 2^y$. Because $2^3 = 8$, $\log_2 8 = 3$.

EXAMPLE 5.6 Graphing a Logarithmic Function

Graph $y = \log_2 x$.

SOLUTION

We may graph $y = \log_2 x$ by graphing $x = 2^y$. The table of values (found by substituting values of y and calculating x) and the graph are shown in Figure 5.6.

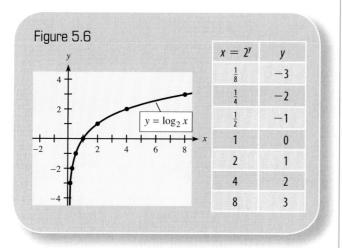

Figure 5.6

$x = 2^y$	y
$\frac{1}{8}$	-3
$\frac{1}{4}$	-2
$\frac{1}{2}$	-1
1	0
2	1
4	2
8	3

From the definition of logarithms, we see that every logarithm has a base. Most applications of logarithms involve logarithms to the base 10 (called **common logarithms**) or logarithms to the base e (called **natural logarithms**). In fact, logarithms to the base 10 and to the base e are the only ones that have function keys on calculators. Thus it is important to be familiar with their names and designations.

Common and Natural Logarithms:

Common logarithms:

$\log x$ means $\log_{10} x.$

Natural logarithms:

$\ln x$ means $\log_e x.$

Values of common and natural logarithmic functions are usually found with a calculator. For example, a calculator gives $\log 2 \approx 0.301$ and $\ln 2 \approx 0.693$.

EXAMPLE 5.7 Doubling Time for an Investment

If $\$P$ is invested for t years at interest rate r, compounded continuously, then the future value of the investment is given by $S = Pe^{rt}$. The doubling time for this investment can be found by solving for t in

$S = Pe^{rt}$ when $S = 2P$. That is, we must solve $2P = Pe^{rt}$, or (equivalently) $2 = e^{rt}$.

(a) Express $2 = e^{rt}$ in logarithmic form and then solve for t to find the doubling-time formula.

(b) If an investment earns 10% compounded continuously, in how many years will it double?

SOLUTION

(a) In logarithmic form, $2 = e^{rt}$ is equivalent to $\log_e 2 = rt$. Solving for t gives the doubling-time formula

$$t = \frac{\log_e 2}{r} = \frac{\ln 2}{r}$$

(b) If the interest rate is $r = 10\%$, compounded continuously, the time required for the investment to double is

$$t = \frac{\ln 2}{0.10} \approx 6.93 \text{ years}$$

Note that we could write the doubling time for this problem as

$$t = \frac{\ln 2}{0.10} \approx \frac{0.693}{0.10} = \frac{69.3}{10}$$

In general we can approximate the doubling time for an investment at $r\%$, compounded continuously, with $\frac{70}{r}$. (In economics, this is called the Rule of 70.)

EXAMPLE 5.8 Natural Logarithm

Graph $y = \ln x$.

SOLUTION

We can graph $y = \ln x$ by evaluating $y = \ln x$ for $x > 0$ (including some values $0 < x < 1$) with a calculator. The graph is shown in Figure 5.7.

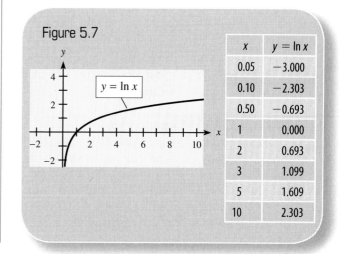

Figure 5.7

x	$y = \ln x$
0.05	-3.000
0.10	-2.303
0.50	-0.693
1	0.000
2	0.693
3	1.099
5	1.609
10	2.303

See the Tech Card for details on graphing base 10 and base e logarithms with a graphing calculator and with Excel.

The shapes of graphs of equations of the form $y = \log_a x$ with $a > 1$ are similar to these two graphs of $y = \log_2 x$ and $y = \ln x$.

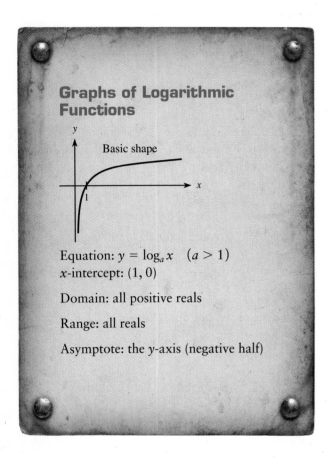

Graphs of Logarithmic Functions

Basic shape

Equation: $y = \log_a x$ $(a > 1)$
x-intercept: $(1, 0)$

Domain: all positive reals

Range: all reals

Asymptote: the y-axis (negative half)

Modeling with Logarithmic Functions

The basic shape of a logarithmic function is important for two reasons. First, when we graph a logarithmic function, we know that the graph should have this shape. Second, when data points have this basic shape, they suggest a logarithmic model. Let's consider the life expectancy example from Chapter 2 again, this time using a different modeling technique.

EXAMPLE 5.9 Life Expectancy

The expected life span of people in the United States depends on their year of birth (see the table and scatter plot, with $x = 0$ in 1900, in Figure 5.8). In Section 2.5, we modeled life span with a linear function and with a quadratic function. However, the scatter plot suggests that the best model may be logarithmic.

(a) Use technology to find a logarithmic equation that models the data.
(b) The National Center for Health Statistics projects the expected life span for people born in 2000 to be 77.0 and that for those born in 2010 to be 78.5. Use your model to project the life spans for those years.

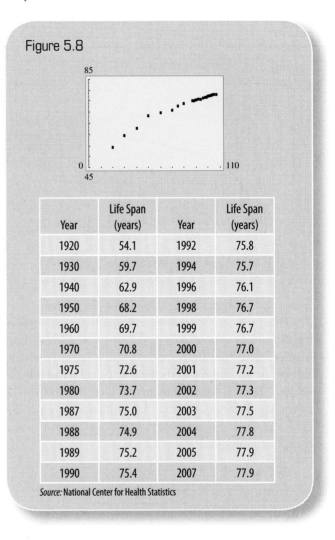

Figure 5.8

Year	Life Span (years)	Year	Life Span (years)
1920	54.1	1992	75.8
1930	59.7	1994	75.7
1940	62.9	1996	76.1
1950	68.2	1998	76.7
1960	69.7	1999	76.7
1970	70.8	2000	77.0
1975	72.6	2001	77.2
1980	73.7	2002	77.3
1987	75.0	2003	77.5
1988	74.9	2004	77.8
1989	75.2	2005	77.9
1990	75.4	2007	77.9

Source: National Center for Health Statistics

SOLUTION

(a) A graphing calculator gives the logarithmic model

$$L(x) = 10.963 + 14.321 \ln x$$

where x is the number of years since 1900. Figure 5.9 shows the scatter plot and graph of the model. See the Tech Card for details on modeling with logarithmic functions.

(b) For the year 2000, the value of x is 100, so the projected life span is $L(100) = 76.9$. For 2010, the model gives 78.3 as the expected life span. These calculations closely approximate the projections from the National Center for Health Statistics.

Figure 5.9

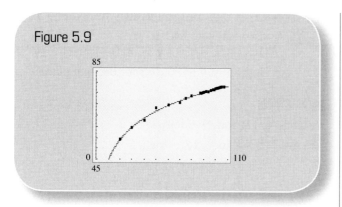

Figure 5.10

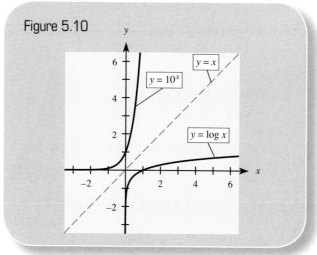

We can also develop logarithmic models with Excel. See the Tech Card for detailed instructions.

Properties of Logarithms

The definition of the logarithmic function and the previous examples suggest a special relationship between the logarithmic function $y = \log_a x$ and the exponential function $y = a^x$ $(a > 0, a \neq 1)$. Because we can write $y = \log_a x$ in exponential form as $x = a^y$, we see that the connection between

$$y = \log_a x \quad \text{and} \quad y = a^x$$

is that x and y have been interchanged from one function to the other.

In general we say that $y = f(x)$ and $y = g(x)$ are **inverse functions** if, whenever the pair (a, b) satisfies $y = f(x)$, the pair (b, a) satisfies $y = g(x)$. Furthermore, because the values of the x- and y-coordinates are interchanged for inverse functions, their graphs are reflections of each other about the line $y = x$.

Thus for $a > 0$ and $a \neq 1$, the logarithmic function $y = \log_a x$ (also written $x = a^y$) and the exponential function $y = a^x$ are inverse functions.

The logarithmic function $y = \log x$ is the inverse of the exponential function $y = 10^x$. Thus the graphs of $y = \log x$ and $y = 10^x$ are reflections of each other about the line $y = x$. Graphs of some values of x and y for these functions are shown in Figure 5.10.

This inverse relationship between exponential and logarithmic functions is a consequence of the definition of the logarithmic function.

The following properties of logarithms are useful in simplifying expressions that contain logarithms.

> **Logarithmic Property I:** If $a > 0$, $a \neq 1$, then $\log_a a^x = x$, for any real number x.

EXAMPLE 5.10 Logarithmic Property I

Use Property I to simplify each of the following.
(a) $\log_4 4^3$ (b) $\ln e^x$

SOLUTION

(a) $\log_4 4^3 = 3$
(b) $\ln e^x = \log_e e^x = x$

We note that two special cases of Property I are used frequently; these are when $x = 1$ and $x = 0$.

> **Special Cases of Logarithmic Property I:** Because $a^1 = a$, we have $\log_a a = 1$. Because $a^0 = 1$, we have $\log_a 1 = 0$.

The logarithmic form of $y = a^{\log_a x}$ is $\log_a y = \log_a x$, so $y = x$. This means that $a^{\log_a x} = x$ and proves Property II.

Logarithmic Property II:

If $a > 0, a \neq 1$ then $a^{\log_a x} = x$, for any positive real number x.

EXAMPLE 5.11 Logarithmic Property II

Use Property II to simplify each of the following.
(a) $2^{\log_2 4}$ (b) $e^{\ln x}$

SOLUTION

(a) $2^{\log_2 4} = 4$
(b) $e^{\ln x} = x$

Logarithmic Property III:

If $a > 0, a \neq 1$, and M and N are positive real numbers, then

$$\log_a (MN) = \log_a M + \log_a N$$

EXAMPLE 5.12 Logarithmic Property III

(a) Find $\log_2 (4 \cdot 16)$, if $\log_2 4 = 2$ and $\log_2 16 = 4$.
(b) Find $\ln 77$, if $\ln 7 = 1.9459$ and $\ln 11 = 2.3979$ (to four decimal places).

SOLUTION

(a) $\log_2 (4 \cdot 16) = \log_2 4 + \log_2 16 = 2 + 4 = 6$
(b) $\ln 77 = \ln (7 \cdot 11) = \ln 7 + \ln 11$
$$= 1.9459 + 2.3979 = 4.3438$$

Logarithmic Property IV: If $a > 0, a \neq 1$, and M and N are positive real numbers, then

$$\log_a (M/N) = \log_a M - \log_a N$$

EXAMPLE 5.13 Logarithmic Property IV

(a) Evaluate $\log_3 \left(\frac{9}{27}\right)$.
(b) Find $\log_{10} \left(\frac{16}{5}\right)$, if $\log_{10} 16 = 1.2041$ and $\log_{10} 5 = 0.6990$ (to four decimal places).

SOLUTION

(a) $\log_3 \left(\frac{9}{27}\right) = \log_3 9 - \log_3 27 = 2 - 3 = -1$
(b) $\log_{10} \left(\frac{16}{5}\right) = \log_{10} 16 - \log_{10} 5$
$$= 1.2041 - 0.6990 = 0.5051$$

Logarithmic Property V:

If $a > 0, a \neq 1$, M is a positive real number, and N is any real number, then

$$\log_a (M^N) = N \log_a M$$

EXAMPLE 5.14 Logarithmic Property V

(a) Simplify $\log_3 (9^2)$.
(b) Simplify $\ln 8^{-4}$, if $\ln 8 = 2.0794$ (to four decimal places).

SOLUTION

(a) $\log_3 (9^2) = 2 \log_3 9 = 2 \cdot 2 = 4$
(b) $\ln 8^{-4} = -4 \ln 8 = -4(2.0794) = -8.3176$

Change of Base

By using a calculator, we can directly evaluate only those logarithms with base 10 or base e. Also, logarithms with base 10 or base e are the only ones that are standard functions on a graphing calculator. Thus, if we had a way to express a logarithmic function with any base in terms of a logarithm with base 10 or base e, we would be able to evaluate the original logarithmic function and graph it with a graphing calculator.

We can use the properties of logarithms to develop the **change-of-base formula** from base b to base a.

Change-of-Base Formulas:

If $a \neq 1, b \neq 1, a > 0, b > 0$, then

$$\log_b x = \frac{\log_a x}{\log_a b}$$

For calculation purposes, we can convert logarithms to base e or base 10.

Base e: $\log_b x = \dfrac{\ln x}{\ln b}$ Base 10: $\log_b x = \dfrac{\log x}{\log b}$

EXAMPLE 5.15 Change-of-Base Formula

→ Evaluate $\log_7 15$ by using a change-of-base formula.

SOLUTION

$$\log_7 15 = \frac{\ln 15}{\ln 7} = \frac{2.70805}{1.94591} = 1.39166 \text{ (approximately)}$$

As a check, a calculator shows that $7^{1.39166} \approx 15$. ✓

See the Tech Card for details on graphing logarithms with other bases with a graphing calculator or with Excel.

Natural logarithms, $y = \ln x$ (and the inverse exponentials with base e), have many practical applications, some of which are considered in the next section. Common logarithms, $y = \log x$, were widely used for computation before computers and calculators became popular. They also have several applications to scaling variables, where the purpose is to reduce the scale of variation when a natural physical variable covers a wide range.

5.2 Exercises

1. Use the definition of a logarithmic function to rewrite the equation in exponential form.

   $$\frac{1}{2} = \log_4 2$$

In Problems 2 and 3, write each equation in exponential form, then solve for x.

2. $\log_3 x = 4$

3. $\log_7 (3x + 1) = 2$

4. Write $2^5 = 32$ in logarithmic form.

5. Write $e^{3x+5} = 0.55$ in logarithmic form and solve for x (to three decimal places).

In Problems 6 and 7, graph each function.

6. $y = \log_3 x$

7. $y = \ln x$

8. Use properties of logarithms or a definition to simplify each expression. Check each result with a change-of-base formula.
   (a) $\log_3 27$
   (b) $\log_5 \left(\dfrac{1}{5}\right)$

9. If $f(x) = e^x$, find $f(\ln 3)$.

10. Evaluate each logarithm by using properties of logarithms and the following facts.

    $$\log_a x = 3.1 \quad \log_a y = 1.8 \quad \log_a z = 2.7$$

    (a) $\log_a (xy)$
    (b) $\log_a \left(\dfrac{x}{z}\right)$
    (c) $\log_a (x^4)$
    (d) $\log_a \sqrt{y}$

11. Write the expression $\log \left(\dfrac{x}{x + 1}\right)$ as the sum or difference of two logarithmic functions.

12. Use the properties of logarithms to write the expression $\log_5 (x + 1) + \dfrac{1}{2} \log_5 x$ as a single logarithm.

13. Use a change-of-base formula to evaluate each logarithm.
    (a) $\log_2 17$
    (b) $\log_5 (0.78)$

Applications

14. *Richter scale* The San Francisco earthquake of 1906 measured 8.25 on the Richter scale, and the San Francisco earthquake of 1989 measured 7.1. How much more intense was the 1906 quake? Use the formula $R = \log (I/I_0)$.

15. *Decibel readings* The loudness of sound (in decibels) perceived by the human ear depends on intensity levels according to $L = 10 \log (I/I_0)$, where I_0 is the threshold of hearing for the average human ear. Find the loudness when I is 10,000 times I_0. This is the intensity level of the average voice.

16. *Doubling time* The formula $2 = \left(1 + \dfrac{r}{100\,n}\right)^{nt}$ is used to find the doubling time t, in years, for an investment at $r\%$ compounded n times per year. Write this exponential statement in logarithmic form. Then use a change-of-base formula to find the doubling time for an investment at 8% compounded quarterly.

17. *Life span update* In Example 5.9 we used data from 1920 to 2007 and found that the life span in the United States depended on the year of birth according to the equation

    $$L(x) = 10.963 + 14.321 \ln x$$

    where x is the number of years after 1900. Using data from 1920 to 1989, the model

    $$\ell(x) = 11.6164 + 14.1442 \ln x$$

where x is the number of years after 1900, predicts life span as a function of birth year. Use both models to predict the life spans for people born in 1999 and in 2012. Did adding data from 1990 to 2007 give predictions that were quite different from or very similar to predictions based on the model found earlier?

Need more practice?
Find more here: cengagebrain.com

5.3 Exponential Equations and Applications

Solving Exponential Equations Using Logarithmic Properties

Many application problems can be answered by solving an equation in which the variable appears in the exponent, called an **exponential equation**.

Solving Exponential Equations

Procedure
To solve an exponential equation by using properties of logarithms:

1. Isolate the exponential by rewriting the equation with a base raised to a power on one side.

2. Take the logarithm (either base e or base 10) of both sides.

3. Use a property of logarithms to remove the variable from the exponent.

4. Solve for the variable.

Unless the exponential in the equation has base e or base 10, the choice of a logarithm in Step 2 makes no difference. However, if the exponential in the equation has base e, then choosing a base e logarithm is slightly easier, and similarly for choosing a base 10 logarithm when the equation has a base 10 exponential.

EXAMPLE 5.16 Solution of an Exponential Equation

 Solve the equation $6(4^{3x-2}) = 120$.

SOLUTION
We first divide both sides by 6 to isolate the exponential.

$$4^{3x-2} = 20$$

Taking the logarithm, base 10, of both sides leads to the solution.

$$\log 4^{3x-2} = \log 20$$

$$(3x - 2) \log 4 = \log 20$$

$$3x - 2 = \frac{\log 20}{\log 4}$$

$$x = \frac{1}{3}\left(\frac{\log 20}{\log 4} + 2\right) \approx 1.387$$

An alternate method of solving the equation is to write $4^{3x-2} = 20$ in logarithmic form.

$$\log_4 20 = 3x - 2$$

$$x = \frac{\log_4 20 + 2}{3} \approx 1.387$$

The change-of-base formula can be used to verify that these solutions are the same.

Growth and Decay

Some applications that use **exponential decay models** are radioactive decay, demand curves, the blood pressure in the aorta, and sales decay. The models for applications such as those can be described by a function of the following form.

> **Decay Models:** $f(x) = C(a^{-x})$ with $a > 1$ and $C > 0$ or
> $f(x) = C(b^x)$ with $0 < b < 1$ and $C > 0$

EXAMPLE 5.17 Sales Decay

A company finds that its daily sales begin to fall after the end of an advertising campaign, and the decline is such that the number of sales is $S = 2000(2^{-0.1x})$, where x is the number of days after the campaign ends.

(a) How many sales will be made 10 days after the end of the campaign?

(b) If the company does not want sales to drop below 350 per day, when should it start a new campaign?

SOLUTION

(a) If $x = 10$, sales are given by $S = 2000(2^{-1}) = 1000$.

(b) Setting $S = 350$ and solving for x will give us the number of days after the end of the campaign when sales will reach 350.

$$350 = 2000(2^{-0.1x})$$

$$\frac{350}{2000} = 2^{-0.1x} \qquad \text{Isolate the exponential.}$$

$$0.175 = 2^{-0.1x}$$

With the base 2 exponential isolated, we take logarithms of both sides. We choose base 10 logarithms and complete the solution as follows.

$$\log 0.175 = \log (2^{-0.1x})$$

$$\log 0.175 = (-0.1x)(\log 2) \qquad \text{Property V}$$

$$\frac{\log 0.175}{\log 2} = -0.1x$$

$$-2.515 = -0.1x \quad \text{so} \quad x = 25.15$$

Thus sales will be 350 at day 25.15, or during the 26th day after the end of the campaign. If a new campaign isn't begun on or before the 26th day, sales will drop below 350. (See Figure 5.11.)

Figure 5.11

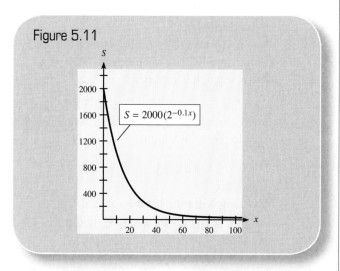

In business, economics, biology, and the social sciences, **exponential growth models** describe the growth of money, bacteria, or population.

> **Growth Models:** $f(x) = C(a^x)$ with $a > 1$ and $C > 0$

The function $f(x) = \ln (3e^x - 5)$ may be used to model population growth for humans, insects, or bacteria.

EXAMPLE 5.18 Population

The population of a certain city was 30,000 in 1990 and 40,500 in 2000. If the formula $P = P_0 e^{ht}$ applies to the growth of the city's population, find the formula for this city's population.

SOLUTION

We can first use the data from 1990 ($t = 0$) and 2000 ($t = 10$) to find the value of h in the formula. Letting $P_0 = 30,000$ and $P = 40,500$, we get

$$40,500 = 30,000e^{h(10)}$$

$$1.35 = e^{10h} \qquad \text{Isolate the exponential.}$$

Taking the natural logarithms of both sides and using logarithmic Property I give

$$\ln 1.35 = \ln e^{10h} = 10h$$

$$0.3001 = 10h$$

$$h = 0.0300 \qquad \text{(approximately)}$$

Thus the formula for this population is $P = P_0 e^{0.03t}$.

Economic and Management Applications

Sometimes cost, revenue, demand, and supply may be modeled by exponential or logarithmic equations. For example, suppose the demand for a product is given by $p = 30(3^{-q/2})$, where q is the number of thousands of units demanded at a price of p dollars per unit. Then the graph of the demand curve is as given in Figure 5.12.

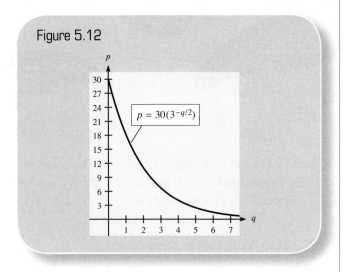

Figure 5.12

$$p = 30(3^{-q/2})$$

EXAMPLE 5.19 Total Revenue

Suppose the demand function for lipstick is given by $p = 100e^{-x/10}$, where p is the price per tube when x tubes are sold.

(a) What is the total revenue for lipstick?

(b) What would be the total revenue if 30 tubes were demanded and supplied?

SOLUTION

(a) The total revenue can be computed by multiplying the quantity sold and the price per tube. The demand function gives the price per tube when x units are sold, so the total revenue for x tubes is $R(x) = x \cdot p = x(100e^{-x/10})$. Thus the total revenue function is
$$R(x) = 100xe^{-x/10}$$

(b) If 30 tubes are sold, the total revenue is

$$R(30) = 100(30)e^{-30/10} \approx 100(30)(0.0498)$$
$$= 149.40 \text{ (dollars)}$$

Two families of curves that are used to model growth that is limited are Gompertz curves and logistic functions. See the Enrichment material on CourseMate for MATH APPS at cengagebrain.com for more about these models.

5.3 Exercises

In Problems 1–5, solve each equation. Give answers correct to 3 decimal places.

1. $8^{3x} = 32{,}768$

2. $10{,}000 = 1500e^{0.10x}$

3. $78 = 100 - 100e^{-0.01x}$

4. $\log x = 5$

5. $\log_4(9x + 1) = 3$

Applications

6. **Sales decay** The sales decay for a product is given by $S = 50{,}000e^{-0.8x}$, where S is the monthly sales and x is the number of months that have passed since the end of a promotional campaign.
   (a) What will be the sales 4 months after the end of the campaign?
   (b) How many months after the end of the campaign will sales drop below 1000, if no new campaign is initiated?

7. **Inflation** The purchasing power P (in dollars) of an annual amount of A dollars after t years of 5% inflation decays according to
$$P = Ae^{-0.05t}$$
   (Source: Viewpoints, VALIC)
   (a) How long will it be before a pension of $60,000 per year has a purchasing power of $30,000?
   (b) How much pension A would be needed so that the purchasing power P is $50,000 after 15 years?

8. **Population growth** If the population of a certain county was 100,000 in 1998 and 110,517 in 2008, and if the formula $y = P_0e^{bt}$ applies to the growth

of the county's population, estimate the population of the county in 2023.

9. **Health care** For selected years from 2001 to 2015, the national health care expenditures H, in billions of dollars, can be modeled by

$$H = 1403e^{0.0712t}$$

where t is the number of years past 2000. (*Source:* U.S. Department of Health and Human Services) According to this model, when will national health care expenditures reach \$4.0 trillion (that is, \$4000 billion)?

10. **Demand** The demand function for a certain commodity is given by $p = 100e^{-q/2}$.
    (a) At what price per unit will the quantity demanded equal 6 units?
    (b) If the price is \$1.83 per unit, how many units will be demanded, to the nearest unit?

11. **Total revenue** If the demand function for a product is given by $p = 200e^{-0.02x}$, where p is the price per unit when x units are demanded, what is the total revenue when 100 units are demanded and supplied?

12. **Compound interest** If \$5000 is invested at 9% per year compounded monthly, the future value S at any time t (in months) is given by $S = 5000(1.0075)^t$.
    (a) What is the amount after 1 year?
    (b) How long before the investment doubles?

....

Need more practice?
Find more here: cengagebrain.com

Chapter Exercises

1. Write each statement in logarithmic form.
   (a) $2^x = y$ (b) $3^y = 2x$

2. Write each statement in exponential form.
   (a) $\log_7 \left(\dfrac{1}{49} \right) = -2$
   (b) $\log_4 x = -1$

In Problems 3–8, graph each function.

3. $y = e^x$

4. $y = e^{-x}$

5. $y = \log_2 x$

6. $y = \frac{1}{2} (4^x)$

7. $y = \ln x$

8. $y = \log_4 x$

In Problems 9–16, evaluate each logarithm without using a calculator. In Problems 9–13, check with the change-of-base formula.

9. $\log_5 1$

10. $\log_2 8$

11. $\log_{25} 5$

12. $\log_3 \left(\dfrac{1}{3} \right)$

13. $\log_3 3^8$

14. $\ln e$

15. $e^{\ln 5}$

16. $10^{\log 3.15}$

In Problems 17–20, if $\log_a x = 1.2$ and $\log_a y = 3.9$, find each of the following by using the properties of logarithms.

17. $\log_a \left(\dfrac{x}{y} \right)$

18. $\log_a \sqrt{x}$

19. $\log_a (xy)$

20. $\log_a (y^4)$

In Problems 21 and 22, use the properties of logarithms to write each expression as the sum or difference of two logarithmic functions containing no exponents.

21. $\log (yz)$ 22. $\ln \sqrt{\dfrac{x+1}{x}}$

23. Is it true that $\ln x + \ln y = \ln (x + y)$ for all values of x?

24. If $f(x) = 2^x + \log (7x - 4)$, find $f(2)$.

25. If $f(x) = \ln (3e^x - 5)$, find $f(\ln 2)$.

In Problems 26 and 27, use a change-of-base formula to evaluate each logarithm.

26. $\log_9 2158$

27. $\log_{12} (0.0195)$

In Problems 28 and 29, rewrite each logarithm by using a change-of-base formula, then graph the function.

28. $y = \log_{\sqrt{3}} x$

29. $f(x) = \log_{11}(2x - 5)$

In Problems 30–33, solve each equation.

30. $6^{4x} = 46{,}656$

31. $11{,}000 = 45{,}000e^{-0.05x}$

32. $312 = 300 + 300e^{-0.08x}$

33. $\log_3(4x - 5) = 3$

Applications

34. *Medicare spending* The graph in the figure shows the projected federal spending for Medicare as a percent of the gross domestic product. If these expenditures were modeled as a function of time, which of a growth exponential, a decay exponential, or a logarithm would be the best model? Justify your answer.

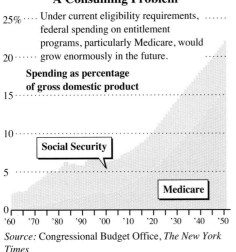

A Consuming Problem

25% ···· Under current eligibility requirements, ······ federal spending on entitlement programs, particularly Medicare, would grow enormously in the future.

Source: Congressional Budget Office, *The New York Times*

35. *Inflation* The purchasing power P of a $60,000 pension after t years of 3% annual inflation is modeled by

$$P(t) = 60{,}000(0.97)^t$$

   (a) What is the purchasing power after 20 years?
   (b) Graph this function for $t = 0$ to $t = 25$ with a graphing utility.

36. **Modeling** *Consumer credit* The data in the table show the total consumer credit (in billions of dollars) that was outstanding in the United States for various years from 1980 to 2005.
   (a) Find an exponential function that models these data. Use $x = 0$ in 1975.
   (b) What does the model predict for the total consumer credit outstanding in 2010?
   (c) In what year did the total consumer credit reach $3089.5 billion, according to the model?

Year	Amount (in billions)
1980	$349.4
1985	$593.2
1990	$789.1
1995	$1095.8
2000	$1560.2
2001	$1666.9
2002	$1725.7
2003	$2119.9
2004	$2232.3
2005	$2323.4

Source: Federal Reserve

37. *Poverty threshold* The average poverty threshold for 1987–2006 for a single individual can be modeled by

$$y = -961.20 + 3293.05 \ln(x)$$

where x is the number of years past 1980 and y is the annual income in dollars. (*Source:* U.S. Bureau of the Census)
   (a) What does the model predict as the poverty threshold in 2015?
   (b) Graph this function for $x = 5$ to $x = 40$.

38. **Modeling** *Percent of paved roads* The table shows various years from 1950 to 2005 and, for those years, the percents of U.S. roads and streets that were paved.
   (a) Find a logarithmic equation that models these data. Use $x = 0$ in 1940.
   (b) What percent of paved roads does the model predict for 2010?

(c) When does the model predict that 70% of U.S. roads and streets will be paved?

Year	Percent	Year	Percent	Year	Percent
1950	23.5	1995	60.8	2003	65.5
1960	34.7	1999	62.4	2004	64.5
1970	44.5	2000	63.4	2005	64.9
1980	53.7	2001	63.7		
1990	58.4	2002	64.8		

Source: U.S. Department of Transportation

39. **Sales decay** The sales decay for a product is given by $S = 50{,}000e^{-0.1x}$, where S is the weekly sales (in dollars) and x is the number of weeks that have passed since the end of an advertising campaign.
   (a) What will sales be 6 weeks after the end of the campaign?
   (b) How many weeks will pass before sales drop below 15,000?

40. **Compound interest** If $1000 is invested at 12%, compounded monthly, the future value S at any time t (in years) is given by

$$S = 1000(1.01)^{12t}$$

How long will it take for the amount to double?

Need more practice?
Find more here: cengagebrain.com

chapter 6

Mathematics
of Finance

Regardless of whether your career is in business, understanding how interest is computed on investments and loans is important to you as a consumer. The proliferation of personal finance and money management software attests to this importance. The goal of this chapter is to provide some understanding of the methods used to determine the interest and future value (principal plus interest) resulting from savings plans and the methods used in repayment of debts.

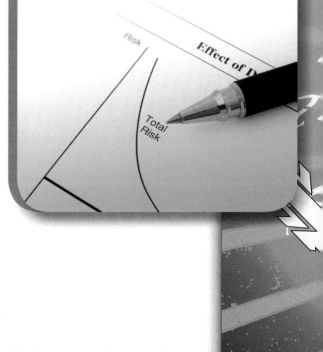

6.1 Simple Interest and Arithmetic Sequences

IN THIS SECTION WE BEGIN OUR STUDY OF THE MATHEMATICS OF FINANCE BY CONSIDERING SIMPLE INTEREST. SIMPLE INTEREST FORMS THE BASIS FOR ALL CALCULATIONS INVOLVING INTEREST THAT IS PAID ON AN INVESTMENT OR THAT IS DUE ON A LOAN.

Simple Interest: If a sum of money P (called the **principal**) is invested for a time period t (frequently in years) at an interest rate r per period, the **simple interest** I is given by

$$I = Prt$$

where I = interest (in dollars)

P = principal (in dollars)

r = annual interest rate (written as a decimal)

t = time (in years)*

Note that the time measurements for r and t must agree.

*Periods of time other than years can be used, as can other monetary systems.

Simple interest is paid on investments involving time certificates issued by banks and on certain types of bonds, such as U.S. government series H bonds and municipal bonds. The interest for a given period is paid to the investor, and the principal remains the same.

If you borrow money from a friend or a relative, interest on your loan might be calculated with the simple interest formula. We'll consider some simple interest loans in this section, but interest on loans from banks and other lending institutions is calculated using methods discussed later.

$I = Prt$

Shutterstock

EXAMPLE 6.1 Simple Interest

a) If $8000 is invested for 2 years at an annual interest rate of 9%, how much interest will be received at the end of the 2-year period?

(b) If $4000 is borrowed for 39 weeks at an annual interest rate of 15%, how much interest is due at the end of the 39 weeks?

Shutterstock

math alpas

SOLUTION

(a) The interest is $I = Prt = \$8000(0.09)(2) = \$1440$.

(b) Use $I = Prt$ with $t = 39/52 = 0.75$ year. Thus

$$I = \$4000(0.15)(0.75) = \$450$$

Future Value

The **future amount of an investment**, or its **future value**, at the end of an interest period is the sum of the principal and the interest. Thus, in Example 6.1(a), the future value is

$$S = \$8000 + \$1440 = \$9440$$

Similarly, the **future amount of a loan**, or its future value, is the amount of money that must be repaid. In Example 6.1(b), the future value of the loan is the principal plus the interest, or

$$S = \$4000 + \$450 = \$4450$$

> **Future Value:** If we use the letter S to denote the future value of an investment or a loan, then we have
>
> Future value of investment or loan: $S = P + I$
>
> where P is the principal (in dollars) and I is the interest (in dollars).
>
> The principal P of a loan is also called the **face value** or the **present value** of the loan.

EXAMPLE 6.2 Future Value and Present Value

(a) If \$2000 is borrowed for one-half year at a simple interest rate of 12% per year, what is the future value of the loan at the end of the half-year?

(b) An investor wants to have \$20,000 in 9 months. If the best available interest rate is 6.05% per year, how much must be invested now to yield the desired amount?

SOLUTION

(a) The interest for the half-year period is $I = \$2000(0.12)(0.5) = \$120$. Thus the future value of the investment for the period is

$$S = P + I = \$2000 + \$120 = \$2120$$

(b) We know that $S = P + I = P + Prt$. In this case, we must solve for P, the present value. Also, the time 9 months is $(9/12)$ of a year.

$$\$20,000 = P + P(0.0605)(9/12) = P + 0.045375P$$

$$\$20,000 = 1.045375P \text{ so } P = \$19,131.89$$

It is also possible to solve simple interest problems that ask for the rate. For example, suppose that Mary Spaulding bought Wind-Gen Electric stock for \$6125.00 and that after 6 months, the value of her shares had risen by \$138.00 and dividends totaling \$144.14 had been paid. Then to find the simple interest rate she earned on this investment if she sold the stock at the end of the 6 months, we find the rate that would yield an amount of simple interest equal to all of Mary's gains (that is, equal to the rise in the stock's price plus the dividends she received). Thus the principal is \$6125.00, the time is 1/2 year, and the interest earned is the total of all gains (that is, interest $I = \$138.00 + \$144.14 = \$282.14$). Using these values in $I = Prt$ gives \$282.14 = $(\$6125)r(0.5)$. Solving for r gives $r \approx 0.092 = 9.2\%$. Thus Mary's return was equivalent to an annual simple interest rate of about 9.2%.

Similarly, we could find how long it would take if \$1000 is invested at 5.8% simple interest to grow to \$1100 by using $P = \$1000$, $S = \$1100$, and $r = 0.058$ in $S = P + Prt$ and solving for t.

Sequences

If a \$2000 investment earns 1% simple interest per month, then the monthly future values for each of the first 5 months would be \$2020, \$2040, \$2060, \$2080, and \$2100. These future values are outputs that result from positive integer inputs corresponding to the number of months of the investment. These outputs form a **sequence**, which is a special type of function defined by outputs that arise uniquely from positive integer inputs.

> **Sequence:** A function whose domain is the set of positive integers is called a **sequence function**. The set of function outputs of a sequence function
>
> $$f(1) = a_1, f(2) = a_2, \ldots, f(n) = a_n, \ldots$$
>
> forms an ordered list called a **sequence**. The outputs $a_1, a_2, a_3, \ldots$ are called **terms** of the sequence, with a_1 the first term, a_2 the second term, and so on.

Because calculations involving interest often result from using positive integer inputs, sequences are the basis for most of the financial formulas derived in this chapter.

Arithmetic Sequences

The sequence

$$2020, 2040, 2060, 2080, 2100, \ldots$$

can also be described in the following way:

$$a_1 = 2020, \quad a_n = a_{n-1} + 20 \quad \text{for } n > 1$$

This sequence is an example of a special kind of sequence called an **arithmetic sequence**. In such a sequence, each term after the first can be found by adding a constant to the preceding term.

Arithmetic Sequence: A sequence is called an **arithmetic sequence** (progression) if there exists a number d, called the **common difference,** such that

$$a_n = a_{n-1} + d \quad \text{for } n > 1$$

EXAMPLE 6.3 Arithmetic Sequences

Write the next three terms of each of the following arithmetic sequences.

(a) $1, 3, 5, \ldots$
(b) $\frac{1}{2}, \frac{5}{6}, \frac{7}{6}, \ldots$

SOLUTION

(a) The common difference is 2, so the next three terms are 7, 9, 11.
(b) The common difference is $\frac{1}{3}$, so the next three terms are $\frac{3}{2}, \frac{11}{6}, \frac{13}{6}$.

Because each term after the first term in an arithmetic sequence is obtained by adding d to the preceding term, the second term is $a_1 + d$, the third is $(a_1 + d) + d = a_1 + 2d, \ldots$

nth Term of an Arithmetic Sequence: The **nth term of an arithmetic sequence** (progression) is given by

$$a_n = a_1 + (n - 1)d$$

where a_1 is the first term and d is the common difference between successive terms.

EXAMPLE 6.4 nth Term of an Arithmetic Sequence

a) Find the 11th term of the arithmetic sequence with first term 3 and common difference -2.
(b) If the first term of an arithmetic sequence is 4 and the 9th term is 20, find the 75th term.

SOLUTION

(a) The 11th term is $a_{11} = 3 + (11 - 1)(-2) = -17$.
(b) Substituting the values $a_1 = 4$, $a_n = 20$, and $n = 9$ in $a_n = a_1 + (n - 1)d$ gives $20 = 4 + (9 - 1)d$. Solving this equation gives $d = 2$. Therefore, the 75th term is $a_{75} = 4 + (75 - 1)(2) = 152$.

Sum of an Arithmetic Sequence: The **sum of the first n terms of an arithmetic sequence** is given by the formula

$$s_n = \frac{n}{2}(a_1 + a_n)$$

where a_1 is the first term of the sequence and a_n is the nth term.

EXAMPLE 6.5 Sum of an Arithmetic Sequence

Find the sum of
(a) the first 10 terms of the arithmetic sequence with first term 2 and common difference 4.
(b) the first 91 terms of the arithmetic sequence $\frac{1}{4}, \frac{7}{12}, \frac{11}{12}, \ldots$.

SOLUTION

(a) We are given the values $n = 10$, $a_1 = 2$, and $d = 4$. Thus the 10th term is $a_{10} = 2 + (10 - 1)4 = 38$, and the sum of the first 10 terms is

$$s_{10} = \frac{10}{2}(2 + 38) = 200$$

(b) The first term is $\frac{1}{4}$ and the common difference is $\frac{1}{3}$. Therefore, the 91st term is $a_{91} = \frac{1}{4} + (91 - 1)(\frac{1}{3}) = 30\frac{1}{4} = \frac{121}{4}$. The sum of the first 91 terms is

$$s_{91} = \frac{91}{2}\left(\frac{1}{4} + \frac{121}{4}\right) = \frac{91}{2}\left(\frac{122}{4}\right) = \frac{91(61)}{4} = \frac{5551}{4}$$
$$= 1387\frac{3}{4}$$

6.1 Exercises

Simple Interest

1. $1000 is invested for 3 months at an annual simple interest rate of 12%.
   (a) How much interest will be earned?
   (b) What is the future value of the investment after 3 months?

2. If you lend $3500 to a friend for 15 months at 8% annual simple interest, find the future value of the loan.

3. (a) To buy a Treasury bill (T-bill) that matures to $10,000 in 6 months, you must pay $9750. What rate does this earn?
   (b) If the bank charges a fee of $40 to buy a T-bill, what is the actual interest rate you earn?

4. If you want to earn 15% annual simple interest on an investment, how much should you pay for a note that will be worth $13,500 in 10 months?

5. If $5000 is invested at 8% annual simple interest, how long does it take to be worth $9000?

6. Bill Casler bought a $2000, 9-month certificate of deposit (CD) that would earn 8% annual simple interest. Three months before the CD was due to mature, Bill needed his CD money, so a friend agreed to lend him money and receive the value of the CD when it matured.
   (a) What is the value of the CD when it matures?
   (b) If their agreement allowed the friend to earn a 10% annual simple interest return on his loan to Bill, how much did Bill receive from his friend?

Sequences

7. Write the first six terms of the sequence whose nth term is $(-1)^n / (2n + 1)$.

8. For the arithmetic sequence 2, 5, 8, . . . (a) identify d and a_1 and (b) write the next three terms.

9. Find the 83rd term of the arithmetic sequence with first term 6 and common difference $-\frac{1}{2}$.

10. Find the 100th term of the arithmetic sequence with first term 5 and eighth term 19.

11. Find the sum of the first 150 terms of the arithmetic sequence 6, $\frac{9}{2}$, 3, ...

Application

12. *Salaries* Suppose you are offered two identical jobs: one paying a starting salary of $40,000 with yearly raises of $2000 and one paying a starting salary of $36,000 with yearly raises of $2400. Which job will pay you more for your tenth year on the job?

Need more practice?
Find more here: cengagebrain.com

6.2 Compound Interest and Geometric Sequences

IN THE PREVIOUS SECTION WE DISCUSSED SIMPLE INTEREST. A SECOND METHOD OF PAYING INTEREST IS THE **COMPOUND INTEREST METHOD,** WHERE THE INTEREST FOR EACH PERIOD IS ADDED TO THE PRINCIPAL BEFORE INTEREST IS CALCULATED FOR THE NEXT PERIOD. UNLIKE THE ARITHMETIC SEQUENCE ASSOCIATED WITH SIMPLE INTEREST, COMPOUND INTEREST IS DEFINED BY A GEOMETRIC SEQUENCE.

Compound Interest

With compound interest, both the interest added *and* the principal earn interest for the next period. With this method, the principal grows as the interest is added to it. Think of it like building a snowman. You start with a snowball the size of your fist (your investment), and as you roll the snowball around the yard, the snowball picks up additional snow (interest). The more you roll it, the snowball itself (principal) keeps growing and as it does, the amount of snow it picks up along the way (interest) also grows. Compound interest is used in investments such as savings accounts and U.S. government series E bonds.

There are several types of compound interest—annual, periodic, and continuous—and not only the interest rate but the type of compound interest affects the annual percentage yield of an investment.

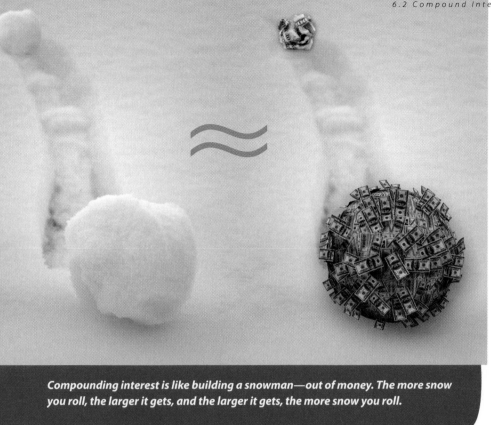

Compounding interest is like building a snowman—out of money. The more snow you roll, the larger it gets, and the larger it gets, the more snow you roll.

© iStockphoto.com/Hakan German, Shutterstock

Because \$3000 of this amount was the original investment, the interest earned is

$$\$4234.74 - \$3000 = \$1234.74.$$

Periodic Compounding

Some accounts have the interest compounded semiannually, quarterly, monthly, or daily. Unless specifically stated otherwise, a stated interest rate, called the **nominal annual rate,** is the rate per year and is denoted by r. The interest rate *per period*, denoted by i, is the nominal rate divided by the number of interest periods per year. The interest periods are also called *conversion periods*, and the number of periods is denoted by n. Thus, if \$100 is invested for 5 years at 6% compounded semiannually (twice a year), it has been invested for $n = 10$ periods (5 years × 2 periods per year) at $i = 3\%$ per period (6% per year ÷ 2 periods per year). The future value of an investment of this type is found using the following formula.

Annual Compounding

The simplest type of compound interest is when the compounding is annual.

> **Future Value (Annual Compounding):** If \$$P$ is invested at an interest rate of r per year, compounded annually, the future value S at the end of the nth year is
>
> $$S = P(1 + r)^n$$

EXAMPLE 6.6 Annual Compounding

➡ If \$3000 is invested for 4 years at 9% compounded annually, how much interest is earned?

SOLUTION

The future value is

$$S = \$3000(1 + 0.09)^4$$
$$= \$3000(1.4115816)$$
$$= \$4234.7448$$
$$= \$4234.74 \text{ (to the nearest cent)}$$

> **Future Value (Periodic Compounding):** If \$$P$ is invested for t years at a nominal interest rate r, compounded m times per year, then the total number of compounding periods is
>
> $$n = mt$$
>
> the interest rate per compounding period is
>
> $$i = \frac{r}{m} \text{ (expressed as a decimal)}$$
>
> and the future value is
>
> $$S = P(1 + i)^n = P\left(1 + \frac{r}{m}\right)^{mt}$$

EXAMPLE 6.7 Periodic Compounding

➡ For the following investment, find the interest rate per period, i, and the number of compounding periods, n:

12% compounded monthly for 7 years

SOLUTION

If the compounding is monthly and $r = 12\% = 0.12$, then $i = 0.12/12 = 0.01$. The number of compounding periods is $n = (7 \text{ yr})(12 \text{ periods/yr}) = 84$.

Once we know i and n, we can calculate the future value from the formula with a calculator.

EXAMPLE 6.8 Future Value

Jim and Eden want to have $200,000 in Maura's college fund on her 18th birthday, and they want to know the impact on this goal of having $10,000 invested at 9.8%, compounded quarterly, on her first birthday. To advise Jim and Eden regarding this, find
(a) the future value of the $10,000 investment,
(b) the amount of compound interest that the investment earns, and
(c) the impact this would have on their goal.

SOLUTION

(a) For this situation, $i = 0.098/4 = 0.0245$ and $n = 4(17) = 68$. Thus the future value of the $10,000 is given by

$$S = P(1 + i)^n = \$10{,}000(1 + 0.0245)^{68}$$
$$= \$10{,}000(5.18577) = \$51{,}857.70$$

(b) The amount of interest earned is
$51,857.70 − $10,000 = $41,857.70.
(c) Thus $10,000 invested by Maura's first birthday grows to an amount that is slightly more than 25% of their goal. This rather large early investment has a substantial impact on their goal.

See the Tech Card for details on how to use a graphing calculator or Excel to compute the future value of a compound interest investment and to solve other compound interest problems.

We saw previously that compound interest calculations are based on those for simple interest, except that interest payments are added to the principal. In this way, interest is earned on both principal and previous interest payments. Let's examine the effect of this compounding by comparing compound interest and simple interest. If the investment in Example 6.8 had been at simple interest, the interest earned would have been $Prt = \$10{,}000(0.098)(17) = \$16{,}660$. This is almost $25,200 less than the amount of compound interest earned. And this difference would have been magnified over a longer period of time. Einstein characterized compound interest as the "most powerful force in the Universe."

EXAMPLE 6.9 Present Value

What amount must be invested now in order to have $12,000 after 3 years if money is worth 6% compounded semiannually?

SOLUTION

We need to find the present value P, knowing that the future value is $S = \$12{,}000$. Use $i = 0.06/2 = 0.03$ and $n = 3(2) = 6$.

$$S = P(1 + i)^n$$
$$\$12{,}000 = P(1 + 0.03)^6 = P(1.03)^6$$
$$P = \frac{\$12{,}000}{1.1940523} = \$10{,}049.81 \text{ (to the nearest cent)}$$

EXAMPLE 6.10 Rate Earned

As Figure 6.1 shows, three years after Google stock was first sold publicly, its share price had risen 650%. The figure also shows that this growth far exceeded Microsoft's performance at the same point since its stock was first publicly traded. Google's 650% increase means that $10,000 invested in Google stock at its initial public offering (IPO) was worth $65,000 three years later. What interest rate compounded annually does this represent? (For more about Microsoft's annual compounding performance, see Problem 7 in the exercises for this section.)

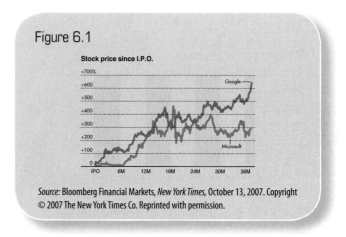

Figure 6.1

Stock price since I.P.O.

Source: Bloomberg Financial Markets, *New York Times*, October 13, 2007. Copyright © 2007 The New York Times Co. Reprinted with permission.

SOLUTION

We use $P = \$10{,}000$, $S = \$65{,}000$, and $n = 3$ in the formula $S = P(1 + i)^n$, and solve for i.

$$\$65{,}000 = \$10{,}000(1 + i)^3$$
$$6.5 = (1 + i)^3$$

At this point we take the cube root (third root) of both sides (or, equivalently, raise both sides to the 1/3 power).

$$6.5^{(1/3)} = [(1 + i)^3]^{(1/3)}$$
$$1.8663 \approx 1 + i, \text{ so } 0.8663 = i$$

Thus, this investment earned about 86.63% compounded annually.

Continuous Compounding

Continuous compounding (compounding every instant) uses a formula different from the one for periodic compounding. This new formula is developed from the one for periodic compounding (see the Enrichment material on CourseMate for MATH APPS at cengagebrain.com).

> **Future Value (Continuous Compounding):** In general, if P is invested for t years at a nominal rate r compounded continuously, then the future value is given by the exponential function
> $$S = Pe^{rt}$$

EXAMPLE 6.11 Continuous Compounding

(a) Find the future value if $1000 is invested for 20 years at 8%, compounded continuously.

(b) What amount must be invested at 6.5%, compounded continuously, so that it will be worth $25,000 after 8 years?

SOLUTION

(a) The future value is

$$S = \$1000e^{(0.08)(20)} = \$1000e^{1.6}$$
$$= \$1000(4.95303) \text{ (because } e^{1.6} \approx 4.95303)$$
$$= \$4953.03$$

(b) Solve for the present value P in $25,000 = Pe^{(0.065)(8)}$.

$$\$25,000 = Pe^{(0.065)(8)} = Pe^{0.52} = P(1.68202765)$$

$$\frac{\$25,000}{1.68202765} = P$$

so $P = \$14,863.01$ (to the nearest cent)

EXAMPLE 6.12 Comparing Investments

How much more will you earn if you invest $1000 for 5 years at 8% compounded continuously instead of at 8% compounded quarterly?

SOLUTION

If the interest is compounded continuously, the future value at the end of the 5 years is

$$S = \$1000e^{(0.08)(5)} = \$1000e^{0.4} \approx \$1000(1.49182)$$
$$= \$1491.82$$

If the interest is compounded quarterly, the future value at the end of the 5 years is

$$S = \$1000(1.02)^{20} \approx \$1000(1.485947)$$
$$= \$1485.95 \text{ (to the nearest cent)}$$

Thus the extra interest earned by compounding continuously is

$$\$1491.82 - \$1485.95 = \$5.87$$

Annual Percentage Yield

As Example 6.12 shows, when we invest money at a given compound interest rate, the method of compounding affects the amount of interest we earn. As a result, a rate of 8% can earn more than 8% interest if compounding is more frequent than annually.

Banks acknowledge this difference between stated nominal interest rates and annual percentage yields by posting both rates for their investments. Note that the annual percentage yield is equivalent to the stated rate when compounding is annual. In general, the annual percentage yield equals I/P, or just I if $P = \$1$. Hence we can calculate the APY with the following formulas.

Shutterstock; Getty Images

COMPOUNDING INTEREST
=
POWERFUL FORCE

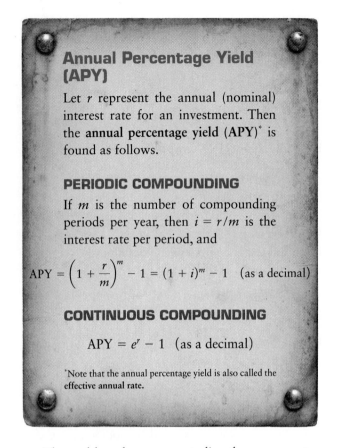

Annual Percentage Yield (APY)

Let r represent the annual (nominal) interest rate for an investment. Then the **annual percentage yield (APY)**[*] is found as follows.

PERIODIC COMPOUNDING

If m is the number of compounding periods per year, then $i = r/m$ is the interest rate per period, and

$$\text{APY} = \left(1 + \frac{r}{m}\right)^m - 1 = (1 + i)^m - 1 \quad \text{(as a decimal)}$$

CONTINUOUS COMPOUNDING

$$\text{APY} = e^r - 1 \quad \text{(as a decimal)}$$

[*]Note that the annual percentage yield is also called the **effective annual rate**.

Thus, although we cannot directly compare two nominal rates with different compounding periods, we can compare their corresponding APYs.

EXAMPLE 6.13 Comparing Yields

Suppose a young couple found three different investment companies that offered college savings plans: (a) one at 10% compounded annually, (b) another at 9.8% compounded quarterly, and (c) a third at 9.65% compounded continuously. Find the annual percentage yield (APY) for each of these three plans in order to discover which plan is best.

SOLUTION

(a) For annual compounding, the stated rate is the APY. Thus, APY = 10%.

(b) Because the number of periods per year is $m = 4$ and the nominal rate is $r = 0.098$, the rate per period is $i = r/m = 0.098/4 = 0.0245$. Thus,

$$\text{APY} = (1 + 0.0245)^4 - 1 = 1.10166 - 1$$
$$= 0.10166 = 10.166\%$$

(c) For continuous compounding and a nominal rate of 9.65%, we have

$$\text{APY} = e^{0.0965} - 1 = 1.10131 - 1$$
$$= 0.10131 = 10.131\%$$

Hence we see that of these three choices, 9.8% compounded quarterly is best. Furthermore, even 9.65% compounded continuously has a higher APY than 10% compounded annually.

EXAMPLE 6.14 Doubling Time

How long does it take an investment of $10,000 to double if it is invested at
(a) 8%, compounded annually?
(b) 8%, compounded continuously?

SOLUTION

(a) We solve for n in $\$20,000 = \$10,000(1 + 0.08)^n$.

$$2 = 1.08^n$$

Taking the logarithm, base e, of both sides of the equation gives

$$\ln 2 = \ln 1.08^n$$
$$\ln 2 = n \ln 1.08 \qquad \text{Logarithm Property V}$$
$$n = \frac{\ln 2}{\ln 1.08} \approx 9.0 \ (\text{years})$$

(b) Solve for t in $\$20,000 = \$10,000e^{0.08t}$.

$$2 = e^{0.08t} \qquad \text{Isolate the exponential.}$$
$$\ln 2 = \ln e^{0.08t} \qquad \text{Take "ln" of both sides.}$$
$$\ln 2 = 0.08t \qquad \text{Logarithm Property I}$$
$$t = \frac{\ln 2}{0.08} \approx 8.7 \ (\text{years})$$

Geometric Sequences

If $\$P$ is invested at an interest rate of i per period, compounded at the end of each period, the future value at the end of each succeeding period is

$$P(1 + i), P(1 + i)^2, P(1 + i)^3, \ldots, P(1 + i)^n, \ldots$$

The future values for each of the succeeding periods form a sequence in which each term (after the first) is found by multiplying the previous term by the same number. Such a sequence is called a **geometric sequence**.

Geometric Sequence: A sequence is called a **geometric sequence** (progression) if there exists a number r, called the **common ratio**, such that

$$a_n = ra_{n-1} \quad \text{for } n > 1$$

Geometric sequences form the foundation for other applications involving compound interest.

EXAMPLE 6.15 Geometric Sequence

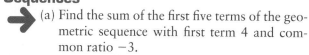

Write the next three terms of the following geometric sequences.

(a) 1, 3, 9, . . .   (b) 3, −6, 12, . . .

SOLUTION

(a) The common ratio is 3, so the next three terms are 27, 81, 243.

(b) The common ratio is −2, so the next three terms are −24, 48, −96.

Because each term after the first in a geometric sequence is obtained by multiplying the previous term by r, the second term is $a_1 r$, the third is $a_1 r^2$, and so on.

nth Term of a Geometric Sequence:
The **nth term of a geometric sequence** (progression) is given by

$$a_n = a_1 r^{n-1}$$

where a_1 is the first term of the sequence and r is the common ratio.

EXAMPLE 6.16 nth Term of a Geometric Sequence

Find the seventh term of the geometric sequence with first term 5 and common ratio −2.

SOLUTION

The seventh term is $a_7 = 5(-2)^{7-1} = 5(64) = 320$.

The sum of a geometric sequence plays an important role in our study of annuities.

Sum of a Geometric Sequence: The **sum of the first n terms of the geometric sequence** with first term a_1 and common ratio r is

$$s_n = \frac{a_1(1 - r^n)}{1 - r} \quad \text{provided that } r \neq 1$$

Development of this formula can be found in the Enrichment material on CourseMate for MATH APPS at cengagebrain.com.

EXAMPLE 6.17 Sums of Geometric Sequences

(a) Find the sum of the first five terms of the geometric sequence with first term 4 and common ratio −3.

(b) Find the sum of the first six terms of the geometric sequence $\frac{1}{4}, \frac{1}{8}, \frac{1}{16}, \dots.$

SOLUTION

(a) We are given that $n = 5$, $a_1 = 4$, and $r = -3$. Thus

$$s_5 = \frac{4[1 - (-3)^5]}{1 - (-3)} = \frac{4[1 - (-243)]}{4} = 244$$

(b) We know that $n = 6$, $a_1 = \frac{1}{4}$, and $r = \frac{1}{2}$. Thus

$$s_6 = \frac{\frac{1}{4}[1 - (\frac{1}{2})^6]}{1 - \frac{1}{2}} = \frac{\frac{1}{4}(1 - \frac{1}{64})}{\frac{1}{2}} = \frac{1 - \frac{1}{64}}{2} = \frac{64 - 1}{128} = \frac{63}{128}$$

6.2 Exercises

Compound Interest

1. For an investment at 8% compounded quarterly for 7 years, identify (a) the annual interest rate, (b) the length of the investment in years, (c) the periodic interest rate, and (d) the number of periods of the investment.

2. What is the future value if $3200 is invested for 5 years at 8% compounded quarterly?

3. What present value amounts to $10,000 if it is invested for 10 years at 6% compounded annually?

4. What is the compound interest if $410 is invested for 10 years at 8% compounded continuously?

5. Which investment will earn more money, a $1000 investment for 5 years at 8% compounded annually or a $1000 investment for 5 years compounded continuously at 7%?

6. Find the annual percentage yield for an investment at
   (a) 7.3% compounded monthly.
   (b) 6% compounded continuously.

7. Microsoft's stock price peaked at 6118% of its IPO price more than 13 years after the IPO. (*Source:* Bloomberg Financial Markets) Suppose that $10,000 invested in Microsoft at its IPO price had been worth $600,000 (6000% of the IPO price) after exactly 13 years. What interest rate, compounded annually, does this represent?

8. How long (in years) would $700 have to be invested at 11.9%, compounded continuously, to earn $300 interest?

9. A couple needs $15,000 as a down payment for a home. If they invest the $10,000 they have at 8% compounded quarterly, how long will it take for the money to grow into $15,000?

10. Mary Stahley invested $2500 in a 36-month certificate of deposit (CD) that earned 8.5% annual simple interest. When the CD matured, she invested the full amount in a mutual fund that had an annual growth equivalent to 18% compounded annually. How much was the mutual fund worth after 9 years?

Geometric Sequences

11. For each geometric sequence given, write the next three terms.
    (a) 3, 6, 12, . . .   (b) 81, 54, 36, . . .

In Problems 12–14, write an expression that gives the requested term or sum.

12. The 13th term of the geometric sequence with first term 10 and common ratio 2

13. The sum of the first 17 terms of the geometric sequence with first term 6 and common ratio 3

14. The sum of the first 18 terms of the geometric sequence $6, 4, \frac{8}{3}, \ldots$

Applications of Sequences

15. *Population growth* Suppose a country has a population of 20 million and projects a growth rate of 2% per year for the next 20 years. What will the population of this country be in 10 years?

16. *Profit* If changing market conditions cause a company earning $8,000,000 in 2005 to project a loss of 2% of its profit in each of the next 5 years, what profit does it project in 2010?

⋯⋯⋯⋯⋯⋯⋯⋯⋯⋯⋯⋯⋯⋯⋯⋯⋯⋯⋯⋯⋯⋯⋯⋯⋯⋯⋯

Need more practice?
Find more here: cengagebrain.com

6.3 Future Values of Annuities

AN **ANNUITY** IS A FINANCIAL PLAN CHARACTERIZED BY REGULAR PAYMENTS. WE CAN VIEW AN ANNUITY AS A SAVINGS PLAN IN WHICH THE REGULAR PAYMENTS ARE CONTRIBUTIONS TO THE ACCOUNT, AND THEN WE CAN ASK WHAT THE TOTAL VALUE OF THE ACCOUNT WILL BECOME. ALSO, WE CAN VIEW AN ANNUITY AS A PAYMENT PLAN (SUCH AS FOR RETIREMENT) IN WHICH REGULAR PAYMENTS ARE MADE FROM AN ACCOUNT, OFTEN TO AN INDIVIDUAL.

Ordinary Annuities

Most people save (or invest) money by depositing relatively small amounts at different times. If a depositor makes equal deposits at regular intervals, he or she is contributing to an annuity. The payments (deposits) may be made weekly, monthly, quarterly, yearly, or over any other interval of time. The sum of all payments plus all interest earned is called the **future amount of the annuity** or its **future value.**

In this text, we will deal with annuities in which the payments begin and end on fixed dates, but we will deal first with annuities in which the payments are made at the end of each of the equal payment intervals. This type of annuity is called an **ordinary annuity** (and also an **annuity immediate**). The ordinary annuities we will consider have payment intervals that coincide with the compounding period of the interest, and we use the following formula for problems involving future value (see the Enrichment material for the formula development).

Future Value of an Ordinary Annuity: If $R is deposited at the end of each period for n periods in an annuity that earns interest at a rate of i per period, the future value of the annuity will be

$$S = R \cdot s_{\overline{n}|i} = R \cdot \left[\frac{(1 + i)^n - 1}{i} \right]$$

where $s_{\overline{n}|i}$ is read "s, n angle i" and represents the future value of an ordinary annuity of $1 per period for n periods with an interest rate of i per period.

The value of $s_{\overline{n}|i}$ can be computed directly with a calculator.

EXAMPLE 6.18 Future Value

Richard Lloyd deposits \$200 at the end of each quarter in an account that pays 4%, compounded quarterly. How much money will he have in his account in $2\frac{1}{4}$ years?

SOLUTION

The number of periods is $n = (4)(2.25) = 9$, and the rate *per period* is $i = 0.04/4 = 0.01$. At the end of $2\frac{1}{4}$ years the future value of the annuity will be

$$S = \$200 \cdot s_{\overline{9}|0.01} = \$200\left[\frac{(1 + 0.01)^9 - 1}{0.01}\right]$$

$$= \$200(9.368527) = \$1873.71 \text{ (to the nearest cent)}$$

See the Tech Card for details on computing future value of an ordinary annuity (and solving other annuity problems) with a graphing calculator or with Excel.

Twins graduate from college together and start their careers. Twin 1 invests \$2000 at the end of 8 years in an account that earns 10%, compounded annually. After the initial 8 years, no additional contributions are made, but the investment continues to earn 10%, compounded annually. Twin 2 invests no money for 8 years but then contributes \$2000 at the end of each year for a period of 36 years (to age 65) to an account that pays 10%, compounded annually. How much money does each twin have at age 65? In the next two examples, we want to answer the questions posed about the twins' savings plans just described.

EXAMPLE 6.19 Future Value for Twin 2

After college, Twin 2 invests nothing for 8 years. Then, at the end of each year for the next 36 years (until age 65), Twin 2 invests \$2000 in an account that pays 10%, compounded annually. How much does Twin 2 have at age 65?

SOLUTION

This savings plan is an ordinary annuity with $i = 0.10$, $n = 36$, and $R = \$2000$. The future value (to the nearest dollar) is

$$S = R\left[\frac{(1 + i)^n - 1}{i}\right] = \$2000\left[\frac{(1.10)^{36} - 1}{0.10}\right] = \$598,254$$

Let's now complete the solution to the twin problem.

EXAMPLE 6.20 Future Value for Twin 1

Twin 1 invests \$2000 at the end of each of 8 years in an account that earns 10%, compounded annually. After the initial 8 years, no additional contri-

butions are made, but the investment continues to earn 10%, compounded annually, for 36 more years (until Twin 1 is age 65). How much does Twin 1 have at age 65?

SOLUTION

We seek the future values of two different investments. The first is an ordinary annuity with $R = \$2000$, $n = 8$ periods, and $i = 0.10$. The second is a compound interest investment with $n = 36$ periods and $i = 0.10$ and whose principal (that is, its present value) is the future value of this twin's ordinary annuity.

We first find the future value of the ordinary annuity.

$$S = R\left[\frac{(1 + i)^n - 1}{i}\right] = \$2000\left[\frac{(1 + 0.10)^8 - 1}{0.10}\right]$$

$$= \$22,871.78 \text{ (to the nearest cent)}$$

This amount is the principal of the compound interest investment. If no deposits or withdrawals were made for the next 36 years, the future value of this investment would be

$$S = P(1 + i)^n = \$22,871.78(1 + 0.10)^{36}$$

$$= \$707,028.03 \text{ (to the nearest cent)}$$

Thus, at age 65, Twin 1's investment is worth about \$707,028.

Would You Rather Be Twin 1 or Twin 2?

\$2000/yr **Yrs. 1–8** + **\$0/yr** **Yrs. 9–44** @ **10%** **compounded annually** = **\$707,028 at age 65**	**\$0/yr** **Yrs. 1–8** + **\$2000/yr** **Yrs. 9–44** @ **10%** **compounded annually** = **\$598,254 at age 65**
SHE SPENT \$16,000 TO GET \$707,028.	**SHE SPENT \$72,000 TO GET \$598,254.**

Shutterstock

Looking back at the previous examples, we can extract the following summary and see which twin was the wiser.

	Contributions	Account Value at Age 65
Twin 1	$2000/year for 8 years = $16,000	$707,028
Twin 2	$2000/year for 36 years = $72,000	$598,254

Note that Twin 1 contributed $56,000 less than Twin 2 but had $108,774 more at age 65. This illustrates the powerful effect that time and compounding have on investments.

Sometimes we want to know how long it will take for an annuity to reach a desired future value.

EXAMPLE 6.21 Time to Reach a Goal

A small business invests $1000 at the end of each month in an account that earns 6% compounded monthly. How long will it take until the business has $100,000 toward the purchase of its own office building?

SOLUTION

This is an ordinary annuity with $S = \$100,000$, $R = \$1000$, $i = 0.06/12 = 0.005$, and $n =$ the number of months. To answer the question of how long, solve for n.

Use $S = R\left[\dfrac{(1 + i)^n - 1}{i}\right]$ and solve for n in

$$100,000 = 1000\left[\frac{(1 + 0.005)^n - 1}{0.005}\right].$$

This is an exponential equation. Hence, we isolate $(1.005)^n$, take the natural logarithm of both sides, and then solve for n as follows.

$$100,000 = \frac{1000}{0.005}[(1.005)^n - 1]$$

$$0.5 = (1.005)^n - 1$$

$$1.5 = (1.005)^n$$

$$\ln(1.5) = \ln[(1.005)^n] = n[\ln(1.005)]$$

$$n = \frac{\ln(1.5)}{\ln(1.005)} \approx 81.3$$

Because this investment is monthly, after 82 months the company will be able to purchase its own office building.

We can also find the periodic payment needed to obtain a specified future value.

EXAMPLE 6.22 Payment for an Ordinary Annuity

A young couple wants to save $50,000 over the next 5 years and then to use this amount as a down payment on a home. To reach this goal, how much money must they deposit at the end of each quarter in an account that earns interest at a rate of 5%, compounded quarterly?

SOLUTION

This plan describes an ordinary annuity with a future value of $50,000 whose payment size, R, is to be determined. Quarterly compounding gives $n = 5(4) = 20$ and $i = 0.05/4 = 0.0125$.

$$S = R\left[\frac{(1 + i)^n - 1}{i}\right]$$

$$\$50,000 = R\left[\frac{(1 + 0.0125)^{20} - 1}{0.0125}\right] = R(22.56297854)$$

$$R = \frac{\$50,000}{22.56297854} = \$2216.02 \text{ (to the nearest cent)}$$

Sinking Funds

Just as the couple in Example 6.22 was saving for a future purchase, some borrowers, such as municipalities, may have a debt that must be paid in a single large sum on a specified future date. If these borrowers make periodic deposits that will produce that sum on a specified date, we say that they have established a **sinking fund.** If the deposits are all the same size and are made regularly, they form an ordinary annuity whose future value (on a specified date) is the amount of the debt. To find the size of these periodic deposits, we solve for R in the equation for the future value of an annuity:

$$S = R\left[\frac{(1 + i)^n - 1}{i}\right]$$

Annuities Due

Deposits in savings accounts, rent payments, and insurance premiums are examples of **annuities due.** Unlike an ordinary annuity, an annuity due has the periodic payments made at the *beginning* of the period. The *term* of an annuity due is from the first payment to the end of one period after the last payment. Thus an annuity due draws interest for one period more than the ordinary annuity.

The formula for the future value of an annuity due is as follows.

Future Value of an Annuity Due:

$$S_{due} = Rs_{\overline{n}|i} = R\left[\frac{(1 + i)^n - 1}{i}\right](1 + i)$$

EXAMPLE 6.23 Future Value

Find the future value of an investment if $150 is deposited at the beginning of each month for 9 years and the interest rate is 7.2%, compounded monthly.

SOLUTION

Because deposits are made at the *beginning* of each month, this is an annuity due with $R = \$150$, $n = 9(12) = 108$, and $i = 0.072/12 = 0.006$.

$$S_{due} = R\left[\frac{(1 + i)^n - 1}{i}\right](1 + i)$$

$$= \$150\left[\frac{(1 + 0.006)^{108} - 1}{0.006}\right](1 + 0.006)$$

$$= \$150(151.3359308)(1.006)$$

$$= \$22,836.59 \text{ (to the nearest cent)}$$

We can also use the formula for the future value of an annuity due to determine the payment size required to reach an investment goal.

....

6.3 Exercises

Ordinary Annuities

1. Find the future value of an annuity of $1300 paid at the end of each year for 5 years, if interest is earned at a rate of 6%, compounded annually.

2. The Weidmans want to save $40,000 in 2 years for a down payment on a house. If they make monthly deposits in an account paying 12%, compounded monthly, what is the size of the payments that are required to meet their goal?

3. If $2500 is deposited at the end of each quarter in an account that earns 5% compounded quarterly, after how many quarters will the account contain $80,000?

Annuities Due

4. Find the future value of an annuity due of $200 paid at the beginning of each 6-month period for 8 years if the interest rate is 6%, compounded semiannually.

5. How much must be deposited at the beginning of each year in an account that pays 8%, compounded annually, so that the account will contain $24,000 at the end of 5 years?

Miscellaneous Problems

In Problems 6–10, (a) state whether the problem relates to an ordinary annuity or an annuity due, and then (b) solve the problem.

6. Parents agree to invest $500 (at 10%, compounded semiannually) for their son on the December 31 or June 30 following each semester that he makes the dean's list during his 4 years in college. If he makes the dean's list in each of the 8 semesters, how much money will his parents have to give him when he graduates?

7. How much will have to be invested at the beginning of each year at 10%, compounded annually, to pay off a debt of $50,000 in 8 years?

8. Mr. Gordon plans to invest $300 at the end of each month in an account that pays 9%, compounded monthly. After how many months will the account be worth $50,000?

9. Jane Adele deposits $500 in an account at the beginning of each 3-month period for 9 years. If the account pays interest at the rate of 8%, compounded quarterly, how much will she have in her account after 9 years?

10. A sinking fund is established by a working couple so that they will have $60,000 to pay for part of their daughter's education when she enters college. If they make deposits at the end of each 3-month period for 10 years, and if interest is paid at 12%, compounded quarterly, what size deposits must they make?

11. Suppose a recent college graduate's first job allows her to deposit $100 at the end of each month in a savings plan that earns 9%, compounded monthly. This savings plan continues for 8 years before new obligations make it impossible to continue. If the accrued amount remains in the plan for the next 15 years without deposits or withdrawals, how much

money will be in the account 23 years after the plan began?

Need more practice?
Find more here: cengagebrain.com

6.4 Present Values of Annuities

WE HAVE DISCUSSED HOW CONTRIBUTING TO AN ANNUITY PROGRAM WILL RESULT IN A SUM OF MONEY, AND WE HAVE CALLED THAT SUM THE FUTURE VALUE OF THE ANNUITY. JUST AS THE TERM *ANNUITY* IS USED TO DESCRIBE AN ACCOUNT IN WHICH A PERSON MAKES EQUAL PERIODIC PAYMENTS (DEPOSITS), THIS TERM IS ALSO USED TO DESCRIBE AN ACCOUNT FROM WHICH A PERSON RECEIVES EQUAL PERIODIC PAYMENTS (WITHDRAWALS). THAT IS, IF YOU INVEST A LUMP SUM OF MONEY IN AN ACCOUNT TODAY, SO THAT AT REGULAR INTERVALS YOU WILL RECEIVE A FIXED SUM OF MONEY, YOU HAVE ESTABLISHED AN ANNUITY. THE SINGLE SUM OF MONEY REQUIRED TO PURCHASE AN ANNUITY THAT WILL PROVIDE THESE PAYMENTS AT REGULAR INTERVALS IS THE **PRESENT VALUE** OF THE ANNUITY.

Ordinary Annuities

Suppose we wish to invest a lump sum of money (denoted by A_n) in an annuity that earns interest at rate i per period in order to receive (withdraw) payments of size R from this account at the end of each of n periods (after which time the account balance will be $0). Recall that receiving payments at the end of each period means that this is an ordinary annuity.

The following formula for the present value of an ordinary annuity is developed in the Enrichment material on CourseMate for MATH APPS at cengagebrain.com.

Present Value of an Ordinary Annuity: If a payment of R is to be made at the end of each period for n periods from an account that earns interest at a rate of i per period, then the account is an ordinary annuity, and the present value is

$$A_n = R \cdot a_{\overline{n}|i} = R \cdot \left[\frac{1 - (1 + i)^{-n}}{i} \right]$$

where $a_{\overline{n}|i}$ represents the present value of an ordinary annuity of $1 per period for n periods, with an interest rate of i per period.

EXAMPLE 6.24 Present Value

What is the present value of an annuity of $1500 payable at the end of each 6-month period for 2 years if money is worth 8%, compounded semiannually?

SOLUTION

We are given that $R = \$1500$ and $i = 0.08/2 = 0.04$. Because a payment is made twice a year for 2 years, the number of periods is $n = (2)(2) = 4$. Thus,

$$A_n = R \cdot a_{\overline{4}|0.04} = \$1500 \left[\frac{1 - (1 + 0.04)^{-4}}{0.04} \right]$$

$$= \$1500(3.629895)$$

$$= \$5444.84 \text{ (to the nearest cent)}$$

EXAMPLE 6.25 Present Value

Find the lump sum that one must invest in an annuity in order to receive $1000 at the end of each month for the next 16 years, if the annuity pays 9%, compounded monthly.

SOLUTION

The sum we seek is the present value of an ordinary annuity, A_n, with $R = \$1000$, $i = 0.09/12 = 0.0075$, and $n = (16)(12) = 192$.

$$A_n = R \left[\frac{1 - (1 + i)^{-n}}{i} \right]$$

$$= \$1000 \left[\frac{1 - (1.0075)^{-192}}{0.0075} \right] = \$1000(101.5727689)$$

$$= \$101,572.77$$

Thus the required lump sum, to the nearest dollar, is $101,573.

See the Tech Card for details on finding the present value of an annuity with a graphing calculator or with Excel.

It is important to note that all annuities involve both periodic payments and a lump sum of money. It is whether this lump sum is in the present or in the future that distinguishes problems that use formulas for present values of annuities from those that use formulas for future values of annuities. In Example 6.25, the lump sum was needed now (in the present) to generate the $1000 payments, so we used the present value formula.

EXAMPLE 6.26 Payments from an Annuity

➡ Suppose that a couple plans to set up an ordinary annuity with a $100,000 inheritance they received. What is the size of the quarterly payments they will receive for the next 6 years (while their children are in college) if the account pays 7%, compounded quarterly?

SOLUTION

The $100,000 is the amount the couple has now, so it is the present value of an ordinary annuity whose payment size, R, we seek. Using present value $A_n = \$100,000$, $n = 6(4) = 24$, and $i = 0.07/4 = 0.0175$, we solve for R.

$$A_n = R\left[\frac{1 - (1 + i)^{-n}}{i}\right]$$

$$\$100,000 = R\left[\frac{1 - (1 + 0.0175)^{-24}}{0.0175}\right]$$

$$\$100,000 = R(19.46068565)$$

$$R = \frac{\$100,000}{19.46068565} = \$5138.57 \text{ (to the nearest cent)}$$

EXAMPLE 6.27 Number of Payments from an Annuity

➡ An inheritance of $250,000 is invested at 9%, compounded monthly. If $2500 is withdrawn at the end of each month, how long will it be until the account balance is $0?

SOLUTION

The regular withdrawals form an ordinary annuity with present value $A_n = \$250,000$, payment $R = \$2500$, $i = 0.09/12 = 0.0075$, and $n =$ the number of months.

Use $A_n = R\left[\dfrac{1 - (1 + i)^{-n}}{i}\right]$ and solve for n in

$$250,000 = 2500\left[\frac{1 - (1.0075)^{-n}}{0.0075}\right].$$

This is an exponential equation, and to solve it we isolate $(1.0075)^{-n}$ and then take the natural logarithm of both sides.

$$\frac{250,000(0.0075)}{2500} = 1 - (1.0075)^{-n}$$

$$0.75 = 1 - (1.0075)^{-n}$$

$$(1.0075)^{-n} = 0.25$$

$$\ln\left[(1.0075)^{-n}\right] = \ln(0.25) \Rightarrow -n\left[\ln(1.0075)\right] = \ln(0.25)$$

$$-n = \frac{\ln(0.25)}{\ln(1.0075)} \approx -185.5, \quad \text{so} \quad n \approx 185.5$$

Thus, the account balance will be $0 in 186 months.

Bond Pricing: Bonds represent a relatively safe investment similar to a bank certificate of deposit (CD), but unlike CDs (and like stocks), bonds can be traded. And, as is also true of stocks, the trading or market price of a bond may fluctuate. Bond pricing is an application of present value of ordinary annuities (see the Enrichment material).

Annuities Due

Recall that an annuity due is one in which payments are made at the beginning of each period. This means that the present value of an annuity due of n payments (denoted $A_{(n,due)}$) of $R at interest rate i per period can be viewed as an initial payment of $R plus the payment program for an ordinary annuity of $n - 1$ payments of $R at interest rate i per period. The following formula for the present value of an annuity due is developed in the Enrichment material on CourseMate for MATH APPS at cengagebrain.com.

Present Value of an Annuity Due:
If a payment of $R is to be made at the beginning of each period for n periods from an account that earns interest rate i per period, then the account is an annuity due, and its present value is given by

$$A_{(n,due)} = R\left[\frac{1 - (1 + i)^{-n}}{i}\right](1 + i) = Ra_{\overline{n}|i}(1 + i)$$

where $a_{\overline{n}|i}$ denotes the present value of an ordinary annuity of $1 per period for n periods at interest rate i per period.

With a payout of a $1.2 million prize in monthly installments of $10,000, your real prize is only about $836,800—or roughly 70% of the total prize.

EXAMPLE 6.28 Lottery Prize

A lottery prize worth $1,200,000 is awarded in payments of $10,000 at the beginning of each month for 10 years. Suppose money is worth 7.8%, compounded monthly. What is the *real* value of the prize?

SOLUTION

The *real* value of this prize is its present value when it is awarded. That is, it is the present value of an annuity due with $R = \$10,000$, $i = 0.078/12 = 0.0065$, and $n = 12(10) = 120$. Thus

$$A_{(120,due)} = \$10,000\left[\frac{1 - (1 + 0.0065)^{-120}}{0.0065}\right](1 + 0.0065)$$

$$= \$10,000(83.1439199)(1.0065)$$

$$= \$836,843.55 \text{ (to the nearest cent)}$$

This means the lottery operator needs this amount to generate the 120 monthly payments of $10,000 each.

EXAMPLE 6.29 Court Settlement Payments

Suppose that a court settlement results in a $750,000 award. If this is invested at 9%, compounded semiannually, how much will it provide at the beginning of each half-year for a period of 7 years?

SOLUTION

Because payments are made at the beginning of each half-year, this is an annuity due. We seek the payment size, R, and use the present value $A_{(n,due)} = \$750,000$, $n = 2(7) = 14$, and $i = 0.09/2 = 0.045$.

$$A_{(n,due)} = R\left[\frac{1 - (1 + i)^{-n}}{i}\right](1 + i)$$

$$\$750,000 = R\left[\frac{1 - (1 + 0.045)^{-14}}{0.045}\right](1 + 0.045)$$

$$\$750,000 \approx R(10.22282528)(1.045) \approx R(10.682852)$$

$$R = \frac{\$750,000}{10.682852} = \$70,205.97 \text{ (to the nearest cent)}$$

Deferred Annuities

A **deferred annuity** is one in which the first payment is made not at the beginning or end of the first period, but at some later date. An annuity that is deferred for k periods and then has payments of R per period at the end of each of the next n periods is an ordinary deferred annuity and can be found with the following formula. (See the Enrichment material on CourseMate for MATH APPS at cengagebrain.com for the formula development.)

> **Present Value of a Deferred Annuity:**
> The present value of a deferred annuity of R per period for n periods, deferred for k periods with interest rate i per period, is given by
>
> $$A_{(n,k)} = R\left[\frac{1 - (1 + i)^{-n}}{i}\right](1 + i)^{-k}$$
>
> $$= Ra_{\overline{n}|i}(1 + i)^{-k}$$

For example, suppose a deferred annuity is purchased that will pay $10,000 per quarter for 15 years after being deferred for 5 years, and money is worth 6% compounded quarterly. To find the present value of this annuity, we use $R = \$10,000$, $n = 4(15) = 60$, $k = 4(5) = 20$, and $i = 0.06/4 = 0.015$ in the formula for the present value of a deferred annuity.

$$A_{(60,20)} = \$10,000\left[\frac{1 - (1.015)^{-60}}{0.015}\right](1.015)^{-20}$$

$$= \$292,386.85 \text{ (to the nearest cent)}$$

We can also use this formula to find the payment size from a deferred annuity.

6.4 Exercises

Ordinary Annuities

1. Suppose a state lottery prize of $5 million is to be paid in 20 payments of $250,000 each at the end of each of the next 20 years. If money is worth 10%, compounded annually, what is the present value of the prize?

2. With a present value of $135,000, what is the size of the withdrawals that can be made at the end of each quarter for the next 10 years if money is worth 6.4%, compounded quarterly?

3. A personal account earmarked as a retirement supplement contains $242,400. Suppose $200,000 is used to establish an annuity that earns 6%, compounded quarterly, and pays $4500 at the end of each quarter. How long will it be until the account balance is $0?

Annuities Due

4. Explain the difference between an ordinary annuity and an annuity due.

5. What amount must be set aside now to generate payments of $50,000 at the beginning of each year for the next 12 years if money is worth 5.92%, compounded annually?

6. A year-end bonus of $25,000 will generate how much money at the beginning of each month for the next year, if it can be invested at 6.48%, compounded monthly?

Miscellaneous Problems for Ordinary Annuities and Annuities Due

In Problems 7–10, (a) decide whether the problem relates to an ordinary annuity or an annuity due, and then (b) solve the problem.

7. An insurance settlement of $1.5 million must replace Trixie Eden's income for the next 40 years. What income will this settlement provide at the end of each month if it is invested in an annuity that earns 8.4%, compounded monthly?

8. A company wants to have $40,000 at the beginning of each 6-month period for the next $4\frac{1}{2}$ years. If an annuity is set up for this purpose, how much must be invested now if the annuity earns 6.68%, compounded semiannually?

9. Recent sales of some real estate and record profits make it possible for a manufacturer to set aside $800,000 in a fund to be used for modernization and remodeling. How much can be withdrawn from this fund at the beginning of each half-year for the next 3 years if the fund earns 7.7%, compounded semiannually?

10. Juanita Domingo's parents want to establish a college trust for her. They want to make 16 quarterly withdrawals of $2000, with the first withdrawal 3 months from now. If money is worth 7.2%, compounded quarterly, how much must be deposited now to provide for this trust?

11. Suppose an individual makes an initial investment of $2500 in an account that earns 7.8%, compounded monthly, and makes additional contributions of $100 at the end of each month for a period of 12 years. After these 12 years, this individual wants to make withdrawals at the end of each month for the next 5 years (so that the account balance will be reduced to $0).
    (a) How much is in the account after the last deposit is made?
    (b) How much was deposited?
    (c) What is the amount of each withdrawal?
    (d) What is the total amount withdrawn?

Deferred Annuities

12. The semiannual tuition payment at a major university is expected to be $30,000 for the 4 years beginning 18 years from now. What lump sum payment should the university accept now, in lieu of tuition payments beginning 18 years, 6 months from now? Assume that money is worth 7%, compounded semiannually, and that tuition is paid at the end of each half-year for 4 years.

13. Danny Metzger's parents invested $1600 when he was born. This money is to be used for Danny's college education and is to be withdrawn in four equal annual payments beginning when Danny is age 19. Find the amount that will be available each year, if money is worth 6%, compounded annually.

Need more practice?
Find more here: cengagebrain.com

6.5 Loans and Amortization

WHEN BUSINESSES, INDIVIDUALS, OR FAMILIES BORROW MONEY TO MAKE A MAJOR PURCHASE, FOUR QUESTIONS COMMONLY ARISE:

1. What will the payments be?

2. How much can be borrowed and still fit a budget?

3. What is the payoff amount of the loan before the final payment is due?

4. What is the total of all payments needed to pay off the loan?

In this section we discuss the most common way that consumers and businesses discharge debts—regular payments of fixed size made on a loan (called amortization)—and determine ways to answer the questions just posed.

Amortization

Just as we invest money to earn interest, banks and lending institutions lend money and collect interest for its use.

Federal law now requires that the full cost of any loan and the **true annual percentage rate (APR)** be disclosed with the loan. Because of this law, loans now are usually paid off by a series of partial payments with interest charged on the unpaid balance at the end of each payment period.

This type of loan is usually repaid by making all payments (including principal and interest) of equal size. This process of repaying the loan is called **amortization**.

When a bank makes a loan of this type, it is purchasing from the borrower an ordinary annuity that pays a fixed return each payment period. The lump sum the bank gives to the borrower (the principal of the loan) is the present value of the ordinary annuity, and each payment the bank receives from the borrower is a payment from the annuity. Hence, to find the size of these periodic payments, we solve for R in the formula for the present value of an ordinary annuity, to obtain the following.

> **Amortization Formula:** If the debt of $\$A_n$, with interest at a rate of i per period, is amortized by n equal periodic payments (each payment being made at the end of a period), the size of each payment is
>
> $$R = A_n \cdot \left[\frac{i}{1 - (1 + i)^{-n}} \right]$$

EXAMPLE 6.30 Buying a Home

A man buys a house for $200,000. He makes a $50,000 down payment and agrees to amortize the rest of the debt with quarterly payments over the next 10 years. If the interest on the debt is 12%, compounded quarterly, find (a) the size of the quarterly payments, (b) the total amount of the payments, and (c) the total amount of interest paid.

SOLUTION

(a) We know that $A_n = \$200{,}000 - \$50{,}000 = \$150{,}000$, $n = 4(10) = 40$, $i = 0.12/4 = 0.03$. Thus, for the quarterly payment, we have

$$R = \$150{,}000 \left[\frac{0.03}{1 - (1.03)^{-40}} \right]$$

$$= (\$150{,}000)(0.043262378)$$

$$= \$6489.36 \text{ (to the nearest cent)}$$

(b) The man made 40 payments of $6489.36, so his payments totaled

$$(40)(\$6489.36) = \$259{,}574.40$$

plus the $50,000 down payment, or $309,574.40.

(c) Of the $309,574.40 paid, $200,000 was for payment of the house. The remaining $109,574.40 was the total amount of interest paid.

See the Tech Card for details on using a graphing calculator or Excel to find the payment required to amortize a loan.

EXAMPLE 6.31 Affordable Home

Carol and Mike have $30,000 for a down payment, and their budget can accommodate a monthly mortgage payment of $1200.00. What is the most expensive home they can buy if they can borrow money for 30 years at 7.8%, compounded monthly?

SOLUTION

We seek the amount that Carol and Mike can borrow, or A_n, knowing that $R = \$1200$, $n = 30(12) = 360$ and $i = 0.078/12 = 0.0065$. We can use these values in the amortization formula and solve for A_n (or use the formula for the present value of an ordinary annuity).

$$R = A_n\left[\frac{i}{1 - (1 + i)^{-n}}\right]$$

$$\$1200 = A_n\left[\frac{0.0065}{1 - (1.0065)^{-360}}\right]$$

$$A_n = \$1200\left[\frac{1 - (1.0065)^{-360}}{0.0065}\right]$$

$$A_n = \$1200(138.9138739)$$

$$= \$166{,}696.65 \text{ (to the nearest cent)}$$

Thus, if they borrow \$166,697 (to the nearest dollar) and put down \$30,000, the most expensive home they can buy would cost \$166,697 + \$30,000 = \$196,697.

Amortization Schedule

We can construct an **amortization schedule** that summarizes all the information regarding the amortization of a loan.

For example, a loan of \$10,000 with interest at 10% could be repaid in 5 equal annual payments of size

$$R = \$10{,}000\left[\frac{0.10}{1 - (1 + 0.10)^{-5}}\right]$$

$$= \$10{,}000(0.263797) = \$2637.97$$

Each time this \$2637.97 payment is made, some is used to pay the interest on the unpaid balance, and some is used to reduce the principal. For the first payment, the unpaid balance is \$10,000, so the interest payment is 10% of \$10,000, or \$1000. The remaining \$1637.97 is applied to the principal. Hence, after this first payment, the unpaid balance is \$10,000 − \$1637.97 = \$8362.03.

For the second payment of \$2637.97, the amount used for interest is 10% of \$8362.03, or \$836.20; the remainder, \$1801.77, is used to reduce the principal.

This information for these two payments and for the remaining payments is summarized in the following amortization schedule.

Period	Payment	Interest	Balance Reduction	Unpaid Balance
				\$10,000.00
1	\$2637.97	\$1000.00	\$1637.97	8362.03
2	2637.97	836.20	1801.77	6560.26
3	2637.97	656.03	1981.94	4578.32
4	2637.97	457.83	2180.14	2398.18
5	2638.00	239.82	2398.18	0.00
Total	13,189.88	3189.88	10,000.00	

Note that the last payment was increased by 3¢ so the balance was reduced to \$0 at the end of the 5 years. Such an adjustment is normal in amortizing a loan.

Unpaid Balance of a Loan

Many people who borrow money, such as for a car or a home, do not pay on the loan for its entire term. Rather, they pay off the loan early by making a final lump sum payment. The **unpaid balance of a loan** (also called the **payoff amount** and the **outstanding principal of the loan**) is the present value needed to generate all the remaining payments.

Unpaid Balance or Payoff Amount of a Loan: For a loan of n payments of \$$R$ per period at interest rate i per period, the unpaid balance after k payments have been made is the present value of an ordinary annuity with $n - k$ payments. That is, with $n - k$ payments remaining,

Unpaid balance =

$$A_{n-k} = R\left[\frac{1 - (1 + i)^{-(n-k)}}{i}\right]$$

EXAMPLE 6.32 Unpaid Balance

In Example 6.30, we found that the monthly payment for a loan of \$150,000 at 12%, compounded quarterly, for 10 years is \$6489.36 (to the nearest cent). Find the unpaid balance immediately after the 15th payment.

SOLUTION

The unpaid balance after the 15th payment is the present value of the annuity with $40 - 15 = 25$ payments remaining. Thus we use $R = \$6489.36$, $i = 0.03$, and $n - k = 25$ in the formula for the unpaid balance of a loan.

$$A_{n-k} = R\left[\frac{1 - (1 + i)^{-(n-k)}}{i}\right]$$

$$= \$6489.36\left[\frac{1 - (1.03)^{-25}}{0.03}\right]$$

$$= (\$6489.36)(17.4131477)$$

$$= \$113{,}000.18 \text{ (to the nearest cent)}$$

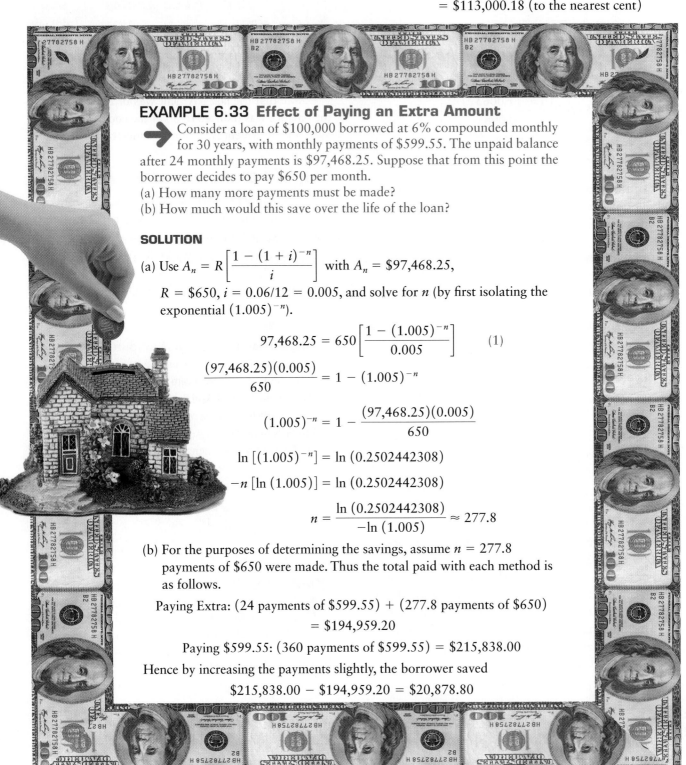

EXAMPLE 6.33 Effect of Paying an Extra Amount

Consider a loan of $100,000 borrowed at 6% compounded monthly for 30 years, with monthly payments of $599.55. The unpaid balance after 24 monthly payments is $97,468.25. Suppose that from this point the borrower decides to pay $650 per month.
(a) How many more payments must be made?
(b) How much would this save over the life of the loan?

SOLUTION

(a) Use $A_n = R\left[\dfrac{1 - (1 + i)^{-n}}{i}\right]$ with $A_n = \$97{,}468.25$,

$R = \$650$, $i = 0.06/12 = 0.005$, and solve for n (by first isolating the exponential $(1.005)^{-n}$).

$$97{,}468.25 = 650\left[\frac{1 - (1.005)^{-n}}{0.005}\right] \qquad (1)$$

$$\frac{(97{,}468.25)(0.005)}{650} = 1 - (1.005)^{-n}$$

$$(1.005)^{-n} = 1 - \frac{(97{,}468.25)(0.005)}{650}$$

$$\ln\left[(1.005)^{-n}\right] = \ln(0.2502442308)$$

$$-n\left[\ln(1.005)\right] = \ln(0.2502442308)$$

$$n = \frac{\ln(0.2502442308)}{-\ln(1.005)} \approx 277.8$$

(b) For the purposes of determining the savings, assume $n = 277.8$ payments of $650 were made. Thus the total paid with each method is as follows.

Paying Extra: (24 payments of $599.55) + (277.8 payments of $650)
$$= \$194{,}959.20$$

Paying $599.55: (360 payments of $599.55) = \$215{,}838.00

Hence by increasing the payments slightly, the borrower saved
$$\$215{,}838.00 - \$194{,}959.20 = \$20{,}878.80$$

See the Tech Card for details on using a graphing calculator to find the number of payments to amortize a loan.

....

6.5 Exercises

1. Answer parts (a) and (b) with respect to loans for 10 years and 25 years. Assume the same interest rate.
   (a) Which loan results in more of each payment being directed toward principal? Explain.
   (b) Which loan results in a lower periodic payment? Explain.

2. A recent graduate's student loans total $18,000. If these loans are at 4.2%, compounded quarterly, for 10 years, what are the quarterly payments?

3. A homeowner planning a kitchen remodeling can afford a $200 monthly payment. How much can the homeowner borrow for 5 years at 6%, compounded monthly, and still stay within the budget?

4. Develop an amortization schedule for a loan of $20,000 for 1 year at 12% compounded quarterly.

5. When Maria Acosta bought a car years ago, she borrowed $14,000 for 48 months at 8.1% compounded monthly. Her monthly payments are $342.44, but she'd like to pay off the loan early. How much will she owe just after her payment at the $2\frac{1}{2}$-year mark?

6. John Fare purchased $10,000 worth of equipment by making a $2000 down payment and promising to pay the remainder of the cost in semiannual payments over the next 4 years. The interest rate on the debt is 10%, compounded semiannually. Find
   (a) the size of each payment,
   (b) the total amount paid over the life of the loan, and
   (c) the total interest paid over the life of the loan.

7. A man buys a car for $36,000. If the interest rate on the loan is 12%, compounded monthly, and if he wants to make monthly payments of $900 for 36 months, how much must he put down?

8. A couple who borrow $90,000 for 30 years at 7.2%, compounded monthly, must make monthly payments of $610.91.
   (a) Find their unpaid balance after 1 year.
   (b) During that first year, how much interest do they pay?

9. A recent college graduate buys a new car by borrowing $18,000 at 8.4%, compounded monthly, for 5 years. She decides to pay an extra $15 per payment.
   (a) What is the monthly payment required by the loan, and how much does she decide to pay each month?
   (b) How many payments (that include the extra $15) will she make?
   (c) How much will she save by paying the extra $15?

10. During four years of college, Nolan MacGregor's student loans are $4000, $3500, $4400, and $5000 for freshman year through senior year, respectively. Each loan amount gathers interest of 1%, compounded quarterly, while Nolan is in school and 3%, compounded quarterly, during a 6-month grace period after graduation.
    (a) What is the loan balance after the grace period? Assume the freshman-year loan earns 1% interest for $\frac{3}{4}$ year during the first year, then for 3 full years until graduation. Make similar assumptions for the loans for the other years.
    (b) After the grace period, the loan is amortized over the next 10 years at 3%, compounded quarterly. Find the quarterly payment.
    (c) If Nolan decides to pay an additional $90 per payment, how many payments of this size will amortize the debt?
    (d) How much will Nolan save by paying the extra $90 with each payment?

....

> **Need more practice?**
> **Find more here: cengagebrain.com**

Chapter Exercises

1. Find the first 4 terms of the sequence with *n*th term
$$a_n = \frac{1}{n^2}$$

2. Identify any arithmetic sequences and find the common differences.
   (a) 12, 7, 2, −3, . . .
   (b) 1, 3, 6, 10, . . .
   (c) $\frac{1}{6}, \frac{1}{3}, \frac{1}{2}, \frac{2}{3}, \ldots$

3. Find the 80th term of the arithmetic sequence with first term −2 and common difference 3.

4. Find the 36th term of the arithmetic sequence with third term 10 and eighth term 25.

5. Find the sum of the first 60 terms of the arithmetic sequence $\frac{1}{3}, \frac{1}{2}, \frac{2}{3}, \ldots$

6. Identify any geometric sequences and their common ratios.
   (a) $\frac{1}{4}, 2, 16, 128, \ldots$
   (b) $16, -12, 9, -\frac{27}{4}, \ldots$
   (c) 4, 16, 36, 64, . . .

7. Find the fourth term of the geometric sequence with first term 64 and eighth term $\frac{1}{2}$.

8. Find the sum of the first 16 terms of the geometric sequence $\frac{1}{9}, \frac{1}{3}, 1, \ldots$.

Applications

Finance In Problems 9–12, identify which of the following formulas applies to each situation.

$$S = P(1 + i)^n \qquad S = Pe^{rt}$$

$$S = R\left[\frac{(1 + i)^n - 1}{i}\right] \qquad A_n = R\left[\frac{1 - (1 + i)^{-n}}{i}\right]$$

9. The future value of a series of payments, with interest compounded when the payments are made

10. The present value of a series of payments, with interest compounded when the payments are made

11. The future value of an investment that earns interest compounded periodically

12. The future value of an investment that earns interest compounded continuously

13. **Loans** If $8000 is borrowed at 12% simple interest for 3 years, what is the future value of the loan at the end of the 3 years?

14. **Loan rate** Mary Toy borrowed $2000 from her parents and repaid them $2100 after 9 months. What simple interest rate did she pay?

15. **Tuition** How much summer earnings must a college student deposit on August 31 in order to have $3000 for tuition and fees on December 31 of the same year, if the investment earns 6% simple interest?

16. **Salaries** Suppose you are offered two identical jobs: one paying a starting salary of $40,000 with yearly raises of $2000 and one paying a starting salary of $36,000 with yearly raises of $2500. Which job will pay more money over a 10-year period?

17. **Interest** If $1000 is invested for 4 years at 8%, compounded quarterly, how much interest will be earned?

18. **Savings goal** How much must one invest now in order to have $18,000 in 4 years if the investment earns 5.4%, compounded monthly?

19. **Future value** What is the future value if $1000 is invested for 6 years at 8%, compounded continuously?

20. **College fund** A couple received an inheritance and plan to invest some of it for their grandchild's college education. How much must they invest if they would like the fund to have $100,000 after 15 years, and their investment earns 10.31%, compounded continuously?

21. **Investments** If $15,000 is invested at 6%, compounded quarterly, how long will it be before it grows to $25,000?

22. **Investment rates** (a) If an initial investment of $35,000 grows to $257,000 in 15 years, what annual interest rate, continuously compounded, was earned? (b) What is the annual percentage yield on this investment?

23. **Comparing yields** Find the annual percentage yield equivalent to a nominal rate of 7.2% (a) compounded quarterly and (b) compounded continuously.

24. **Sinking fund** How much would have to be invested at the end of each year at 6%, compounded annually, to pay off a debt of $80,000 in 10 years?

25. **Annuity** Find the future value of an annuity due of $800 paid at the beginning of every 6-month period for 10 years, if it earns interest at 12%, compounded semiannually.

26. **Construction fund** A company wants to have $250,000 available in $4\frac{1}{2}$ years for new construction. How much must be deposited at the beginning of each quarter to reach this goal if the investment earns 10.2%, compounded quarterly?

27. **Time to reach a goal** If $1200 is deposited at the end of each quarter in an account that earns 7.2%, compounded quarterly, how long will it be until the account is worth $60,000?

28. **Annuity** What lump sum would have to be invested at 9%, compounded semiannually, to provide an annuity of $10,000 at the end of each half-year for 10 years?

29. **Powerball lottery** Winners of lotteries receive the jackpot distributed over a period of years, usually 20 or 25 years. The winners of the Powerball lottery on July 29, 1998, elected to take a one-time cash payout rather than receive the $295.7 million jackpot in 25 annual payments beginning on the date the lottery was won.
    (a) How much money would the winners have received at the beginning of each of the 25 years?
    (b) If the value of money was 5.91%, compounded annually, what onetime payout did they receive in lieu of the annual payments?

30. **Payment from an annuity** A recent college graduate's gift from her grandparents is $20,000. How much will this provide at the end of each month for the next 12 months while the graduate travels? Assume that money is worth 6.6%, compounded monthly.

31. **Payment from an annuity** An IRA of $250,000 is rolled into an annuity that pays a retired couple at the beginning of each quarter for the next 20 years. If the annuity earns 6.2%, compounded quarterly, how much will the couple receive each quarter?

32. **Duration of an annuity** A retirement account that earns 6.8%, compounded semiannually, contains $488,000. For how long can $40,000 be withdrawn at the end of each half-year until the account balance is $0?

33. **Amortization** A debt of $1000 with interest at 12%, compounded monthly, is amortized by 12 monthly payments (of equal size). What is the size of each payment?

34. **Loan payoff** A debt of $8000 is amortized with eight semiannual payments of $1288.29 each. If money is worth 12%, compounded semiannually, find the unpaid balance after five payments have been made.

35. **Cash value** A woman paid $90,000 down for a house and agreed to pay 18 quarterly payments of $4500 each. If money is worth 4%, compounded quarterly, how much would the house have cost if she had paid cash?

36. **Amortization schedule** Complete the next two lines of the amortization schedule for a $100,000 loan for 30 years at 7.5%, compounded monthly, with monthly payments of $699.22.

Payment Number	Payment Amount	Interest	Balance Reduction	Unpaid Balance
56	$699.22	$594.67	$104.55	$95,042.20

Miscellaneous Financial Problems

37. If an initial investment of $2500 grows to $38,000 in 18 years, what annual interest rate, compounded annually, did this investment earn?

38. How much must be deposited at the end of each month in an account that earns 8.4%, compounded monthly, if the goal is to have $40,000 after 10 years?

39. What is the future value if $1000 is invested for 6 years at 8% (a) simple interest and (b) compounded semiannually?

40. Quarterly payments of $500 are deposited in an account that pays 8%, compounded quarterly. How much will have accrued in the account at the end of 4 years if each payment is made at the end of each quarter?

41. If $8000 is invested at 7%, compounded continuously, how long will it be before it grows to $22,000?

42. An investment broker bought some stock at $87.89 per share and sold it after 3 months for $105.34 per share. What was the annual simple interest rate earned on this transaction?

43. Kevin Patrick paid off the loan he took out to buy his car, but once the loan was paid, he continued to deposit $400 on the first of each month in an account that paid 5.4%, compounded monthly. After four years of making these deposits, Kevin was ready to buy a new car. How much did he have in the account?

44. A young couple receive an inheritance of $72,000 that they want to set aside for a college fund for

their two children. How much will this provide at the end of each half-year for a period of 9 years if it is deferred for 11 years and can be invested at 7.3%, compounded semiannually?

45. A bank is trying to decide whether to advertise some new 18-month certificates of deposit (CDs) at 6.52%, compounded quarterly, or at 6.48%, compounded continuously. Which rate is a better investment for the consumer who buys such a CD? Which rate is better for the bank?

46. If $8000 is invested at 12%, compounded continuously, for 3 years, what is the total interest earned at the end of the 3 years?

47. A couple borrowed $184,000 to buy a condominium. Their loan was for 25 years and money is worth 6%, compounded monthly.
    (a) Find their monthly payment size.
    (b) Find the total amount they would pay over 25 years.
    (c) Find the total interest they would pay over 25 years.
    (d) Find the unpaid balance after 7 years.

48. Suppose a salesman invests his $12,500 bonus in a fund that earns 10.8%, compounded monthly.

Suppose also that he makes contributions of $150 at the end of each month to this fund.
    (a) Find the future value after $12\frac{1}{2}$ years.
    (b) If after the $12\frac{1}{2}$ years, the fund is used to set up an annuity, how much will it pay at the end of each month for the next 10 years?

49. At age 22, Aruam Sdlonyer receives a $2000 IRA from her parents. At age 30, she decides that at age 67 she'd like to have a retirement fund that would pay $10,000 at the end of each month for 20 years. Suppose all investments earn 8.4%, compounded monthly. How much does Aruam need to deposit at the end of each month from ages 30 to 67 to realize her goal?

50. A company borrows $2.6 million for 15 years at 5.6%, compounded quarterly. After 2 years of regular payments, the company's profits are such that management feels they can increase the quarterly payments to $70,000 each. How long will it take to pay off the loan, and how much interest will be saved?

Need more practice?
Find more here: cengagebrain.com

STUDY YOUR WAY

At no additional cost, you have access to online learning resources that include **tutorial videos, printable flashcards, quizzes,** and more!

Watch videos that offer step-by-step conceptual explanations and guidance for each chapter in the text.

Along with the printable flashcards and other online resources, you will have a multitude of ways to check your comprehension of key mathematical concepts.

You can find these resources at **login.cengagebrain.com.**

chapter 7

Introduction
to Probability

7.1 Probability and Odds

AN ECONOMIST CANNOT PREDICT EXACTLY HOW THE GROSS NATIONAL PRODUCT WILL CHANGE, A PHYSICIAN CANNOT DETERMINE EXACTLY THE CAUSE OF LUNG CANCER, AND A PSYCHOLOGIST CANNOT DETERMINE THE EXACT EFFECT OF ENVIRONMENT ON BEHAVIOR. BUT EACH OF THESE DETERMINATIONS CAN BE MADE WITH VARYING PROBABILITIES. THUS, AN UNDERSTANDING OF THE MEANING AND DETERMINATION OF THE PROBABILITIES OF EVENTS OCCURRING IS IMPORTANT TO SUCCESS IN BUSINESS, ECONOMICS, THE LIFE SCIENCES, AND THE SOCIAL SCIENCES. PROBABILITY WAS INITIALLY DEVELOPED TO SOLVE GAMBLING PROBLEMS, BUT IT IS NOW THE BASIS FOR SOLVING PROBLEMS IN A WIDE VARIETY OF AREAS.

Sample Spaces and Single Events

When we toss a coin, it can land in one of two ways, heads or tails. If the coin is a "fair" coin, these two possible **outcomes** have an equal chance of occurring, and we say the outcomes of this **probability experiment** are **equally likely**. If we seek the probability that an experiment has a certain result, that result is called an **event**.

Suppose that an experiment can have a total of n equally likely outcomes and that k of these outcomes would be considered successes. Then the probability of achieving a success in the experiment is k/n. That is, the probability of a success in an experiment is the number of ways the experiment can result in a success divided by the total number of possible outcomes.

Probability of a Single Event: If an event E can happen in k ways out of a total of n equally likely possibilities, the probability of the occurrence of the event is denoted by

$$\Pr(E) = \frac{\text{Number of successes}}{\text{Number of possible outcomes}} = \frac{k}{n}$$

EXAMPLE 7.1 Probability

If we draw a ball from a bag containing 4 white balls and 6 black balls, what is the probability of
(a) getting a white ball? (b) getting a black ball? (c) not getting a white ball?

SOLUTION

(a) A white ball can occur (be drawn) in 4 ways out of a total of 10 equally likely possibilities. Thus the probability of drawing a white ball is

$$\Pr(W) = \frac{4}{10} = \frac{2}{5}$$

Shutterstock

(b) We have 6 chances to succeed at drawing a black ball out of a total of 10 possible outcomes. Thus the probability of getting a black ball is

$$\Pr(B) = \frac{6}{10} = \frac{3}{5}$$

(c) The probability of not getting a white ball is $\Pr(B)$, because not getting a white ball is the same as getting a black ball. Thus $\Pr(\text{not } W) = \Pr(B) = 3/5$.

A set that contains all the possible outcomes of an experiment is called a **sample space.** For the experiment of randomly drawing a numbered card from a box of cards numbered 1 through 12, a sample space is

$$S = \{1, 2, 3, 4, 5, 6, 7, 8, 9, 10, 11, 12\}$$

Each element of the sample space is called a **sample point,** and an event is a subset of the sample space. Each sample point in the sample space is assigned a nonnegative **probability measure** or **probability weight** such that the sum of the weights in the sample space is 1. The probability of an event is the sum of the weights of the sample points in the event's **subspace** (subset of S).

Some experiments have an infinite number of outcomes, but we will concern ourselves only with experiments having finite sample spaces. We can usually construct more than one sample space for an experiment. The sample space in which each sample point is equally likely is called an **equiprobable sample space.** Suppose the number of elements in a sample space is $n(S) = n$, with each element representing an equally likely outcome of a probability experiment. Then the weight (probability) assigned to each element is

$$\frac{1}{n(S)} = \frac{1}{n}$$

If an event E that is a subset of an equiprobable sample space contains $n(E) = k$ elements of S, then we can restate the probability of an event E in terms of the number of elements of E.

Probability of an Event: If an event E can occur in $n(E) = k$ ways out of $n(S) = n$ equally likely ways, then

$$\Pr(E) = \frac{n(E)}{n(S)} = \frac{k}{n}$$

EXAMPLE 7.2 Probability of an Event

→ If a number is to be selected at random from the integers 1 through 12, what is the probability that it is (a) divisible by 3? (b) even and divisible by 3? (c) even or divisible by 3?

SOLUTION

The set $S = \{1, 2, 3, 4, 5, 6, 7, 8, 9, 10, 11, 12\}$ is an equiprobable sample space for this experiment (as mentioned previously).

(a) The set $E = \{3, 6, 9, 12\}$ contains the numbers that are divisible by 3. Thus

$$\Pr(\text{divisible by 3}) = \frac{n(E)}{n(S)} = \frac{4}{12} = \frac{1}{3}$$

(b) The numbers 1 through 12 that are even *and* divisible by 3 are 6 and 12, so

$$\Pr(\text{even and divisible by 3}) = \frac{2}{12} = \frac{1}{6}$$

(c) The numbers that are even *or* divisible by 3 are 2, 3, 4, 6, 8, 9, 10, 12, so

$$\Pr(\text{even or divisible by 3}) = \frac{8}{12} = \frac{2}{3}$$

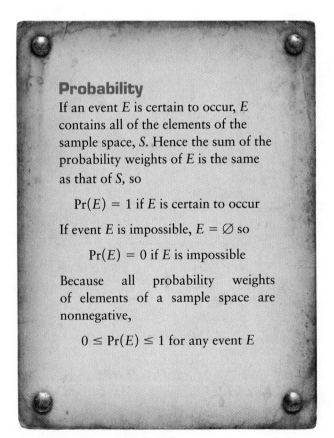

Probability

If an event E is certain to occur, E contains all of the elements of the sample space, S. Hence the sum of the probability weights of E is the same as that of S, so

$$\Pr(E) = 1 \text{ if } E \text{ is certain to occur}$$

If event E is impossible, $E = \varnothing$ so

$$\Pr(E) = 0 \text{ if } E \text{ is impossible}$$

Because all probability weights of elements of a sample space are nonnegative,

$$0 \le \Pr(E) \le 1 \text{ for any event } E$$

EXAMPLE 7.3 Sample Spaces

→ Suppose a coin is tossed 3 times.
(a) Construct an equiprobable sample space for the experiment.
(b) Find the probability of obtaining 0 heads.
(c) Find the probability of obtaining 2 heads.

SOLUTION

(a) Perhaps the most obvious way to record the possibilities for this experiment is to list the number of heads that could result: {0, 1, 2, 3}. But the probability of obtaining 0 heads is different from the probability of obtaining 2 heads, so this sample space is not equiprobable. A sample space in which each outcome is equally likely is {HHH, HHT, HTH, THH, HTT, THT, TTH, TTT}, where HHT indicates that the first two tosses were heads and the third was a tail.

(b) Because there are 8 equally likely possible outcomes, $n(S) = 8$. Only one of the eight possible outcomes, $E = \{TTT\}$, gives 0 heads, so $n(E) = 1$. Thus

$$\Pr(0 \text{ heads}) = \frac{n(E)}{n(S)} = \frac{1}{8}$$

(c) The event "two heads" is $F = \{HHT, HTH, THH\}$, so

$$\Pr(2 \text{ heads}) = \frac{n(F)}{n(S)} = \frac{3}{8}$$

To find the probability of obtaining a given sum when a pair of dice is rolled, we need to determine the outcomes. However, the sums {2, 3, 4, 5, 6, 7, 8, 9, 10, 11, 12} are not equally likely; there is only one way to obtain a sum of 2, but there are several ways to obtain a sum of 6. If we distinguish between the two dice we are rolling, and we record all the possible outcomes for each die, we see that there are 36 possibilities, each of which is equally likely (see Table 7.1).

This list of possible outcomes for finding the sum of two dice is an equiprobable sample space for the experiment. Because the 36 elements in the sample space are equally likely, we can find the probability that a given

Table 7.1

First Die	Second Die					
	1	2	3	4	5	6
1	(1, 1)	(1, 2)	(1, 3)	(1, 4)	(1, 5)	(1, 6)
2	(2, 1)	(2, 2)	(2, 3)	(2, 4)	(2, 5)	(2, 6)
3	(3, 1)	(3, 2)	(3, 3)	(3, 4)	(3, 5)	(3, 6)
4	(4, 1)	(4, 2)	(4, 3)	(4, 4)	(4, 5)	(4, 6)
5	(5, 1)	(5, 2)	(5, 3)	(5, 4)	(5, 5)	(5, 6)
6	(6, 1)	(6, 2)	(6, 3)	(6, 4)	(6, 5)	(6, 6)

sum results by determining the number of ways this sum can occur and dividing that number by 36. Thus the probability that sum 6 will occur is 5/36, and the probability that sum 9 will occur is 4/36 = 1/9. (See the four ways the sum 9 can occur in Table 7.1.) We can see that the probabilities of events can be found more easily if an equiprobable sample space is used to determine the possibilities.

Empirical Probability

Up to this point, the probabilities assigned to events were determined *theoretically,* either by assumption (the probability of obtaining a head in one toss of a fair coin is 1/2) or by knowing the entire population under consideration (if 1000 people are in a room and 400 of them are males, the probability of picking a person at random and getting a male is 400/1000 = 2/5). We can also determine probabilities **empirically,** where the assignment of the probability of an event is frequently derived from our experiences or from data that have been gathered. Probabilities developed in this way are called **empirical probabilities.** An empirical probability is formed by conducting an experiment, or by observing a situation a number of times, and noting how many times a certain event occurs.

Empirical Probability of an Event:

$$\Pr(\text{Event}) = \frac{\text{Number of observed occurrences of the event}}{\text{Total number of trials or observations}}$$

Shutterstock

In some cases empirical probability is the only way to assign a probability to an event. For example, life insurance premiums are based on the probability that an individual will live to a certain age. These probabilities are developed empirically and organized into mortality tables.

In other cases, empirical probabilities may have a theoretical framework. For example, for tossing a coin,

$$Pr(\text{Head}) = Pr(\text{Tail}) = \frac{1}{2}$$

gives the theoretical probability. If a coin was actually tossed 1000 times, it might come up heads 517 times and tails 483 times, giving empirical probabilities for these 1000 tosses of

$$Pr(\text{Head}) = \frac{517}{1000} = 0.517$$

$$Pr(\text{Tail}) = \frac{483}{1000} = 0.483$$

In general, empirical probabilities can differ from theoretical probabilities, but they are likely to reflect the theoretical probabilities when the number of observations is large. Moreover, when this is not the case, there may be reason to suspect that the theoretical model does not fit the event (for example, that a coin may be biased).

$$\frac{1}{2} \qquad \frac{1}{2}$$

Shutterstock

EXAMPLE 7.4 Quality Control

The General Standard Company can use information about a large sample of its faucets to compute the empirical probability that a faucet chosen at random will function properly (not leak) for 3 years. Of a sample of 12,316 faucets produced in 2007, 137 were found to fail within 3 years. Find the empirical probability that any faucet the company produces will work properly for 3 years.

SOLUTION

If 137 of these faucets fail to work properly, then $12{,}316 - 137 = 12{,}179$ will work properly. Thus the probability that any faucet selected at random will function properly is

$$Pr(\text{not leak}) = \frac{12{,}179}{12{,}316} \approx 0.989$$

Odds

We sometimes use **odds** to describe the probability that an event will occur. We say that its odds were 2 to 1, or 2:1. The odds in favor of an event E occurring and the odds against E occurring are found as follows.

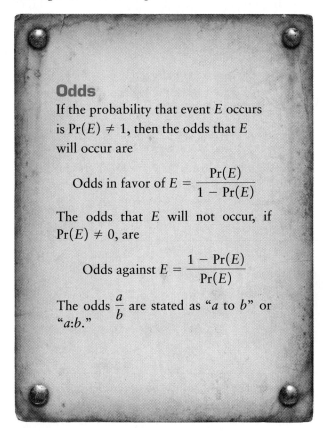

Odds

If the probability that event E occurs is $Pr(E) \neq 1$, then the odds that E will occur are

$$\text{Odds in favor of } E = \frac{Pr(E)}{1 - Pr(E)}$$

The odds that E will not occur, if $Pr(E) \neq 0$, are

$$\text{Odds against } E = \frac{1 - Pr(E)}{Pr(E)}$$

The odds $\dfrac{a}{b}$ are stated as "a to b" or "$a{:}b$."

EXAMPLE 7.5 Odds

If the probability of drawing a queen from a deck of playing cards is $1/13$, what are the odds
(a) in favor of drawing a queen?
(b) against drawing a queen?

SOLUTION

(a) $\dfrac{1/13}{1 - 1/13} = \dfrac{1/13}{12/13} = \dfrac{1}{12}$

The odds in favor of drawing a queen are 1 to 12, which we write as 1:12.

(b) $\dfrac{12/13}{1/13} = \dfrac{12}{1}$

The odds against drawing a queen are 12 to 1, denoted 12:1.

Odds and Probability

If the odds in favor of an event E occurring are $a{:}b$, the probability that E will occur is

$$Pr(E) = \frac{a}{a + b}$$

If the odds against E are $b{:}a$, then

$$Pr(E) = \frac{a}{a + b}$$

EXAMPLE 7.6 Robbery

The day after O.J. Simpson was arrested for armed robbery in September 2007, the odds against his being convicted were given in Las Vegas as 6 to 5. (*Source:* Fox News Network) What probability of conviction did this indicate?

SOLUTION

The odds against the conviction are 6:5, so the probability of conviction is

$$Pr(\text{convicted}) = \frac{5}{5 + 6} = \frac{5}{11}$$

7.1 Exercises

1. One ball is drawn at random from a bag containing 4 red balls and 6 white balls. What is the probability that the ball is
   (a) red?   (b) green?   (c) red or white?

2. If you draw one card at random from a deck of 12 cards numbered 1 through 12, inclusive, what is the probability that the number you draw is divisible by 4?

3. A die is rolled. Find the probability of getting a number greater than 0.

4. An urn contains three red balls numbered 1, 2, 3; four white balls numbered 4, 5, 6, 7; and three black balls numbered 8, 9, 10. A ball is drawn from the urn. What is the probability that
   (a) it is red?
   (b) it is odd-numbered?
   (c) it is red and odd-numbered?
   (d) it is red or odd-numbered?
   (e) it is not black?

5. From a deck of 52 ordinary playing cards, one card is drawn. Find the probability that it is
   (a) a queen.
   (b) a red card.
   (c) a spade.

6. Suppose a fair coin is tossed two times. Construct an equiprobable sample space for the experiment, and determine each of the following probabilities:
   (a) Pr(0 heads)   (b) Pr(1 head)   (c) Pr(2 heads)

7. Suppose a die is tossed 1200 times and a 6 comes up 431 times.
   (a) Find the empirical probability for a 6 to occur.
   (b) On the basis of a comparison of the empirical probability and the theoretical probability, do you think the die is fair or biased?

8. If the probability that an event will occur is 2/5, what are the odds
   (a) in favor of the event occurring?
   (b) against the event occurring?

9. If the odds that a particular horse will win a race are 1:20, what is the probability
   (a) that the horse will win the race?
   (b) that the horse will lose the race?

Applications

10. *Drug use* Forty-six percent of marijuana use among youth occurs in the inner cities. (*Source:* Partnership for a Drugfree America) If an instance of such marijuana use is chosen at random, what is the probability that the use occurs in an inner city?

11. *Car maintenance* A car rental firm has 425 cars. Sixty-three of these cars have defective turn signals and 32 have defective tires. What is the probability that one of these cars selected at random has
    (a) defective turn signals?
    (b) defective tires?

12. *Voting* The table gives the average number of voters in each of three political parties during the last 12 years, along with the average number that voted in presidential elections during this period.

(a) For each political party, use these data to find the probability that a person selected at random from the registered voters in the party will vote in the next election.

(b) For which party is the probability highest?

	Political Party		
	Republican	Democratic	Independent
Registered voters	4500	6100	2200
Voted	2835	2501	1122

13. ***Sales promotion*** In a sales promotion, a clothing store gives its customers a chance to draw a ticket from a box that contains a discount on their next purchase. The box contains 3000 tickets giving a 10% discount, 500 giving a 30% discount, 100 giving a 50% discount, and 1 giving a 100% discount. What is the probability that a given customer will randomly draw a ticket giving

(a) a 100% discount?

(b) a 50% discount?

(c) a discount of less than 50%?

(d) Is a given customer more likely to get a ticket with a 30% discount or with a discount higher than 30%?

14. ***Blood types*** Human blood is classified by blood type, which indicates the presence or absence of the antigens A, B, and Rh, as follows.

A present	Type A
B present	Type B
Both A and B present	Type AB
Neither A nor B present	Type O

Each of these types is combined with a + or a − sign to indicate whether the Rh antigen is present. Write a sample space containing all possible blood types. Do you think this is an equiprobable sample space?

15. ***Education*** A professor assigns 6 homework projects and then, during the next class, rolls a die to determine which one of the 6 projects to grade. If a student has completed 2 of the assignments, what is the probability that the professor will grade a project that the student has completed?

Need more practice?
Find more here: cengagebrain.com

7.2 Unions, Intersections, and Complements of Events

AS WE STATED IN THE PREVIOUS SECTION, EVENTS DETERMINE SUBSETS OF THE SAMPLE SPACE FOR A PROBABILITY EXPERIMENT. THE INTERSECTION, UNION, AND COMPLEMENT OF EVENTS ARE DEFINED AS FOLLOWS.

Unions, Intersections, and Complements of Events

If E and F are two events in a sample space S, then the intersection of E and F is

$$E \cap F = \{a: a \in E \quad \text{and} \quad a \in F\}$$

the union of E and F is

$$E \cup F = \{a: a \in E \quad \text{or} \quad a \in F\}$$

and the complement of E is

$$E' = \{a: a \in S \quad \text{and} \quad a \notin E\}$$

Using these definitions, we can write the probabilities that certain events will occur as follows:

$$Pr(E \textbf{ and } F \text{ both occur}) = Pr(E \cap F)$$
$$Pr(E \textbf{ or } F \text{ occurs}) = Pr(E \cup F)$$
$$Pr(E \text{ does } \textbf{not} \text{ occur}) = Pr(E')$$

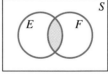 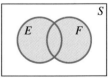 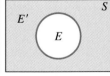

$E \cap F$ $E \cup F$

Let's see how these definitions work in practice by considering a card drawing.

EXAMPLE 7.7 Unions and Intersections of Events

→ A card is drawn from a box containing 15 cards numbered 1 to 15. What is the probability that the card is

(a) even and divisible by 3?

(b) even or divisible by 3?   (c) not even?

SOLUTION

The sample space S contains the 15 numbers.

$$S = \{1, 2, 3, 4, 5, 6, 7, 8, 9, 10, 11, 12, 13, 14, 15\}$$

If we let E represent "even-numbered" and D represent "number divisible by 3," we have

$$E = \{2, 4, 6, 8, 10, 12, 14\} \quad \text{and} \quad D = \{3, 6, 9, 12, 15\}$$

These sets are shown in the Venn diagram in Figure 7.1.

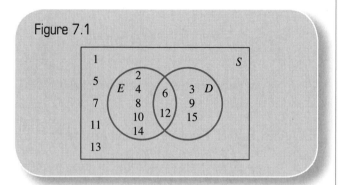

Figure 7.1

(a) The event "even and divisible by 3" is $E \cap D = \{6, 12\}$, so

$$\text{Pr(even and divisible by 3)} = \text{Pr}(E \cap D) = \frac{n(E \cap D)}{n(S)} = \frac{2}{15}$$

(b) The event "even or divisible by 3" is $E \cup D = \{2, 3, 4, 6, 8, 9, 10, 12, 14, 15\}$, so

$$\text{Pr(even or divisible by 3)} = \frac{n(E \cup D)}{n(S)} = \frac{10}{15} = \frac{2}{3}$$

(c) The event "not even," or E', contains all elements of S not in E, so $E' = \{1, 3, 5, 7, 9, 11, 13, 15\}$, and the probability that the card is not even is

$$\text{Pr(not }E) = \text{Pr}(E') = \frac{n(E')}{n(S)} = \frac{8}{15}$$

Probability of the Complement:

Because the complement of an event contains all the elements of *S except* for those elements in *E* (see Figure 7.2), the number of elements in *E′* is $n(S) - n(E)$, so

$$\text{Pr}(E') = \frac{n(S) - n(E)}{n(S)} = 1 - \frac{n(E)}{n(S)} = 1 - \text{Pr}(E)$$

$$\text{Pr(not }E) = \text{Pr}(E') = 1 - \text{Pr}(E)$$

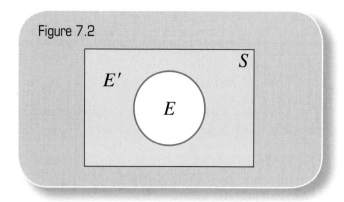

Figure 7.2

EXAMPLE 7.8 Heart Attack Risk

Suppose that a 50-year-old man has systolic blood pressure of 110 and that 60% of his cholesterol is HDL. The risk of heart attack in the next 10 years is given in Table 7.2. If a man satisfying these conditions is selected at random and has a cholesterol level of 200, what is the probability that he will not have a heart attack by age 60?

Table 7.2

Total Cholesterol	Risk of Heart Attack
130	1%
150	1
175	2
200	2
250	3
300	5

SOLUTION

If the man selected at random satisfies the given conditions and has a cholesterol level of 200, the probability that he will have a heart attack in the next 10 years (that is, by age 60) is 2% = 0.02. Thus the probability that he will not have a heart attack by age 60 is

$$\text{Pr(no attack)} = 1 - \text{Pr(attack)} = 1 - 0.02 = 0.98$$

Inclusion-Exclusion Principle

The number of elements in the set $E \cup F$ is the number of elements in E plus the number of elements in F, *minus* the number of elements in $E \cap F$ [because this number was counted twice in $n(E) + n(F)$]. Thus,

$$n(E \cup F) = n(E) + n(F) - n(E \cap F)$$

Thus,

$$\Pr(E \cup F) = \frac{n(E \cup F)}{n(S)} = \frac{n(E)}{n(S)} + \frac{n(F)}{n(S)} - \frac{n(E \cap F)}{n(S)}$$

$$= \Pr(E) + \Pr(F) - \Pr(E \cap F)$$

This establishes what is called the **inclusion-exclusion principle**, which is summarized as follows.

> **Inclusion-Exclusion Principle:**
> If *E* and *F* are any two events, then the probability that one event or the other will occur, denoted Pr(*E* or *F*), is given by
>
> Pr(*E* or *F*) = Pr(*E* ∪ *F*) = Pr(*E*) + Pr(*F*) − Pr(*E* ∩ *F*)

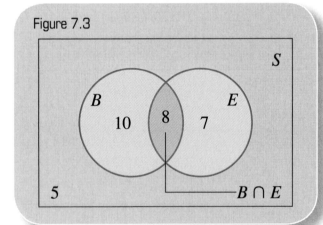

Figure 7.3

Suppose that of 30 people interviewed for a position, 18 had a business degree, 15 had previous experience, and 8 of those with experience also had a business degree. The Venn diagram in Figure 7.3 illustrates this situation.

Because 8 of the 30 people have both a business degree and experience,

$$\Pr(B \cap E) = \frac{8}{30}$$

We also see that

$$\Pr(B) = \frac{18}{30} \quad \text{and} \quad \Pr(E) = \frac{15}{30}$$

Thus the probability of choosing a person at random who has a business degree or experience is

$$\Pr(B \cup E) = \Pr(B) + \Pr(E) - \Pr(B \cap E)$$

$$= \frac{18}{30} + \frac{15}{30} - \frac{8}{30} = \frac{25}{30} = \frac{5}{6}$$

Mutually Exclusive Events

We say that events *E* and *F* are **mutually exclusive** if and only if $E \cap F = \emptyset$. Thus,

$$\Pr(E \cup F) = \Pr(E) + \Pr(F) - 0 = \Pr(E) + \Pr(F)$$

if and only if *E* and *F* are mutually exclusive.

> **Mutually Exclusive Events:** If *E* and *F* are **mutually exclusive,** then
> Pr(*E* ∩ *F*) = 0, and
>
> Pr(*E* or *F*) = Pr(*E* ∪ *F*) = Pr(*E*) + Pr(*F*).

EXAMPLE 7.9 Property Development

A firm is considering three possible locations for a new factory. The probability that site A will be selected is 1/3 and the probability that site B will be selected is 1/5. Only one location will be chosen.

(a) What is the probability that site A or site B will be chosen?

(b) What is the probability that neither site A nor site B will be chosen?

SOLUTION

(a) The two events are mutually exclusive, so

$$\text{Pr(site A or site B)} = \text{Pr(site A} \cup \text{site B)}$$

$$= \text{Pr(site A)} + \text{Pr(site B)} = \frac{1}{3} + \frac{1}{5} = \frac{8}{15}$$

(b) The probability that neither will be chosen is $1 - 8/15 = 7/15$.

The formula for the probability of the union of mutually exclusive events can be extended beyond two events.

$$\text{Pr}(E_1 \cup E_2 \cup \cdots \cup E_n) = \text{Pr}(E_1) + \text{Pr}(E_2) + \cdots + \text{Pr}(E_n)$$

for mutually exclusive events $E_1, E_2, \ldots, E_n$.

7.2 Exercises

1. If you draw one card at random from a deck of 12 cards numbered 1 through 12, what is the probability that the number you draw is divisible by 3 and even?

2. An ordinary die is tossed. What is the probability of getting a number divisible by 3 or an odd number?

3. A bag contains 5 red balls numbered 1, 2, 3, 4, 5 and 9 white balls numbered 6, 7, 8, 9, 10, 11, 12, 13, 14. If a ball is drawn, what is the probability that
   (a) the ball is red and even-numbered?
   (b) the ball is red or even-numbered?

4. If you draw one card from a deck of 12 cards numbered 1 through 12, what is the probability that the card will be odd or divisible by 3?

5. A cube has 2 faces painted red, 2 painted white, and 2 painted blue. What is the probability of getting a red face or a white face in one roll?

6. A ball is drawn from a bag containing 13 red balls numbered 1–13 and 5 white balls numbered 14–18. What is the probability that
   (a) the ball is not even-numbered?
   (b) the ball is red and even-numbered?
   (c) the ball is red or even-numbered?
   (d) the ball is neither red nor even-numbered?

Applications

7. **Drug use** Forty-six percent of marijuana use among youth occurs in the inner cities. If an instance of such marijuana use is chosen at random, what is the probability that the use does not occur in an inner city? (*Source:* Partnership for a Drugfree America)

8. **Linguistics** Of 100 students, 24 can speak French, 18 can speak German, and 8 can speak both French and German. If a student is picked at random, what is the probability that he or she can speak French or German?

9. **Salaries** The table gives the percent of employees of the Ace Company in each of three salary brackets, categorized by the sex of the employees. An employee is selected at random.
   (a) What is the probability that the person selected is female and makes less than $30,000?
   (b) What is the probability that the person selected makes at least $50,000?
   (c) What is the probability that the person is male or makes less than $30,000?

	Earns Less Than $30,000	Earns at Least $30,000 and Less Than $50,000	Earns at Least $50,000
Male	25%	18%	5%
Female	35%	14%	3%

10. **AIDS cases** The table gives the numbers of new AIDS cases in 2003 for people over age 13 in various categories. Use the table to find the probability that a person who contracted AIDS in 2003
    (a) was female.
    (b) was female or was an injected drug user.
    (c) was male or had sex with a drug user.

	Total	Sex with Drug User	Heterosexual	Injected Drug User	Hemophilia Disorder
Male	33,250	477	3371	4866	74
Female	11,561	985	5234	2262	11

Source: HIV/AIDS Surveillance Report, 2003

11. **Education** A mathematics class consists of 16 engineering majors, 12 science majors, and 4 liberal arts majors.
    (a) What is the probability that a student selected at random will be a science or liberal arts major?
    (b) What is the probability that a student selected at random will be an engineering or science major?
    (c) Five of the engineering students, 6 of the science majors, and 2 of the liberal arts majors

are female. What is the probability that a student selected at random is an engineering major or is female?

Need more practice?
Find more here: cengagebrain.com

7.3 Conditional Probability: The Product Rule

EACH PROBABILITY THAT WE HAVE COMPUTED HAS BEEN RELATIVE TO THE SAMPLE SPACE FOR THE EXPERIMENT. SOMETIMES INFORMATION IS GIVEN THAT REDUCES THE SAMPLE SPACE NEEDED TO SOLVE A STATED PROBLEM. FOR EXAMPLE, IF A DIE IS ROLLED, THE PROBABILITY THAT A 5 OCCURS, GIVEN THAT AN ODD NUMBER IS ROLLED, IS DENOTED BY

PR(5 ROLLED | ODD NUMBER ROLLED)

The knowledge that an odd number occurs reduces the sample space from the numbers on a die, $S = \{1, 2, 3, 4, 5, 6\}$, to the sample space for odd numbers on a die, $S_1 = \{1, 3, 5\}$. Thus if E is the event "a 5 is rolled," then we have

Pr(5 rolled | odd number rolled) =

$$\frac{n(5 \text{ and odd number rolled})}{n(\text{odd number rolled})} = \frac{n(E \cap S_1)}{n(S_1)} = \frac{1}{3}$$

To find $\Pr(A|B)$, we seek the probability that A occurs, given that B occurs. Figure 7.4 shows the original sample space S with the reduced sample space B shaded. Because all elements must be contained in B, we evaluate $\Pr(A|B)$ by dividing the number of elements in $A \cap B$ by the number of elements in B.

The probability that A occurs, given that B occurs, is denoted by $\Pr(A|B)$ and is given by

$$\Pr(A|B) = \frac{n(A \cap B)}{n(B)}$$

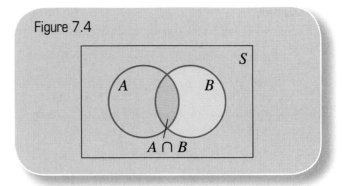

Figure 7.4

EXAMPLE 7.10 Public Opinion

→ Suppose the following table summarizes the opinions of various groups on the issue of increased enforcement of immigration laws. What is the probability that a person selected at random from this group of people favors increased enforcement of immigration laws, given that the person is nonwhite?

| Opinion | Whites | | Nonwhites | | |
	Republicans	Democrats	Republicans	Democrats	Totals
Favor	300	100	25	10	435
Oppose	100	250	25	190	565
Total	400	350	50	200	1000

SOLUTION

There are $50 + 200 = 250$ nonwhite and $25 + 10 = 35$ people who both are nonwhite and favor increased enforcement, so

$$\Pr(\text{favor} \mid \text{nonwhite}) =$$

$$\frac{n(\text{favor and nonwhite})}{n(\text{nonwhite})} = \frac{35}{250} = \frac{7}{50}$$

The probability found in Example 7.10 is called **conditional probability**.

> **Conditional Probability:** Let A and B be events in the sample space S with $\Pr(B) > 0$. The conditional probability that event A occurs, given that event B occurs, is given by
>
> $$\Pr(A|B) = \frac{\Pr(A \cap B)}{\Pr(B)}$$

EXAMPLE 7.11 Conditional Probability

→ A red die and a green die are rolled. What is the probability that the sum rolled on the dice is 6, given that the sum is less than 7?

SOLUTION

Using the conditional probability formula and the sample space in Table 7.1 in Section 7.1, we have

$$\Pr(\text{sum is 6}\,|\,\text{sum} < 7) = \frac{\Pr(\text{sum is 6} \cap \text{sum} < 7)}{\Pr(\text{sum} < 7)}$$

Because

$$\Pr(\text{sum is 6} \cap \text{sum} < 7) = \Pr(\text{Sum is 6}) = \frac{5}{36}$$

the probability is

$$\Pr(\text{sum is 6}\,|\,\text{sum} < 7) =$$
$$= \frac{5/36}{1/36 + 2/36 + 3/36 + 4/36 + 5/36}$$
$$= \frac{5/36}{15/36} = \frac{1}{3}$$

The Product Rule

We can use the formulas for $\Pr(A|B)$ and $\Pr(B|A)$ to obtain the following for $\Pr(A \cap B)$.

> **Product Rule:** If A and B are probability events, then the probability of the event "A and B" is $\Pr(A \text{ and } B)$, and it can be found by one or the other of these two formulas.
> $$\Pr(A \text{ and } B) = \Pr(A \cap B) = \Pr(A) \cdot \Pr(B|A)$$
> or
> $$\Pr(A \text{ and } B) = \Pr(A \cap B) = \Pr(B) \cdot \Pr(A|B)$$

EXAMPLE 7.12 Cards

→ Suppose that from a deck of 52 cards, two cards are drawn in succession, without replacement. Find the probability that both cards are kings.

SOLUTION

We seek the probability that the first card will be a king and the probability that the second card will be a king given that the first card was a king. The probability that the first card is a king is

$$\Pr(\text{1st card is a king}) = \frac{4}{52}$$

and the probability that the second card is a king is

$$\Pr(\text{2nd king}\,|\,\text{1st king}) = \frac{3}{51}$$

because one king has been removed from the deck on the first draw. Thus

$$\Pr(\text{2 kings}) = \Pr(\text{1st king}) \cdot \Pr(\text{2nd king}\,|\,\text{1st king})$$
$$= \frac{4}{52} \cdot \frac{3}{51} = \frac{1}{221}$$

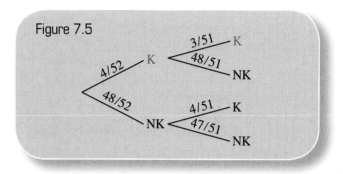

Figure 7.5

We can represent the outcomes in Example 7.12 with a **probability tree,** as shown in Figure 7.5. The outcome of the first draw is either a king or not a king; it is illustrated by the beginning two branches of the tree. The possible outcomes of the second trial can be treated as branches that emanate from each of the outcomes of the first trial. The tree shows that if a king occurs on the first trial, the outcome of the second trial is either a king or not a king, and the probabilities of these outcomes depend on the fact that a king occurred on the first trial. On the other hand, the probabilities of getting or not getting a king are different if the first draw does not yield a king, so the branches illustrating these possible outcomes have different probabilities.

Shutterstock

How does the number of sides on a die—for example, 6 sides, 10 sides, 21 sides—affect probability calculations?

There is only one "path" through the tree that illustrates getting kings on both draws, and the probability that both draws result in kings is the product of the probabilities on the branches of this path. Note that this gives $\frac{4}{52} \cdot \frac{3}{51} = \frac{1}{221}$, which is the same Product Rule probability found in Example 7.12.

EXAMPLE 7.13 Delinquent Accounts

Polk's Department Store has observed that 80% of its charge accounts have men's names on them and that 16% of the accounts with men's names on them have been delinquent at least once, while 5% of the accounts with women's names on them have been delinquent at least once. What is the probability that an account selected at random
(a) is in a man's name and is delinquent?
(b) is in a woman's name and is delinquent?

SOLUTION

(a) Pr(man's account) = 0.80
    Pr(delinquent | man's) = 0.16
    Pr(man's and delinquent) = Pr(man's ∩ delinquent)
    = Pr(man's) · Pr(delinquent | man's)
    = (0.80)(0.16) = 0.128
(b) Pr(woman's account) = 0.20
    Pr(delinquent | woman's) = 0.05
    Pr(woman's and delinquent) = (0.20)(0.05) = 0.01

Independent Events

If a coin is tossed twice, the sample space is $S = \{HH, HT, TH, TT\}$, so

Pr(2nd is H | 1st is H) =

$$\frac{\text{Pr(1st is H} \cap \text{2nd is H)}}{\text{Pr(1st is H)}} = \frac{\text{Pr(Both H)}}{\text{Pr(1st is H)}} = \frac{1/4}{1/2} = \frac{1}{2}$$

We can see that the probability that the second toss of a coin is a head, given that the first toss was a head, was 1/2, which is the same as the probability that the second toss is a head, regardless of what happened on the first toss. This is because the results of the coin tosses are **independent events**. We can define *independent events* as follows.

Independent Events: The events A and B are independent if and only if

Pr($A | B$) = Pr(A)   and   Pr($B | A$) = Pr(B)

This means that the occurrence or non-occurrence of one event does not affect the other.

If A and B are independent events, then this definition can be used to simplify the Product Rule:

$$\text{Pr}(A \cap B) = \text{Pr}(A) \cdot \text{Pr}(B \mid A) = \text{Pr}(A) \cdot \text{Pr}(B)$$

Thus we have the following.

Product Rule for Independent Events: If A and B are independent events, then

Pr(A and B) = Pr($A \cap B$) = Pr(A) · Pr(B)

EXAMPLE 7.14 Independent Events

A die is rolled and a coin is tossed. Find the probability of getting a 4 on the die and a head on the coin.

SOLUTION

Let E_1 be "4 on the die" and E_2 be "head on the coin." The events are independent because what occurs on the die does not affect what happens to the coin.

$$\text{Pr}(E_1) = \text{Pr}(4 \text{ on die}) = \frac{1}{6} \quad \text{and}$$

$$\text{Pr}(E_2) = \text{Pr}(\text{head on coin}) = \frac{1}{2}$$

Thus, $\text{Pr}(E_1 \cap E_2) = \text{Pr}(E_1) \cdot \text{Pr}(E_2) = \frac{1}{6} \cdot \frac{1}{2} = \frac{1}{12}$

The Product Rules can be expanded to include more than two events, as the next example shows.

EXAMPLE 7.15 Identity Theft

Identity theft can occur when an unscrupulous individual learns someone else's Social Security number. Suppose a person learns the first five digits of your Social Security number. Calculate the probability that the person guesses the last four digits and thus gains access to your Social Security number.

SOLUTION

The digits in your Social Security number can repeat, so the probability that each digit is guessed correctly is 1/10. Thus the probability that the person guesses the last four digits is

$$\frac{1}{10} \cdot \frac{1}{10} \cdot \frac{1}{10} \cdot \frac{1}{10} = \frac{1}{10,000}$$

In general, we can calculate the probability of at least one success in n trials as follows.

Pr(at least 1 success in n trials) = 1 − Pr(0 successes)

7.3 Exercises

1. A card is drawn from a deck of 52 playing cards. Given that it is a red card, what is the probability that
   (a) it is a heart?   (b) it is a king?

2. A die has been "loaded" so that the probability of rolling any even number is 2/9 and the probability of rolling any odd number is 1/9. What is the probability of
   (a) rolling a 6, given that an even number is rolled?
   (b) rolling a 3, given that a number divisible by 3 is rolled?

3. A bag contains 9 red balls numbered 1, 2, 3, 4, 5, 6, 7, 8, 9 and 6 white balls numbered 10, 11, 12, 13, 14, 15. One ball is drawn from the bag. What is the probability that the ball is red, given that the ball is even-numbered?

4. A bag contains 4 red balls and 6 white balls. Two balls are drawn without replacement.
   (a) What is the probability that the second ball is white, given that the first ball is red?
   (b) What is the probability that the second ball is red, given that the first ball is white?
   (c) Answer part (a) if the first ball is replaced before the second is drawn.

5. A die is thrown twice. What is the probability that a 3 will result the first time and a 6 the second time?

6. (a) A box contains 3 red balls, 2 white balls, and 5 black balls. Two balls are drawn at random from the box (with replacement of the first before the second is drawn). What is the probability of getting a red ball on the first draw and a white ball on the second?
   (b) Answer part (a) if the first ball is not replaced before the second is drawn.
   (c) Are the events in part (a) or in part (b) independent? Explain.

7. One card is drawn at random from a deck of 52 cards. The first card is not replaced, and a second card is drawn. Find the probability that
   (a) both cards are spades.
   (b) the first card is a heart and the second is a club.

Applications

8. *Blood types* In the pretrial hearing of the O.J. Simpson case, the prosecution stated that Mr. Simpson's blood markers included type A blood, which 33.7% of the population has; blood SD subtype 1, which 79.6% of the population has; and PGM 2+2−, which 1.6% of the population has. If these blood markers are independent, what is the probability that a person selected at random will have the same blood markers as O.J. Simpson?

9. *Quality control* Each computer component that the Peggos Company produces is tested twice before it is shipped. There is a 0.7 probability that a defective component will be so identified by the first test and a 0.8 probability that it will be identified as being defective by the second test. What is the probability that a defective component will not be identified as defective before it is shipped?

10. *Lactose intolerance* Lactose intolerance affects about 50% of Hispanic Americans, and 9% of the residents of the United States are Hispanic. (*Source:* Jean Carper, "Eat Smart," *USA Weekend,* September 4, 1994) If a U.S. resident is selected at random, what is the probability that the person will be Hispanic and have lactose intolerance?

11. *Quality control* To test its shotgun shells, a company fires 5 of them. What is the probability that all 5 will fire properly if 5% of the company's shells are defective?

12. *Birth control* Suppose a birth control pill is 99% effective in preventing pregnancy.
    (a) What is the probability that none of 100 women using the pill will become pregnant?
    (b) What is the probability that at least one woman per 100 users will become pregnant?

Need more practice?
Find more here: cengagebrain.com

7.4 Probability Trees and Bayes' Formula

IN SECTION 7.3, "CONDITIONAL PROBABIL-ITY: THE PRODUCT RULE," WE INTRODUCED PROBABILITY TREES TO DESCRIBE THE

PROBABILITIES THAT ARISE IN EXPERIMENTS INVOLVING MORE THAN ONE TRIAL. TREES PROVIDE A SYSTEMATIC WAY TO ANALYZE PROBABILITY EXPERIMENTS THAT HAVE TWO OR MORE TRIALS. IN THIS SECTION, WE WILL USE TREES TO SOLVE SOME ADDITIONAL PROBABILITY PROBLEMS AND TO SOLVE BAYES PROBLEMS.

Probability Trees

Recall that in a probability tree, each set of branches on the tree represents one trial, and we find the probability of an event as follows.

Probability Trees

The probability attached to each branch is the conditional probability that the specified event will occur, given that the events on the preceding branches have occurred.

1. For each event that can be described by a single path through the tree, the probability that that event will occur is the product of the probabilities on the branches along the path that represents the event.

2. If an event can be described by two or more paths through a probability tree, its probability is found by adding the probabilities from the paths.

EXAMPLE 7.16 Quality Control

A security system is manufactured with a "fail-safe" provision so that it functions properly if any two or more of its three main components, A, B, and C, are functioning properly. The probabilities that components A, B, and C are functioning properly are 0.95, 0.90, and 0.92, respectively. What is the probability that the system functions properly?

SOLUTION

The probability tree in Figure 7.6 shows all of the possibilities for the components of the system. Each stage of the tree corresponds to a component, and G

indicates that the component is good; N indicates it is not good. Four paths through the probability tree give at least two good (functioning) components, so there are four mutually exclusive possible outcomes. The probability that the system functions properly is

$$\Pr(\text{at least } 2G) = \Pr(GGG \cup GGN \cup GNG \cup NGG)$$
$$= (0.95)(0.90)(0.92) + (0.95)(0.90)(0.08)$$
$$+ (0.95)(0.10)(0.92) + (0.05)(0.90)(0.92)$$
$$= 0.9838$$

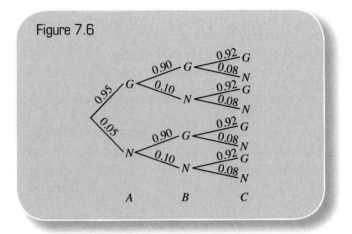

Figure 7.6

Bayes' Formula

If a probability experiment has n possible outcomes in its first stage given by $E_1, E_2, \ldots, E_n$, and if F_1 is an event in the second stage, then the probability that event E_1 occurs in the first stage, given that F_1 has occurred in the second stage, is

$$\Pr(E_1|F_1) = \frac{\Pr(E_1) \cdot \Pr(F_1|E_1)}{\Pr(E_1) \cdot \Pr(F_1|E_1) + \Pr(E_2) \cdot \Pr(F_1|E_2) + \cdots + \Pr(E_n) \cdot \Pr(F_1|E_n)}$$

Note that Bayes problems can be solved either by this formula or by the use of a probability tree. Of the two methods, many students find using the tree easier because it is less abstract than the formula.

The development of Bayes' formula can be found in the Enrichment material on CourseMate for MATH APPS at cengagebrain.com.

Bayes' Formula and Trees:

$$\Pr(E_1|F_1) = \frac{\text{Product of branch probabilities on path leading to } F_1 \text{ through } E_1}{\text{Sum of all branch products on paths leading to } F_1}$$

EXAMPLE 7.17 Medical Tests

Suppose a test for diagnosing a certain serious disease is successful in detecting the disease in 95% of all people infected but incorrectly diagnoses 4% of all healthy people as having the serious disease. Suppose also that it incorrectly diagnoses 12% of all people having another minor disease as having the serious disease. It is known that 2% of the population have the serious disease, 90% of the population are healthy, and 8% have the minor disease. Find the probability that a person selected at random has the serious disease if the test indicates that he or she does. Use H to represent healthy, M to represent having the minor disease, and D to represent having the serious disease.

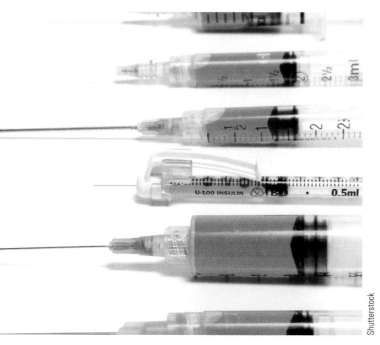

Shutterstock

SOLUTION

The tree that represents the health condition of a person chosen at random and the results of the test on that person are shown in Figure 7.7. (A test that indicates that a person has the disease is called a *positive* test.)

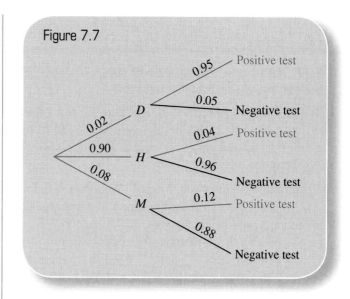

Figure 7.7

We seek $\Pr(D \mid \text{pos. test})$. Using the tree and the conditional probability formula gives

$$\Pr(D \mid \text{pos. test}) = \frac{\Pr(D \cap \text{pos. test})}{\Pr(\text{pos. test})}$$

$$= \frac{(0.02)(0.95)}{(0.90)(0.04) + (0.02)(0.95) + (0.08)(0.12)}$$

$$= \frac{0.0190}{0.0360 + 0.0190 + 0.0096}$$

$$\approx 0.2941$$

Using Bayes' formula directly gives the same result.

$$\Pr(D \mid \text{pos. test}) = \Pr(D \mid +)$$

$$= \frac{\Pr(D) \cdot \Pr(+ \mid D)}{\Pr(H) \cdot \Pr(+ \mid H) + \Pr(D) \cdot \Pr(+ \mid D) + \Pr(M) \cdot \Pr(+ \mid M)}$$

$$= \frac{(0.02)(0.95)}{(0.90)(0.04) + (0.02)(0.95) + (0.08)(0.12)} \approx 0.2941$$

Table 7.3 gives a summary of probability formulas and when they are used.

Table 7.3 **Summary of Probability Formulas**

		One Trial	Two Trials	More Than Two Trials
Pr(A and B)	Independent	Sample space	$\Pr(A) \cdot \Pr(B)$	Product of probabilities
	Dependent	Sample space	$\Pr(A) \cdot \Pr(B \mid A)$	Product of conditional probabilities
Pr(A or B)	Mutually exclusive	$\Pr(A) + \Pr(B)$	Probability tree	Probability tree
	Not mutually exclusive	$\Pr(A) + \Pr(B) - \Pr(A \text{ and } B)$	Probability tree	Probability tree

7.4 Exercises

In Problems 1–4, use probability trees to find the probabilities of the indicated outcomes.

1. A bag contains 5 coins, 4 of which are fair and 1 that has a head on each side. If a coin is selected from the bag and tossed twice, what is the probability of obtaining 2 heads?

2. An urn contains 4 red, 5 white, and 6 black balls. One ball is drawn from the urn, it is not replaced, and a second ball is drawn.
   (a) What is the probability that both balls are white?
   (b) What is the probability that one ball is white and one is red?
   (c) What is the probability that at least one ball is black?

3. Three balls are drawn, without replacement, from a bag containing 4 red balls and 5 white balls. Find the probability that
   (a) three white balls are drawn.
   (b) two white balls and one red ball are drawn.
   (c) the third ball drawn is red.

4. A bag contains 5 coins, of which 4 are fair; the remaining coin has a head on both sides. If a coin is selected at random from the bag and tossed three times, what is the probability that heads will occur at least twice?

In Problems 5 and 6, use a probability tree or Bayes' formula to find the probabilities.

5. There are 3 urns containing balls. Urns I and II each contain 5 red balls, 3 white balls, and 2 green balls. Urn III contains 1 red ball, 1 white ball, and 8 green balls. An urn is selected at random and a ball is drawn from the urn. If the ball is green, find the probability that urn III was selected.

6. There are 3 urns containing coins. Urn I contains 3 gold coins, urn II contains 1 gold coin and 1 silver coin, and urn III contains 2 silver coins. An urn is selected and a coin is drawn from the urn. If the coin is gold, what is the probability that urn I was selected?

Applications

7. *Lactose intolerance* Lactose intolerance affects about 20% of non-Hispanic white Americans, 50% of Hispanic Americans, and 75% of African, Asian, and Native Americans. (*Source:* Jean Carper, "Eat Smart," *USA Weekend*, September 4, 1994)

Seventy-six percent of U.S. residents are non-Hispanic whites, 9% are Hispanic, and 15% are African, Asian, or Native American. If a person is selected from this group of people, what is the probability that the person will have lactose intolerance?

8. *Quality control* Suppose a box contains 3 defective transistors and 12 good transistors. If 2 transistors are drawn from the box without replacement, what is the probability that
   (a) the first transistor is good and the second transistor is defective?
   (b) the first transistor is defective and the second one is good?
   (c) one of the transistors drawn is good and one of them is defective?

9. *Alcoholism* A small town has 8000 adult males and 6000 adult females. A sociologist conducted a survey and found that 40% of the males and 30% of the females drink heavily. An adult is selected at random from the town.
   (a) What is the probability the person is a male?
   (b) What is the probability the person drinks heavily?
   (c) What is the probability the person is a male or drinks heavily?
   (d) What is the probability the person is a male, if it is known that the person drinks heavily?

10. *Drinking age* A survey questioned 1000 people regarding raising the legal drinking age from 18 to 21. Of the 560 who favored raising the age, 390 were female. Of the 440 opposition responses, 160 were female. If a person selected at random from this group is a man, what is the probability that the person favors raising the drinking age?

Need more practice?
Find more here: cengagebrain.com

7.5 Counting: Permutations and Combinations

SUPPOSE THAT YOU DECIDE TO DINE AT A RESTAURANT THAT OFFERS 3 APPETIZERS,

8 ENTREES, AND 6 DESSERTS. HOW MANY DIFFERENT MEALS CAN YOU HAVE? WE CAN ANSWER QUESTIONS OF THIS TYPE USING THE **FUNDAMENTAL COUNTING PRINCIPLE.**

> **Fundamental Counting Principle:** If there are $n(A)$ ways in which an event A can occur, and if there are $n(B)$ ways in which a second event B can occur after the first event has occurred, then the two events can occur in $n(A) \cdot n(B)$ ways.

The Fundamental Counting Principle can be extended to any number of events as long as they are independent. Thus the total number of possible meals at the restaurant just mentioned is

$$n(\text{appetizers}) \cdot n(\text{entrees}) \cdot n(\text{desserts}) = 3 \cdot 8 \cdot 6 = 144$$

Let's consider an example using vanity license plates to illustrate how the Fundamental Counting Principle can help solve more complex counting problems.

EXAMPLE 7.18 License Plates

Pennsylvania offers a special "Save Wild Animals" license plate, which gives Pennsylvania zoos $15 for each plate sold. These plates have the 2 letters P and Z, followed by 4 numbers. When plates like these have all been purchased, additional plates will replace the final digit with a letter other than O. What is the maximum amount that the zoos could receive from the sale of the plates?

SOLUTION

The following numbers represent the number of possible digits or letters in each of the spaces on the license plates.

$$\underline{1} \quad \underline{1} \quad \underline{10} \quad \underline{10} \quad \underline{10} \quad \underline{35}$$

Note that the last space can be a number from 0 to 9 or a letter other than O. By the Fundamental Counting Principle, the product of these numbers gives the number of license plates that can be made satisfying the given conditions.

$$1 \cdot 1 \cdot 10 \cdot 10 \cdot 10 \cdot 35 = 35{,}000$$

Thus 35,000 automobiles can be licensed with these plates, so the zoos could receive $(\$15)(35{,}000) = \$525{,}000$.

Sometimes applying The Fundamental Counting Principle results in a product of sequential integers, such as $6 \cdot 5 \cdot 4 \cdot 3 \cdot 2 \cdot 1$. Because special products such as this frequently occur in counting theory, we use special notation to denote them. We write 6! (read "6 factorial") to denote $6 \cdot 5 \cdot 4 \cdot 3 \cdot 2 \cdot 1$. Likewise, $4! = 4 \cdot 3 \cdot 2 \cdot 1$.

> **Factorial:** For any positive integer n, we define n **factorial**, written $n!$, as
> $$n! = n(n - 1)(n - 2) \cdots 3 \cdot 2 \cdot 1$$
> We define $0! = 1$.

See the Tech Card for instructions on evaluating factorials with a graphing calculator.

Permutations

Suppose an event planner knows that 8 people feel they should be at the head table, but only 6 spaces are available. There are 6 spaces to fill, but any of 8 people can be placed in the first space, any of 7 people in the second space, and so on. Thus the total number of possible arrangements is

$$8 \cdot 7 \cdot 6 \cdot 5 \cdot 4 \cdot 3 = 20{,}160$$

The number of possible arrangements—20,160 in this case—is called the number of **permutations** of 8 things taken 6 at a time, and it is denoted $_8P_6$. Note that $_8P_6$ gives the first 6 factors of 8!, so we can use factorial notation to write the product.

$$_8P_6 = \frac{8!}{2!} = \frac{8!}{(8 - 6)!}$$

> **Permutations:** The number of possible ordered arrangements of r objects chosen from a set of n objects is called the number of **permutations** of n objects taken r at a time, and it equals
> $$_nP_r = \frac{n!}{(n - r)!}$$
> Note that $_nP_n = n!$

Imagine you're trying to determine how many ways you can select a president, a vice president, a secretary,

$$_nP_r = \frac{n!}{(n-r)!}$$

Shutterstock

Note that the fundamental difference between permutations and combinations is that permutations are used when order is a factor in the selection, and combinations are used when it is not. Because $_nP_r = n!/(n-r)!$, we have the following:

> **Combination Formula:** The number of combinations of n objects taken r at a time is also denoted by $\binom{n}{r}$; that is,
>
> $$\binom{n}{r} = \,_nC_r = \frac{n!}{r!(n-r)!}$$

To see how the combination formula works, let's apply it to sales promotions for cars. An auto dealer is offering a promotion: For a single price on a specially equipped car, you can select any 4 of 6 special options. How many different choices of specially equipped cars do you have?

The order in which you choose the options is not relevant, so we seek the number of combinations of 6 things taken 4 at a time. Thus the number of possible choices is

$$_6C_4 = \frac{6!}{4!(6-4)!} = \frac{6!}{4!2!} = 15$$

See the Tech Card for details on evaluating combinations with a graphing calculator.

We can also think about combinations in selecting people for membership in teams. Suppose 6 men and 8 women are qualified to serve on a project team to revise operational procedures, but you need to have 3 men and 3 women on the team. Three men can be selected from 6 in $\binom{6}{3}$ ways, and 3 women can be selected from 8 in $\binom{8}{3}$ ways. Therefore, 3 men *and* 3 women can be selected in

$$\binom{6}{3}\binom{8}{3} = \frac{6!}{3!3!} \cdot \frac{8!}{5!3!} = 1120 \text{ ways}$$

and a treasurer from an organization with 20 members. What you're really looking for is the number of orders (arrangements) in which 4 people can be selected from a group of 20. This number is

$$_{20}P_4 = \frac{20!}{(20-4)!} = 116{,}280$$

See the Tech Card for steps for evaluating permutations with a graphing calculator.

Combinations

When we talk about arrangements of people at a table or arrangements of digits on a license plate, order is important. We now consider counting problems in which order is not important.

To find the number of ways to select 3 people from 5 without regard to order, we would first find $_5P_3$ and then divide by the number of ways the 3 people could be ordered (3!). Thus 3 people can be selected from 5 (without regard for order) in $_5P_3/3! = 10$ ways. We say that the number of **combinations** of 5 things taken 3 at a time is 10.

> **Combinations:** The number of ways in which r objects can be chosen from a set of n objects without regard to the order of selection is called the number of **combinations** of n objects taken r at a time, and it equals
>
> $$_nC_r = \frac{_nP_r}{r!}$$

7.5 Exercises

Permutations

1. Compute $_6P_4$.

2. Compute $_{10}P_6$.

3. Compute $_5P_0$.

4. How many 4-digit numbers can be formed from the digits 1, 3, 5, 7, 8, and 9
   (a) if each digit may be used once in each number?
   (b) if each digit may be used repeatedly in each number?

Combinations

5. Compute $_{100}C_{98}$.

6. Compute $_4C_4$.

7. Compute $\binom{5}{0}$.

Applications

8. *Binz, not Benz* Mercedes Benz E-Class sedans are converted to limousines by a coach company. The limousines, named Binz, come in two models; XL Six Door and XL Vis-à-Vis are available with a choice of 2 gasoline engines (a V6 or a V8) or 2 diesel engines.
   (a) If no other options are considered, how many different Binz limos are available?
   (b) If, in addition, 5 option packages and 6 colors are available, how many different Binz limos are available?

9. *Signaling* A sailboat owner received 6 different signal flags with his new sailboat. If the order in which the flags are arranged on the mast determines the signal being sent, how many 3-flag signals can be sent?

10. *Molecules* Biologists have identified 4 kinds of small molecules: adenine, cytosine, guanine, and thymine, which link together to form larger molecules in genes. How many 3-molecule chains can be formed if the order of linking is important and each small molecule can occur more than once in a chain?

11. *Poker* A flush (5 cards from the same suit) is an excellent hand in poker. If there are 13 cards in each of four suits in a deck used in the game and there are 5 cards in a hand, how many flush hands are possible?

12. *License plates* South Carolina previously had license plates containing 2 letters and 4 digits but now has plates with 3 letters and 3 digits. How many more plates are now available?

13. *Politics* To determine voters' feelings regarding an issue, a candidate asks a sample of people to pick 4 words out of 10 that they feel best describe the issue. How many different groupings of 4 words are possible?

14. *Committees* In how many ways can a committee consisting of 6 men and 6 women be selected from a group consisting of 20 men and 22 women?

> **Need more practice?**
> Find more here: cengagebrain.com

7.6 Permutations, Combinations, and Probability

NOW THAT YOU'VE SEEN HOW THE FUNDAMENTAL COUNTING PRINCIPLE, PERMUTATIONS, AND COMBINATIONS WORK, LET'S SEE HOW THEY CAN BE APPLIED TO THE SOLUTION OF PROBABILITY PROBLEMS.

For example, suppose a psychologist claims she can teach 4-year-old children to spell 3-letter words very quickly. To test her claim, an examiner gave one of the psychologist's students cards with the letters A, C, D, K, and T on them and told the student to spell CAT.

Even if the child does spell CAT, it could occur by chance. In fact, the probability that the child spells CAT by chance is 1/60. The number of different 3-letter "words" that can be formed using these 5 cards is $n(S) = {}_5P_3 = 5!/(5-3)! = 60$. Because only one of those 60 arrangements will be CAT, $n(E) = 1$, and the probability that the child will be successful by guessing is 1/60.

EXAMPLE 7.19 License Plates

→ Suppose that a state issues license plates with 3 letters followed by 3 digits and that there are 12 words of 3 letters that are not permitted on license plates. If all possible plates are produced and a plate is selected at random, what is the probability that the plate is unacceptable and will not be issued?

SOLUTION

The total number of license plates that could be produced with 3 letters followed by 3 digits is

$$N(S) = 26^3 \cdot 10^3 = 17,576,000$$

and the number that will be unacceptable is

$$N(E) = 12 \cdot 10^3 = 12,000$$

Thus the probability that a plate selected at random will be unacceptable is

$$\frac{12,000}{17,576,000} = \frac{12}{17,576} = \frac{3}{4394} \approx 0.0006827$$

EXAMPLE 7.20 Quality Control

A manufacturing process for computer chips is such that 5 out of 100 chips are defective. If 10 chips are chosen at random from a box containing 100 newly manufactured chips, what is the probability that none of the chips will be defective?

SOLUTION

The number of ways in which any 10 chips can be chosen from 100 is

$$n(S) = {}_{100}C_{10} = \binom{100}{10} = \frac{100!}{10!90!}$$

The number of ways in which 10 chips can be chosen with none of them being defective is the same as the number of ways 10 good chips can be chosen from the 95 good chips in the box. Thus

$$n(E) = {}_{95}C_{10} = \binom{95}{10} = \frac{95!}{10!85!}$$

Thus the probability that none of the chips chosen will be defective is

$$\Pr(E) = \frac{n(E)}{n(S)} = \frac{\dfrac{95!}{10!85!}}{\dfrac{100!}{10!90!}} \approx 0.58375$$

EXAMPLE 7.21 Jury Selection

Suppose that of the 20 prospective jurors for a trial, 12 favor the death penalty and 8 do not. If 12 jurors are chosen at random from these 20, what is the probability that 7 of the jurors will favor the death penalty?

SOLUTION

The number of ways in which 7 jurors of the 12 will favor the death penalty is the number of ways in which 7 favor and 5 do not favor, or

$$n(E) = {}_{12}C_7 \cdot {}_8C_5 = \frac{12!}{7!5!} \cdot \frac{8!}{5!3!}$$

The total number of ways in which 12 jurors can be selected from 20 people is

$$n(S) = {}_{20}C_{12} = \frac{20!}{12!8!}$$

Thus the probability we seek is

$$\Pr(E) = \frac{n(E)}{n(S)} = \frac{\dfrac{12!}{7!5!} \cdot \dfrac{8!}{5!3!}}{\dfrac{20!}{12!8!}} \approx 0.3521$$

EXAMPLE 7.22 Auto Keys

Beginning in 1990, General Motors began to use a theft deterrent key on some of its cars. The key has 6 parts, with 3 patterns for each part, plus an electronic chip containing a code from 1 to 15. What is the probability that one of these keys selected at random will start a GM car requiring a key of this type?

SOLUTION

Each of the 6 parts has 3 patterns, and the chip can have any of 15 codes, so there are

$$3 \cdot 3 \cdot 3 \cdot 3 \cdot 3 \cdot 3 \cdot 15 = 10,935$$

possible keys. Because there is only one key that will start the car, the probability that a key selected at random will start a given car is $1/10,935$.

Shutterstock

A graphing calculator can also be used to find probabilities that use permutations or combinations. See the Tech Card for details.

Markov Chains

See the Enrichment material on CourseMate for MATH APPS at cengagebrain.com for a discussion of how probability applies to the study of Markov Chains.

....

7.6 Exercises

Applications

1. *Politics* A poll asks voters to rank Social Security, economics, the war on terror, health care, and education in the order of importance.
   (a) How many rankings are possible?
   (b) What is the probability that one reply chosen at random has the issues ranked in the order they appear on the survey?

2. *License plates* Suppose that all license plates in a state have 3 letters and 3 digits. If a plate is chosen at random, what is the probability that all 3 letters and all 3 numbers on the plate will be different?

3. *ATMs*
   (a) An automatic teller machine requires that each customer enter a 4-digit personal identification number (PIN) when he or she inserts a bank card. If a person finds a bank card and guesses at a PIN to use the card fraudulently, what is the probability that the person will succeed in one attempt?
   (b) If the person knows that the PIN will not have any digit repeated, what is the probability the person will succeed in guessing in one attempt?

4. *Telephones* If the first digit of a 7-digit telephone number cannot be a 0 or a 1, what is the probability that a number chosen at random will have all 7 digits the same?

5. *Rewards* As a reward for a record year, the Ace Software Company is randomly selecting 4 people from its 500 employees for a free trip to Hawaii, but it will not pay for a traveling companion. If John and Jill are married and both are employees, what is the probability that they will both win?

6. *Quality control* A box of 12 transistors has 3 defective ones. If 2 transistors are drawn from the box together, what is the probability
   (a) that both transistors are defective?
   (b) that neither transistor is defective?
   (c) that one transistor is defective?

7. *Banking* To see whether a bank has enough minority construction company loans, a social agency selects 30 loans to construction companies at random and finds that 2 of them are loans to minority companies. If the bank's claim that 10 of every 100 of its loans to construction companies are minority loans is true, what is the probability that 2 loans out of 30 are minority loans? Leave your answer with combination symbols.

8. *Diversity* Suppose that an employer plans to hire 4 people from a group of 9 equally qualified people, of whom 3 are minority candidates. If the employer does not know which candidates are minority candidates, and if she selects her employees at random, what is the probability that
   (a) no minority candidates are hired?
   (b) all 3 minority candidates are hired?
   (c) 1 minority candidate is hired?

9. *Management* Suppose that a children's basketball coach knows the best 6 players to use on his team, but pressure from parents to give everyone a chance to start in a game causes him to pick the starting team by choosing 5 players at random from the 10 team members. What is the probability that this will give him a team with 5 of his 6 best players?

....

Need more practice?
Find more here: cengagebrain.com

Chapter Exercises

1. If one ball is drawn from a bag containing 9 balls numbered 1 through 9, what is the probability that the ball's number is
   (a) odd?
   (b) divisible by 3?
   (c) odd and divisible by 3?

2. Suppose one ball is drawn from a bag containing 9 red balls numbered 1 through 9 and three white balls numbered 10, 11, 12. What is the probability that
   (a) the ball is red?
   (b) the ball is odd-numbered?
   (c) the ball is white and even-numbered?
   (d) the ball is white or odd-numbered?

3. If the probability that an event E occurs is 3/7, what are the odds that
   (a) E will occur?   (b) E will not occur?

4. Suppose that a fair coin is tossed two times. Construct an equiprobable sample space for the experiment and determine each of the following probabilities.
   (a) Pr(0 heads)   (b) Pr(1 head)   (c) Pr(2 heads)

5. Suppose that a fair coin is tossed three times. Construct an equiprobable sample space for the experiment and determine each of the following probabilities.
   (a) Pr(2 heads)   (b) Pr(3 heads)   (c) Pr(1 head)

6. A card is drawn at random from an ordinary deck of 52 playing cards. What is the probability that it is a queen or a jack?

7. A deck of 52 cards is shuffled. A card is drawn, it is replaced, the pack is again shuffled, and a second card is drawn. What is the probability that each card drawn is an ace, king, queen, or jack?

8. A card is drawn from a deck of 52 playing cards. What is the probability that it is an ace or a 10?

9. A card is drawn at random from a deck of playing cards. What is the probability that it is a king or a red card?

10. A bag contains 4 red balls numbered 1, 2, 3, 4 and 5 white balls numbered 5, 6, 7, 8, 9. A ball is drawn. What is the probability that the ball
    (a) is red and even-numbered?
    (b) is red or even-numbered?
    (c) is white or odd-numbered?

11. A bag contains 4 red balls and 3 black balls. Two balls are drawn at random from the bag without replacement. Find the probability that both balls are red.

12. A box contains 2 red balls and 3 black balls. Two balls are drawn from the box without replacement. Find the probability that the second ball is red, given that the first ball is black.

13. A bag contains 8 white balls, 5 red balls, and 7 black balls. If three balls are drawn at random from the bag, with replacement, what is the probability that the first two balls are red and the third is black?

14. A bag contains 8 white balls, 5 red balls, and 7 black balls. If three balls are drawn at random from the bag, without replacement, what is the probability that the first two balls are red and the third is black?

15. An urn contains 4 red and 6 white balls. One ball is drawn, it is not replaced, and a second ball is drawn. What is the probability that one ball is white and one is red?

16. Urn I contains 3 red and 4 white balls and urn II contains 5 red and 2 white balls. An urn is selected and a ball is drawn from it.
    (a) What is the probability that urn I is selected and a red ball is drawn?
    (b) What is the probability that a red ball is selected?
    (c) If a red ball is selected, what is the probability that urn I was selected?

17. Bag A contains 3 red, 6 black, and 5 white balls, and bag B contains 4 red, 5 black, and 7 white balls. A bag is selected at random and a ball is drawn. If the ball is white, what is the probability that bag B was selected?

18. Compute $_6P_2$.

19. Compute $_7C_3$.

20. How many 3-letter sets of initials are possible?

21. How many combinations of 8 things taken 5 at a time are possible?

Applications

22. **Senior citizens** In a certain city, 30,000 citizens out of 80,000 are over age 50. What is the probability that a citizen selected at random will be 50 years old or younger?

23. **World Series** Suppose the preseason odds that the Chicago Cubs would win the 2010 World Series were 1 to 6. What is the associated probability that the Cubs would win this series?

24. **United Nations** Of 100 job applicants to the United Nations, 30 speak French, 40 speak German, and 12 speak both French and German. If an applicant is chosen at random, what is the probability that

the applicant speaks French or German?

Productivity In Problems 25–27, suppose that a study of leadership style versus industrial productivity obtained the following data.

Productivity	Leadership Style			
	Democratic	Authoritarian	Laissez-faire	Total
Low	40	15	40	95
Medium	25	75	10	110
High	25	30	20	75
Total	90	120	70	280

25. Find the probability that a person chosen at random has a democratic style and medium productivity.

26. Find the probability that an individual chosen at random has high productivity or an authoritarian style.

27. Find the probability that an individual chosen at random has an authoritarian style, given that he or she has medium productivity.

28. **Quality control** A product must pass an initial inspection, where the probability that it will be rejected is 0.2. If it passes this inspection, it must also pass a second inspection, where the probability that it will be rejected is 0.1. What is the probability that it will pass both inspections?

29. **Color blindness** Sixty men out of 1000 and 3 women out of 1000 are color blind. A person is picked at random from a group containing 10 men and 10 women.
    (a) What is the probability that the person is color blind?
    (b) What is the probability that the person is a man if the person is color blind?

30. **Purchasing** A regional survey found that 70% of all families who indicated an intention to buy a new car bought a new car within 3 months, that 10% of families who did not indicate an intention to buy a new car bought one within 3 months, and that 22% indicated an intention to buy a new car. If a family chosen at random bought a car, find the probability that the family had not previously indicated an intention to buy a car.

31. **Management** A personnel director ranks 4 applicants for a job. How many rankings are possible?

32. **Management** An organization wants to select a president, vice president, secretary, and treasurer. If 8 people are willing to serve and each of them is eligible for any of the offices, in how many different ways can the offices be filled?

33. **Utilities** A utility company sends teams of 4 people each to perform repairs. If it has 12 qualified people, how many different ways can people be assigned to a team?

34. **Juries** A jury can be deadlocked if one person disagrees with the rest. There are 12 ways in which a jury can be deadlocked if one person disagrees, because any one of the 12 jurors could disagree.
    (a) In how many ways can a jury be deadlocked if 2 people disagree?
    (b) If 3 people disagree?

35. **Blood types** In a book describing the mass murder of 18 people in northern California, a policewoman was quoted as stating that there are blood groups O, A, B, and AB, positive and negative, blood types M, N, MN, and P, and 8 Rh blood types, so there are 288 unique groups. (*Source*: Joseph Harrington and Robert Berger, *Eye of Evil*, St. Martin's Press, 1993) Comment on this conclusion.

36. **Scheduling** A college registrar adds 4 new courses to the list of offerings for the spring semester. If he added the course names in random order at the end of the list, what is the probability that these 4 courses are listed in alphabetical order?

37. **Lottery** Pennsylvania's Daily Number pays 500 to 1 to people who play the winning 3-digit number exactly as it is drawn. However, players can "box" a number so they can win $80 from a $1 bet if the 3 digits they pick come out in any order. What is the probability that a "boxed" number with 3 different digits will be a winner?

38. **Quality control** A supplier has 200 compact discs, of which 10% are known to be defective. If a music store purchases 10 of these discs, what is the probability that
    (a) none of the discs is defective?
    (b) 2 of the discs are defective?

39. **Stocks** Mr. Way must sell stocks from 3 of the 5 companies whose stocks he owns so that he can send his children to college. If he chooses the companies at random, what is the probability that the 3 companies will be the 3 with the best future earnings?

40. **Quality control** A sample of 6 fuses is drawn from a lot containing 10 good fuses and 2 defective fuses. Find the probability that the number of defective fuses is
    (a) exactly 1.   (b) at least 1.

Need more practice?
Find more here: cengagebrain.com

Probability
and Data Description

In this chapter, we discuss how a set of data can be described with **descriptive statistics,** including mode, median, mean, and standard deviation. Descriptive statistics are used by businesses to summarize data about advertising effectiveness, production costs, and profit. We continue our discussion of probability by considering **binomial probability distributions, discrete probability distributions,** and **normal probability distributions.** Social and behavioral scientists collect data about carefully selected **samples** and use probability distributions to reach conclusions about the populations from which the samples were drawn.

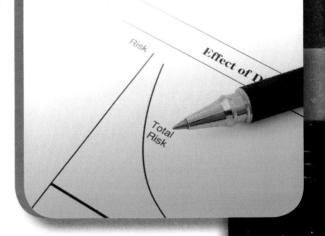

objectives

8.1 **Binomial Probability Experiments**

8.2 **Describing Data**

8.3 **Discrete Probability Distributions**

8.4 **Normal Probability Distribution**

8.1 Binomial Probability Experiments

SUPPOSE YOU HAVE A COIN THAT IS BIASED SO THAT WHEN THE COIN IS TOSSED, THE PROBABILITY OF GETTING A HEAD IS 2/3 AND THE PROBABILITY OF GETTING A TAIL IS 1/3. WHAT IS THE PROBABILITY OF GETTING 2 HEADS IN 3 TOSSES OF THIS COIN?

We must consider all orders that will give us 2 heads and 1 tail. We can use a tree to find all the possibilities (see Figure 8.1). We can see that there are 3 paths through the tree that correspond to 2 heads and 1 tail in 3 tosses. Because the probability for each successful path is 4/27,

$$\Pr(2H \text{ and } 1T) = 3 \cdot \frac{4}{27} = \frac{4}{9}$$

In this problem we can find the probability of 2 heads in 3 tosses by finding the probability for any one path of the tree (such as HHT) and then multiplying that probability by the number of paths that result in 2 heads. Notice that the 3 paths in the tree that result in 2 heads correspond to the different orders in which 2 heads and 1 tail can occur.

This is the number of ways that we can pick 2 positions out of 3 for the heads to occur, which is $_3C_2 = \binom{3}{2}$. Thus the probability of 2 heads resulting from 3 tosses is

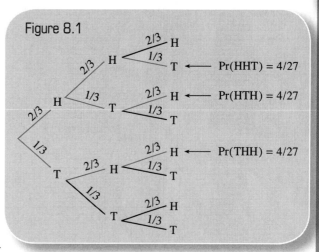

Figure 8.1

$$\Pr(2H \text{ and } 1T) = \binom{3}{2}\frac{2}{3}\cdot\frac{2}{3}\cdot\frac{1}{3} = \binom{3}{2}\left(\frac{2}{3}\right)^2\left(\frac{1}{3}\right)^1 = \frac{4}{9}$$

Shutterstock

math alps

HHT

HTH

THH

Shutterstock

The experiment just discussed (tossing a coin 3 times) is an example of a general class of experiments called **binomial probability experiments**. These experiments are also called **Bernoulli experiments,** or **Bernoulli trials,** after the 18th-century mathematician Jacob Bernoulli.

A **binomial probability experiment** satisfies the following properties:

1. There are n repeated trials of the experiment.

2. Each trial results in one of two outcomes, with one denoted success (S) and the other failure (F).

3. The trials are independent, with the probability of success, p, the same for every trial. The probability of failure is $q = 1 - p$ for every trial.

4. The probability of x successes in n trials is

$$\Pr(x) = \binom{n}{x} p^x q^{n-x}$$

$\Pr(x)$ is called a **binomial probability.** Binomial probability can be used to determine the likelihood of real-world outcomes, including birth defects.

EXAMPLE 8.1 Birth Defects

It has been found that the probability is 0.01 that any child whose mother lives in an area near a chemical dump will be born with a birth defect. Suppose that 10 children whose mothers live in the area are born in a given month.
(a) What is the probability that 2 of the 10 will be born with birth defects?
(b) What is the probability that at least 2 of them will have birth defects?

SOLUTION

(a) Note that we may consider a child having a birth defect as a success for this problem even though it is certainly not true in reality. Each of the 10 births is independent, with probability of success $p = 0.01$ and probability of failure $q = 0.99$. Therefore, the experiment is a binomial experiment, and

$$\Pr(2) = \binom{10}{2}(0.01)^2(0.99)^8 = 0.00415$$

(b) The probability of at least 2 of the children having birth defects is

$$1 - [\Pr(0) + \Pr(1)] =$$

$$1 - \left[\binom{10}{0}(0.01)^0(0.99)^{10} + \binom{10}{1}(0.01)^1(0.99)^9\right]$$

$$= 1 - [0.90438 + 0.09135] = 0.00427$$

We interpret this to mean that there is a 0.427% likelihood that at least 2 of these children will have birth defects.

Binomial probabilities can also be calculated with a graphing calculator or Excel. See the Tech Card for details.

8.1 Exercises

1. Suppose a fair die is rolled 18 times.
   (a) What is the probability p that a 4 will occur each time the die is rolled?
   (b) What is the probability q that a 4 will not occur each time the die is rolled?
   (c) What is n for this experiment?
   (d) What is the probability that a 4 will occur 6 times in the 18 rolls?

2. A bag contains 6 red balls and 4 black balls. We draw 5 balls, with each one replaced before the next is drawn.
   (a) What is the probability that 2 balls drawn will be red?
   (b) What is the probability that 5 black balls will be drawn?
   (c) What is the probability that at least 3 black balls will be drawn?

3. Suppose a pair of dice is thrown 4 times. What is the probability that a sum of 9 occurs exactly 2 times?

4. Suppose the probability that a marksman will hit a target each time he shoots is 0.85. If he fires 10 shots at a target, what is the probability that he will hit it 8 times?

Applications

5. *Management* The manager of a store buys portable radios in lots of 12. Suppose that, on the average, 2 out of each group of 12 are defective. The manager randomly selects 4 radios out of the group to test.
   (a) What is the probability that he will find 2 defective radios?
   (b) What is the probability that he will find no defective radios?

6. *Genetics* The probability that a certain couple will have a blue-eyed child is 1/4, and they have 4 children. What is the probability that
   (a) 1 of their children has blue eyes?
   (b) 2 of their children have blue eyes?
   (c) none of their children has blue eyes?

7. *Health care* Suppose that 10% of the patients who have a certain disease die from it. If 5 patients have the disease, what is the probability that
   (a) exactly 2 patients will die from it?
   (b) no patients will die from it?
   (c) no more than 2 patients will die from it?

8. *Genetics* If the ratio of boys born to girls born is 105 to 100, and if 6 children are born in a certain hospital in a day, what is the probability that 4 of them are boys?

9. *Quality control* A company produces shotgun shells in batches of 300. A sample of 10 is tested from each batch, and if more than one defect is found, the entire batch is tested. If 1% of the shells are actually defective,
   (a) what is the probability of 0 defective shells in the sample?
   (b) what is the probability of 1 defective shell?
   (c) what is the probability of more than 1 defective shell?

10. *Testing* A quiz consists of 10 multiple-choice questions with 5 choices for each question. Suppose a student is sure of the first 5 answers and has each of the last 5 questions narrowed to 3 of the possible 5 choices. If the student guesses among the narrowed choices on the last 5 questions, find
    (a) the probability of passing the quiz (getting at least 60%).

(b) the probability of getting at least a B (at least 80%).

> **Need more practice?**
> Find more here: cengagebrain.com

8.2 Describing Data

IN MODERN SOCIETY, A VAST AMOUNT OF DATA IS COLLECTED FOR MULTIPLE END USES, SUCH AS BUSINESS, SPORTS, AND CONSUMER BEHAVIOR. INTERPRETATIONS DRAWN FROM THE DATA DEPEND AS MUCH ON HOW THE DATA ARE DESCRIBED AS ON THE DATA THEMSELVES.

Before we examine the multiple ways that data can be described and how to identify variability and deviations in the data, let's first review several techniques for graphically depicting those data.

Statistical Graphs

Statistical data that have been collected can be visualized with a graph that shows the relationships among various quantities. Statistical graphs include **pie charts** (see Figure 8.2), **bar graphs,** and **frequency histograms.**

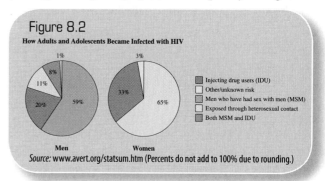

Figure 8.2
How Adults and Adolescents Became Infected with HIV

Men — Women

- Injecting drug users (IDU)
- Other/unknown risk
- Men who have had sex with men (MSM)
- Exposed through heterosexual contact
- Both MSM and IDU

Source: www.avert.org/statsum.htm (Percents do not add to 100% due to rounding.)

Bar graphs are useful in showing qualitative data as well as quantitative data. Bar graphs are often used to represent frequencies of data in different periods of time. Figure 8.3 on the following page is a bar graph that shows the industry-wide net income in the property-casualty industry for the years 2002–2007.

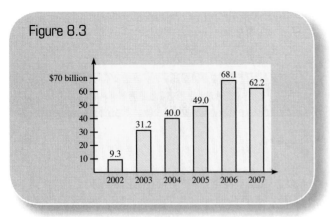

Figure 8.3

Data can also be organized in a **frequency table**, which lists data values (individually or grouped in intervals called classes) along with their corresponding frequencies. A **frequency histogram** is a special bar graph that gives a graphical view of a frequency table. It is constructed by putting the data values or data classes along the horizontal axis and putting the frequencies with which they occur along the vertical axis.

EXAMPLE 8.2 Frequency Histograms

Construct a histogram from the following frequency table.

Interval	Frequency
0–5	0
6–10	2
11–15	5
16–20	1
21–25	3

SOLUTION

The histogram is shown in Figure 8.4.

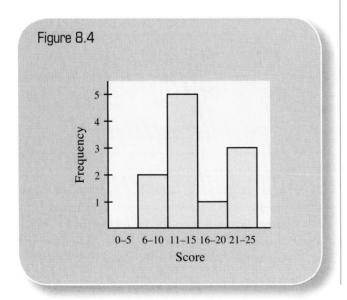

Figure 8.4

Graphing calculators, spreadsheets, and computer programs can be used to plot the histogram of a set of data. The width of each bar of the histogram can be determined by the calculator or adjusted by the user when the histogram is plotted. See the Tech Card for details on making histograms with a graphing calculator or Excel.

Types of Averages

A set of data can be described by listing all the scores or by drawing a frequency histogram. But we often wish to describe a set of scores by giving the *average score* or the typical score in the set. Three types of measures are called *averages*, or measures of central tendency: the **mode,** the **median,** and the **mean.**

Mode

To determine what value represents the most "typical" score in a set of scores, we use the **mode** of the scores.

> **Mode:** The **mode** of a set of scores is the score that occurs most frequently. The mode of a set of scores may not be unique.

That is, the mode is the most popular score—the one that is most likely to occur. The mode can be readily determined from a frequency table or frequency histogram because it is determined according to the frequencies of the scores. The score associated with the highest bar on a histogram is the mode. (See Figure 8.5.)

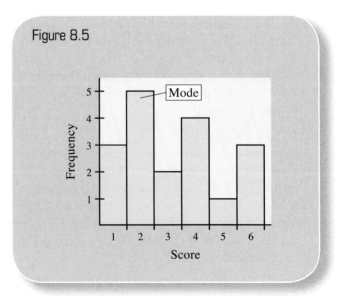

Figure 8.5

Consider the following scores: 10, 4, 3, 6, 4, 2, 3, 4, 5, 6, 8, 10, 2, 1, 4, 3. Arranging the scores in order gives: 1, 2, 2, 3, 3, 3, 4, 4, 4, 4, 5, 6, 6, 8, 10, 10. The mode is 4 because it occurs more frequently than any other score.

Let's consider another set of data: 4, 3, 2, 4, 6, 5, 5, 7, 6, 5, 7, 3, 1, 7, 2. Again, the first step is to arrange the scores in order: 1, 2, 2, 3, 3, 4, 4, 5, 5, 5, 6, 6, 7, 7, 7. Both 5 and 7 occur three times. Thus they are both modes. This set of data is said to be *bimodal* because it has two modes.

Median

Although the mode of a set of scores tells us the most frequent score, it does not always represent the typical performance, or central tendency, of the set of scores. For example, the scores 2, 3, 3, 3, 5, 8, 9, 10, 13 have a mode of 3. But 3 is not a good measure of central tendency for this set of scores, for it is nowhere near the middle of the distribution.

One value that gives a better measure of central tendency is the **median**.

> **Median:** The **median** is the score or point above and below which an equal number of the scores lie when the scores are arranged in ascending or descending order. If there is an odd number of scores, we find the median by ranking the scores from smallest to largest and picking the middle score.
>
> If the number of scores is even, there will be no middle score when the scores are ranked, so the median is the point that is the average of the two middle scores.

EXAMPLE 8.3 Median

➡ Find the median of each of the following sets of scores.

(a) 2, 3, 16, 5, 15, 38, 18, 17, 12
(b) 3, 2, 6, 8, 12, 4, 3, 2, 1, 6

SOLUTION

(a) We rank the scores from smallest to largest: 2, 3, 5, 12, 15, 16, 17, 18, 38. The median is the middle score, which is 15. Note that there are 4 scores above and 4 scores below 15.

(b) We rank the scores from smallest to largest: 1, 2, 2, 3, 3, 4, 6, 6, 8, 12. The two scores that lie in the

middle of the distribution are 3 and 4. Thus the median is a point that is midway between 3 and 4; that is, it is the arithmetic average of 3 and 4. The median is

$$\frac{3+4}{2} = \frac{7}{2} = 3.5$$

In this case we say the median is a point because there is no *score* that is 3.5.

Some calculators have keys or functions for finding the median of a set of data. Such calculators, or computers, are valuable when the set of data is large.

Mean

The median is the most easily interpreted measure of central tendency, and it is the best indicator of central tendency when the set of scores contains a few extreme values. However, the most frequently used measure of central tendency is the **mean**.

> **Mean:** The symbol $\bar{x}$ is used to represent the **mean** of a set of scores in a sample. Thus, if n is the number of scores, the formula for the mean is
>
> $$\bar{x} = \frac{\text{sum of scores}}{\text{number of scores}} = \frac{\Sigma x}{n}$$

The mean is used most often as a measure of central tendency because it is more useful than the median in the general applications of statistics. Let's see how to calculate it using the following sample of numbers: 12, 8, 7, 10, 6, 14, 7, 6, 12, 9. Applying the formula for the mean produces this solution:

$$\bar{x} = \frac{\Sigma x}{10} = \frac{12 + 8 + 7 + 10 + 6 + 14 + 7 + 6 + 12 + 9}{10}$$

$$= \frac{91}{10}$$

so $\bar{x} = 9.1$.

Note that the mean need not be one of the numbers (scores) given.

If the data are given in a frequency table or a frequency histogram, we can use a more efficient formula for finding the mean.

Mean of Grouped Data

If a set of n scores is grouped into k classes with x representing the score in a given class and $f(x)$

representing the number of scores in that class, then $n = \Sigma f(x)$ and the mean of the data is given by

$$\bar{x} = \frac{\Sigma(x \cdot f(x))}{\Sigma f(x)}$$

with each value of x used once.

Let's calculate the mean of a set of test scores using this formula.

EXAMPLE 8.4 Test Scores

Find the mean of the following sample of test scores for a math class.

Scores	Class Marks	Frequencies
40–49	44.5	2
50–59	54.5	0
60–69	64.5	6
70–79	74.5	12
80–89	84.5	8
90–99	94.5	2

Shutterstock

SOLUTION

For the purpose of finding the mean of interval data, we assume that all scores within an interval are represented by the class mark (midpoint of the interval). Also note, the sum of the frequencies is 30. Thus

$$\Sigma(x \cdot f(x)) = (44.5)(2) + (54.5)(0) + (64.5)(6)$$
$$+ (74.5)(12) + (84.5)(8) + (94.5)(2) = 2235.6$$

$$\bar{x} = \frac{\Sigma(x \cdot f(x))}{\Sigma f(x)} = \frac{2235.6}{30} = 74.52$$

Range

Although the mean of a set of data is useful in locating the center of the distribution of the data, it doesn't tell us as much about the distribution as we might think at first. For example, a basketball team with five 6-ft players is quite different from a team with one 6-ft, two 5-ft, and two 7-ft players. The distributions of the heights of these teams differ not in the mean height but in how the heights *vary* from the mean.

One measure of how a distribution varies is the **range** of the distribution.

> **Range:** The **range** of a set of numbers is the difference between the largest and smallest numbers in the set.

Consider the following set of heights of players on a basketball team, in inches: 69, 70, 75, 69, 73, 78, 74, 73, 78, 71. The range of this set of heights is $78 - 69 = 9$ inches. Note that this range is determined by only two numbers and does not give any information about how the other heights vary.

Variance and Standard Deviation

If we calculated the deviation of each score from the mean, $(x - \bar{x})$, and then attempted to average the deviations, we would see that the deviations add to 0. In an effort to find a meaningful measure of dispersion about the mean, statisticians developed the **variance**, which squares the deviations (to make them all positive) and then averages the squared deviations.

Frequently, we do not have all the data for a population and must use a sample of these data. To estimate the variance of a population from a sample, statisticians compensate for the fact that there is usually less variability in a sample than in the population itself by summing the squared deviations and dividing them by $n - 1$ rather than averaging them.

To get a measure comparable to the original deviations before they were squared, the **standard deviation**, which is the square root of the variance, was introduced. The formulas for the variance and standard deviation of sample data follow.

> **Variance and Standard Deviation:**
>
> Sample Variance $\qquad s^2 = \dfrac{\Sigma(x - \bar{x})^2}{n - 1}$
>
> Sample Standard Deviation $\qquad s = \sqrt{\dfrac{\Sigma(x - \bar{x})^2}{n - 1}}$

The standard deviation of a sample is a measure of the concentration of the scores about their mean. The smaller the standard deviation, the closer the scores lie to the mean.

EXAMPLE 8.5 Standard Deviation

→ Suppose two classes of fourth graders are polled as to how many television sets there are in their house. The sample data from the first class are represented in the histogram labeled Figure 8.6(a): 1, 1, 1, 3, 3, 4, 4, 5, 6, 6, 7, 7. Figure 8.6(b) is the histogram for the sample data from the second class: 1, 2, 3, 3, 4, 4, 4, 4, 5, 5, 6, 7. Both sets of data have a mean of 4 and a range of 6. Find the standard deviations of these samples.

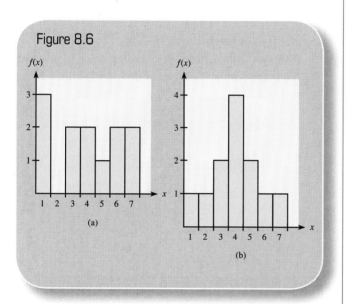

Figure 8.6

(a)

(b)

SOLUTION

For the sample data in Figure 8.6(a), we have

$$\Sigma(x - \bar{x})^2 = (1 - 4)^2 + (1 - 4)^2 + (1 - 4)^2$$
$$+ (3 - 4)^2 + (3 - 4)^2 + (4 - 4)^2 + (4 - 4)^2$$
$$+ (5 - 4)^2 + (6 - 4)^2 + (6 - 4)^2$$
$$+ (7 - 4)^2 + (7 - 4)^2$$
$$= 56$$
$$s^2 = \frac{\Sigma(x - \bar{x})^2}{n - 1} = \frac{56}{11} \approx 5.0909$$
$$s = \sqrt{s^2} = \sqrt{5.0909} \approx 2.26$$

For the sample data in Figure 8.6(b), we have

$$\Sigma(x - \bar{x})^2 = (1 - 4)^2 + (2 - 4)^2 + (3 - 4)^2$$
$$+ (3 - 4)^2 + (4 - 4)^2 + (4 - 4)^2$$
$$+ (4 - 4)^2 + (4 - 4)^2 + (5 - 4)^2$$
$$+ (5 - 4)^2 + (6 - 4)^2 + (7 - 4)^2$$
$$= 30$$
$$s^2 = \frac{30}{11} \approx 2.7273$$
$$s = \sqrt{\frac{30}{11}} \approx 1.65$$

Calculator Note: Most calculators have keys or built-in functions that compute the mean and standard deviation of a set of data. The ease of using calculators to find the standard deviation makes it the preferred method. See the Tech Card for details on finding means and standard deviations with a graphing calculator or Excel.

Shutterstock (both)

8.2 Exercises

In Problems 1 and 2, construct frequency histograms for the data given in the frequency tables.

1.

Score	Frequency
12	2
13	3
14	4
15	3
16	1

2.

Interval	Frequency
1–4	1
5–8	0
9–12	2
13–16	3
17–20	1

3. Construct a frequency histogram for the following data: 3, 2, 5, 6, 3, 2, 6, 5, 4, 2, 1, 6

4. Find the mode of the following set of scores: 3, 4, 3, 2, 2, 3, 5, 7, 6, 2, 3

In Problems 5 and 6, find the median of each set of scores.

5. 1, 3, 6, 7, 5

6. 4, 7, 9, 18, 36, 14, 3, 12

In Problems 7 and 8, find the mode, median, and mean of each set of scores.

7. 3, 2, 1, 6, 8, 12, 14, 2

8. 2.8, 6.4, 5.3, 5.3, 6.8

9. Use class marks to find the mean, mode, and median of the data in Problem 2.

10. Find the range of the following set of numbers: 3, 5, 7, 8, 2, 11, 6, 5

In Problems 11 and 12, find the mean, variance, and standard deviation of each set of sample data.

11. 5, 7, 1, 3, 0, 8, 6, 2

12. 11, 12, 13, 14, 15, 16, 17

Applications

13. **Unemployment rates** The table gives the U.S. unemployment rates for civilian workers for the decade-beginning years 1960–2000.
    (a) Display these data with a bar graph.
    (b) Summarize the data by finding the mean and standard deviation of the unemployment rate.

Year	1960	1970	1980	1990	2000
Unemployment rate	5.5	4.9	7.1	5.6	4.0

Source: Bureau of Labor Statistics, U.S. Department of Labor

14. **Educational expenditures** The table gives the per-pupil expenditures (in thousands of dollars) and their frequencies for the 50 states and the District of Columbia.
    (a) Display these data with a frequency histogram.
    (b) Find the mean per-pupil expenditure.

Expenditure ($ thousands)	Frequency
1–3	0
4–6	6
7–9	28
10–12	14
13–15	3

Source: National Education Association

15. **Birth weights** The birth weights (in kilograms) of a sample of 160 children are given in the table. Find the mean and standard deviation of the weights.

Weight (kg)	Frequency	Weight (kg)	Frequency
2.0	4	3.5	26
2.3	12	3.8	20
2.6	20	4.1	16
2.9	26	4.4	10
3.2	20	4.7	6

> **Need more practice?**
> Find more here: cengagebrain.com

8.3 Discrete Probability Distributions

UNDERSTANDING THE MATHEMATICS BEHIND DISTRIBUTIONS ALLOWS DECISION MAKERS TO COMPUTE EXPECTED VALUES AND DETERMINE THE RELIABILITY OF THAT CALCULATION. IN THIS SECTION, WE'LL EXAMINE DISCRETE PROBABILITY DISTRIBU-

TIONS, MEASURES OF DISPERSION, AND THE BINOMIAL DISTRIBUTION.

Discrete Probability Distributions

If a player rolls a die and receives $1 for each dot on the face she rolls, the amount of money won on one roll can be represented by the variable x. If the die is rolled once, the following table gives the possible outcomes of the experiment and their probabilities.

Shutterstock

x	1	2	3	4	5	6
Pr(x)	1/6	1/6	1/6	1/6	1/6	1/6

Note that we have used the variable x to denote the possible numerical outcomes of this experiment. Because x results from a probability experiment, we call x a **random variable,** and because there are a finite number of possible values of x, we say that x is a discrete random variable. Whenever all possible values of a random variable can be listed (or counted), the random variable is a **discrete random variable.**

Discrete Probability Distribution

A table, graph, or formula that assigns to each value of a discrete random variable x a probability Pr(x) describes a **discrete probability distribution** if the following two conditions hold.

(i) $0 \leq \text{Pr}(x) \leq 1$, for any value of x.

(ii) The sum of all the probabilities is 1. We use Σ to denote "the sum of" and write $\Sigma\text{Pr}(x) = 1$ where the sum is taken over all values of x.

Let's consider an experiment that illustrates this definition.

EXAMPLE 8.6 Discrete Probability Distribution

An experiment consists of selecting a ball from a bag containing 15 balls, each with a number 1 through 5. If the probability of selecting a ball with the number x on it is $\text{Pr}(x) = x/15$, where x is an integer and $1 \leq x \leq 5$, verify that $f(x) = \text{Pr}(x) = x/15$ describes a discrete probability distribution for the random variable x.

SOLUTION

For each integer $x(1 \leq x \leq 5)$, $\text{Pr}(x)$ satisfies $0 \leq \text{Pr}(x) \leq 1$.

$$\sum \text{Pr}(x) = \text{Pr}(1) + \text{Pr}(2) + \text{Pr}(3) + \text{Pr}(4) + \text{Pr}(5)$$

$$= \frac{1}{15} + \frac{2}{15} + \frac{3}{15} + \frac{4}{15} + \frac{5}{15} = 1$$

Hence $\text{Pr}(x) = x/15$ describes a discrete probability distribution.

Probability Density Histograms

We can visualize the possible values of a discrete random variable and their associated probabilities by constructing a graph. This graph, called a **probability density histogram**, is designed so that centered over each value of the discrete random variable x along the horizontal axis is a bar having width equal to 1 unit and height (and thus, area) equal to Pr(x). Figure 8.7 gives the probability density histogram for the experiment described in Example 8.6.

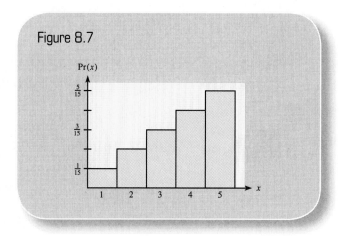
Figure 8.7

Mean and Expected Value

We found the **mean** of a set of n numbers grouped into k classes by using the formula

$$\frac{\sum [x \cdot f(x)]}{n}, \text{ or with } \sum \left[x \cdot \frac{f(x)}{n} \right]$$

We can also find the **theoretical mean** of a probability distribution by using the values x of the random variable, and the probabilities that those values will occur. This mean is also called the **expected value** of x because it gives the average outcome of a probability experiment.

> **Mean and Expected Value:** If x is a discrete random variable with values $x_1, x_2, \ldots, x_n$, then the **mean**, μ, of the distribution is the **expected value of x,** denoted by $E(x)$, and is given by
>
> $$\mu = E(x) = \sum [xPr(x)] = x_1Pr(x_1) + x_2Pr(x_2) + \cdots + x_nPr(x_n)$$

Many decisions in business and science are made on the basis of what the outcomes of specific decisions will be, so expected value is very important in decision making and planning.

EXAMPLE 8.7 Lottery

A fire company sells chances to win a new car, with each ticket costing $10. If the car is worth $36,000 and the company sells 6000 tickets, what is the expected value for a person buying one ticket?

SOLUTION

For a person buying one ticket, there are two possible outcomes from the drawing: winning or losing. The probability of winning is $1/6000$ and the amount won would be $36,000 - 10 (the cost of the ticket). The probability of losing is $5999/6000$ and the amount lost is written as a winning of $-$10$. Thus the expected value is

$$35,990\left(\frac{1}{6000}\right) + (-10)\left(\frac{5999}{6000}\right) = \frac{35,990}{6000} - \frac{59,990}{6000}$$

$$= -4$$

Thus, on the average, a person can expect to lose $4 on every ticket. It is not possible to lose exactly $4 on a ticket—the person will either win $36,000 or lose $10; the expected value means that the fire company will make $4 on each ticket it sells if it sells all 6000.

Measures of Dispersion

Figure 8.8 shows two probability histograms that have the same mean (expected value), 5. However, these histograms look very different. This is because the histograms differ in the **dispersion**, or **variation**, of the values of the random variables.

Thus we need a way to indicate how distributions with the same mean, such as the ones in Figure 8.8, differ. As with sample data, we can find the **variance** of a probability distribution by finding the mean of the *squares* of these deviations. The square root of this measure, called the **standard deviation**, is useful in describing how the values of x are concentrated about the mean of the probability distribution.

Variance and Standard Deviation

If x is a discrete random variable with values $x_1, x_2, \ldots, x_n$, and mean μ, then the variance of the distribution of x is

$$\sigma^2 = \sum (x - \mu)^2 Pr(x)$$

The standard deviation is

$$\sigma = \sqrt{\sigma^2} = \sqrt{\sum (x - \mu)^2 Pr(x)}$$

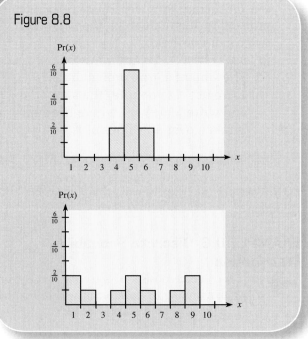

Figure 8.8

EXAMPLE 8.8 Standard Deviation of a Distribution

➡ Find the standard deviations of the two distributions described in Figure 8.8.

SOLUTION

(a) $\sigma^2 = \sum (x - \mu)^2 \Pr(x)$

$= (4-5)^2 \cdot \dfrac{2}{10} + (5-5)^2 \cdot \dfrac{6}{10} + (6-5)^2 \cdot \dfrac{2}{10}$

$= 0.4$

$\sigma = \sqrt{0.4} \approx 0.632$

(b) $\sigma^2 = (1-5)^2 \cdot \dfrac{2}{10} + (2-5)^2 \cdot \dfrac{1}{10} + (4-5)^2 \cdot \dfrac{1}{10}$

$+ (5-5)^2 \cdot \dfrac{2}{10} + (6-5)^2 \cdot \dfrac{1}{10}$

$+ (8-5)^2 \cdot \dfrac{1}{10} + (9-5)^2 \cdot \dfrac{2}{10}$

$= 8.4$

$\sigma = \sqrt{8.4} \approx 2.898$

As we can see by referring to Figure 8.8 and to the standard deviations calculated in Example 8.8, smaller values of σ indicate that the values of x are clustered nearer μ (as is true with sample mean and standard deviation).

The Binomial Distribution

A special discrete probability distribution is the **binomial probability distribution**, which describes all possible outcomes of a binomial experiment with their probabilities. Table 8.1 gives the binomial probability distribution for the experiment of tossing a fair coin 6 times, with x equal to the number of heads.

Table 8.1

x	Pr(x)	x	Pr(x)
0	$\binom{6}{0}\left(\frac{1}{2}\right)^0\left(\frac{1}{2}\right)^6 = \frac{1}{64}$	4	$\binom{6}{4}\left(\frac{1}{2}\right)^4\left(\frac{1}{2}\right)^2 = \frac{15}{64}$
1	$\binom{6}{1}\left(\frac{1}{2}\right)^1\left(\frac{1}{2}\right)^5 = \frac{6}{64}$	5	$\binom{6}{5}\left(\frac{1}{2}\right)^5\left(\frac{1}{2}\right)^1 = \frac{6}{64}$
2	$\binom{6}{2}\left(\frac{1}{2}\right)^2\left(\frac{1}{2}\right)^4 = \frac{15}{64}$	6	$\binom{6}{6}\left(\frac{1}{2}\right)^6\left(\frac{1}{2}\right)^0 = \frac{1}{64}$
3	$\binom{6}{3}\left(\frac{1}{2}\right)^3\left(\frac{1}{2}\right)^3 = \frac{20}{64}$		

Binomial Probability Distribution:
If x is a variable that assumes the values $0, 1, 2, \ldots, r, \ldots, n$ with probabilities

$$\binom{n}{0}p^0q^n, \binom{n}{1}p^1q^{n-1}, \binom{n}{2}p^2q^{n-2}, \ldots, \binom{n}{r}p^rq^{n-r}, \ldots, \binom{n}{n}p^nq^0$$

respectively, where p is the probability of success and q is the probability of failure, then x is called a **binomial variable.**

The values of x and their corresponding probabilities just described form the **binomial probability distribution.**

Mean and Standard Deviation

The expected value of the number of successes (heads) for the coin toss experiment in Table 8.1 is given by

$$E(x) = 0 \cdot \frac{1}{64} + 1 \cdot \frac{6}{64} + 2 \cdot \frac{15}{64} + 3 \cdot \frac{20}{64}$$
$$+ 4 \cdot \frac{15}{64} + 5 \cdot \frac{6}{64} + 6 \cdot \frac{1}{64} = \frac{192}{64} = 3$$

This expected number of successes seems reasonable. If the probability of success is $1/2$, we would expect to succeed on half of the 6 trials. We could also use the formula for expected value to see that the expected number of heads in 100 tosses is $100(1/2) = 50$. For any binomial distribution the expected number of successes is given by np, where n is the number of trials and p is the probability of success.

Mean of a Binomial Distribution:
The theoretical mean of any binomial distribution is

$$\mu = np$$

where n is the number of trials in the corresponding binomial experiment and p is the probability of success on each trial.

A simple formula can also be developed for the standard deviation of a binomial distribution (see next page). That formula gives the standard deviation of the binomial distribution corresponding to the number of heads resulting when a coin is tossed 16 times as

$$\sigma = \sqrt{16 \cdot \frac{1}{2} \cdot \frac{1}{2}} = \sqrt{4} = 2$$

Standard Deviation of a Binomial Distribution

The standard deviation of a binomial distribution is

$$\sigma = \sqrt{npq}$$

where n is the number of trials, p is the probability of success on each trial, and $q = 1 - p$.

EXAMPLE 8.9 Nursing Homes

According to the *New England Journal of Medicine*, there is a 43% chance that a person age 65 or older will enter a nursing home sometime in his or her lifetime. Of 100 people in this age group, what is the expected number of people who will eventually enter a nursing home, and what is the standard deviation?

SOLUTION

The number of people who will eventually enter a nursing home follows a binomial distribution, with the probability that any one person in this group will enter a home equal to $p = 0.43$ and the probability that he or she will not equal to $q = 0.57$. Thus the expected number out of 100 that will eventually enter a nursing home is

$$E(x) = np = 100(0.43) = 43$$

The standard deviation of this distribution is

$$\sigma = \sqrt{npq} = \sqrt{(100)(0.43)(0.57)} \approx 4.95$$

Binomial Formula

The binomial probability distribution is closely related to the powers of a binomial. For example, if a binomial experiment has 3 trials with probability of success p on each trial, then the binomial probability distribution is given in Table 8.2, with the values of x written from 3 to 0. Compare this with the expansion of $(p + q)^3$.

$$(p + q)^3 = p^3 + 3p^2q + 3pq^2 + q^3$$

Table 8.2

x	Pr(x)
3	$\binom{3}{3}p^3q^0 = p^3$
2	$\binom{3}{2}p^2q = 3p^2q$
1	$\binom{3}{1}pq^2 = 3pq^2$
0	$\binom{3}{0}p^0q^3 = q^3$

The formula we can use to expand a binomial $(a + b)$ to any positive integer power n is as follows.

Binomial Formula:

$$(a + b)^n = \binom{n}{n}a^n + \binom{n}{n-1}a^{n-1}b + \binom{n}{n-2}a^{n-2}b^2 + \cdots$$

$$+ \binom{n}{2}a^2b^{n-2} + \binom{n}{1}ab^{n-1} + \binom{n}{0}b^n$$

EXAMPLE 8.10 Binomial Formula

Expand $(x + y)^4$.

SOLUTION

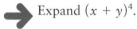

$$(x + y)^4 = \binom{4}{4}x^4 + \binom{4}{3}x^3y + \binom{4}{2}x^2y^2 + \binom{4}{1}xy^3 + \binom{4}{0}y^4$$

$$= x^4 + 4x^3y + 6x^2y^2 + 4xy^3 + y^4$$

8.3 Exercises

In Problems 1 and 2, determine whether each table describes a discrete probability distribution. Explain.

1.

x	Pr(x)
1	−1/5
2	1/4
3	5/10
4	9/20

2.

x	Pr(x)
−1	1/4
1	3/8
2	1/4
4	1/8

3. The table defines a discrete probability distribution. Find the expected value of the distribution.

x	0	1	2	3
Pr(x)	1/8	1/4	1/4	3/8

4. Find the mean, variance, and standard deviation for the following probability distribution.

x	0	1	2	3
Pr(x)	1/4	1/4	1/8	3/8

5. Suppose an experiment has five possible outcomes for x: 0, 1, 2, 3, 4. The probability that each of these outcomes occurs is $x/10$. What is the expected value of x for the experiment?

6. A die is rolled 3 times, and success is rolling a 5.
   (a) Construct the binomial distribution that describes this experiment.
   (b) Find the mean of this distribution.
   (c) Find the standard deviation of this distribution.

7. A variable x has a binomial distribution with probability of success 0.7 for each trial. For a total of 60 trials, what are
   (a) the mean and
   (b) the standard deviation of the distribution?

8. Expand $(a + b)^6$.

Applications

9. *Animal relocation* In studying endangered species, scientists have found that when animals are relocated, it takes x years without offspring before the first young are born, where x and the probability of x are given in the table. What is the expected number of years before the first young are born?

x	0	1	2	3	4
Pr(x)	0.04	0.35	0.38	0.18	0.05

10. *Raffle* Living Arrangements for the Developmentally Disabled (LADD), Inc., a nonprofit organization, sells chances for a $40,000 Corvette at $100 per ticket. It sells 1500 tickets and offers 4 prizes, summarized in the table that follows. What are the expected winnings (or loss) for each ticket?

Prize	Amount
First	$40,000
Second	5,000
Third	2,500
Fourth	1,500

    Source: Automobile

11. *Gambling* Suppose a student is offered a chance to draw a card from an ordinary deck of 52 playing cards and win $15 for an ace, $10 for a king, and $1 for a queen. If $4 must be paid to play the game, what are the expected winnings every time the game is played by the student?

12. *Insurance* A car owner must decide whether she should take out a $100-deductible collision policy in addition to her liability insurance policy. Records show that each year, in her area, 8% of the drivers have an accident that is their fault or for which no fault is assigned, and that the average cost of repairs for these types of accidents is $1000. If the $100-deductible collision policy costs $100 per year, would she save money in the long run by buying the insurance or "taking the chance"? (*Hint:* Find the expected values if she has the policy and if she doesn't have the policy and compare them.)

13. *Voting* A candidate claims that 60% of the voters in his district will vote for him.
    (a) If his district contains 100,000 voters, how many votes does he expect to get from his district?
    (b) What is the standard deviation of the number of these votes?

14. *Quality control* A certain calculator circuit board is manufactured in lots of 200. If 1% of the boards are defective, find the mean and standard deviation of the number of defects in each lot.

Need more practice?
Find more here: cengagebrain.com

8.4 Normal Probability Distribution

DISCRETE PROBABILITY DISTRIBUTIONS ARE IMPORTANT, BUT THEY DO NOT APPLY TO MANY KINDS OF MEASUREMENTS. FOR EXAMPLE, THE WEIGHTS OF PEOPLE, THE HEIGHTS OF TREES, AND THE IQ SCORES OF COLLEGE STUDENTS CANNOT BE MEASURED WITH WHOLE NUMBERS BECAUSE EACH OF THEM CAN ASSUME ANY ONE OF AN INFINITE NUMBER OF VALUES ON A MEASURING SCALE, AND THE VALUES CANNOT BE LISTED OR COUNTED, AS IS POSSIBLE FOR A DISCRETE RANDOM VARIABLE. THESE MEASUREMENTS FOLLOW A SPECIAL CONTINUOUS PROBABILITY DISTRIBUTION CALLED THE NORMAL DISTRIBUTION.

The normal distribution is perhaps the most important probability distribution, because so many measurements that occur in nature follow this particular distribution.

The **normal distribution** has the following properties.

1. Its graph is a bell-shaped curve like that in Figure 8.9.* The graph is called the **normal curve**. It approaches but never touches the horizontal axis as it extends in both directions.

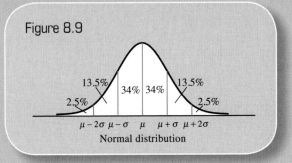

Figure 8.9

Normal distribution

2. The curve is symmetric about a vertical center line that passes through the value that is the mean, the median, *and* the mode of the distribution. That is, the mean, median, and mode are the same for a normal distribution. (That is why they are all called average.)

3. A normal distribution is completely determined when its mean μ and its standard deviation σ are known.

4. Approximately 68% of all scores lie within 1 standard deviation of the mean. Approximately 95% of all scores lie within 2 standard deviations of the mean. More than 99.5% of all scores will lie within 3 standard deviations of the mean.

*Percents shown are approximate.

EXAMPLE 8.11 IQ Scores

→ IQ scores follow a normal distribution with mean 100 and standard deviation 15. Find the probability that a person picked at random has an IQ score

(a) between 85 and 115.
(b) greater than 130.

SOLUTION

The approximate percents associated with these values are shown on the graph of a normal distribution in Figure 8.10.

The total area under the normal curve is 1. The area under the curve from value x_1 to value x_2 repre-

sents the percent of the scores that lie between x_1 and x_2. Thus the percent of the scores between x_1 and x_2 equals the area under the curve from x_1 to x_2 and both *represent the probability* that a score chosen at random will lie between x_1 and x_2. Thus, (a) the probability that a person chosen at random has an IQ between 85 and 115 is 0.68, and (b) the probability that a person chosen at random has an IQ greater than 130 is 0.025.

We have seen that 34% of the scores lie between 100 and 115 in the normal distribution graph in Figure 8.10. We can write this as

$$\Pr(100 \le x \le 115) = 0.34$$

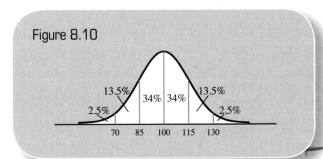

Figure 8.10

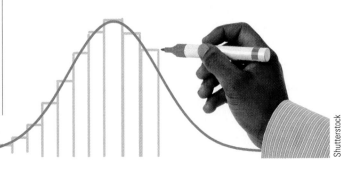

z-Scores

Because the probability of obtaining a score from a normal distribution is always related to how many standard deviations the score is away from the mean, it is desirable to convert all scores from a normal distribution to **standard scores**, or **z-scores**. The z-score for any score x is found by determining how many standard deviations x is from the mean μ. The formula for converting scores to z-scores follows.

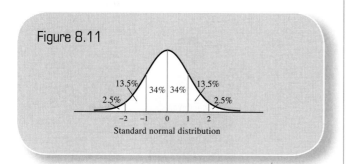

z-Scores:

If σ is the standard deviation of the population data, then the number of standard deviations that x is from the mean μ is given by the z-score

$$z = \frac{x - \mu}{\sigma}$$

This formula enables us to convert scores from any normal distribution to a distribution of z-scores with no units of measurement.

The distribution of z-scores will always be a normal distribution with mean 0 and standard deviation 1. This distribution is called the **standard normal distribution**. Figure 8.11 shows the graph of the standard normal distribution, with approximate percents shown. The total area under the curve is 1, with 0.5 on either side of the mean 0.

Figure 8.11

Standard normal distribution

By comparing Figure 8.11 with Figure 8.9, we see that each unit from 0 in the standard normal distribution corresponds to 1 standard deviation from the mean of the normal distribution.

We can use a table to determine more accurately the area under the standard normal curve between two z-scores. Table 8.3 at the end of the chapter (pages 196–197) gives the area under the normal curve from $z = 0$ to $z = z_0$, for values of z_0 from 0 to 3. As with the normal curve, the area under the curve from $z = 0$ to $z = z_0$ is the probability that a z-score lies between 0 and z_0.

EXAMPLE 8.12 Standard Normal Distribution

→ (a) Find the area under the standard normal curve from $z = 0$ to $z = 1.50$.

(b) Find $\Pr(0 \le z \le 1.50)$.

SOLUTION

(a) Looking in the column headed by z_0 in Table 8.3, we see 1.50. Across from 1.50 in the column headed by A is 0.4332. Thus the area under the standard normal curve between $z = 0$ and $z = 1.50$ is $A_{1.50} = 0.4332$. (See Figure 8.12.)

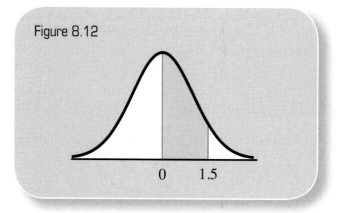

Figure 8.12

(b) The area under the curve from 0 to 1.50 equals the probability that z lies between 0 and 1.50. Thus

$$\Pr(0 \le z \le 1.50) = A_{1.50} = 0.4332$$

The following facts, which are direct results of the symmetry of the normal curve about μ, are useful in calculating probabilities using Table 8.3.

1. $\Pr(-z_0 \le z \le 0) = \Pr(0 \le z \le z_0) = A_{z_0}$

2. $\Pr(z \ge 0) = \Pr(z \le 0) = 0.5$

EXAMPLE 8.13 Probabilities with z-Scores

→ Find the following probabilities for the random variable z with standard normal distribution.

(a) $\Pr(-1 \le z \le 0)$ (b) $\Pr(-1 \le z \le 1.5)$

(c) $\Pr(1 \le z \le 1.5)$ (d) $\Pr(z > 2)$

(e) $\Pr(z < 1.35)$

SOLUTION

(a) $\Pr(-1 \le z \le 0) = \Pr(0 \le z \le 1) = A_1 = 0.3413$

(b) We find $\Pr(-1 \le z \le 1.5)$ by using A_1 and $A_{1.5}$ as follows:

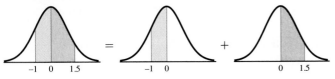

$$\Pr(-1 \le z \le 1.5) = \Pr(-1 \le z \le 0) + \Pr(0 \le z \le 1.5)$$
$$= \quad A_1 \quad + \quad A_{1.5}$$
$$= \quad 0.3413 \quad + \quad 0.4332$$
$$= 0.7745$$

(c) We find $\Pr(1 \le z \le 1.5)$ by using A_1 and $A_{1.5}$ as follows:

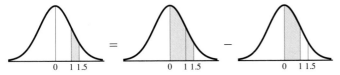

$$\Pr(1 \le z \le 1.5) = \Pr(0 \le z \le 1.5) - \Pr(0 \le z \le 1)$$
$$= \quad A_{1.5} \quad - \quad A_1$$
$$= \quad 0.4332 \quad - \quad 0.3413$$
$$= 0.0919$$

(d) We find $\Pr(z > 2)$ by using 0.5 and A_2 as follows:

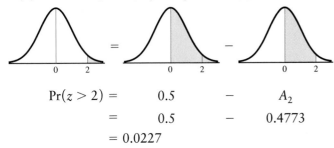

$$\Pr(z > 2) = \quad 0.5 \quad - \quad A_2$$
$$= \quad 0.5 \quad - \quad 0.4773$$
$$= 0.0227$$

(e) We find $\Pr(z < 1.35)$ by using 0.5 and $A_{1.35}$ as follows:

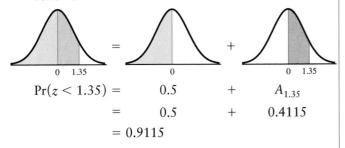

$$\Pr(z < 1.35) = \quad 0.5 \quad + \quad A_{1.35}$$
$$= \quad 0.5 \quad + \quad 0.4115$$
$$= 0.9115$$

Recall that if a population follows a normal distribution, but with mean and/or standard deviation different from 0 and 1, respectively, we can convert the scores to **standard scores,** or **z-scores,** by using the formula

$$z = \frac{x - \mu}{\sigma}$$

Thus

$$\Pr(a \le x \le b) = \Pr(z_a \le z \le z_b)$$

where

$$z_a = \frac{a - \mu}{\sigma} \text{ and } z_b = \frac{b - \mu}{\sigma}$$

and we can use Table 8.3, the table of standard scores (z-scores), to find these probabilities. For example, the normal distribution of IQ scores has mean $\mu = 100$ and standard deviation $\sigma = 15$. Thus the z-score for $x = 115$ is

$$z = \frac{115 - 100}{15} = 1$$

A z-score of 1 indicates that 115 is 1 standard deviation above the mean (see Figure 8.13), and

$$\Pr(100 \le x \le 115) = \Pr(0 \le z \le 1) = 0.3413$$

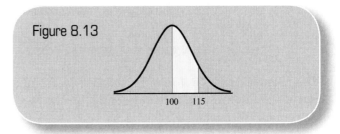

Figure 8.13

EXAMPLE 8.14 Height

➡ If the mean height of a population of students is $\mu = 68$ in. with standard deviation $\sigma = 3$ in., what is the probability that a person chosen at random from the population will be between
(a) 68 and 74 in. tall? (b) 65 and 74 in. tall?

Shutterstock (both)

SOLUTION

(a) To find $\Pr(68 \le x \le 74)$, we convert 68 and 74 to z-scores.

$$\text{For } 68: z = \frac{68 - 68}{3} = 0$$

$$\text{For } 74: z = \frac{74 - 68}{3} = 2$$

Thus

$$\Pr(68 \le x \le 74) = \Pr(0 \le z \le 2) = A_2 = 0.4773$$

(b) The z-score for 65 is

$$z = \frac{65 - 68}{3} = -1$$

and the z-score for 74 is 2. Thus

$$\Pr(65 \le x \le 74) = \Pr(-1 \le z \le 2)$$
$$= A_1 + A_2$$
$$= 0.3413 + 0.4773 = 0.8186$$

Normal probabilities can also be calculated with a graphing calculator or Excel. See the Tech Card for details.

Another application of the normal distribution is in approximating binomial probabilities when a large number of trials are involved (see the Enrichment material on CourseMate for MATH APPS at cengagebrain .com to see this application).

8.4 Exercises

In Problems 1–6, use Table 8.3 to find the probability that a z-score from the standard normal distribution will lie within each interval.

1. $0 \le z \le 1.8$

2. $-0.6 \le z \le 0$

3. $-1.5 \le z \le 2.1$

4. $2.1 \le z \le 3.0$

5. $z > 2$

6. $z < 1.2$

In Problems 7 and 8, suppose a population of scores x is normally distributed with $\mu = 20$ and $\sigma = 5$. Use the standard normal distribution to find the probabilities indicated.

7. $\Pr(20 \le x \le 22.5)$

8. $\Pr(13.75 \le x \le 20)$

9. Suppose a population of scores x is normally distributed with $\mu = 50$ and $\sigma = 10$. Use the standard normal distribution to find $\Pr(45 \le x \le 55)$.

10. Suppose a population of scores x is normally distributed with $\mu = 110$ and $\sigma = 12$. Use the standard normal distribution to find $\Pr(x > 128)$.

Applications

11. *Growth* The heights of a certain species of plant are normally distributed, with mean $\mu = 20$ cm and standard deviation $\sigma = 4$ cm. What is the probability that a plant chosen at random will be between 10 and 30 cm tall?

12. *Mileage* A certain model of automobile has its gas mileage (in miles per gallon, or mpg) normally distributed, with a mean of 28 mpg and a standard deviation of 4 mpg. Find the probability that a car selected at random has gas mileage
(a) less than 22 mpg.
(b) greater than 30 mpg.
(c) between 26 and 30 mpg.

13. *Blood pressure* Systolic blood pressure for a group of women is normally distributed, with a mean of 120 and a standard deviation of 12. Find the probability that a woman selected at random has blood pressure
(a) greater than 140.   (b) less than 110.
(c) between 110 and 130.

> **Need more practice?**
> Find more here: cengagebrain.com

Chapter Exercises

1. If the probability of success on each trial of an experiment is 0.4, what is the probability of 5 successes in 7 trials?

2. A bag contains 5 black balls and 7 white balls. If we draw 4 balls, with each one replaced before the next is drawn, what is the probability that
(a) 2 balls are black?
(b) at least 2 balls are black?

3. If a die is rolled 4 times, what is the probability that a number greater than 4 is rolled at least 2 times?

In Problems 4–7, use the data in the following frequency table.

Score	Frequency
1	4
2	6
3	8
4	3
5	5

4. Construct a frequency histogram for the data.

5. What is the mode of the data?

6. What is the mean of the data?

7. What is the median of the data?

8. Construct a frequency histogram for the following set of test scores: 14, 16, 15, 14, 17, 16, 12, 12, 13, 14.

9. What is the median of the scores in Problem 8?

10. What is the mode of the scores in Problem 8?

11. What is the mean of the scores in Problem 8?

12. Find the mean, variance, and standard deviation of the following data: 4, 3, 4, 6, 8, 0, 2.

13. Find the mean, variance, and standard deviation of the following data: 3, 2, 1, 5, 1, 4, 0, 2, 1, 1.

14. For the following probability distribution, find the expected value $E(x)$.

x	Pr(x)
1	0.2
2	0.3
3	0.4
4	0.1

In Problems 15 and 16, determine whether each function or table represents a discrete probability distribution.

15. $\Pr(x) = x/15$, $x = 1, 2, 3, 4, 5$

16.

x	Pr(x)
2	1/2
3	1/4
4	−1/4
5	1/2

17. For the following probability distribution, find the expected value $E(x)$.

x	1	2	3	4
Pr(x)	0.4	0.3	0.2	0.1

18. For the discrete probability distribution described by $\Pr(x) = x/16$, $x = 1, 2, 3, 4, 6$, find
    (a) the mean.
    (b) the variance.
    (c) the standard deviation.

19. For the probability distribution described by the table, find
    (a) the mean.
    (b) the variance.
    (c) the standard deviation.

x	Pr(x)
1	1/12
2	1/6
3	1/3
4	5/12

20. A coin has been altered so that the probability that a head will occur is $2/3$. If the coin is tossed 6 times, give the mean and standard deviation of the distribution of the number of heads.

21. Suppose a pair of dice is thrown 18 times. How many times would we expect a sum of 7 to occur?

22. Expand $(x + y)^5$.

23. What is the area under the standard normal curve between $z = -1.6$ and $z = 1.9$?

24. If z is a standard normal score, find $\Pr(-1 \le z \le -0.5)$.

25. Find $\Pr(1.23 \le z \le 2.55)$.

26. If a variable x is normally distributed, with $\mu = 25$ and $\sigma = 5$, find $\Pr(25 \le x \le 30)$.

27. If a variable x is normally distributed, with $\mu = 25$ and $\sigma = 5$, find $\Pr(20 \le x \le 30)$.

28. If a variable x is normally distributed, with $\mu = 25$ and $\sigma = 5$, find $\Pr(30 \le x \le 35)$.

Applications

29. **Genetics** Suppose the probability that a certain couple will have a blond child is $1/4$. If they have 6 children, what is the probability that 2 of them will be blond?

30. **Sampling** Suppose 70% of a population opposes a proposal and a sample of size 5 is drawn from the population. What is the probability that the majority of the sample will favor the proposal?

31. **Disease** One person in 100,000 develops a certain disease. Calculate
    (a) Pr(exactly 1 person in 100,000 has the disease).
    (b) Pr(at least 1 person in 100,000 has the disease).

Farm families In Problems 32–34, use the data on the distribution of farm families (as a percent of the population) in a 50-county survey given in the following table.

Percent	Number of Counties
10–19	5
20–29	16
30–39	25
40–49	3
50–59	1

32. Make a frequency histogram for these data.

33. Find the mean percent of farm families in these 50 counties.

34. Find the standard deviation of the percent of farm families in these 50 counties.

35. **Cancer testing** The probability that a man with prostate cancer will test positive with the prostate specific antigen (PSA) test is 0.91. If 500 men with prostate cancer are tested, how many would we expect to test positive?

36. **Fraud** A company selling substandard drugs to developing countries sold 2,000,000 capsules with 60,000 of them empty. (*Source: 60 Minutes*, episode aired January 18, 1998.) If a person gets 100 randomly chosen capsules from this company, what is the expected number of empty capsules that this person will get?

37. **Testing** Suppose the mean SAT score for students admitted to a university is 1000, with a standard deviation of 200. Suppose that a student is selected at random. If the scores are normally distributed, find the probability that the student's SAT score is
    (a) between 1000 and 1400.
    (b) between 1200 and 1400.
    (c) greater than 1400.

38. **Net worth** Suppose the mean net worth of the residents of Sun City, a retirement community, is $611,000. (*Source: The Island Packet*) If their net worth is normally distributed with a standard deviation of $96,000, what percent of the residents have a net worth between $700,000 and $800,000?

39. **Quality control** Records over 10 years indicate that 5% of Acer televisions will have some defective components. If the latest models have the same number of defects, what is the probability that 6 or more of 100 sets will have defective components?

> ### Need more practice?
> Find more here: cengagebrain.com

Table 8.3   Areas Under the Standard Normal Curve

The value of A is the area under the standard normal curve between $z = 0$ and $z = z_0$, for $z_0 \geq 0$. Areas for negative values of z_0 are obtained by symmetry.

z_0	A	z_0	A	z_0	A	z_0	A
0.00	0.0000	0.46	0.1772	0.92	0.3212	1.38	0.4162
0.01	0.0040	0.47	0.1808	0.93	0.3238	1.39	0.4177
0.02	0.0080	0.48	0.1844	0.94	0.3264	1.40	0.4192
0.03	0.0120	0.49	0.1879	0.95	0.3289	1.41	0.4207
0.04	0.0160	0.50	0.1915	0.96	0.3315	1.42	0.4222
0.05	0.0199	0.51	0.1950	0.97	0.3340	1.43	0.4236
0.06	0.0239	0.52	0.1985	0.98	0.3365	1.44	0.4251
0.07	0.0279	0.53	0.2019	0.99	0.3389	1.45	0.4265
0.08	0.0319	0.54	0.2054	1.00	0.3413	1.46	0.4279
0.09	0.0359	0.55	0.2088	1.01	0.3438	1.47	0.4292
0.10	0.0398	0.56	0.2123	1.02	0.3461	1.48	0.4306
0.11	0.0438	0.57	0.2157	1.03	0.3485	1.49	0.4319
0.12	0.0478	0.58	0.2190	1.04	0.3508	1.50	0.4332
0.13	0.0517	0.59	0.2224	1.05	0.3531	1.51	0.4345
0.14	0.0557	0.60	0.2258	1.06	0.3554	1.52	0.4357
0.15	0.0596	0.61	0.2291	1.07	0.3577	1.53	0.4370
0.16	0.0636	0.62	0.2324	1.08	0.3599	1.54	0.4382
0.17	0.0675	0.63	0.2357	1.09	0.3621	1.55	0.4394
0.18	0.0714	0.64	0.2389	1.10	0.3643	1.56	0.4406
0.19	0.0754	0.65	0.2422	1.11	0.3665	1.57	0.4418
0.20	0.0793	0.66	0.2454	1.12	0.3686	1.58	0.4430
0.21	0.0832	0.67	0.2486	1.13	0.3708	1.59	0.4441
0.22	0.0871	0.68	0.2518	1.14	0.3729	1.60	0.4452
0.23	0.0910	0.69	0.2549	1.15	0.3749	1.61	0.4463
0.24	0.0948	0.70	0.2580	1.16	0.3770	1.62	0.4474
0.25	0.0987	0.71	0.2612	1.17	0.3790	1.63	0.4485
0.26	0.1026	0.72	0.2642	1.18	0.3810	1.64	0.4495
0.27	0.1064	0.73	0.2673	1.19	0.3830	1.65	0.4505
0.28	0.1103	0.74	0.2704	1.20	0.3849	1.66	0.4515
0.29	0.1141	0.75	0.2734	1.21	0.3869	1.67	0.4525
0.30	0.1179	0.76	0.2764	1.22	0.3888	1.68	0.4535
0.31	0.1217	0.77	0.2794	1.23	0.3907	1.69	0.4545
0.32	0.1255	0.78	0.2823	1.24	0.3925	1.70	0.4554
0.33	0.1293	0.79	0.2852	1.25	0.3944	1.71	0.4564
0.34	0.1331	0.80	0.2881	1.26	0.3962	1.72	0.4573
0.35	0.1368	0.81	0.2910	1.27	0.3980	1.73	0.4582
0.36	0.1406	0.82	0.2939	1.28	0.3997	1.74	0.4591
0.37	0.1443	0.83	0.2967	1.29	0.4015	1.75	0.4599
0.38	0.1480	0.84	0.2996	1.30	0.4032	1.76	0.4608
0.39	0.1517	0.85	0.3023	1.31	0.4049	1.77	0.4616
0.40	0.1554	0.86	0.3051	1.32	0.4066	1.78	0.4625
0.41	0.1591	0.87	0.3079	1.33	0.4082	1.79	0.4633
0.42	0.1628	0.88	0.3106	1.34	0.4099	1.80	0.4641
0.43	0.1664	0.89	0.3133	1.35	0.4115	1.81	0.4649
0.44	0.1700	0.90	0.3159	1.36	0.4131	1.82	0.4656
0.45	0.1736	0.91	0.3186	1.37	0.4147	1.83	0.4664

z_0	A	z_0	A	z_0	A	z_0	A
1.84	0.4671	2.36	0.4909	2.88	0.4980	3.39	0.4997
1.85	0.4678	2.37	0.4911	2.89	0.4981	3.40	0.4997
1.86	0.4686	2.38	0.4913	2.90	0.4981	3.41	0.4997
1.87	0.4693	2.39	0.4916	2.91	0.4982	3.42	0.4997
1.88	0.4700	2.40	0.4918	2.92	0.4983	3.43	0.4997
1.89	0.4706	2.41	0.4920	2.93	0.4983	3.44	0.4997
1.90	0.4713	2.42	0.4922	2.94	0.4984	3.45	0.4997
1.91	0.4719	2.43	0.4925	2.95	0.4984	3.46	0.4997
1.92	0.4726	2.44	0.4927	2.96	0.4985	3.47	0.4997
1.93	0.4732	2.45	0.4929	2.97	0.4985	3.48	0.4998
1.94	0.4738	2.46	0.4931	2.98	0.4986	3.49	0.4998
1.95	0.4744	2.47	0.4932	2.99	0.4986	3.50	0.4998
1.96	0.4750	2.48	0.4934	3.00	0.4987	3.51	0.4998
1.97	0.4756	2.49	0.4936	3.01	0.4987	3.52	0.4998
1.98	0.4762	2.50	0.4938	3.02	0.4987	3.53	0.4998
1.99	0.4767	2.51	0.4940	3.03	0.4988	3.54	0.4998
2.00	0.4773	2.52	0.4941	3.04	0.4988	3.55	0.4998
2.01	0.4778	2.53	0.4943	3.05	0.4989	3.56	0.4998
2.02	0.4783	2.54	0.4945	3.06	0.4989	3.57	0.4998
2.03	0.4788	2.55	0.4946	3.07	0.4989	3.58	0.4998
2.04	0.4793	2.56	0.4948	3.08	0.4990	3.59	0.4998
2.05	0.4798	2.57	0.4949	3.09	0.4990	3.60	0.4998
2.06	0.4803	2.58	0.4951	3.10	0.4990	3.61	0.4999
2.07	0.4808	2.59	0.4952	3.11	0.4991	3.62	0.4999
2.08	0.4812	2.60	0.4953	3.12	0.4991	3.63	0.4999
2.09	0.4817	2.61	0.4955	3.13	0.4991	3.64	0.4999
2.10	0.4821	2.62	0.4956	3.14	0.4992	3.65	0.4999
2.11	0.4826	2.63	0.4957	3.15	0.4992	3.66	0.4999
2.12	0.4830	2.64	0.4959	3.16	0.4992	3.67	0.4999
2.13	0.4834	2.65	0.4960	3.17	0.4992	3.68	0.4999
2.14	0.4838	2.66	0.4961	3.18	0.4993	3.69	0.4999
2.15	0.4842	2.67	0.4962	3.19	0.4993	3.70	0.4999
2.16	0.4846	2.68	0.4963	3.20	0.4993	3.71	0.4999
2.17	0.4850	2.69	0.4964	3.21	0.4993	3.72	0.4999
2.18	0.4854	2.70	0.4965	3.22	0.4994	3.73	0.4999
2.19	0.4857	2.71	0.4966	3.23	0.4994	3.74	0.4999
2.20	0.4861	2.72	0.4967	3.24	0.4994	3.75	0.4999
2.21	0.4865	2.73	0.4968	3.25	0.4994	3.76	0.4999
2.22	0.4868	2.74	0.4969	3.26	0.4994	3.77	0.4999
2.23	0.4871	2.75	0.4970	3.27	0.4995	3.78	0.4999
2.24	0.4875	2.76	0.4971	3.28	0.4995	3.79	0.4999
2.25	0.4878	2.77	0.4972	3.29	0.4995	3.80	0.4999
2.26	0.4881	2.78	0.4973	3.30	0.4995	3.81	0.4999
2.27	0.4884	2.79	0.4974	3.31	0.4995	3.82	0.4999
2.28	0.4887	2.80	0.4974	3.32	0.4996	3.83	0.4999
2.29	0.4890	2.81	0.4975	3.33	0.4996	3.84	0.4999
2.30	0.4893	2.82	0.4976	3.34	0.4996	3.85	0.4999
2.31	0.4896	2.83	0.4977	3.34	0.4996	3.86	0.4999
2.32	0.4898	2.84	0.4977	3.35	0.4996		
2.33	0.4901	2.85	0.4978	3.36	0.4996		
2.34	0.4904	2.86	0.4979	3.37	0.4996		
2.35	0.4906	2.87	0.4980	3.38	0.4996		

Derivatives

9.1 Limits

IF A FIRM RECEIVES $30,000 IN REVENUE DURING A 30-DAY MONTH, ITS AVERAGE REVENUE PER DAY IS $30,000/30 = $1000. THIS DOES NOT NECESSARILY MEAN THE ACTUAL REVENUE WAS $1000 ON ANY ONE DAY, JUST THAT THE AVERAGE WAS $1000 PER DAY. SIMILARLY, IF A PERSON DROVE 50 MILES IN ONE HOUR, THE AVERAGE VELOCITY WAS 50 MILES PER HOUR, BUT THE DRIVER COULD STILL HAVE RECEIVED A SPEEDING TICKET FOR TRAVELING 70 MILES PER HOUR.

The smaller the time interval, the nearer the average velocity will be to the instantaneous velocity (the speedometer reading). Similarly, changes in revenue over a smaller number of units can give information about the instantaneous rate of change of revenue. The mathematical bridge from average rates of change to instantaneous rates of change is the **limit**.

This chapter is concerned with *limits* and *rates of change*. We will see that the *derivative* of a function can be used to determine instantaneous rates of change.

Notion of a Limit

The concept of *limit* is essential to the study of calculus. We have used the notation $f(c)$ to indicate the value of a function $f(x)$ at $x = c$. If we need to discuss a value that $f(x)$ approaches as x approaches c, we use the idea of a **limit**. For example, if

$$f(x) = \frac{x^2 - x - 6}{x + 2}$$

then we know that $x = -2$ is not in the domain of $f(x)$, so $f(-2)$ does not exist even though $f(x)$ exists for every value of $x \neq -2$. Figure 9.1 shows the graph of $y = f(x)$ with an open circle where

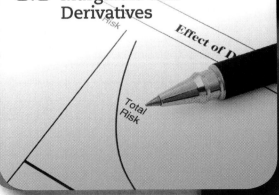

$x = -2$. The open circle indicates that $f(-2)$ does not exist but shows that points near $x = -2$ have functional values that lie on the line on either side of the open circle. Even though $f(-2)$ is not defined, the figure shows that as x approaches -2 from either side of -2, the graph approaches the open circle at $(-2, -5)$ and the values of $f(x)$ approach -5. Thus -5 is the limit of $f(x)$ as x approaches -2, and we write

$$\lim_{x \to -2} f(x) = -5, \quad \text{or} \quad f(x) \to -5 \quad \text{as} \quad x \to -2$$

This conclusion is fairly obvious from the graph, but it is not so obvious from the equation for $f(x)$.

Figure 9.1

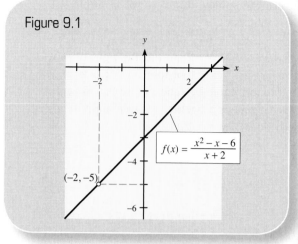

$$f(x) = \frac{x^2 - x - 6}{x + 2}$$

$(-2, -5)$

math apps

We can use the values near $x = -2$ in Table 9.1 to help verify that $f(x) \to -5$ as $x \to -2$. Note that to the left of -2, the values of $f(x)$ get very close to -5 as x gets very close to -2, and to the right of -2, the values of $f(x)$ get very close to -5 as x gets very close to -2. Hence, Table 9.1 and Figure 9.1 indicate that the value of $f(x)$ approaches -5 as x approaches -2 from both sides of $x = -2$.

Table 9.1

Left of -2	
x	$f(x) = \dfrac{x^2 - x - 6}{x + 2}$
-3.000	-6.000
-2.500	-5.500
-2.100	-5.100
-2.010	-5.010
-2.001	-5.001
Right of -2	
x	$f(x) = \dfrac{x^2 - x - 6}{x + 2}$
-1.000	-4.000
-1.500	-4.500
-1.900	-4.900
-1.990	-4.990
-1.999	-4.999

From our discussion of the graph in Figure 9.1 and Table 9.1, we see in general that as x approaches a value c from either side of c, the limit of a function $f(x)$ is the value L that the function approaches. This limit L is not necessarily the value of the function at at $x = c$. This leads to our intuitive definition of *limit*.

Limit: Let $f(x)$ be a function defined on an open interval containing c, except perhaps at c. Then
$$\lim_{x \to c} f(x) = L$$
is read "the limit of $f(x)$ as x approaches c equals L." The number L exists if we can make values of $f(x)$ as close to L as we desire by choosing values of x sufficiently close to c. When the values of $f(x)$ do not approach a single finite value L as x approaches c, we say the limit does not exist.

As the definition states, a limit as $x \to c$ can exist only if the function approaches a single finite value as x approaches c from both the left and right of c.

EXAMPLE 9.1 Limits

Figure 9.2 shows three functions for which the limit exists as x approaches 2. Use this figure to find the following.

(a) $\lim\limits_{x \to 2} f(x)$ and $f(2)$ (if it exists)

(b) $\lim\limits_{x \to 2} g(x)$ and $g(2)$ (if it exists)

(c) $\lim\limits_{x \to 2} h(x)$ and $h(2)$ (if it exists)

Figure 9.2

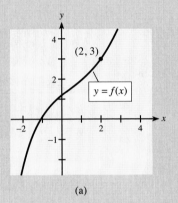

(a)

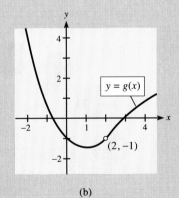

(b)

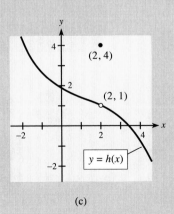

(c)

SOLUTION

(a) From the graph in Figure 9.2(a), we see that as x approaches 2 from both the left and the right, the graph approaches the point (2, 3). Thus $f(x)$ approaches the single value 3. That is,

$$\lim_{x \to 2} f(x) = 3$$

The value of $f(2)$ is the y-coordinate of the point on the graph at $x = 2$. Thus $f(2) = 3$.

(b) Figure 9.2(b) shows that as x approaches 2 from both the left and the right, the graph approaches the open circle at $(2, -1)$. Thus

$$\lim_{x \to 2} g(x) = -1$$

The figure also shows that at $x = 2$ there is no point on the graph. Thus $g(2)$ is undefined.

(c) Figure 9.2(c) shows that

$$\lim_{x \to 2} h(x) = 1$$

The figure also shows that at $x = 2$ there is a point on the graph at $(2, 4)$. Thus $h(2) = 4$, and we see that $\lim_{x \to 2} h(x) \neq h(2)$.

In Example 9.1 we saw that the limit as x approaches 2 meant the limit as x approaches 2 from both the left and the right. We can also consider limits only from the left or only from the right; these are called one-sided limits.

One-Sided Limits:

Limit from the Right: $\lim_{x \to c^+} f(x) = L$ means the values of $f(x)$ approach the value L as $x \to c$ but $x > c$.

Limit from the Left: $\lim_{x \to c^-} f(x) = M$ means the values of $f(x)$ approach the value M as $x \to c$ but $x < c$.

Note that when one or both one-sided limits fail to exist, then the limit does not exist. Also, when the one-sided limits differ, such as if $L \neq M$, then the values of $f(x)$ do not approach a *single* value as x approaches c, and $\lim_{x \to c} f(x)$ does not exist.

EXAMPLE 9.2 One-Sided Limits

Using the functions graphed in Figure 9.3, determine why the limit as $x \to 2$ does not exist for

(a) $f(x)$.
(b) $h(x)$.

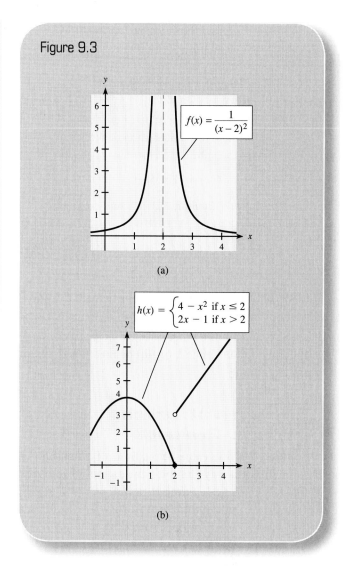

Figure 9.3

$$f(x) = \frac{1}{(x-2)^2}$$

(a)

$$h(x) = \begin{cases} 4 - x^2 & \text{if } x \leq 2 \\ 2x - 1 & \text{if } x > 2 \end{cases}$$

(b)

SOLUTION

(a) As $x \to 2$ from the left side and the right side of $x = 2$, $f(x)$ increases without bound, which we denote by saying that $f(x)$ approaches $+\infty$ as $x \to 2$. In this case, $\lim_{x \to 2} f(x)$ does not exist [denoted by $\lim_{x \to 2} f(x)$ DNE] because $f(x)$ does not approach a finite value as $x \to 2$. In this case, we write

$$f(x) \to +\infty \text{ as } x \to 2$$

The graph has a vertical asymptote at $x = 2$.

(b) As $x \to 2$ from the left, the graph approaches the point at $(2, 0)$, so $\lim_{x \to 2^-} h(x) = 0$. As $x \to 2$ from the right, the graph approaches the open circle at $(2, 3)$, so $\lim_{x \to 2^+} h(x) = 3$. Because these one-sided limits differ, $\lim_{x \to 2} h(x)$ does not exist.

Examples 9.1 and 9.2 illustrate the following.

The limit is said to exist only if the following conditions are satisfied:

(a) The limit L is a finite value (real number).

(b) The limit as x approaches c from the left equals the limit as x approaches c from the right. That is, we must have

$$\lim_{x \to c^-} f(x) = \lim_{x \to c^+} f(x)$$

Figure 9.3 and Example 9.2 illustrate cases where $\lim_{x \to c} f(x)$ does not exist.

The limit of a function as x approaches c is independent of the value of the function at c. When $\lim_{x \to c} f(x)$ exists, the value of the function at c may be (i) the same as the limit, (ii) undefined, or (iii) defined but different from the limit.

Properties of Limits, Algebraic Evaluation

We've seen that the value of the limit of a function as $x \to c$ will not always be the same as the value of the function at $x = c$. However, there are many functions for which the limit and the functional value agree [see Figure 9.2(a)], and for these functions we can easily evaluate limits.

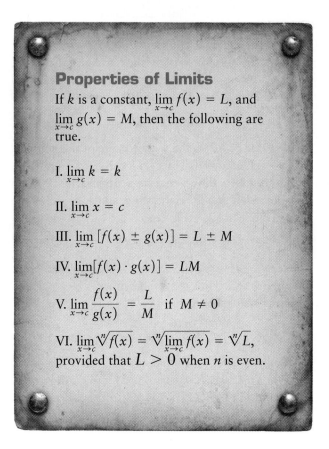

Properties of Limits

If k is a constant, $\lim_{x \to c} f(x) = L$, and $\lim_{x \to c} g(x) = M$, then the following are true.

I. $\lim_{x \to c} k = k$

II. $\lim_{x \to c} x = c$

III. $\lim_{x \to c} [f(x) \pm g(x)] = L \pm M$

IV. $\lim_{x \to c} [f(x) \cdot g(x)] = LM$

V. $\lim_{x \to c} \dfrac{f(x)}{g(x)} = \dfrac{L}{M}$ if $M \neq 0$

VI. $\lim_{x \to c} \sqrt[n]{f(x)} = \sqrt[n]{\lim_{x \to c} f(x)} = \sqrt[n]{L}$, provided that $L > 0$ when n is even.

EXAMPLE 9.3 Limits

Find the following limits, if they exist.

(a) $\lim\limits_{x \to -1} (x^3 - 2x)$

(b) $\lim\limits_{x \to 4} \dfrac{x^2 - 4x}{x - 2}$

For polynomial and rational functions, the properties of limits give the following:

Function	Definition	Limit
Polynomial function	The function $f(x) = a_n x^n + a_{n-1} x^{n-1} + \cdots + a_1 x + a_0,$ where $a \neq 0$ and n is a positive integer, is called a **polynomial function** of degree n.	$\lim_{x \to c} f(x) = f(c)$ for all values c (by Properties I–IV)
Rational function	The function $h(x) = \dfrac{f(x)}{g(x)}$ where both $f(x)$ and $g(x)$ are polynomial functions, is called a **rational function**.	$\lim_{x \to c} h(x) = \lim_{x \to c} \dfrac{f(x)}{g(x)} = \dfrac{f(c)}{g(c)}$ when $g(c) \neq 0$ (by Property V)

Limits of Polynomial and Rational Functions

SOLUTION

(a) Note that $f(x) = x^3 - 2x$ is a polynomial, so

$$\lim_{x \to -1} f(x) = f(-1) = (-1)^3 - 2(-1) = 1$$

Figure 9.4(a) shows the graph of $f(x) = x^3 - 2x$.

(b) Note that this limit has the form

$$\lim_{x \to c} \frac{f(x)}{g(x)}$$

where $f(x)$ and $g(x)$ are polynomials and $g(c) \neq 0$. Therefore, we have

$$\lim_{x \to 4} \frac{x^2 - 4x}{x - 2} = \frac{4^2 - 4(4)}{4 - 2} = \frac{0}{2} = 0$$

Figure 9.4(b) shows the graph of $g(x) = \dfrac{x^2 - 4x}{x - 2}$.

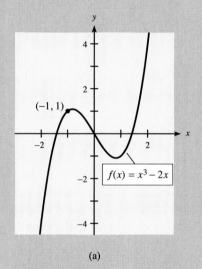

Figure 9.4

$f(x) = x^3 - 2x$

(−1, 1)

(a)

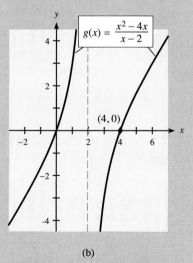

$g(x) = \dfrac{x^2 - 4x}{x - 2}$

(4, 0)

(b)

We have seen that we can find the limit of a rational function $f(x)/g(x)$ as long as the denominator is *not* zero. If the limit of the denominator of $f(x)/g(x)$ *is* zero, then there are two possible cases.

Rational Functions: Evaluating Limits of the Form

$$\lim_{x \to c} \frac{f(x)}{g(x)} \text{ where } \lim_{x \to c} g(x) = 0$$

Type I. If $\lim\limits_{x \to c} f(x) = 0$ and $\lim\limits_{x \to c} g(x) = 0$, then the fractional expression has the **0/0 indeterminate form** at $x = c$. We can factor $x - c$ from $f(x)$ and $g(x)$, reduce the fraction, and then find the limit of the resulting expression, if it exists.

Type II. If $\lim\limits_{x \to c} f(x) \neq 0$ and $\lim\limits_{x \to c} g(x) = 0$, then $\lim\limits_{x \to c} \dfrac{f(x)}{g(x)}$ does not exist. In this case, the values of $f(x)/g(x)$ are unbounded near $x = c$; the line $x = c$ is a vertical asymptote.

EXAMPLE 9.4 0/0 Indeterminate Form

Evaluate the following limits, if they exist.

(a) $\lim\limits_{x \to 2} \dfrac{x^2 - 4}{x - 2}$

(b) $\lim\limits_{x \to 1} \dfrac{x^2 + 3x + 2}{x - 1}$

SOLUTION

(a) We cannot find the limit by using Property V because the denominator is zero at $x = 2$. The numerator is also zero at $x = 2$, so the expression

$$\frac{x^2 - 4}{x - 2}$$

has the 0/0 indeterminate form at $x = 2$. Thus we can factor $x - 2$ from both the numerator and the denominator and reduce the fraction. (We can divide by $x - 2$ because $x - 2 \neq 0$ while $x \to 2$.)

$$\lim_{x \to 2} \frac{x^2 - 4}{x - 2} = \lim_{x \to 2} \frac{(x - 2)(x + 2)}{x - 2} = \lim_{x \to 2} (x + 2) = 4$$

The graph of $f(x) = (x^2 - 4)/(x - 2)$ is shown in Figure 9.5. Note the open circle at $(2, 4)$.

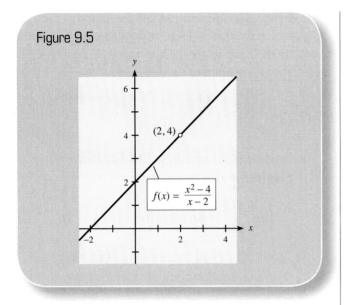

Figure 9.5

$$f(x) = \frac{x^2 - 4}{x - 2}$$

In part (b) of Example 9.4, even though the left-hand and right-hand limits do not exist (see Table 9.2), knowledge that the functional values are unbounded (that is, that they become infinite) is helpful in graphing. The graph is shown in Figure 9.6. We see that $x = 1$ is a vertical asymptote.

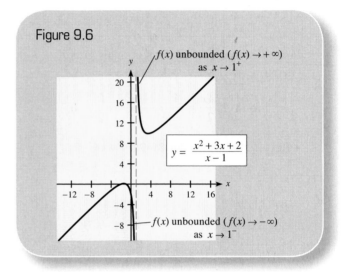

Figure 9.6

$f(x)$ unbounded ($f(x) \to +\infty$) as $x \to 1^+$

$$y = \frac{x^2 + 3x + 2}{x - 1}$$

$f(x)$ unbounded ($f(x) \to -\infty$) as $x \to 1^-$

(b) Substituting 1 for x in the function results in $6/0$, so this limit has the form $a/0$, with $a \neq 0$, and is like Type II discussed previously. Hence the limit does not exist. Because the numerator is not zero when $x = 1$, we know that $x - 1$ is *not* a factor of the numerator, and we cannot divide numerator and denominator as we did in part (a). Table 9.2 confirms that this limit does not exist, because the values of the expression become unbounded as x approaches 1.

Table 9.2

Left of $x = 1$		Right of $x = 1$	
x	$\dfrac{x^2 + 3x + 2}{x - 1}$	x	$\dfrac{x^2 + 3x + 2}{x - 1}$
0	-2	2	12
0.5	-7.5	1.5	17.5
0.7	-15.3	1.2	35.2
0.9	-55.1	1.1	65.1
0.99	-595.01	1.01	605.01
0.999	$-5,995.001$	1.001	6005.01
0.9999	$-59,995.0001$	1.0001	60,005.0001
$\displaystyle\lim_{x \to 1^-} \frac{x^2 + 3x + 2}{x - 1}$ DNE		$\displaystyle\lim_{x \to 1^+} \frac{x^2 + 3x + 2}{x - 1}$ DNE	
$f(x) \to -\infty$ as $x \to 1^-$		$f(x) \to +\infty$ as $x \to 1^+$	

The left-hand and right-hand limits do not exist. Thus $\displaystyle\lim_{x \to 1} \frac{x^2 + 3x + 2}{x - 1}$ does not exist.

Let's think about these concepts using a concrete scenario. USA Steel has shown that the cost C of removing p percent of the particulate pollution from the smokestack emissions at one of its plants is

$$C = C(p) = \frac{7300p}{100 - p}$$

The cost of removing 100% of the pollution is undefined because the denominator of the function is 0 when $p = 100$. To see what the cost approaches as p approaches 100 from values smaller than 100, we evaluate $\displaystyle\lim_{p \to 100^-} \frac{7300p}{100 - p}$. This limit has the Type II form for rational functions. Thus $\dfrac{7300p}{100 - p} \to +\infty$ as $p \to 100^-$, which means that as the amount of pollution that is removed approaches 100%, the cost increases without bound. (That is, it is impossible to remove 100% of the pollution.)

Limits of Piecewise Defined Functions

As we noted in Section 2.4, "Special Functions and Their Graphs," many applications are modeled by piecewise defined functions. To see how we evaluate a limit involving a piecewise defined function, consider the following example.

©iStockphoto.com/Stefan Baum

AS THE AMOUNT OF

POLLUTION THAT IS

REMOVED APPROACHES 100%, THE
COST INCREASES WITHOUT BOUND.

Table 9.3

Left of 1	
x	$f(x) = x^2 + 1$
0.1	1.01
0.9	1.81
0.99	1.98
0.999	1.998
0.9999	1.9998
Right of 1	
x	$f(x) = x + 2$
1.2	3.2
1.01	3.01
1.001	3.001
1.0001	3.0001
1.00001	3.00001

EXAMPLE 9.5 Limit of a Piecewise Defined Function

Find $\lim_{x \to 1^-} f(x)$, $\lim_{x \to 1^+} f(x)$, and $\lim_{x \to 1} f(x)$, if they exist, for

$$f(x) = \begin{cases} x^2 + 1 \text{ for } x \leq 1 \\ x + 2 \text{ for } x > 1 \end{cases}$$

SOLUTION

Because $f(x)$ is defined by $x^2 + 1$ when $x < 1$,

$$\lim_{x \to 1^-} f(x) = \lim_{x \to 1^-} (x^2 + 1) = 2$$

Because $f(x)$ is defined by $x + 2$ when $x > 1$,

$$\lim_{x \to 1^+} f(x) = \lim_{x \to 1^+} (x + 2) = 3$$

And because

$$2 = \lim_{x \to 1^-} f(x) \neq \lim_{x \to 1^+} f(x) = 3$$

$\lim_{x \to 1} f(x)$ does not exist.

Table 9.3 and Figure 9.7 show these results numerically and graphically.

Figure 9.7

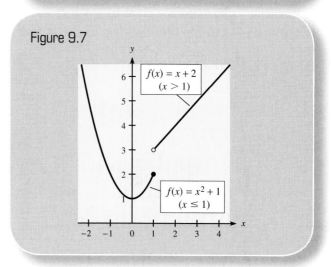

We have used graphical, numerical, and algebraic methods to understand and evaluate limits. Graphing calculators and Excel can be especially effective when we explore limits graphically or numerically. See the Tech Card for details.

9.1 Exercises

In Problems 1–3, a graph of $y = f(x)$ is shown and a c-value is given. For each problem, use the graph to find the following, whenever they exist.

(a) $\lim\limits_{x \to c} f(x)$ and   (b) $f(c)$

1. $c = 4$

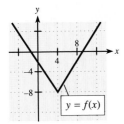

2. $c = 20$

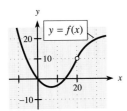

3. $c = -8$

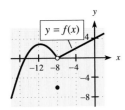

In Problems 4 and 5, use the graph of $y = f(x)$ and the given c-value to find the following, whenever they exist.

(a) $\lim\limits_{x \to c^-} f(x)$ (b) $\lim\limits_{x \to c^+} f(x)$
(c) $\lim\limits_{x \to c} f(x)$ (d) $f(c)$

4. $c = -10$

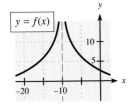

5. $c = -4\frac{1}{2}$

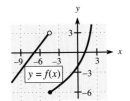

6. Complete the table and predict the limit, if it exists.

$$f(x) = \frac{2 - x - x^2}{x - 1}$$
$$\lim_{x \to 1} f(x) = ?$$

x	$f(x)$
0.9	
0.99	
0.999	
1.001	
1.01	
1.1	

In Problems 7–12, use properties of limits and algebraic methods to find the limits, if they exist.

7. $\lim\limits_{x \to -1} (4x^3 - 2x^2 + 2)$

8. $\lim\limits_{x \to -1/2} \dfrac{4x - 2}{4x^2 + 1}$

9. $\lim\limits_{x \to 3} \dfrac{x^2 - 9}{x - 3}$

10. $\lim\limits_{x \to 7} \dfrac{x^2 - 8x + 7}{x^2 - 6x - 7}$

11. $\lim\limits_{x \to -1} f(x)$, where
$$f(x) = \begin{cases} x^2 + \frac{4}{x} & \text{for } x \le -1 \\ 3x^3 - x - 1 & \text{for } x > -1 \end{cases}$$

12. $\lim\limits_{x \to 2} \dfrac{x^2 + 6x + 9}{x - 2}$

Applications

13. **Revenue** The total revenue for a product is given by
$$R(x) = 1600x - x^2$$
where x is the number of units sold. What is $\lim\limits_{x \to 100} R(x)$?

14. **Productivity** During an 8-hour shift, the rate of change of productivity (in units per hour) of children's phonographs assembled after t hours on the job is
$$r(t) = \frac{128(t^2 + 6t)}{(t^2 + 6t + 18)^2}, \quad 0 \le t \le 8$$

(a) Find $\lim\limits_{t \to 4} r(t)$.
(b) Find $\lim\limits_{t \to 8^-} r(t)$.
(c) Is the rate of productivity higher near the lunch break (at $t = 4$) or near quitting time (at $t = 8$)?

15. **Cost-benefit** Suppose that the cost C of obtaining water that contains p percent impurities is given by

$$C(p) = \frac{120,000}{p} - 1200$$

(a) Find $\lim\limits_{p \to 100^-} C(p)$, if it exists. Interpret this result.

(b) Find $\lim\limits_{p \to 0^+} C(p)$, if it exists.

(c) Is complete purity possible? Explain.

16. **Municipal water rates** The Corner Water Corp. of Shippenville, Pennsylvania, has the following rates per 1000 gallons of water used.

Usage (x)	Cost per 1000 Gallons ($C(x)$)
First 10,000 gallons	$7.98
Next 110,000 gallons	6.78
Over 120,000 gallons	5.43

If Corner Water has a monthly service fee of $3.59, write a function $C = C(x)$ that models the charges (where x is thousands of gallons) and find $\lim\limits_{x \to 10} C(x)$ (that is, as usage approaches 10,000 gallons).

Need more practice?
Find more here: cengagebrain.com

9.2 Continuous Functions and Limits at Infinity

IN THIS SECTION, WE WILL SHOW HOW TO DETERMINE WHETHER A FUNCTION IS CONTINUOUS, AND WE WILL INVESTIGATE SOME DIFFERENT TYPES OF DISCONTINUOUS FUNCTIONS. WE WILL ALSO DISCUSS LIMITS AT INFINITY AND THEIR RELATIONSHIP TO HORIZONTAL ASYMPTOTES.

Continuous Functions

We have found that $f(c)$ is the same as the limit as $x \to c$ for any polynomial function $f(x)$ and any real number c.

Any function for which this special property holds is called a **continuous function**. The graphs of continuous functions can be drawn without lifting the pencil from the paper, and graphs of others may have holes, vertical asymptotes, or jumps that make it impossible to draw them without lifting the pencil. In general, we define continuity of a function at the value $x = c$ as follows.

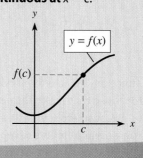

Continuity at a Point: The function f is **continuous at $x = c$** if all of the following conditions are satisfied.

1. $f(c)$ exists
2. $\lim\limits_{x \to c} f(x)$ exists
3. $\lim\limits_{x \to c} f(x) = f(c)$

The figure below illustrates these three conditions.

If one or more of the preceding conditions do not hold, we say the function is **discontinuous at $x = c$.**

If a function is discontinuous at one or more points, it is called a **discontinuous function**. Figure 9.8 shows graphs of some functions that are discontinuous at $x = 2$.

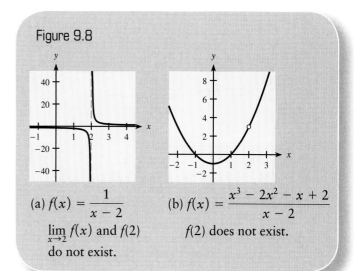

Figure 9.8

(a) $f(x) = \dfrac{1}{x - 2}$
$\lim\limits_{x \to 2} f(x)$ and $f(2)$ do not exist.

(b) $f(x) = \dfrac{x^3 - 2x^2 - x + 2}{x - 2}$
$f(2)$ does not exist.

In the previous section, we saw that if f is a polynomial function, then $\lim_{x \to c} f(x) = f(c)$ for every real number c, and also that $\lim_{x \to c} h(x) = h(c)$ if $h(x) = \dfrac{f(x)}{g(x)}$ is a rational function and $g(c) \neq 0$. Thus, by definition, we have the following.

> Every polynomial function is continuous for all real numbers.
>
> Every rational function is continuous at all values of x except those that make the denominator 0.

EXAMPLE 9.6 Discontinuous Functions

For what values of x, if any, is the following function continuous?

$$h(x) = \frac{3x + 2}{4x - 6}$$

SOLUTION

This is a rational function, so it is continuous for all values of x except for those that make the denominator, $4x - 6$, equal to 0. Because $4x - 6 = 0$ at $x = 3/2$, $h(x)$ is continuous for all real numbers except $x = 3/2$. Figure 9.9 shows a vertical asymptote at $x = 3/2$.

If the pieces of a piecewise defined function are polynomials, the only values of x where the function might be discontinuous are those at which the definition of the function changes.

EXAMPLE 9.7 Piecewise Defined Functions

Determine the values of x, if any, for which the following function is discontinuous.

$$f(x) = \begin{cases} 4 - x^2 & \text{if } x < 2 \\ x - 2 & \text{if } x \geq 2 \end{cases}$$

SOLUTION

The function $f(x)$ is continuous everywhere except perhaps at $x = 2$, where the definition of $f(x)$ changes. Because $x = 2$ satisfies $x \geq 2$, $f(2) = 2 - 2 = 0$. The left- and right-hand limits are

$$\lim_{x \to 2^-} f(x) = \lim_{x \to 2^-} (4 - x^2) = 4 - 2^2 = 0$$

and

$$\lim_{x \to 2^+} f(x) = \lim_{x \to 2^+} (x - 2) = 2 - 2 = 0$$

Because the right- and left-hand limits are equal, we conclude that $\lim_{x \to 2} f(x) = 0$. The limit is equal to the functional value

$$\lim_{x \to 2} f(x) = f(2)$$

so we conclude that f is continuous at $x = 2$ and thus f is continuous for all values of x. This result is confirmed by the graph of f, shown in Figure 9.10.

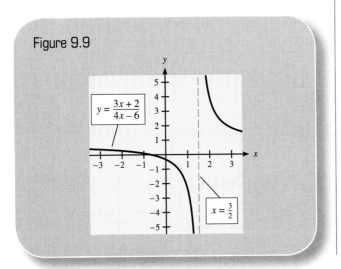

Figure 9.9

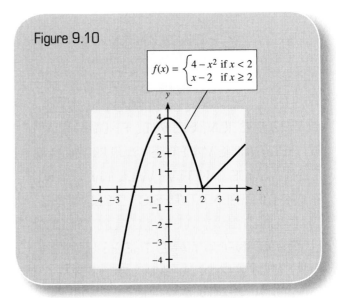

Figure 9.10

[TAX BRACKET]

Taxing Question

Suppose that a friend of yours and her husband have a taxable income of $137,050, and she tells you that she doesn't want to make any more money because that would put them in a higher tax bracket. She makes this statement because the tax rate schedule for married taxpayers filing a joint return (shown in the table) appears to have a jump in taxes for taxable income at $137,050.

Schedule Y-1—If your filing status is
Married filing jointly or **Qualifying widow(er)**

If taxable income is over—	But not over—	The tax is:
$0	$16,700	10% of the amount over $0
$16,700	$67,900	$1,670.00 plus 15% of the amount over 16,700
$67,900	$137,050	$9,350.00 plus 25% of the amount over 67,900
$137,050	$208,850	$26,637.50 plus 28% of the amount over 137,050
$208,850	$372,950	$46,741.50 plus 33% of the amount over 208,850
$372,950	no limit	$100,894.50 plus 35% of the amount over 372,950

Source: Internal Revenue Service, 2010, Form 1040 Instructions

To see whether the couple's taxes would jump to some higher level, let's write the function that gives income tax for married taxpayers as a function of taxable income and see whether the function is continuous. That is, we will see whether the tax paid does not jump at \$137,050, even though the tax on income above \$137,050, is collected at a higher rate.

The function that gives the tax due for married taxpayers is

$$T(x) =$$

$$\begin{cases} 0.10x & \text{if} & 0 \le x \le 16{,}700 \\ 1670.00 + 0.15(x - 16{,}700) & \text{if} & 16{,}700 < x \le 67{,}900 \\ 9350.00 + 0.25(x - 67{,}900) & \text{if} & 67{,}900 < x \le 137{,}050 \\ 26{,}637.50 + 0.28(x - 137{,}050) & \text{if} & 137{,}050 < x \le 208{,}850 \\ 46{,}741.50 + 0.33(x - 208{,}850) & \text{if} & 208{,}850 < x \le 372{,}950 \\ 100{,}894.50 + 0.35(x - 372{,}950) & \text{if} & x > 372{,}950 \end{cases}$$

If this function is continuous at $x = 137{,}050$, there is no jump in taxes at \$137,050. We examine the three conditions for continuity at $x = 137{,}050$

(a) $T(137{,}050) = 26{,}637.50$, so $T(137{,}050)$ exists.

(b) Because the function is piecewise defined near 137,050, we evaluate $\lim\limits_{x \to 137{,}050} T(x)$ by evaluating one-sided limits:

From the left evaluate $\lim\limits_{x \to 137{,}050^-} T(x)$:

$$\lim\limits_{x \to 137{,}050^-} [9350.00 + 0.25(x - 67{,}900)] = 26{,}637.50$$

From the right evaluate $\lim\limits_{x \to 137{,}050^+} T(x)$:

$$\lim\limits_{x \to 137{,}050^+} [26{,}637.50 + 0.28(x - 137{,}050)] = 26{,}637.50$$

Because these one-sided limits agree, the limit exists and is $\lim\limits_{x \to 137{,}050} T(x) = 26{,}637.50$.

(c) Because $\lim\limits_{x \to 137{,}050} T(x) = T(137{,}050) = 26{,}637.50$, the function is continuous at 137,050.

If your friend earned more than \$137,050, she and her husband would pay taxes at a higher rate on the money earned *above* the \$137,050, but it would not increase the tax rate on any income *up to* \$137,050. Thus she should take any raise that's offered.

Limits at Infinity

We noted in Section 2.4, "Special Functions and Their Graphs," that the graph of $y = 1/x$ has a vertical asymptote at $x = 0$ (shown in Figure 9.11(a)). By graphing $y = 1/x$ and evaluating the function for very large x-values, we can see that $y = 1/x$ never becomes negative for positive x-values regardless of how large the x-value is. Although no value of x makes $1/x$ equal to 0, it is easy to see that $1/x$ approaches 0 as x gets very large. This is denoted by

$$\lim\limits_{x \to +\infty} \frac{1}{x} = 0$$

and means that the line $y = 0$ (the x-axis) is a horizontal asymptote for $y = 1/x$. We also see that $y = 1/x$ approaches 0 as x decreases without bound, and we denote this by

$$\lim\limits_{x \to -\infty} \frac{1}{x} = 0$$

These limits for $f(x) = 1/x$ can also be established with numerical tables.

x	$f(x) = 1/x$	x	$f(x) = 1/x$
100	0.01	−100	−0.01
100,000	0.00001	−100,000	−0.00001
100,000,000	0.00000001	−100,000,000	−0.00000001
↓	↓	↓	↓
$+\infty$	0	$-\infty$	0
$\lim\limits_{x \to +\infty} \dfrac{1}{x} = 0$		$\lim\limits_{x \to -\infty} \dfrac{1}{x} = 0$	

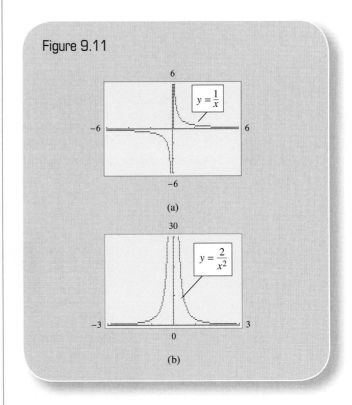

Figure 9.11

(a)

(b)

We can use the graph of $y = 2/x^2$ in Figure 9.11(b) to see that the x-axis ($y = 0$) is a horizontal asymptote and that

$$\lim\limits_{x \to +\infty} \frac{2}{x^2} = 0 \quad \text{and} \quad \lim\limits_{x \to -\infty} \frac{2}{x^2} = 0$$

By using graphs and/or tables of values, we can generalize the results for the functions shown in Figure 9.11 and conclude the following.

Limits at Infinity

If c is any constant, then

1. $\lim\limits_{x \to +\infty} c = c$ and $\lim\limits_{x \to -\infty} c = c.$

2. $\lim\limits_{x \to +\infty} \dfrac{c}{x^p} = 0,$ where $p > 0.$

3. $\lim\limits_{x \to -\infty} \dfrac{c}{x^n} = 0,$ where $n > 0$ is any integer.

In order to use these properties for finding the limits of rational functions as x approaches $+\infty$ or $-\infty$, we first divide each term of the numerator and denominator by the highest power of x present and then determine the limit of the resulting expression.

EXAMPLE 9.8 Limits at Infinity

Find $\lim\limits_{x \to +\infty} \dfrac{2x - 1}{x + 2}$ if it exists.

SOLUTION

The highest power of x present is x^1, so we divide each term in the numerator and denominator by x and then use the properties for limits at infinity.

$$\lim_{x \to +\infty} \frac{2x - 1}{x + 2} = \lim_{x \to +\infty} \frac{\dfrac{2x}{x} - \dfrac{1}{x}}{\dfrac{x}{x} + \dfrac{2}{x}} = \lim_{x \to +\infty} \frac{2 - \dfrac{1}{x}}{1 + \dfrac{2}{x}}$$

$$= \frac{2 - 0}{1 + 0} = 2 \text{ (by Properties 1 and 2)}$$

Figure 9.12 shows the graph of this function with the y-coordinates of the graph approaching 2 as x approaches $+\infty$ and as x approaches $-\infty$. That is, $y = 2$ is a horizontal asymptote. Note also that there is a discontinuity (vertical asymptote) where $x = -2$.

In our work with limits at infinity, we have mentioned horizontal asymptotes several times. The connection between these concepts is shown at right.

Graphing calculators and Excel can be used to help locate and investigate discontinuities and limits and infinity. See the Tech Card for details.

Summary

The following information is useful in discussing continuity of functions.

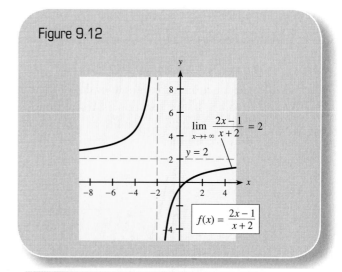

Figure 9.12

$$\lim_{x \to +\infty} \frac{2x - 1}{x + 2} = 2$$

$$y = 2$$

$$f(x) = \frac{2x - 1}{x + 2}$$

Limits at Infinity and Horizontal Asymptotes

If $\lim\limits_{x \to \infty} f(x) = b$ or $\lim\limits_{x \to -\infty} f(x) = b$, where b is a constant, then the line $y = b$ is a horizontal asymptote for the graph of $y = f(x)$. Otherwise, $y = f(x)$ has no horizontal asymptotes.

A. A polynomial function is continuous everywhere.

B. A rational function is a function of the form $\dfrac{f(x)}{g(x)}$, where $f(x)$ and $g(x)$ are polynomials.

1. If $g(x) \neq 0$ at any value of x, the function is continuous everywhere.
2. If $g(c) = 0$, the function is discontinuous at $x = c$.
   (a) If $g(c) = 0$ and $f(c) \neq 0$, then there is a vertical asymptote at $x = c$.
   (b) If $g(c) = 0$ and $\lim\limits_{x \to c} \dfrac{f(x)}{g(x)} = L$, then the graph has a missing point at (c, L).

C. A piecewise defined function *may* have a discontinuity at any x-value where the function changes its formula. One-sided limits must be used to see whether the limit exists.

The following steps are useful when we are evaluating limits at infinity for a rational function $f(x) = p(x)/q(x)$.

1. Divide both $p(x)$ and $q(x)$ by the highest power of x found in either polynomial.

2. Use the properties of limits at infinity to complete the evaluation.

9.2 Exercises

1. For each given x-value, use the figure to determine whether the function is continuous or discontinuous at that x-value. If the function is discontinuous, state which of the three conditions that define continuity is not satisfied.

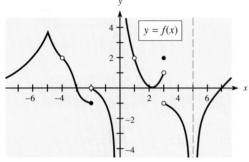

(a) $x = -5$ (b) $x = 1$
(c) $x = 3$ (d) $x = 0$

In Problems 2–4, determine whether each function is continuous or discontinuous at the given x-value. Examine the three conditions in the definition of continuity.

2. $f(x) = \dfrac{x^2 - 4}{x - 2}, \quad x = -2$

3. $y = \dfrac{x^2 - x - 12}{x^2 + 3x}, \quad x = -3$

4. $f(x) = \begin{cases} x - 3 & \text{if } x \le 2 \\ 4x - 7 & \text{if } x > 2 \end{cases}, \quad x = 2$

In Problems 5–7, determine whether the given function is continuous. If it is not, identify where it is discontinuous and which condition fails to hold. A graphing utility can be used to verify your conclusions.

5. $f(x) = 4x^2 - 1$

6. $g(x) = \dfrac{4x^2 + 3x + 2}{x + 2}$

7. $f(x) = \begin{cases} 3 & \text{if } x \le 1 \\ x^2 + 2 & \text{if } x > 1 \end{cases}$

8. Consider the function $f(x) = \dfrac{8}{x + 2}$ and its graph, presented in the figure.

(a) Use the graph to determine, as well as you can,
  (i) any vertical asymptotes.
  (ii) $\displaystyle \lim_{x \to +\infty} f(x)$.
  (iii) $\displaystyle \lim_{x \to -\infty} f(x)$.

(iv) any horizontal asymptotes.
(b) Check your conclusions in part (a) by using the function to determine items (i)–(iv) analytically.

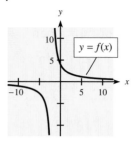

In Problems 9–11, complete (a) and (b).
(a) Use analytic methods to evaluate each limit.
(b) What does the result from part (a) tell you about horizontal asymptotes?

A graphing utility can be used to verify your conclusions.

9. $\displaystyle \lim_{x \to +\infty} \frac{3}{x + 1}$

10. $\displaystyle \lim_{x \to -\infty} \frac{5x^3 - 4x}{3x^3 - 2}$

11. $\displaystyle \lim_{x \to +\infty} \frac{3x^2 + 5x}{6x + 1}$

Applications

12. **Sales volume** Suppose that the weekly sales volume (in thousands of units) for a product is given by

$$y = \frac{32}{(p + 8)^{2/5}}$$

where p is the price in dollars per unit. Is this function continuous
(a) for all values of p?   (b) at $p = 24$?
(c) for all $p \ge 0$?
(d) What is the domain for this application?

13. **Cost-benefit** Suppose that the cost C of removing p percent of the impurities from the waste water in a manufacturing process is given by

$$C(p) = \frac{9800p}{101 - p}$$

Is this function continuous for all those p-values for which the problem makes sense?

14. **Pollution** The percent p of particulate pollution that can be removed from the smokestacks of an industrial plant by spending C dollars is given by

$$p = \frac{100C}{7300 + C}$$

Find the percent of the pollution that could be removed if spending C were allowed to increase without bound. Can 100% of the pollution be removed? Explain.

15. **Electrical usage costs** The monthly charge in dollars for x kilowatt-hours (kWh) of electricity used by a residential consumer of Excelsior Electric Membership Corporation from November through June is given by the function

$$C(x) = \begin{cases} 10 + 0.094x & \text{if} \quad 0 \le x \le 100 \\ 19.40 + 0.075(x - 100) & \text{if} \quad 100 < x \le 500 \\ 49.40 + 0.05(x - 500) & \text{if} \quad x > 500 \end{cases}$$

(a) What is the monthly charge if 1100 kWh of electricity is consumed in a month?

(b) Find $\lim\limits_{x \to 100} C(x)$ and $\lim\limits_{x \to 500} C(x)$, if the limits exist.

(c) Is C continuous at $x = 100$ and at $x = 500$?

....

Need more practice?
Find more here: cengagebrain.com

9.3 Rates of Change and the Derivative

FOR LINEAR FUNCTIONS, WE HAVE SEEN THAT THE SLOPE OF THE LINE MEASURES THE AVERAGE RATE OF CHANGE OF THE FUNCTION AND CAN BE FOUND FROM ANY TWO POINTS ON THE LINE. HOWEVER, FOR A FUNCTION THAT IS NOT LINEAR, THE SLOPE BETWEEN DIFFERENT PAIRS OF POINTS NO LONGER ALWAYS GIVES THE SAME NUMBER, BUT IT CAN BE INTERPRETED AS AN AVERAGE RATE OF CHANGE.

Average Rate of Change: The **average rate of change** of a function $y = f(x)$ from $x = a$ to $x = b$ is defined by

$$\text{Average rate of change} = \frac{f(b) - f(a)}{b - a}$$

The figure shows that this average rate is the same as the slope of the segment (or secant line) joining the points $(a, f(a))$ and $(b, f(b))$.

EXAMPLE 9.9 Total Cost

Suppose a company's total cost in dollars to produce x units of its product is given by

$$C(x) = 0.01x^2 + 25x + 1500$$

Find the average rate of change of total cost for the first 100 units produced (from $x = 0$ to $x = 100$).

SOLUTION

The average rate of change of total cost from $x = 0$ to $x = 100$ units is

$$\frac{C(100) - C(0)}{100 - 0}$$

$$= \frac{[0.01(100)^2 + 25(100) + 1500] - (1500)}{100}$$

$$= \frac{4100 - 1500}{100} = \frac{2600}{100} = 26 \text{ dollars per unit}$$

Instantaneous Rates of Change: Velocity

A common rate of change is velocity. For instance, if we travel 200 miles in our car over a 4-hour period, we know that we averaged 50 mph. However, during that trip there may have been times when we were traveling on an interstate at faster than 50 mph and times when we were stopped at a traffic light. Thus, for the trip we have not only an average velocity but also instantaneous velocities (or instantaneous speeds as displayed on the speedometer). Instantaneous velocity, commonly called *velocity,* is an example of an instantaneous rate of change. (To see how average velocity leads to instantaneous velocity, see the Enrichment material.)

The instantaneous rate of change of any function is commonly called *rate of change.* The function that gives this instantaneous rate of change of a function f is called the **derivative** of f.

> **Derivative:** If f is a function defined by $y = f(x)$, then the **derivative** of $f(x)$ at any value x, denoted $f'(x)$, is
>
> $$f'(x) = \lim_{h \to 0} \frac{f(x + h) - f(x)}{h}$$
>
> if this limit exists. If $f'(c)$ exists, we say that f is **differentiable** at c.

The following procedure illustrates how to find the derivative of a function $y = f(x)$ at any value x.

Derivative Using the Definition

Procedure

To find the derivative of $y = f(x)$ at any value x:

1. Let h represent the change in x from x to x + h.

2. The corresponding change in $y = f(x)$ is

    $$f(x + h) - f(x)$$

3. Form the difference quotient
    $\dfrac{f(x + h) - f(x)}{h}$ and simplify.

4. Find $\lim\limits_{h \to 0} \dfrac{f(x + h) - f(x)}{h}$ to determine $f'(x)$, the derivative of $f(x)$.

EXAMPLE 9.10

 Find the derivative of $f(x) = 4x^2$.

SOLUTION

Following the steps in the Procedure we have the following:

1. The change in x from x to $x + h$ is h.

2. The change in $f(x)$ is

$$f(x + h) - f(x) = 4(x + h)^2 - 4x^2$$
$$= 4(x^2 + 2xh + h^2) - 4x^2$$
$$= 4x^2 + 8xh + 4h^2 - 4x^2$$
$$= 8xh + 4h^2$$

3.
$$\frac{f(x + h) - f(x)}{h} = \frac{8xh + 4h^2}{h}$$
$$= 8x + 4h$$

4.
$$f'(x) = \lim_{h \to 0} \frac{f(x + h) - f(x)}{h}$$
$$f'(x) = \lim_{h \to 0} (8x + 4h) = 8x$$

Note that in the preceding example, we could have found the derivative of the function $f(x) = 4x^2$ at a particular value of x, say $x = 3$, by evaluating the derivative formula at that value:

$$f'(x) = 8x \quad \text{so} \quad f'(3) = 8(3) = 24$$

In addition to $f'(x)$, the derivative at any point x may be denoted by

$$\frac{dy}{dx}, \quad y', \quad \frac{d}{dx}f(x), \quad D_x y, \quad \text{or} \quad D_x f(x)$$

We can, of course, use variables other than x and y to represent functions and their derivatives. For example, we can represent the derivative of the function defined by $p = 2q^2 - 1$ by dp/dq.

Tangent to a Curve

Just as average rates of change are connected with slopes, so are instantaneous rates (derivatives). In fact, the slope of the graph of a function at any point is the same as the derivative at that point. In order to show this, we must define the slope of a curve at a point on the curve. We will define the slope of a curve at a point as the slope of the line tangent to the curve at the point.

In geometry, a **tangent** to a circle is defined as a line that has one point in common with the circle. (See Figure 9.13(a).) This definition does not apply to all curves, as Figure 9.13(b) shows. Many lines can be drawn through the point A that touch the curve only at A. One of the lines, line l, looks like it is tangent to the curve.

Figure 9.13

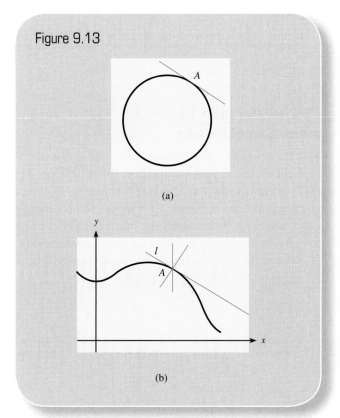

(a)

(b)

Figure 9.14

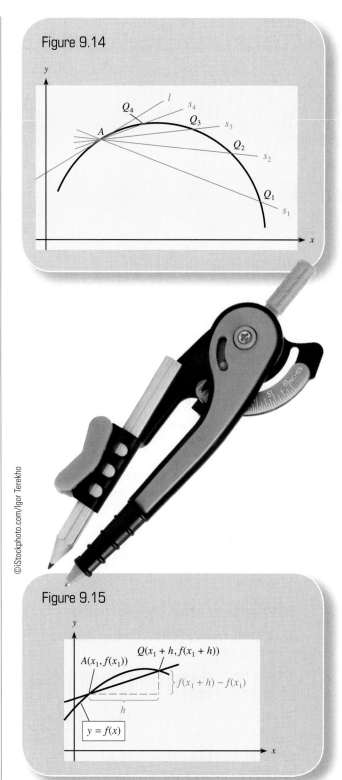

In Figure 9.14, the line l represents the tangent line to the curve at point A and shows that secant lines (s_1, s_2, etc.) through A approach line l as the second points (Q_1, Q_2, etc.) approach A. (For points and secant lines to the left of A, there would be a similar figure and conclusion.) This means that as we choose points on the curve closer and closer to A (from both sides of A), the limiting position of the secant lines through A is the **tangent line** to the curve at A.

From Figure 9.15, we see that the slope of an arbitrary secant line through $A(x_1, f(x_1))$ is given by

$$m_{AQ} = \frac{f(x_1 + h) - f(x_1)}{h}$$

Thus, we have the following.

Slope of the Tangent: The **slope of the tangent** to the graph of $y = f(x)$ at point $A(x_1, f(x_1))$ is

$$m = \lim_{h \to 0} \frac{f(x_1 + h) - f(x_1)}{h}$$

if this limit exists. That is, $m = f'(x_1)$, the derivative at $x = x_1$.

Figure 9.15

EXAMPLE 9.11 Tangent Line

Given $y = f(x) = 3x^2 + 2x + 11$, find
(a) the derivative of $f(x)$ at any point $(x, f(x))$.
(b) the slope of the curve at $(1, 16)$.
(c) the equation of the line tangent to $y = 3x^2 + 2x + 11$ at $(1, 16)$.

SOLUTION

(a) The derivative of $f(x)$ at any value x is denoted by $f'(x)$ and is

$$y' = f'(x) = \lim_{h \to 0} \frac{f(x + h) - f(x)}{h}$$

$$= \lim_{h \to 0} \frac{[3(x + h)^2 + 2(x + h) + 11] - (3x^2 + 2x + 11)}{h}$$

$$= \lim_{h \to 0} \frac{3(x^2 + 2xh + h^2) + 2x + 2h + 11 - 3x^2 - 2x - 11}{h}$$

$$= \lim_{h \to 0} \frac{6xh + 3h^2 + 2h}{h}$$

$$= \lim_{h \to 0} (6x + 3h + 2)$$

$$= 6x + 2$$

(b) The derivative is $f'(x) = 6x + 2$, so the slope of the tangent to the curve at $(1, 16)$ is $f'(1) = 6(1) + 2 = 8$. We can also say "The slope of the curve at $(1, 6)$ is 8."

(c) The equation of the tangent line uses the given point $(1, 16)$ and the slope $m = 8$. Using $y - y_1 = m(x - x_1)$ gives $y - 16 = 8(x - 1)$, or $y = 8x + 8$.

The discussion in this section indicates that the derivative of a function has several interpretations.

Interpretations of the Derivative

For a given function, each of the following means "find the **derivative**."

1. Find the **velocity** of an object moving in a straight line.

2. Find the **instantaneous rate of change** of a function.

3. Find the **slope of the tangent** to the graph of a function.

That is, all the terms printed in boldface are mathematically the same, and the answers to questions about any one of them give information about the others.

TECH NOTE

Graphing calculators and Excel can be used to approximate the value of a derivative at a point. See the Tech Card for details.

Differentiability and Continuity

So far we have talked about how the derivative is defined, what it represents, and how to find it. However, there are functions for which derivatives do not exist at every value of x. Figure 9.16 shows some common cases where $f'(c)$ does not exist but where $f'(x)$ exists for all other values of x. These cases occur where there is a discontinuity, a corner, or a vertical tangent line.

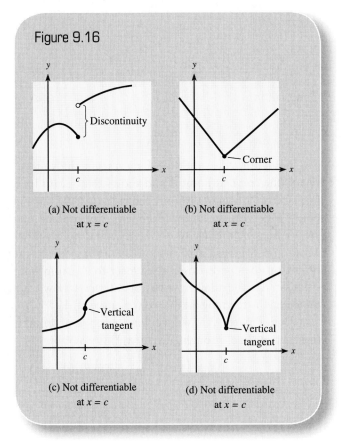

Figure 9.16

(a) Not differentiable at $x = c$

(b) Not differentiable at $x = c$

(c) Not differentiable at $x = c$

(d) Not differentiable at $x = c$

From Figure 9.16 we see that a function may be continuous at $x = c$ even though $f'(c)$ does not exist. Thus continuity does not imply differentiability at a point. However, differentiability does imply continuity.

Differentiability Implies Continuity:
If a function f is differentiable at $x = c$, then f is continuous at $x = c$.

9.3 Exercises

1. For the function $f(x) = x^2 + x - 12$, find the average rate of change over (a) $[0, 5]$ and (b) $[-3, 10]$.

2. Let $f(x) = 3x^2 - 2x$.
   (a) Use the definition of derivative and the Procedure given in this section to verify that $f'(x) = 6x - 2$.
   (b) Find the instantaneous rate of change of $f(x)$ at $x = -1$.
   (c) Find the slope of the tangent to the graph of $y = f(x)$ at $x = -1$.
   (d) Find the point on the graph of $y = f(x)$ at $x = -1$.

3. The tangent line to the graph of $f(x)$ at $x = 1$ is shown in the figure. On the tangent line, P is the point of tangency and A is another point on the line.
   (a) Find the coordinates of the points P and A.
   (b) Use the coordinates of P and A to find the slope of the tangent line.
   (c) Find $f'(1)$.
   (d) Find the instantaneous rate of change of $f(x)$ at P.

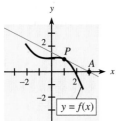

In Problems 4 and 5, for each function, find
   (a) the derivative, by using the definition.
   (b) the instantaneous rate of change of the function at any value and at the given value.
   (c) the slope of the tangent at the given value.

4. $f(x) = 5x^2 + 6x - 11$; $x = -2$

5. $p(q) = 2q^2 + q + 5$; $q = 10$

6. In the figure, at each point A and B draw an approximate tangent line and then use it to complete parts (a) and (b).
   (a) Is $f'(x)$ greater at point A or at point B? Explain.

(b) Estimate $f'(x)$ at point B.

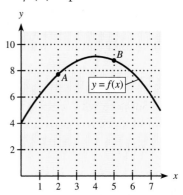

7. Because the derivative of a function represents both the slope of the tangent to the curve and the instantaneous rate of change of the function, it is possible to use information about one to gain information about the other. Consider the graph of the function $y = f(x)$ given in the figure.

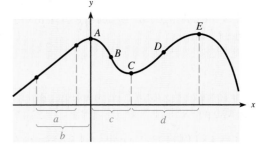

   (a) Over what interval(s) (*a*) through (*d*) is the rate of change of $f(x)$ positive?
   (b) Over what interval(s) (*a*) through (*d*) is the rate of change of $f(x)$ negative?
   (c) At what point(s) A through E is the rate of change of $f(x)$ equal to zero?

8. Given the graph of $y = f(x)$ in the figure, determine for which x-values A, B, C, D, or E the function is
   (a) continuous.   (b) differentiable.

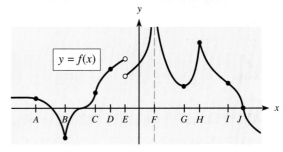

9. Consider the function $f(x) = x^2 + x$ and the point $(2, 6)$.
   (a) Find the slope of the tangent to the graph of $f(x)$ at any point.
   (b) Find the slope of the tangent at the given point.

(c) Write the equation of the line tangent to the graph of $f(x)$ at the given point.

(d) Graph both $f(x)$ and its tangent line (use a graphing utility if one is available).

Applications

10. **Total cost** Suppose total cost in dollars from the production of x printers is given by

$$C(x) = 0.0001x^3 + 0.005x^2 + 28x + 3000$$

Find the average rate of change of total cost when production changes

(a) from 100 to 300 printers.

(b) from 300 to 600 printers.

(c) Interpret the results from parts (a) and (b).

11. **Demand** If the demand for a product is given by

$$D(p) = \frac{1000}{\sqrt{p}} - 1$$

what is the average rate of change of demand when p increases from

(a) 1 to 25?

(b) 25 to 100?

12. **Labor force and output** The monthly output at the Olek Carpet Mill is

$$Q(x) = 15{,}000 + 2x^2 \quad \text{units}, \quad (40 \le x \le 60)$$

where x is the number of workers employed at the mill. If there are currently 50 workers, find the instantaneous rate of change of monthly output with respect to the number of workers. That is, find $Q'(50)$.

Need more practice?
Find more here: cengagebrain.com

9.4   Derivative Formulas

AS WE DISCUSSED IN THE PREVIOUS SECTION, THE DERIVATIVE OF A FUNCTION CAN BE USED TO FIND THE RATE OF CHANGE OF THE FUNCTION. IN THIS SECTION WE WILL DEVELOP FORMULAS THAT WILL MAKE IT EASIER TO FIND CERTAIN DERIVATIVES.

Derivative of $f(x) = x^n$: We can use the definition of derivative to show the following:

If $f(x) = x^2$, then $f'(x) = 2x$.

If $f(x) = x^3$, then $f'(x) = 3x^2$.

If $f(x) = x^4$, then $f'(x) = 4x^3$.

If $f(x) = x^5$, then $f'(x) = 5x^4$.

By using the definition of a derivative and the binomial formula, we can prove the following formula.

Powers of x Rule: If $f(x) = x^n$, where n is a real number, then $f'(x) = nx^{n-1}$.

For the development of this formula, see the Enrichment material on CourseMate for MATH APPS.

EXAMPLE 9.12   Powers of x Rule

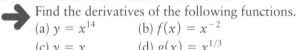   Find the derivatives of the following functions.

(a) $y = x^{14}$ (b) $f(x) = x^{-2}$

(c) $y = x$ (d) $g(x) = x^{1/3}$

SOLUTION

(a) If $y = x^{14}$, then $dy/dx = 14x^{14-1} = 14x^{13}$.

(b) The Powers of x Rule applies for all real values. Thus for $f(x) = x^{-2}$, we have

$$f'(x) = -2x^{-2-1} = -2x^{-3} = \frac{-2}{x^3}$$

(c) If $y = x$, then $dy/dx = 1x^{1-1} = x^0 = 1$. (Note that $y = x$ is a line with slope 1.)

(d) The Powers of x Rule applies to $y = x^{1/3}$.

$$g'(x) = \frac{1}{3}x^{1/3-1} = \frac{1}{3}x^{-2/3} = \frac{1}{3x^{2/3}}$$

In Example 9.12 we took the derivative with respect to x of *both sides* of each equation. We denote the operation "take the derivative with respect to x" by $\dfrac{d}{dx}$. Thus for $y = x^{14}$, in part (a), we can use this notation on both sides of the equation.

$$\frac{d}{dx}(y) = \frac{d}{dx}(x^{14}) \quad \text{gives} \quad \frac{dy}{dx} = 14x^{13}$$

Similarly, we can use this notation to indicate the derivative of an expression.

$$\frac{d}{dx}(x^{-2}) = -2x^{-3}$$

The differentiation rules are stated and proved for the independent variable x, but they also apply to other independent variables. For example, if $u(s) = s^8$, then

$u'(s) = 8s^{8-1} = 8s^7$, and if $s = \frac{1}{\sqrt{t}}$, then

$$s = \frac{1}{t^{1/2}} = t^{-1/2}$$

so

$$\frac{ds}{dt} = -\frac{1}{2}t^{-1/2-1} = -\frac{1}{2}t^{-3/2}$$

Writing the derivative in a form similar to that of the original function gives

$$\frac{ds}{dt} = -\frac{1}{2} \cdot \frac{1}{t^{3/2}} = -\frac{1}{2\sqrt{t^3}}$$

EXAMPLE 9.13 Tangent Line

Write the equation of the tangent line to the graph of $y = x^3$ at $x = 1$.

SOLUTION

Writing the equation of the tangent line to $y = x^3$ at $x = 1$ involves three steps.

1. Evaluate the function to find the point of tangency.

At $x = 1$: $y = (1)^3 = 1$, so the point is $(1, 1)$

2. Evaluate the derivative to find the slope of the tangent.

At any point: $m_{\tan} = y' = 3x^2$

At $x = 1$: $m_{\tan} = y'(1) = y'|_{x=1} = 3(1^2) = 3$

3. Use $y - y_1 = m(x - x_1)$ with the point $(1, 1)$ and slope $m = 3$.

$$y - 1 = 3(x - 1) \Rightarrow y = 3x - 3 + 1 \Rightarrow y = 3x - 2$$

Figure 9.17 shows the graph of $y = x^3$ and the tangent line at $x = 1$.

Additional Formulas

The following formulas can be proved with the definition (see the Enrichment material).

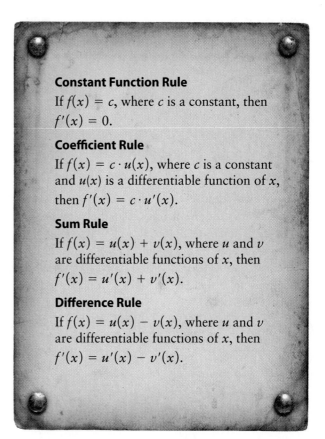

Constant Function Rule

If $f(x) = c$, where c is a constant, then $f'(x) = 0$.

Coefficient Rule

If $f(x) = c \cdot u(x)$, where c is a constant and $u(x)$ is a differentiable function of x, then $f'(x) = c \cdot u'(x)$.

Sum Rule

If $f(x) = u(x) + v(x)$, where u and v are differentiable functions of x, then $f'(x) = u'(x) + v'(x)$.

Difference Rule

If $f(x) = u(x) - v(x)$, where u and v are differentiable functions of x, then $f'(x) = u'(x) - v'(x)$.

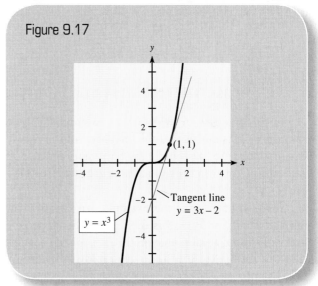

Figure 9.17

Thus from the Constant Function Rule, we see that the derivative of $y = 4$ is $\frac{dy}{dx} = 0$. Also, by the Coefficient Rule, the derivative of $g(t) = \frac{1}{2}t^2$ is $g'(t) = \frac{1}{2}(2t) = t$.

The following example illustrates the sum and difference rules.

EXAMPLE 9.14 Sum and Difference Rules

Find the derivatives of the following functions.

(a) $y = 3x + 5$

(b) $f(x) = 4x^3 - 2x^2 + 5x - 3$

(c) $y = 4x^3 + \sqrt{x}$

(d) $s = 5t^6 - \dfrac{1}{t^2}$

SOLUTION

(a) $y' = 3 \cdot 1 + 0 = 3$

(b) The rules regarding the derivatives of sums and differences of two functions also apply if more than two functions are involved. We may think of the functions that are added and subtracted as terms of the function f. Then it would be correct to say that we may take the derivative of a function term by term. Thus,

$$f'(x) = 4(3x^2) - 2(2x) + 5(1) - 0 = 12x^2 - 4x + 5$$

(c) We may write the function as $y = 4x^3 + x^{1/2}$, so

$$y' = 4(3x^2) + \frac{1}{2}x^{-1/2} = 12x^2 + \frac{1}{2x^{1/2}}$$

$$= 12x^2 + \frac{1}{2\sqrt{x}}$$

(d) We may write $s = 5t^6 - 1/t^2$ as $s = 5t^6 - t^{-2}$, so

$$\frac{ds}{dt} = 5(6t^5) - (-2t^{-3}) = 30t^5 + 2t^{-3}$$

$$= 30t^5 + \frac{2}{t^3}$$

Each derivative in the previous example has been *simplified*. This means that the final form of the derivative contains no negative exponents and the use of radicals or fractional exponents matches the original problem.

Also, in part (a) of the example, we saw that the derivative of $y = 3x + 5$ is 3. Because the slope of a line is the same at all points on the line, it is reasonable that the derivative of a linear equation is a constant. In particular, the slope of the graph of the equation $y = mx + b$ is m at all points on its graph because the derivative of $y = mx + b$ is $y' = f'(x) = m$.

EXAMPLE 9.15 World Tourism

Over the past 20 years, world tourism has grown into one of the world's major industries. Since 1990 the receipts from world tourism y, in billions of dollars, can be modeled by the function

$$y = 113.54x^{0.52584}$$

where x is the number of years past 1985. (*Source:* World Tourism Organization)

(a) Name the function that models the rate of change of the receipts from world tourism.

(b) Find the function from part (a).

(c) Find the rate of change in world tourism in 2010.

SOLUTION

(a) The rate of change of world tourism receipts is modeled by the derivative.

(b) $\dfrac{dy}{dx} = 113.54(0.52584x^{0.52584-1}) \approx 59.704x^{-0.47416}$

(c) $\dfrac{dy}{dx}\bigg|_{x=25} = 59.704(25^{-0.47416}) \approx 12.976$

Thus, the model estimates that world tourism changed by about $12.976 billion per year in 2010.

EXAMPLE 9.16 Personal Income

Suppose that t is the number of years past 1960 and that

$$I = I(t) = 4.88t^2 + 2.68t + 370$$

models the U.S. total personal income in billions of current dollars. For 2010, find the model's prediction for the U.S. total personal income and the rate of change of U.S. total personal income.

SOLUTION

For 2010, $t = 50$ and

$$I(50) = 4.88(50^2) + 2.68(50) + 370 = \$12{,}704$$
$$\text{billion (current dollars)}$$

2011

$490.68 billion

2010

The rate of change of U.S. total personal income is given by

$$I'(t) = 9.76t + 2.68$$

The predicted rate for 2010 is

$$I'(50) = \$490.68 \text{ billion (current dollars) per year}$$

This predicts that U.S. total personal income will change by about $490.68 billion from 2010 to 2011.

Marginal Revenue

One very common application of derivatives is in calculating marginal revenue. We define the marginal revenue for a product as the instantaneous rate of change, or the derivative, of the revenue function. Thus, the marginal revenue is $\overline{MR} = R'(x)$ and is used to estimate the change in revenue caused by the sale of one additional unit.

EXAMPLE 9.17 Revenue

Suppose that the manufacturer of Segaway knows that because of the demand for one type of personal transporters, his revenue is given by

$$R(x) = 1500x - 0.02x^2, \quad 0 \le x \le 1000$$

where x is the number of units sold and $R(x)$ is in dollars.
(a) Find the marginal revenue at $x = 500$.
(b) Find the change in revenue caused by the increase in sales from 500 to 501 units.
(c) Find the difference between the marginal revenue found in (a) and the change in revenue found in (b).

SOLUTION

(a) The marginal revenue for any value of x is

$$R'(x) = 1500 - 0.04x$$

The marginal revenue at $x = 500$ is

$$R'(500) = 1500 - 20 = 1480 \text{ (dollars per unit)}$$

We can interpret this to mean that the approximate revenue from the sale of the 501st unit will be $1480.

(b) The revenue at $x = 500$ is $R(500) = 745,000$, and the revenue at $x = 501$ is $R(501) = 746,479.98$, so the change in revenue is

$$R(501) - R(500) = 746,479.98 - 745,000$$
$$= 1479.98 \text{ (dollars)}$$

(c) The difference is $1480 - 1479.98 = 0.02$. Thus we see that the marginal revenue at $x = 500$ is a good estimate of the revenue from the 501st unit.

See the Tech Card for details on how to use the numerical derivative feature of a graphing calculator to check the derivative of a function.

9.4 Exercises

In Problems 1–3, find the derivative of each function.

1. $y = 4$

2. $f(t) = t$

3. $y = 10x^5 - 3x^3 + 5x - 11$

4. For the function $y = 7x^2 + 2x + 1$, at the point $x = 2$, find
   (a) the slope of the tangent to the curve, and
   (b) the instantaneous rate of change of the function.

In Problems 5 and 6, find the derivative of each function.

5. $y = 3x^{11/3} - 2x^{7/4} - x^{1/2} + 8$

6. $g(x) = \dfrac{3}{x^5} + \dfrac{2}{x^4} + 6\sqrt[3]{x}$

7. Write the equation of the tangent line to $y = x^3 - 5x^2 + 7$ at $x = 1$. As a check, graph both the function and the tangent line.

8. Find the coordinates of points where the graph of $f(x) = -x^3 + 9x^2 - 15x + 6$ has horizontal tangents. As a check, graph $f(x)$ and see whether the points you found look as though they have horizontal tangents.

Applications

9. *Revenue* Suppose that a wholesaler expects that his monthly revenue, in dollars, for small television sets will be

$$R(x) = 100x - 0.1x^2, \quad 0 \le x \le 800$$

where x is the number of units sold. Find his marginal revenue and interpret it when the quantity sold is
   (a) $x = 300$.
   (b) $x = 600$.

10. ***Workers and output*** The number of units of weekly output of a certain product is

$$Q(x) = 200x + 6x^2$$

where x is the number of workers on the assembly line. There are presently 60 workers on the line.

(a) Find $Q'(x)$ and estimate the change in the weekly output caused by the addition of one worker.

(b) Calculate $Q(61) - Q(60)$ to see the actual change in the weekly output.

11. ***Demand*** The demand for q units of a product depends on the price p (in dollars) according to

$$q = \frac{1000}{\sqrt{p}} - 1, \quad \text{for } p > 0$$

Find and explain the meaning of the instantaneous rate of change of demand with respect to price when the price is

(a) \$25.   (b) \$100.

12. ***Cost and average cost*** Suppose that the total cost function, in dollars, for the production of x units of a product is given by

$$C(x) = 4000 + 55x + 0.1x^2$$

Then the average cost of producing x items is

$$\overline{C(x)} = \frac{\text{total cost}}{x} = \frac{4000}{x} + 55 + 0.1x$$

(a) Find the instantaneous rate of change of average cost with respect to the number of units produced, at any level of production.

(b) Find the level of production at which this rate of change equals zero.

(c) At the value found in part (b), find the instantaneous rate of change of cost and find the average cost. What do you notice?

13. ***Cost-benefit*** Suppose that for a certain city the cost C, in dollars, of obtaining drinking water that contains p percent impurities (by volume) is given by

$$C = \frac{120{,}000}{p} - 1200$$

(a) Find the rate of change of cost with respect to p when impurities account for 1% (by volume).

(b) Write a sentence that explains the meaning of your answer in part (a).

⋯⋯⋯⋯⋯⋯⋯⋯⋯⋯⋯⋯⋯⋯⋯⋯⋯⋯⋯⋯⋯⋯⋯⋯⋯

Need more practice?
Find more here: cengagebrain.com

9.5 The Product Rule and the Quotient Rule

LET'S CONTINUE OUR EXPLORATION OF RULES THAT APPLY TO DERIVATIVES BY EXAMINING THE PRODUCT AND QUOTIENT RULES.

Product Rule

We have simple formulas for finding the derivatives of the sums and differences of functions. But we are not so lucky with products. The derivative of a product is *not* the product of the derivatives. To see this, we consider the function $f(x) = x \cdot x$. Because this function is $f(x) = x^2$, its derivative is $f'(x) = 2x$. But the product of the derivatives of x and x would give $1 \cdot 1 = 1 \neq 2x$. Thus we need a different formula to find the derivative of a product. This formula is given by the Product Rule.

> **Product Rule:** If $f(x) = u(x) \cdot v(x)$, where u and v are differentiable functions of x, then
>
> $$f'(x) = u(x) \cdot v'(x) + v(x) \cdot u'(x)$$

Thus the derivative of a product of two functions is the first function times the derivative of the second plus the second function times the derivative of the first. The Product Rule can be proven from the definition of derivative. (See the Enrichment material on Course-Mate for MATH APPS at cengagebrain.com.)

EXAMPLE 9.18 Product Rule

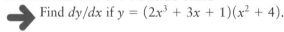 Find dy/dx if $y = (2x^3 + 3x + 1)(x^2 + 4)$.

SOLUTION

Using the Product Rule with $u(x) = 2x^3 + 3x + 1$ and $v(x) = x^2 + 4$, we have

$$\frac{dy}{dx} = (2x^3 + 3x + 1)(2x) + (x^2 + 4)(6x^2 + 3)$$

$$= 4x^4 + 6x^2 + 2x + 6x^4 + 3x^2 + 24x^2 + 12$$

$$= 10x^4 + 33x^2 + 2x + 12$$

EXAMPLE 9.19 Slope of a Tangent Line

Given
$$f(x) = (4x^3 + 5x^2 - 6x + 5)(x^3 - 4x^2 + 1),$$
find the slope of the tangent to the graph of
$y = f(x)$ at $x = 1$.

SOLUTION

$$f'(x) = (4x^3 + 5x^2 - 6x + 5)(3x^2 - 8x)$$
$$+ (x^3 - 4x^2 + 1)(12x^2 + 10x - 6)$$

so $f'(1) = 8(-5) + (-2)(16) = -72$ is the slope of
the curve at $x = 1$.

Quotient Rule

For a quotient of two functions, we might be tempted
to take the derivative of the numerator divided by the
derivative of the denominator; but this is incorrect. With
the example $f(x) = x^3/x$ (which equals x^2 if $x \neq 0$),
this approach would give $3x^2/1 = 3x^2$ as the derivative, rather than $2x$. Thus, finding the derivative of a
function that is the quotient of two functions requires
a new formula.

> **Quotient Rule:** If $f(x) = \dfrac{u(x)}{v(x)}$, where u
> and v are differentiable functions of x, with
> $v(x) \neq 0$, then
> $$f'(x) = \frac{v(x) \cdot u'(x) - u(x) \cdot v'(x)}{[v(x)]^2}$$

The preceding formula says that the derivative of
a quotient is the denominator times the derivative of
the numerator minus the numerator times the derivative of the denominator, all divided by the square of the
denominator.

EXAMPLE 9.20 Quotient Rule

If $f(x) = \dfrac{x^3 - 3x^2 + 2}{x^2 - 4}$, find (a) $f'(x)$ and (b) the
instantaneous rate of change of $f(x)$ at $x = 3$.

SOLUTION

(a) We evaluate $f'(x)$ at $x = 3$ to find the desired
rate of change. Using the Quotient Rule with
$u(x) = x^3 - 3x^2 + 2$ and $v(x) = x^2 - 4$, we get

$$f'(x) = \frac{(x^2 - 4)(3x^2 - 6x) - (x^3 - 3x^2 + 2)(2x)}{(x^2 - 4)^2}$$

$$= \frac{(3x^4 - 6x^3 - 12x^2 + 24x) - (2x^4 - 6x^3 + 4x)}{(x^2 - 4)^2}$$

$$= \frac{x^4 - 12x^2 + 20x}{(x^2 - 4)^2}$$

(b) Thus, the instantaneous rate of change at $x = 3$ is

$$f'(3) = \frac{3^4 - 12(3^2) + 20(3)}{(3^2 - 4)^2} = \frac{33}{25} = 1.32$$

It is not necessary to use the Quotient Rule when the
denominator of the function in question contains only
a constant. For example, the function $y = (x^3 - 3x)/3$
can be written $y = \frac{1}{3}(x^3 - 3x)$, so the derivative is
$y' = \frac{1}{3}(3x^2 - 3) = x^2 - 1$.

Let's reexamine the marginal revenue problem
using the Quotient Rule.

EXAMPLE 9.21 Marginal Revenue

Suppose that the revenue function for a flash
drive is given by

$$R(x) = 10x + \frac{100x}{3x + 5}$$

where x is the number of units sold and R is in dollars.
(a) Find the marginal revenue function.
(b) Find the marginal revenue when $x = 15$.

$$R'(15) = \$10.20$$

SOLUTION

(a) We must use the Quotient Rule to find the marginal revenue (the derivative).

$$\overline{MR} = R'(x) = 10 + \frac{(3x + 5)(100) - 100x(3)}{(3x + 5)^2}$$

$$= 10 + \frac{300x + 500 - 300x}{(3x + 5)^2} = 10 + \frac{500}{(3x + 5)^2}$$

(b) The marginal revenue when $x = 15$ is $R'(15)$.

$$R'(15) = 10 + \frac{500}{[(3)(15) + 5]^2} = 10 + \frac{500}{(50)^2}$$

$$= 10.20 \text{ (dollars per unit)}$$

Recall that $R'(15)$ estimates the revenue from the sale of the 16th item.

The Tech Card techniques for approximating and checking derivatives apply to functions that are products or quotients.

···

9.5 Exercises

1. Find y' if $y = (5x + 3)(x^2 - 2x)$.

2. Find the derivative of $y = (7x^6 - 5x^4 + 2x^2 - 1)(4x^9 + 3x^7 - 5x^2 + 3x)$, but do not simplify your answer.

3. For the function $y = (x^2 + 1)(x^3 - 4x)$, at $(-2, 0)$ find
   (a) the slope of the tangent line.
   (b) the instantaneous rate of change of the function.

In Problems 4–6, find the indicated derivatives and simplify.

4. $\dfrac{dp}{dq}$ for $p = \dfrac{q^2 + 3}{2q - 1}$

5. $\dfrac{dy}{dx}$ for $y = \dfrac{1 - 2x^2}{x^4 - 2x^2 + 5}$

6. $\dfrac{dz}{dx}$ for $z = x^2 + \dfrac{x^2}{1 - x - 2x^2}$

7. For the function $y = \dfrac{x^2 + 1}{x + 3}$ at $(2, 1)$, find
   (a) the slope of the tangent line.
   (b) the instantaneous rate of change of the function.

8. Write the equation of the tangent line to the graph of $y = (9x^2 - 6x + 1)(1 + 2x)$ at $x = 1$. Check the reasonableness of your answer by graphing both the function and the tangent line.

Applications

9. **Cost-benefit** If the cost C (in dollars) of removing p percent of the particulate pollution from the exhaust gases at an industrial site is given by

$$C(p) = \frac{8100p}{100 - p}$$

find the rate of change of C with respect to p.

10. **Revenue** A travel agency will plan a group tour for groups of size 25 or larger. If the group contains exactly 25 people, the cost is $300 per person. If each person's cost is reduced by $10 for each additional person above the 25, then the revenue is given by

$$R(x) = (25 + x)(300 - 10x)$$

where x is the number of additional people above 25. Find the marginal revenue if the group contains 30 people. Interpret your result.

11. **Response to a drug** The reaction R to an injection of a drug is related to the dosage x (in milligrams) according to

$$R(x) = x^2\left(500 - \frac{x}{3}\right)$$

where 1000 mg is the maximum dosage. If the rate of reaction with respect to the dosage defines the sensitivity to the drug, find the sensitivity.

12. **Candidate recognition** Suppose that the proportion P of voters who recognize a candidate's name t months after the start of the campaign is given by

$$P(t) = \frac{13t}{t^2 + 100} + 0.18$$

   (a) Find the rate of change of P when $t = 6$, and explain its meaning.
   (b) Find the rate of change of P when $t = 12$, and explain its meaning.
   (c) One month prior to the election, is it better for $P'(t)$ to be positive or negative? Explain.

13. **Wind chill** According to the National Climatic Data Center, during 1991, the lowest temperature recorded in Indianapolis, Indiana, was 0°F. If x is the wind speed in miles per hour and $x \geq 5$, then the wind chill (in degrees Fahrenheit) for an air

temperature of 0°F can be approximated by the function

$$f(x) = \frac{289.173 - 58.5731x}{x + 1}$$

(a) At what rate is the wind chill changing when the wind speed is 20 mph?

(b) Explain the meaning of your answer to part (a).

....

Need more practice?
Find more here: cengagebrain.com

9.6 The Chain Rule and the Power Rule

SO FAR, WE HAVE DERIVATIVE FORMULAS FOR SEVERAL WAYS THAT FUNCTIONS CAN BE COMBINED. IN THIS SECTION, WE CONSIDER TWO MORE DERIVATIVE FORMULAS, THE CHAIN RULE AND THE POWER RULE. THE CHAIN RULE IS USEFUL IN TAKING THE DERIVATIVE OF A FUNCTION THAT IS THE COMPOSITION OF TWO OTHER FUNCTIONS.

Recall from Section 1.2, "Functions," that if f and g are functions, then the composite functions g of f (denoted $g \circ f$) and f of g (denoted $f \circ g$) are defined as follows:

$$(g \circ f)(x) = g(f(x)) \quad \text{and} \quad (f \circ g)(x) = f(g(x))$$

Chain Rule

We could find the derivative of the function $F(x) = 3(2x - 1)^2$ by expanding the expression $3(2x - 1)^2$. Then

$$F(x) = 3(4x^2 - 4x + 1) = 12x^2 - 12x + 3$$

so $F'(x) = 24x - 12$. But we can also use a very powerful rule, called the **Chain Rule**, to find derivatives of composite functions. If we write the composite function $y = f(g(x))$ in the form $y = f(u)$, where $u = g(x)$, we state the Chain Rule as follows.

Chain Rule: If f and g are differentiable functions with $y = f(u)$, and $u = g(x)$, then y is a differentiable function of x, and

$$\frac{dy}{dx} = f'(u) \cdot g'(x)$$

or, equivalently,

$$\frac{dy}{dx} = \frac{dy}{du} \cdot \frac{du}{dx}$$

Note that dy/du represents the derivative of $y = f(u)$ *with respect to u* and du/dx represents the derivative of $u = g(x)$ *with respect to x*. For example, if $y = 3(2x - 1)^2$, then the outside function, f, is the squaring function, and the inside function, g, is $2x - 1$, so we may write $y = f(u) = 3u^2$, where $u = g(x) = 2x - 1$. Then the derivative is

$$\frac{dy}{dx} = \frac{dy}{du} \cdot \frac{du}{dx} = 6u \cdot 2 = 12u$$

To write this derivative in terms of x, we substitute $2x - 1$ for u. Thus

$$\frac{dy}{dx} = 12(2x - 1) = 24x - 12$$

Note that we get the same result by using the Chain Rule as we did by expanding $f(x) = 3(2x - 1)^2$.

EXAMPLE 9.22 Chain Rule

➡ If $y = \sqrt{x^2 - 1}$, find $\dfrac{dy}{dx}$.

SOLUTION

If we write this function as $y = f(u) = \sqrt{u} = u^{1/2}$, when $u = x^2 - 1$, we can find the derivative.

$$\frac{dy}{dx} = \frac{dy}{du} \cdot \frac{du}{dx} = \frac{1}{2} \cdot u^{-1/2} \cdot 2x = u^{-1/2} \cdot x = \frac{1}{\sqrt{u}} \cdot x = \frac{x}{\sqrt{u}}$$

To write this derivative in terms of x alone, we substitute $x^2 - 1$ for u. Then

$$\frac{dy}{dx} = \frac{x}{\sqrt{x^2 - 1}}$$

Note that we could not find the derivative of a function like that of Example 9.22 by the methods learned previously.

Power Rule

The Chain Rule is very useful and will be extremely important with functions that we will study later. A special case of the Chain Rule, called the **Power Rule**, is useful for the algebraic functions we have studied so far, composite functions where the outside function is a power.

> **Power Rule:** If $y = u^n$, where u is a differentiable function of x, then
>
> $$\frac{dy}{dx} = nu^{n-1} \cdot \frac{du}{dx}$$

EXAMPLE 9.23 Power Rule

Find the derivative of each of the following functions.

(a) $y = (x^2 - 4x)^6$ (b) $p = \dfrac{4}{3q^2 + 1}$

(c) $g(x) = \dfrac{1}{\sqrt{(x^2 + 1)^3}}$

SOLUTION

(a) This has the form $y = u^n = u^6$, with $u = x^2 - 4x$. Thus, by the Power Rule,

$$\frac{dy}{dx} = nu^{n-1} \cdot \frac{du}{dx} = 6u^5(2x - 4)$$

Substituting $x^2 - 4x$ for u gives

$$\frac{dy}{dx} = 6(x^2 - 4x)^5(2x - 4) = (12x - 24)(x^2 - 4x)^5$$

(b) We can use the Power Rule to find dp/dq if we write the equation in the form

$$p = 4(3q^2 + 1)^{-1}$$

Then

$$\frac{dp}{dq} = 4[-1(3q^2 + 1)^{-2}(6q)] = \frac{-24q}{(3q^2 + 1)^2}$$

The derivative of the function in part (b) can also be found by using the Quotient Rule, but the Power Rule provides a more efficient method.

(c) Writing $g(x)$ as a power gives

$$g(x) = (x^2 + 1)^{-3/2}$$

Then

$$g'(x) = -\frac{3}{2}(x^2 + 1)^{-5/2}(2x)$$

$$= -3x \cdot \frac{1}{(x^2 + 1)^{5/2}} = \frac{-3x}{\sqrt{(x^2 + 1)^5}}$$

9.6 Exercises

In Problems 1–4, differentiate each function.

1. $f(x) = (3x^5 - 2)^{20}$

2. $h(x) = \frac{3}{4}(x^5 - 2x^3 + 5)^8$

3. $f(x) = \dfrac{3}{(2x^5 + 1)^4}$

4. $y = \sqrt{3x^2 + 4x + 9}$

5. For the function $y = (x^3 + 2x)^4$, at $x = 2$, find
   (a) the slope of the tangent line.
   (b) the instantaneous rate of change of the function.
   A graphing utility's numerical derivative feature can be used to check your work.

6. Write the equation of the line tangent to the graph of $y = \sqrt{3x^2 - 2}$ at $x = 3$. As a check, graph both the function and the tangent line you found to see whether it looks correct.

Applications

7. **Revenue** The revenue from the sale of a product is

   $$R = 1500x + 3000(2x + 3)^{-1} - 1000 \text{ dollars}$$

   where x is the number of units sold. Find the marginal revenue when 100 units are sold. Interpret your result.

8. **Pricing and sales** Suppose that the weekly sales volume y (in thousands of units sold) depends on the price per unit of the product according to

   $$y = 32(3p + 1)^{-2/5}, \quad p > 0$$

   where p is in dollars.
   (a) What is the rate of change in sales volume when the price is $21?
   (b) Interpret your answer to part (a).

9. **Demand** If the demand for q units of a product priced at p per unit is described by the equation

   $$p = \frac{100}{\sqrt{2q + 1}}$$

   find the rate of change of p with respect to q.

10. **Investments** If an IRA is a variable-rate investment for 20 years at rate r percent per year, compounded monthly, then the future value S that accumulates from an initial investment of $1000 is

    $$S = 1000\left[1 + \frac{0.01r}{12}\right]^{240}$$

What is the rate of change of S with respect to r and what does it tell us if the interest rate is (a) 6%? (b) 12%?

....

Need more practice?
Find more here: cengagebrain.com

9.7 Using Derivative Formulas

WE HAVE USED THE POWER RULE TO FIND THE DERIVATIVES OF FUNCTIONS LIKE

$$y = (x^3 - 3x^2 + x + 1)^5$$

BUT WE HAVE NOT FOUND THE DERIVATIVES OF FUNCTIONS LIKE

$$y = [(x^2 + 1)(x^3 + x + 1)]^5$$

This function is different because the function u (which is raised to the fifth power) is the product of two functions, $(x^2 + 1)$ and $(x^3 + x + 1)$. The equation is of the form $y = u^5$, where $u = (x^2 + 1)(x^3 + x + 1)$. This means that the Product Rule should be used to find du/dx. Then

$$\frac{dy}{dx} = 5u^4 \cdot \frac{du}{dx}$$

$$= 5u^4[(x^2 + 1)(3x^2 + 1) + (x^3 + x + 1)(2x)]$$

$$= 5[(x^2 + 1)(x^3 + x + 1)]^4(5x^4 + 6x^2 + 2x + 1)$$

$$= (25x^4 + 30x^2 + 10x + 5)[(x^2 + 1)(x^3 + x + 1)]^4$$

A different type of problem involving the Power Rule and the Product Rule is finding the derivative of $y = (x^2 + 1)^5(x^3 + x + 1)$. We may think of y as the *product* of two functions, one of which is a power. Thus the fundamental formula we should use is the Product Rule. The two functions are $u(x) = (x^2 + 1)^5$ and $v(x) = x^3 + x + 1$. The Product Rule gives

$$\frac{dy}{dx} = u(x) \cdot v'(x) + v(x) \cdot u'(x)$$

$$= (x^2 + 1)^5(3x^2 + 1) + (x^3 + x + 1)[5(x^2 + 1)^4(2x)]$$

Note that the Power Rule was used to find $u'(x)$, since $u(x) = (x^2 + 1)^5$.

We can simplify dy/dx by factoring $(x^2 + 1)^4$ from both terms:

$$\frac{dy}{dx} = (x^2 + 1)^4[(x^2 + 1)(3x^2 + 1) + (x^3 + x + 1) \cdot 5 \cdot 2x]$$

$$= (x^2 + 1)^4(13x^4 + 14x^2 + 10x + 1)$$

EXAMPLE 9.24 Power of a Quotient

If $y = \left(\dfrac{x^2}{x - 1}\right)^5$, find y'.

SOLUTION

We again have an equation of the form $y = u^n$, but this time u is a quotient. Thus we will need the Quotient Rule to find du/dx.

$$y' = nu^{n-1} \cdot \frac{du}{dx} = 5u^4 \frac{(x - 1) \cdot 2x - x^2 \cdot 1}{(x - 1)^2}$$

Substituting for u and simplifying give

$$y' = 5\left(\frac{x^2}{x - 1}\right)^4 \cdot \frac{2x^2 - 2x - x^2}{(x - 1)^2}$$

$$= \frac{5x^8(x^2 - 2x)}{(x - 1)^6} = \frac{5x^{10} - 10x^9}{(x - 1)^6}$$

EXAMPLE 9.25 Revenue

To see how these rules apply in a concrete situation, let's suppose that the weekly revenue function for a software package is given by

$$R(x) = \frac{36,000,000x}{(2x + 500)^2}$$

where $R(x)$ is the dollars of revenue from the sale of x units.
(a) Find the marginal revenue function.
(b) Find the marginal revenue when 50 units are sold.

SOLUTION

(a) $\overline{MR} = R'(x)$

$$= \frac{(2x + 500)^2(36,000,000) - 36,000,000x[2(2x + 500)^1(2)]}{(2x + 500)^4}$$

$$= \frac{36,000,000(2x + 500)(2x + 500 - 4x)}{(2x + 500)^4}$$

$$= \frac{36,000,000(500 - 2x)}{(2x + 500)^3}$$

(b) $\overline{MR}(50) = R'(50) = \dfrac{36,000,000(500 - 100)}{(100 + 500)^3}$

$$= \frac{36,000,000(400)}{(600)^3} = \frac{200}{3} \approx 66.67$$

The marginal revenue is $66.67 when 50 units are sold.

The following sumary presents examples of different types of functions and the formulas needed to find their derivatives.

Summary of Derivative Formulas

Examples	Formulas

$f(x) = 14$

If $f(x) = c$, then $f'(x) = 0$.

$y = x^4$

If $f(x) = x^n$, then $f'(x) = nx^{n-1}$.

$g(x) = 5x^3$

If $g(x) = cf(x)$, then $g'(x) = cf'(x)$.

$y = 3x^2 + 4x$

If $f(x) = u(x) + v(x)$, then $f'(x) = u'(x) + v'(x)$.

$y = (x^2 - 2)(x + 4)$

If $f(x) = u(x) \cdot v(x)$, then $f'(x) = u(x) \cdot v'(x) + v(x) \cdot u'(x)$.

$f(x) = \dfrac{x^3}{x^2 + 1}$

If $f(x) = \dfrac{u(x)}{v(x)}$, then $f'(x) = \dfrac{v(x) \cdot u'(x) - u(x) \cdot v'(x)}{[v(x)]^2}$.

$y = (x^3 - 4x)^{10}$

If $y = u^n$ and $u = g(x)$, then $\dfrac{dy}{dx} = nu^{n-1} \cdot \dfrac{du}{dx}$.

$y = \left(\dfrac{x - 1}{x^2 + 3}\right)^3$

Power Rule, then Quotient Rule to find $\dfrac{du}{dx}$, where $u = \dfrac{x - 1}{x^2 + 3}$.

$y = (x + 1)\sqrt{x^3 + 1}$

Product Rule, then Power Rule to find $v'(x)$, where $v(x) = \sqrt{x^3 + 1}$.

$y = \dfrac{(x^2 - 3)^4}{x + 1}$

Quotient Rule, then Power Rule to find the derivative of the numerator.

9.7 Exercises

In Problems 1–8, find the derivative of each function. Simplify the answer and express it using positive exponents only.

1. $f(x) = \pi^4$

2. $g(x) = 5x^3 + \dfrac{4}{x}$

3. $y = \dfrac{(x^3 - 4x)^{10}}{10}$

4. $y = \dfrac{5}{3}x^3(4x^5 - 5)^3$

5. $y = \dfrac{(x^2 - 4)^3}{x^2 + 1}$

6. $R(x) = [x^2(x^2 + 3x)]^4$

7. $y = \left(\dfrac{2x - 1}{x^2 + x}\right)^4$

8. $c(x) = 2x\sqrt{x^3 + 1}$

Applications

9. *Revenue* Suppose that the revenue in dollars from the sale of x campers is given by

$$R(x) = 60{,}000x + 40{,}000(10 + x)^{-1} - 4000$$

(a) Find the marginal revenue when 10 units are sold.

(b) How is revenue changing when 10 units are sold?

10. *Advertising and sales* An inferior product with an extensive advertising campaign does well when it is released, but sales decline as people discontinue

use of the product. If the sales S (in thousands of dollars) after t weeks are given by

$$S(t) = \frac{200t}{(t + 1)^2}, \quad t \geq 0$$

what is the rate of change of sales when $t = 9$? Interpret your result.

....

Need more practice?
Find more here: cengagebrain.com

9.8 Higher-Order Derivatives

SINCE CELL PHONES WERE INTRODUCED, THEIR POPULARITY HAS INCREASED ENORMOUSLY. FIGURE 9.18(A) SHOWS A GRAPH OF THE MILLIONS OF WORLDWIDE CELLULAR SUBSCRIBERS (ACTUAL AND PROJECTED) AS A FUNCTION OF THE NUMBER OF YEARS PAST 1995. (*SOURCE:* INTERNATIONAL TELECOMMUNICATIONS UNION) NOTE THAT THE NUMBER OF SUBSCRIBERS IS ALWAYS INCREASING AND THAT THE RATE OF CHANGE OF THAT NUMBER (AS SEEN FROM TANGENT LINES TO THE GRAPH) IS ALWAYS POSITIVE. HOWEVER, THE TANGENT LINES SHOWN IN FIGURE 9.18(B) INDICATE THAT THE RATE OF CHANGE OF THE NUMBER OF SUBSCRIBERS IS GREATER AT *B* THAN AT EITHER *A* OR *C*.

Furthermore, the rate of change of the number of subscribers (the slopes of tangents) increases from *A* to *B* and then decreases from *B* to *C*. To learn how the rate of change of the number of subscribers is changing, we are interested in finding the derivative of the rate of change of the number of subscribers—that is, the derivative of the derivative of the number of subscribers. This is called the second derivative.

©iStockphoto.com

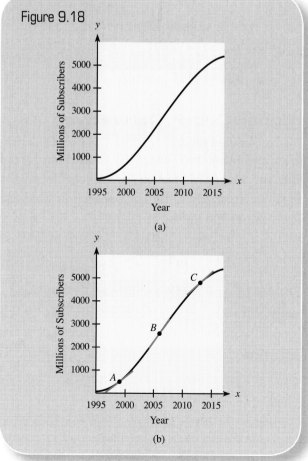

Figure 9.18

(a)

(b)

Second Derivatives

Because the derivative of a function is itself a function, we can take a derivative of the derivative. The derivative of a first derivative is called a **second derivative**. We can find the second derivative of a function f by differentiating it twice. If f' represents the first derivative of a function, then f'' represents the second derivative of that function. It is also common to use $\dfrac{d^2y}{dx^2}$ and $\dfrac{d^2f(x)}{dx^2}$ to denote the second derivative of a function.

EXAMPLE 9.26 Second Derivative

➡️ (a) If $f(x) = 3x^3 - 4x^2 + 5$, find $f''(x)$.
    (b) If $y = \sqrt{2x - 1}$, find d^2y/dx^2.

SOLUTION

(a) The first derivative is $f'(x) = 9x^2 - 8x$.
    The second derivative is $f''(x) = 18x - 8$.
(b) The first derivative is

$$\frac{dy}{dx} = \frac{1}{2}(2x - 1)^{-1/2}(2) = (2x - 1)^{-1/2}$$

The second derivative is

$$\frac{d^2y}{dx^2} = -\frac{1}{2}(2x - 1)^{-3/2}(2) = -(2x - 1)^{-3/2}$$

$$= \frac{-1}{(2x - 1)^{3/2}} = \frac{-1}{\sqrt{(2x - 1)^3}}$$

Higher-Order Derivatives

We can also find third, fourth, fifth, and higher derivatives, continuing indefinitely. The third, fourth, and fifth derivatives of a function f are denoted by f''', $f^{(4)}$, and $f^{(5)}$, respectively. Other notations for the third and fourth derivatives include

$$y''' = \frac{d^3y}{dx^3} = \frac{d^3f(x)}{dx^3}, \quad y^{(4)} = \frac{d^4y}{dx^4} = \frac{d^4f(x)}{dx^4}$$

EXAMPLE 9.27 Higher-Order Derivatives

➡️ Find the first four derivatives of $f(x) = 4x^3 + 5x^2 + 3$.

SOLUTION

$f'(x) = 12x^2 + 10x$, $\quad f''(x) = 24x + 10$,
$f'''(x) = 24$, $\quad f^{(4)}(x) = 0$

Just as the first derivative, $f'(x)$, can be used to determine the rate of change of a function $f(x)$, the second derivative, $f''(x)$, can be used to determine the rate of change of $f'(x)$.

Let $f(x) = 3x^4 + 6x^3 - 3x^2 + 4$. Then:

Because $f'(x) = 12x^3 + 18x^2 - 6x$, we have

$$f'(1) = 12 + 18 - 6 = 24$$

Thus the rate of change of $f(x)$ at $(1, 10)$ is 24 (y units per x unit).

Because $f''(x) = 36x^2 + 36x - 6$, we have

$$f''(1) = 66$$

Thus the rate of change of $f'(x)$ at $(1, 10)$ is 66 (y units per x unit per x unit).

Because $f''(1) = 66 > 0$, $f'(x)$ is increasing at $(1, 10)$.

EXAMPLE 9.28 Acceleration

➡️ Suppose that a particle travels according to the equation

$$s = 100t - 16t^2 + 200$$

where s is the distance in feet and t is the time in seconds. Then ds/dt is the velocity, and $d^2s/dt^2 = dv/dt$ is the acceleration of the particle. Find the acceleration.

SOLUTION

The velocity is $v = ds/dt = 100 - 32t$ feet per second, and the acceleration is

$$\frac{dv}{dt} = \frac{d^2s}{dt^2} = -32 \text{ (ft/sec)/sec} = -32 \text{ ft/sec}^2$$

We can use the numerical derivative feature of a graphing calculator to find the value of the second derivative of a function at a point. See the Tech Card for details.

....

9.8 Exercises

In Problems 1–3, find the second derivative.

1. $f(x) = 2x^{10} - 18x^5 - 12x^3 + 4$

2. $g(x) = x^3 - \dfrac{1}{x}$

3. $y = x^3 - \sqrt{x}$

4. Find the third derivative of $y = x^5 - 16x^3 + 12$.

In Problems 5–7, find the indicated derivative.

5. If $f(x) = \sqrt{x + 1}$, find $f'''(x)$.

6. Find $\dfrac{d^4y}{dx^4}$ if $y = 4x^3 - 16x$.

7. Find $y^{(4)}$ if $y' = \sqrt{4x - 1}$.

Applications

8. **Acceleration** A particle travels as a function of time according to the formula

$$s = 100 + 10t + 0.01t^3$$

where s is in meters and t is in seconds. Find the acceleration of the particle when $t = 2$.

9. **Revenue** The revenue (in dollars) from the sale of x units of a certain product can be described by

$$R(x) = 100x - 0.01x^2$$

Find the instantaneous rate of change of the marginal revenue.

10. **Sensitivity** When medicine is administered, reaction (measured in change of blood pressure or temperature) can be modeled by

$$R = m^2\left(\frac{c}{2} - \frac{m}{3}\right)$$

where c is a positive constant and m is the amount of medicine absorbed into the blood. (*Source*: R. M. Thrall et al., *Some Mathematical Models in Biology*, U.S. Department of Commerce, 1967) The sensitivity to the medication is defined to be the rate of change of reaction R with respect to the amount of medicine m absorbed in the blood.
    (a) Find the sensitivity.
    (b) Find the instantaneous rate of change of sensitivity with respect to the amount of medicine absorbed in the blood.
    (c) Which order derivative of reaction gives the rate of change of sensitivity?

11. **Advertising and sales** The daily sales S (in thousands of dollars) that are attributed to an advertising campaign are given by

$$S = 1 + \frac{3}{t + 3} - \frac{18}{(t + 3)^2}$$

where t is the number of weeks the campaign runs.
    (a) Find the rate of change of sales at any time t.
    (b) Use the second derivative to find how this rate is changing at $t = 15$.
    (c) Interpret your result in part (b).

Need more practice?
Find more here: cengagebrain.com

9.9 Marginals and Derivatives

PREVIOUSLY, WE EXTENDED THE NOTION OF MARGINAL REVENUE TO NONLINEAR TOTAL REVENUE FUNCTIONS BY DEFINING MARGINAL REVENUE AS THE DERIVATIVE OF TOTAL REVENUE. IN THIS SECTION, FOR ANY TOTAL COST OR PROFIT FUNCTION WE CAN EXTEND THE NOTION OF MARGINAL TO NONLINEAR FUNCTIONS.

Marginal Revenue

Recall that the instantaneous rate of change (the derivative) of the revenue function is the marginal revenue.

> **Marginal Revenue:** If $R = R(x)$ is the total revenue function for a commodity, then the **marginal revenue function** is $\overline{MR} = R'(x)$.

Recall that if the demand function for a product in a monopoly market is $p = f(x)$, then the total revenue from the sale of x units is

$$R(x) = px = f(x) \cdot x$$

So if the demand for a product in a monopoly market is $p = 16 - 0.02x$, the total revenue function is $R(x) = 16x - 0.02x^2$, and the marginal revenue function is

$$\overline{MR} = R'(x) = 16 - 0.04x.$$

At $x = 40$, $R'(40) = 16 - 1.6 = 14.40$ dollars per unit. Thus the 41st item sold will increase the total revenue by approximately $14.40.

The marginal revenue is an approximation or estimate of the revenue gained from the sale of 1 additional unit. We have used marginal revenue to find that the revenue from the sale of the 41st item will be approximately

$14.40. The actual increase in revenue from the sale of the 41st item is

$$R(41) - R(40) = 622.38 - 608 = \$14.38$$

Marginal revenue (and other marginals) also can be used to predict for more than one additional unit. For instance, in the preceding situation, $\overline{MR}(40) = \$14.40$ per unit means that the approximate revenue for the 41st through the 45th items sold would be $5(\$14.40) = \$72.00$. The actual revenue for these 5 items is

$$R(45) - R(40) = \$679.50 - \$608 = \$71.50.$$

Marginal Cost

As with marginal revenue, the derivative of the total cost gives marginal cost.

Marginal Cost: If $C = C(x)$ is a total cost function for a commodity, then its derivative, $\overline{MC} = C'(x)$, is the **marginal cost function.**

Notice that the linear total cost function with equation

$$C(x) = 300 + 6x \text{ (in dollars)}$$

has marginal cost $6 per unit because its slope is 6. Taking the derivative of $C(x)$ also gives

$$\overline{MC} = C'(x) = 6$$

which verifies that the marginal cost is $6 per unit at all levels of production.

EXAMPLE 9.29 Marginal Cost

Suppose the daily total cost in dollars for a certain factory to produce x kitchen blenders is given by

$$C(x) = 0.001x^3 - 0.3x^2 + 32x + 2500$$

(a) Find the marginal cost function for these blenders.
(b) Find and interpret the marginal cost when $x = 80$ and $x = 200$.

SOLUTION

(a) The marginal cost function is the derivative of $C(x)$.
$$\overline{MC} = C'(x) = 0.003x^2 - 0.6x + 32$$

(b) $C'(80) = 0.003(80)^2 - 0.6(80) + 32$
$$= 3.2 \text{ dollars per unit}$$

©iStockphoto.com/Sebastien Cote

$C'(200) = 0.003(200)^2 - 0.6(200) + 32$
$$= 32 \text{ dollars per unit}$$

These values for the marginal cost can be used to *estimate* the amount that total cost would change if production were increased by one blender. Thus,

$$C'(80) = 3.2$$

means total cost would increase by *about* $3.20 if an 81st blender were produced. Note that

$$C(81) - C(80) = 3.141 \approx \$3.14$$

is the actual increase in total cost for an 81st blender.

Also, $C'(200) = 32$ means that total cost would increase by *about* $32 if a 201st blender were produced.

Whenever $C'(x)$ is positive, it means that an additional unit produced adds to or increases the total cost. In addition, the value of the derivative (or marginal cost) measures how fast $C(x)$ is increasing. Thus, our preceding calculations indicate that $C(x)$ is increasing faster at $x = 200$ than at $x = 80$. The graphs of the total cost function and the marginal cost function in Figure 9.19(a) and (b) also illustrate these facts.

The graphs of many marginal cost functions tend to be U-shaped; they eventually will rise, even though

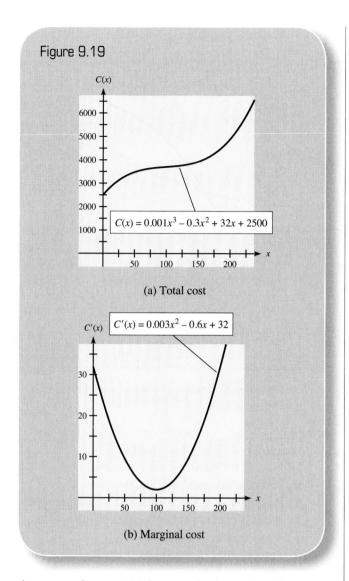

Figure 9.19

$$C(x) = 0.001x^3 - 0.3x^2 + 32x + 2500$$

(a) Total cost

$$C'(x) = 0.003x^2 - 0.6x + 32$$

(b) Marginal cost

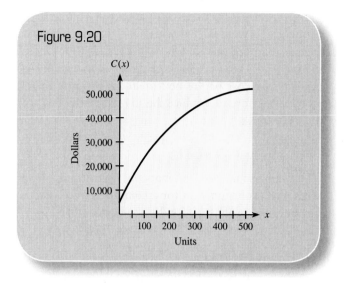

Figure 9.20

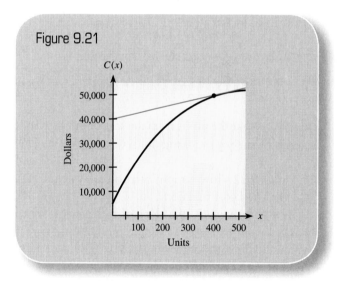

Figure 9.21

there may be an initial interval where they decrease. Looking at the marginal cost graph in Figure 9.19(b), we see that marginal cost reaches its minimum near $x = 100$. We can also see this in Figure 9.19(a) by noting that tangent lines drawn to the total cost graph would have slopes that decrease until about $x = 100$ and then would increase.

Because producing more units can never reduce the total cost of production, the following properties are valid for total cost functions.

1. The total cost can never be negative. If there are fixed costs, the cost of producing 0 units is positive; otherwise, the cost of producing 0 units is 0.

2. The total cost function is always increasing; the more units produced, the higher the total cost. Thus the marginal cost is always positive.

3. There may be limitations on the units produced, such as those imposed by plant space.

Let's look at another example of total and marginal cost. Suppose the graph in Figure 9.20 shows the monthly total cost for producing a product.

Then the total cost of producing 400 items is the height of the total cost graph when $x = 400$, or about $50,000. The approximate cost of the 401st item is the marginal cost when $x = 400$, or the slope of the tangent line drawn to the graph at $x = 400$. Figure 9.21 shows the total cost graph with a tangent line at $x = 400$.

Note that the tangent line passes through the point $(0, 40,000)$, so we can find the slope by using the points $(0, 40,000)$ and $(400, 50,000)$.

$$m = \frac{y_2 - y_1}{x_2 - x_1} = \frac{50,000 - 40,000}{400 - 0} = \frac{10,000}{400} = 25$$

Thus the marginal cost at $x = 400$ is 25 dollars per item, so the approximate cost of the 401st item is $25.

From Figure 9.21 we can see that if a tangent line to the graph were drawn where $x = 150$, it would be steeper than the one at $x = 400$. Because the slope of the tangent is the marginal cost, and the marginal cost predicts the cost of the next item, this means that it would cost more to produce the 151st item than to produce the 401st.

Marginal Profit

As with marginal cost and marginal revenue, the derivative of a profit function for a commodity will give us the marginal profit function for the commodity.

> **Marginal Profit:** If $P = P(x)$ is the profit function for a commodity, then the **marginal profit function** is $\overline{MP} = P'(x)$.

EXAMPLE 9.30 Marginal Profit

➡ If the total profit, in thousands of dollars, for a product is given by $P(x) = 20\sqrt{x + 1} - 2x - 22$, what is the marginal profit at a production level of 15 units?

SOLUTION

The marginal profit function is

$$\overline{MP} = P'(x) = 20 \cdot \frac{1}{2}(x + 1)^{-1/2} - 2 = \frac{10}{\sqrt{x + 1}} - 2$$

If 15 units are produced, the marginal profit is

$$P'(15) = \frac{10}{\sqrt{15 + 1}} - 2 = \frac{1}{2}$$

This means that the profit from the sale of the 16th unit is approximately $\frac{1}{2}$(thousand dollars), or $500.

Figure 9.22 shows graphs of a company's total revenue and total cost functions.

At a given x-value, the slope of the tangent line to the revenue function gives the marginal revenue at that x-value and the slope of the tangent line to the cost function gives the marginal cost there. At $x = 100$, we can see that the graph of $R(x)$ is steeper than the graph of $C(x)$. Thus, the tangent line to $R(x)$ will be steeper than the tangent line to $C(x)$. Hence $\overline{MR}(100) > \overline{MC}(100)$, which means that the revenue from the 101st item will exceed the cost. Therefore, profit will increase when the 101st item is produced. Note that at $x = 100$, total costs are greater than total revenue, so the company is losing money but should still sell the 101st item because it will reduce the amount of loss.

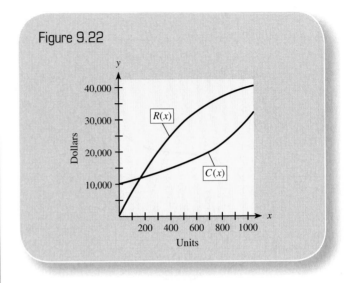

Figure 9.22

9.9 Exercises

Marginal Revenue, Cost, and Profit

In Problems 1 and 2, total revenue is in dollars and x is the number of units.

1. Suppose that the total revenue function for a commodity is $R = 36x - 0.01x^2$.
   (a) Find $R(100)$ and tell what it represents.
   (b) Find the marginal revenue function.
   (c) Find the marginal revenue at $x = 100$, and tell what it predicts about the sale of the next unit and the next 3 units.
   (d) Find $R(101) - R(100)$ and explain what this value represents.

2. Suppose that demand for local cable TV service is given by
   $$p = 80 - 0.4x$$
   where p is the monthly price in dollars and x is the number of subscribers (in hundreds). Recall that revenue = (price per unit)(number of units).
   (a) Find the total revenue as a function of the number of subscribers.
   (b) Find the number of subscribers when the company charges $50 per month for cable service. Then find the total revenue for $p = 50.
   (c) How could the company attract more subscribers?
   (d) Find and interpret the marginal revenue when the price is $50 per month. What does this suggest about the monthly charge to subscribers?

In Problems 3–5, cost is in dollars and x is the number of units. Find the marginal cost functions for the given cost functions.

3. $C(x) = 500 + 13x + x^2$

4. $C = 400 + 27x + x^3$

5. Suppose that the cost function for a commodity is

$$C(x) = 40 + x^2 \quad \text{dollars}$$

   (a) Find the marginal cost at $x = 5$ units and tell what this predicts about the cost of producing 1 additional unit.

   (b) Calculate $C(6) - C(5)$ to find the actual cost of producing 1 additional unit.

6. The graph of a company's total cost function is shown. Use the graph to answer the following questions.

   (a) Will the 101st item or the 501st item cost more to produce? Explain.

   (b) Does this total cost function represent a manufacturing process that is getting more efficient or less efficient? Explain.

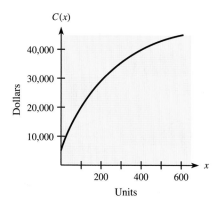

In Problems 7 and 8, cost, revenue, and profit are in dollars and x is the number of units.

7. If the total profit function is $P(x) = 5x - 25$, find the marginal profit. What does this mean?

8. Suppose that the total revenue function for a product is $R(x) = 50x$ and that the total cost function is $C(x) = 1900 + 30x + 0.01x^2$.

   (a) Find the profit from the production and sale of 500 units.

   (b) Find the marginal profit function.

   (c) Find $\overline{MP}$ at $x = 500$ and explain what it predicts.

   (d) Find $P(501) - P(500)$ and explain what this value represents.

9. The graphs of a company's total revenue function and total cost function are shown. Use the graphs to answer the following questions.

   (a) From the sale of 100 items, 400 items, and 700 items, rank from smallest to largest the amount of profit received. Explain your choices and note whether any of these scenarios results in a loss.

   (b) From the sale of the 101st item, the 401st item, and the 701st item, rank from smallest to largest the amount of profit received. Explain your choices, and note whether any of these scenarios results in a loss.

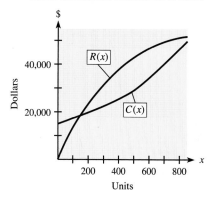

10. The graph of a company's profit function is shown. Use the graph to answer the following questions about points A, B, and C.

   (a) Rank from smallest to largest the amounts of profit received at these points. Explain your choices, and note whether any point results in a loss.

   (b) Rank from smallest to largest the marginal profit at these points. Explain your choices, and note whether any marginal profit is negative and what this means.

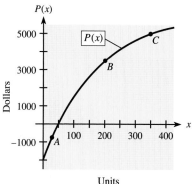

Need more practice?
Find more here: cengagebrain.com

Chapter Exercises

In Problems 1–4, use the graph of $y = f(x)$ in the figure to find the functional values and limits, if they exist.

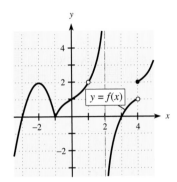

1. (a) $f(-1)$ (b) $\lim_{x \to -1} f(x)$

2. (a) $f(4)$ (b) $\lim_{x \to 4} f(x)$

3. (a) $\lim_{x \to 4^+} f(x)$ (b) $\lim_{x \to 4} f(x)$

4. (a) $f(2)$ (b) $\lim_{x \to 2} f(x)$

In Problems 5–11, find each limit, if it exists.

5. $\lim_{x \to 4} \dfrac{x^2 - 16}{x + 4}$

6. $\lim_{x \to -1} \dfrac{x^2 - 1}{x + 1}$

7. $\lim_{x \to 2} \dfrac{4x^3 - 8x^2}{4x^3 - 16x}$

8. $\lim_{x \to 3} \dfrac{x^2 - 16}{x - 3}$

9. $\lim_{x \to 1} \dfrac{x^2 - 9}{x - 3}$

10. $\lim_{x \to 1} f(x)$ where $f(x) = \begin{cases} 4 - x^2 & \text{if } x < 1 \\ 4 & \text{if } x = 1 \\ 2x + 1 & \text{if } x > 1 \end{cases}$

11. $\lim_{h \to 0} \dfrac{3(x + h)^2 - 3x^2}{h}$

12. Use tables to investigate the limit. Check your result analytically or graphically.

$$\lim_{x \to 2} \dfrac{x^2 + 10x - 24}{x^2 - 5x + 6}$$

13. Refer back to the graph of $y = f(x)$ you used for Problems 1–4. Use that same graph to determine whether $f(x)$ is continuous at
   (a) $x = -1$.   (b) $x = 1$.

In Problems 14–16, suppose that

$$f(x) = \begin{cases} x^2 + 1 & \text{if } x \leq 0 \\ x & \text{if } 0 < x < 1 \\ 2x^2 - 1 & \text{if } x \geq 1 \end{cases}$$

14. What is $\lim_{x \to -1} f(x)$?

15. What is $\lim_{x \to 1} f(x)$, if it exists?

16. Is $f(x)$ continuous at $x = 1$?

In Problems 17 and 18, determine which functions are continuous. Identify discontinuities for those that are not continuous.

17. $y = \dfrac{x^2 + 25}{x - 5}$

18. $f(x) = \begin{cases} x + 2 & \text{if } x \leq 2 \\ 5x - 6 & \text{if } x > 2 \end{cases}$

19. Use the graph to find (a) the points of discontinuity, (b) $\lim_{x \to +\infty} f(x)$, and (c) $\lim_{x \to -\infty} f(x)$.

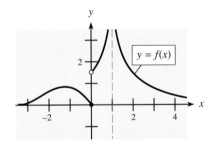

In Problems 20 and 21, evaluate the limits, if they exist. Then state what each limit tells about any horizontal asymptotes.

20. $\lim_{x \to -\infty} \dfrac{2x^2}{1 - x^2}$ 21. $\lim_{x \to +\infty} \dfrac{3x^{2/3}}{x + 1}$

22. Use the definition of derivative to find $f'(x)$ for $f(x) = 3x^2 + 2x - 1$.

23. Use the definition of derivative to find $f'(x)$ if $f(x) = x - x^2$.

24. Let $f(x) = \dfrac{\sqrt[3]{4x}}{(3x^2 - 10)^2}$. Approximate $f'(2)$.

   (a) by using the numerical derivative feature of a graphing utility, and

(b) by evaluating

$$\frac{f(2 + h) - f(2)}{h} \text{ with } h = 0.0001.$$

25. If $c = 4x^5 - 6x^3$, find c'.

26. If $f(x) = 10x^9 - 5x^6 + 4x - 2^7 + 19$, find $f'(x)$.

27. If $f(z) = \sqrt[3]{2^4}$ find $f'(z)$.

28. If $v(x) = 4/\sqrt[3]{x}$, find $v'(x)$.

29. If $y = \dfrac{1}{x} - \dfrac{1}{\sqrt{x}}$, find y'.

30. Write the equation of the line tangent to the graph of $y = 3x^5 - 6$ at $x = 1$.

31. For $f(x) = x^3 - 3x^2 + 1$, find (a) all x-values where the slope of the tangent equals zero, and (b) points (x, y) where the slope of the tangent equals zero, and (c) use a graphing utility to graph the function and label the points found in (b).

32. If $f(x) = (3x - 1)(x^2 - 4x)$, find $f'(x)$.

33. If $p = \dfrac{5q^3}{2q^3 + 1}$, find $\dfrac{dp}{dq}$.

34. Find $\dfrac{ds}{dt}$ if $s = \dfrac{\sqrt{t}}{(3t + 1)}$.

35. Find $\dfrac{dy}{dx}$ for $y = \sqrt{x}(3x + 2)$.

36. Find $\dfrac{dC}{dx}$ for $C = \dfrac{5x^4 - 2x^2 + 1}{x^3 + 1}$.

37. If $y = (x^3 - 4x^2)^3$, find y'.

38. If $y = (2x^4 - 9)^9$, find $\dfrac{dy}{dx}$.

39. Find $g'(x)$ if $g(x) = \dfrac{1}{\sqrt{x^3 - 4x}}$.

40. Find $f'(x)$ if $f(x) = x^2(2x^4 + 5)^8$.

41. Find S' if $S = \dfrac{(3x + 1)^2}{x^2 - 4}$.

42. Find y' if $y = \left(\dfrac{x + 1}{1 - x^2}\right)^3$.

43. Find the second derivative of $y = \sqrt{x} - x^2$.

44. Find the fifth derivative of $y = (2x + 1)^4$.

45. If $\dfrac{dy}{dx} = \sqrt{x^2 - 4}$, find $\dfrac{d^3y}{dx^3}$.

Applications

46. *Cost, revenue, and profit* Assume that a company's monthly total revenue and total cost (both in dollars) are given by

$$R(x) = 140x - 0.01x^2 \quad \text{and} \quad C(x) = 60x + 70{,}000$$

where x is the number of units. If $\overline{R}(x) = \dfrac{R(x)}{x}$ and $\overline{C}(x) = \dfrac{C(x)}{x}$ are, respectively, the company's average revenue per unit and average cost per unit, find
(a) $\lim\limits_{x \to 0^+} \overline{R}(x)$ (b) $\lim\limits_{x \to 0^+} \overline{C}(x)$

47. *Demand* Suppose that the demand for x units of a product is given by $x = (100/p) - 1$, where p is the price per unit of the product. Find and interpret the rate of change of demand with respect to price if the price is
(a) \$10.   (b) \$20.

48. *Severe weather ice makers* Thunderstorms severe enough to produce hail develop when an upper-level low (a pool of cold air high in the atmosphere) moves through a region where there is warm, moist air at the surface. These storms create an updraft that draws the moist air into subfreezing air above 10,000 feet. Data from the National Weather Service indicates that the strength of the updraft, as measured by its speed s in mph, affects the size of the hail according to

$$h = 0.000595s^{1.922}$$

where h is the diameter of the hail (in inches). Find and interpret $h(100)$ and $h'(100)$.

49. *Revenue* In a 100-unit apartment building, when the price charged per apartment rental is $(830 + 30x)$ dollars, then the number of apartments rented is $100 - x$ and the total revenue for the building is

$$R(x) = (830 + 30x)(100 - x)$$

where x is the number of \$30 rent increases (and also the resulting number of unrented apartments). Find the marginal revenue when $x = 10$. Does this tell you that the rent should be raised (causing more vacancies) or lowered? Explain.

50. *Productivity* Suppose the productivity of a worker (in units per hour) after x hours of training and time on the job is given by

$$P(x) = 3 + \dfrac{70x^2}{x^2 + 1000}$$

(a) Find and interpret $P(20)$.
(b) Find and interpret $P'(20)$.

51. **Demand** The demand q for a product at price p is given by

$$q = 10{,}000 - 50\sqrt{0.02p^2 + 500}$$

Find the rate of change of demand with respect to price.

52. **Supply** The number of units x of a product that is supplied at price p is given by

$$x = \sqrt{p} - 1, \quad p \geq 1$$

If the price p is \$10, what is the rate of change of the supply with respect to the price, and what does it tell us?

53. **Acceleration** Suppose an object moves so that its distance to a sensor, in feet, is given by

$$s(t) = 16 + 140t + 8\sqrt{t}$$

where t is the time in seconds. Find the acceleration at time $t = 4$ seconds.

54. **Profit** Suppose a company's profit (in dollars) is given by

$$P(x) = 70x - 0.1x^2 - 5500$$

where x is the number of units. Find and interpret $P'(300)$ and $P''(300)$.

In Problems 55–60, cost, revenue, and profit are in dollars and x is the number of units.

55. **Cost** If the cost function for a particular good is $C(x) = 3x^2 + 6x + 600$, what is
(a) the marginal cost function?
(b) the marginal cost if 30 units are produced?
(c) the interpretation of your answer in part (b)?

56. **Cost** If the total cost function for a commodity is $C(x) = 400 + 5x + x^3$, what is the marginal cost when 4 units are produced, and what does it mean?

57. **Profit** If the total revenue function for a product is given by $R(x) = 60x$ and the total cost function is given by $C = 200 + 10x + 0.1x^2$, what is the marginal profit at $x = 10$? What does the marginal profit at $x = 10$ predict?

58. **Revenue** The total revenue function for a commodity is given by $R = 80x - 0.04x^2$.
(a) Find the marginal revenue function.
(b) What is the marginal revenue at $x = 100$?
(c) Interpret your answer in part (b).

59. **Revenue** If the revenue function for a product is

$$R(x) = \frac{60x^2}{2x + 1}$$

find the marginal revenue.

60. **Profit** A firm has monthly costs given by

$$C = 45{,}000 + 100x + x^3$$

where x is the number of units produced per month. The firm can sell its product in a competitive market for \$4600 per unit. Find the marginal profit.

Need more practice?
Find more here: cengagebrain.com

LEARN YOUR WAY

With **MATH APPS**, you have a multitude of study aids at your fingertips.

The Student Solutions Manual contains worked-out solutions to every odd-numbered exercise, to further reinforce your understanding of mathematical concepts in **MATH APPS**.

In addition to the Student Solutions Manual, Cengage Learning's **CourseMate** offers exercises and questions that correspond to every section and chapter in the text for extra practice.

For access to these study aids sign in at **login.cengagebrain.com**.

Applications
of Derivatives

The derivative can be used to determine where a function has a "turning point" on its graph, so that we can determine where the graph reaches its highest or lowest point within a particular interval. These points are called the **relative maxima** and **relative minima,** respectively, and are useful in sketching the graph of the function. The techniques for finding these points are also useful in solving applied problems, such as finding the maximum profit, the minimum average cost, and the maximum productivity. The second derivative can be used to find **points of inflection** of the graph of a function and to find the point of diminishing returns in certain applications.

objectives

10.1 Relative Maxima and Minima: Curve Sketching

10.2 Concavity: Points of Inflection

10.3 Optimization in Business and Economics

10.4 Applications of Maxima and Minima

10.5 Rational Functions: More Curve Sketching

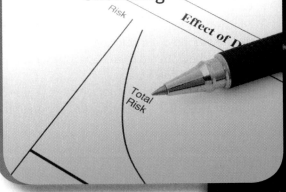

10.1 Relative Maxima and Minima: Curve Sketching

RECALL THAT WE CAN FIND THE MAXIMUM VALUE OR THE MINIMUM VALUE (ALSO CALLED THE OPTIMUM VALUE) OF A QUADRATIC FUNCTION BY FINDING THE VERTEX OF ITS GRAPH. BUT FOR FUNCTIONS OF HIGHER DEGREE, THE SPECIAL FEATURES OF THEIR GRAPHS MAY BE HARDER TO FIND ACCURATELY, EVEN WHEN A GRAPHING UTILITY IS USED. THE FIRST DERIVATIVE IDENTIFIES THE "TURNING POINTS" OF THE GRAPH, WHICH HELP US DETERMINE THE GENERAL SHAPE OF THE GRAPH AND CHOOSE A VIEWING WINDOW THAT INCLUDES THE INTERESTING POINTS OF THE GRAPH.

In Figure 10.1(a) we see that the graph of $y = \frac{1}{3}x^3 - x^2 - 3x + 2$ has two "turning points," at $(-1, \frac{11}{3})$ and $(3, -7)$. The curve has a relative maximum at $(-1, \frac{11}{3})$ because this point is higher than any other point "near" it on the curve; the curve has a relative minimum at $(3, -7)$ because this point is lower than any other point "near" it on the curve. A formal definition follows.

Figure 10.1

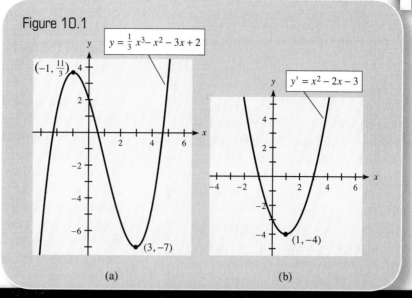

$y = \frac{1}{3}x^3 - x^2 - 3x + 2$

$y' = x^2 - 2x - 3$

(a)   (b)

math apps

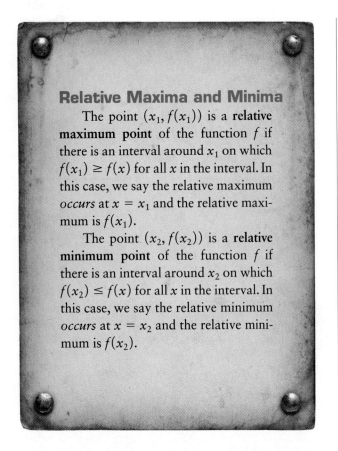

Relative Maxima and Minima

The point $(x_1, f(x_1))$ is a **relative maximum point** of the function f if there is an interval around x_1 on which $f(x_1) \geq f(x)$ for all x in the interval. In this case, we say the relative maximum *occurs* at $x = x_1$ and the relative maximum is $f(x_1)$.

The point $(x_2, f(x_2))$ is a **relative minimum point** of the function f if there is an interval around x_2 on which $f(x_2) \leq f(x)$ for all x in the interval. In this case, we say the relative minimum *occurs* at $x = x_2$ and the relative minimum is $f(x_2)$.

In order to determine whether a turning point of a function is a maximum point or a minimum point, it is frequently helpful to know what the graph of the function does in intervals on either side of the turning point. We say a function is **increasing** on an interval if the values of the function increase as the x-values increase (that is, if the graph rises as we move from left to right on the interval). Similarly, a function is **decreasing** on an interval if the values of the function decrease as the x-values increase (that is, if the graph falls as we move from left to right on the interval).

Because the derivative of the function gives the slope of the tangent to the curve, we see the following.

Increasing and Decreasing Functions:

If f is a function that is differentiable on an interval (a, b), then

if $f'(x) > 0$ for all x in (a, b), f is increasing on (a, b).

if $f'(x) < 0$ for all x in (a, b), f is decreasing on (a, b).

Figure 10.1(a) on the previous page shows the graph of a function, and Figure 10.1(b) shows the graph of its derivative. The figures show that the graph of $y = f(x)$ is increasing for the same x-values for which the graph of $y' = f'(x)$ is above the x-axis (when $f'(x) > 0$). Similarly, the graph of $y = f(x)$ is decreasing for the same x-values $(-1 < x < 3)$ for which the graph of $y' = f'(x)$ is below the x-axis (when $f'(x) < 0$).

The derivative $f'(x)$ can change signs only at values of x at which $f'(x) = 0$ or $f'(x)$ is undefined. We call these values of x **critical values**. The point corresponding to a critical value for x is a **critical point**.* Because a curve changes from increasing to decreasing at a relative maximum and from decreasing to increasing at a relative minimum (see Figure 10.1(a)), we have the following.

Relative Maximum and Minimum:

If f has a relative maximum or a relative minimum at $x = x_0$, then $f'(x_0) = 0$ or $f'(x_0)$ is undefined.

Figure 10.2 shows a function with two relative maxima, one at $x = x_1$ and the second at $x = x_3$, and one relative minimum at $x = x_2$. At $x = x_1$ and $x = x_2$, we see that $f'(x) = 0$, and at $x = x_3$ the derivative does not exist.

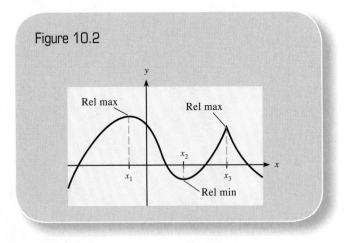

Figure 10.2

*There may be some critical values at which both $f'(x)$ and $f(x)$ are undefined. Critical points do not occur at these values, but studying the derivative on either side of such values may be of interest.

Figure 10.3

Direction of graph of $f(x)$:
Signs and values of $f'(x)$:
x-axis with critical values:

*means $f(x_3)$ is undefined.

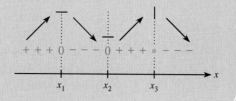

Thus we can find relative maxima and minima for a curve by finding values of x for which the function has critical points. The behavior of the derivative to the left and right of (and near) these points will tell us whether they are relative maxima, relative minima, or neither.

Using the critical values of $f(x)$ and the sign of $f'(x)$ between those critical values, we can create a **sign diagram for** $f'(x)$. The sign diagram for the graph in Figure 10.2 is shown in Figure 10.3. This sign diagram was created from the graph of f, but it is also possible to predict the shape of a graph from a sign diagram.

Figure 10.4 shows the two ways that a function can have a relative maximum at a critical point, and Figure 10.5 shows the two ways for a relative minimum.

The preceding discussion suggests the procedure at right for finding relative maxima and minima of a function.

Figure 10.4

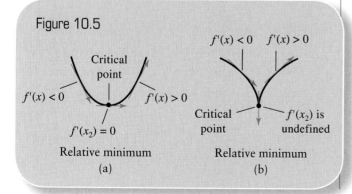

First-Derivative Test Procedure

To find relative maxima and minima of a function:

1. Find the first derivative of the function.

2. Set the derivative equal to 0, and solve for values of x that satisfy $f'(x) = 0$. These are called **critical values.** Values that make $f'(x)$ undefined are also critical values.

3. Substitute the critical values into the *original function* to find the **critical points.**

4. Evaluate $f'(x)$ at a value of x to the left and one to the right of each critical point to develop a sign diagram.
   (a) If $f'(x) > 0$ to the left and $f'(x) < 0$ to the right of the critical value, the critical point is a relative maximum point.
   (b) If $f'(x) < 0$ to the left and $f'(x) > 0$ to the right of the critical value, the critical point is a relative minimum point.

5. Use the information from the sign diagram and selected points to sketch the graph.

EXAMPLE 10.1 First-Derivative Test

Find the relative maxima and minima of $f(x) = \frac{1}{3} x^3 - x^2 - 3x + 2$.

SOLUTION

1. $f'(x) = x^2 - 2x - 3$
2. $0 = x^2 - 2x - 3 = (x + 1)(x - 3)$ has solutions $x = -1, x = 3$. No values of x make $f'(x) = x^2 - 2x - 3$ undefined. Critical values are -1 and 3.
3. $f(-1) = \frac{11}{3}$ $f(3) = -7$
   The critical points are $(-1, \frac{11}{3})$ and $(3, -7)$.
4. $f'(-2) = 5 > 0$ and $f'(0) = -3 < 0$
   Thus $(-1, 11/3)$ is a relative maximum point.

$$f'(2) = -3 < 0 \text{ and } f'(4) = 5 > 0$$

Thus $(3, -7)$ is a relative minimum point. The sign diagram for $f'(x)$ is

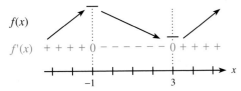

5. The information from this sign diagram is shown in Figure 10.6(a). Plotting additional points gives the graph of the function, shown in Figure 10.6(b).

Because the critical values are the only x-values at which the graph can have turning points, we can test to

Figure 10.6

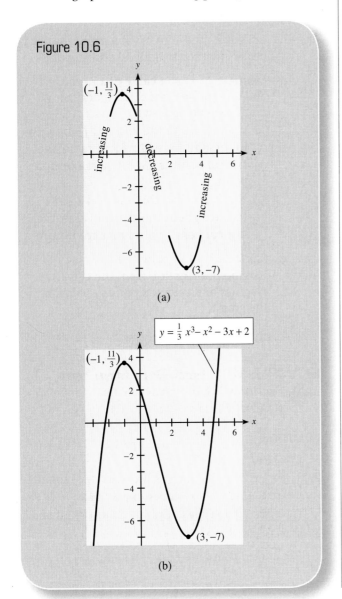

(a)

$$y = \frac{1}{3}x^3 - x^2 - 3x + 2$$

(b)

the left and right of each critical value by testing to the left of the smallest critical value, then testing a value *between* each two successive critical values, and then testing to the right of the largest critical value.

Note that we substitute the critical values into the *original function $f(x)$* to find the y-values of the critical points, but we test for relative maxima and minima by substituting values near the critical values into the *derivative of the function, $f'(x)$*.

If the first derivative of f is 0 at x_0 but does not change from positive to negative or from negative to positive as x passes through x_0, then the critical point at x_0 is neither a relative maximum nor a relative minimum. In this case we say that f has a **horizontal point of inflection** (abbreviated HPI) at x_0.

EXAMPLE 10.2 Maxima, Minima, and Horizontal Points of Inflection

Find the relative maxima, relative minima, and horizontal points of inflection of $h(x) = \frac{1}{4}x^4 - \frac{2}{3}x^3 - 2x^2 + 8x + 4$, and sketch its graph.

SOLUTION

1. $h'(x) = x^3 - 2x^2 - 4x + 8$

2. $0 = x^3 - 2x^2 - 4x + 8$ or
   $0 = x^2(x - 2) - 4(x - 2)$. Therefore, we have
   $0 = (x - 2)(x^2 - 4)$. Thus $x = -2$ and $x = 2$ are solutions.

3. The critical points are $\left(-2, -\frac{32}{3}\right)$ and $\left(2, \frac{32}{3}\right)$.

4. Using test values, such as $x = -3$, $x = 0$, and $x = 3$, gives the sign diagram for $h'(x)$.

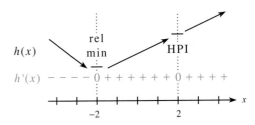

5. Figure 10.7(a) shows the graph of the function near the critical points, and Figure 10.7(b) shows the graph of the function.

See the Tech Card for details on using a graphing calculator or Excel to find critical values and to determine relative maxima and minima.

Figure 10.7

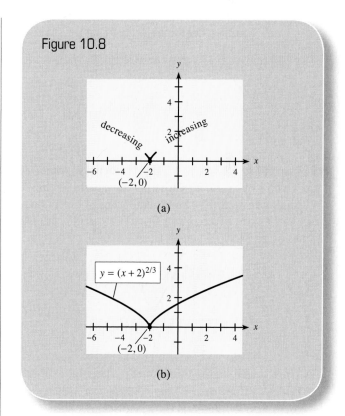

$$h(x) = \frac{1}{4}x^4 - \frac{2}{3}x^3 - 2x^2 + 8x + 4$$

(a)

(b)

EXAMPLE 10.3 Undefined Derivatives

Find the relative maxima and minima (if any) of the graph of $y = (x + 2)^{2/3}$.

SOLUTION

1. $y' = f'(x) = \frac{2}{3}(x + 2)^{-1/3} = \frac{2}{3\sqrt[3]{x + 2}}$

2. $0 = \frac{2}{3\sqrt[3]{x + 2}}$ has no solutions; $f'(x)$ is undefined at $x = -2$.

3. $f(-2) = 0$, so the critical point is $(-2, 0)$.

4. The sign diagram for $f'(x)$ is

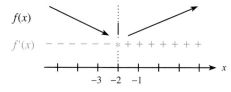

   *means $f'(-2)$ is undefined.

   Thus a relative minimum occurs at $(-2, 0)$.

5. Figure 10.8(a) shows the graph of the function near the critical point, and Figure 10.8(b) shows the graph.

Figure 10.8

$$y = (x + 2)^{2/3}$$

Calculator Note: With a graphing calculator, choosing an appropriate viewing window is the key to understanding the graph of a function. We saw that the derivative can be used to determine the critical values of a function and hence the interesting points of its graph. Therefore, the derivative can be used to determine the viewing window that provides an accurate representation of the graph. The derivative can also be used to discover graphical behavior that might be overlooked in a graph with a standard window. See the Tech Card for details.

Spreadsheet Note: Excel can also be used to develop and graph a model for data. Also, the solver feature of Excel can be used to solve optimization problems from calculus. Details are given on the Tech Card.

10.1 Exercises

1. Use the graph of $y = f(x)$ to identify points at which $f(x)$ has (a) a relative maximum, (b) a relative minimum, and (c) a horizontal point of inflection.

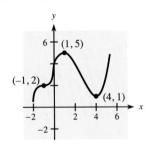

2. Use the sign diagram for $f'(x)$ to determine (a) the critical values of $f(x)$, (b) intervals on which $f(x)$ increases, (c) intervals on which $f(x)$ decreases, (d) x-values at which relative maxima occur, and (e) x-values at which relative minima occur.

$$f'(x) \quad \frac{---\ 0\ +++++\ 0\ ---}{37} \to x$$

3. Given $y = 2x^3 - 12x^2 + 6$. (a) Find the critical values of the function, and (b) make a sign diagram and determine the relative maxima and minima.

In Problems 4 and 5, for each function:
   (a) Find $y' = f'(x)$.
   (b) Find the critical values.
   (c) Find the critical points.
   (d) Find intervals of x-values where the function is increasing and where it is decreasing.
   (e) Classify the critical points as relative maxima, relative minima, or horizontal points of inflection. In each case, you may check your conclusions with a graphing utility.

4. $y = \dfrac{x^3}{3} + \dfrac{x^2}{2} - 2x + 1$

5. $y = x^{2/3}$

6. Let $y = 6 - x - x^2$.
   (a) Use the graph to identify x-values for which $y' > 0$, $y' < 0$, $y' = 0$, and y' does not exist.
   (b) Use the derivative to check your conclusions.

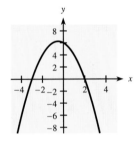

In Problems 7 and 8, for each function, find the relative maxima, relative minima, and horizontal points of inflection, and sketch the graph. You may check your graph with a graphing utility.

7. $y = \frac{1}{3}x^3 - x^2 + x + 1$

8. $y = 3x^5 - 5x^3 + 1$

9. Given $f(x) = x^{2/3}(x - 5)$ and its derivative $f'(x) = \dfrac{5(x - 2)}{3x^{1/3}}$, use these to find critical values, critical points, intervals on which the function is increasing and decreasing, relative maxima, relative minima, and horizontal points of inflection; sketch the graph of the function.

10. Given $f'(x) = x^2 - x - 2$ and its graph, use the graph to determine the critical values of $f(x)$, where $f(x)$ is increasing, where it is decreasing, and where it has relative maxima, relative minima,

and horizontal points of inflection. Then sketch a possible graph for $f(x)$ that passes through $(0, 0)$.

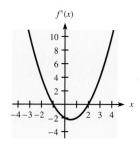

Applications

11. *Productivity* A time study showed that, on average, the productivity of a worker after t hours on the job can be modeled by

$$P(t) = 27t + 6t^2 - t^3, \quad 0 \le t \le 8$$

where P is the number of units produced per hour.
(a) Find the critical values of this function.
(b) Which critical value makes sense in this model?
(c) For what values of t is P increasing?
(d) Graph the function for $0 \le t \le 8$.

12. *Production costs* Suppose that the average cost, in dollars, of producing a shipment of a certain product is

$$\overline{C} = 5000x + \frac{125{,}000}{x}, \quad x > 0$$

where x is the number of machines used in the production process.
(a) Find the critical values of this function.
(b) Over what interval does the average cost decrease?
(c) Over what interval does the average cost increase?

13. *Marginal revenue* Suppose the weekly marginal revenue function for selling x units of a product is given by the graph in the figure.
(a) At each of $x = 150$, $x = 250$, and $x = 350$, what is happening to revenue?
(b) Over what interval is revenue increasing?
(c) How many units must be sold to maximize revenue?

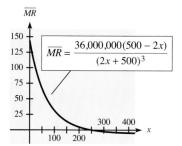

14. *Revenue* The weekly revenue of a certain recently released film is given by

$$R(t) = \frac{50t}{t^2 + 36}, \quad t \ge 0$$

where R is in millions of dollars and t is in weeks.
(a) Find the critical values.
(b) For how many weeks will weekly revenue increase?

Need more practice?
Find more here: cengagebrain.com

10.2 Concavity: Points of Inflection

JUST AS WE USED THE FIRST DERIVATIVE TO DETERMINE WHETHER A CURVE WAS INCREASING OR DECREASING ON A GIVEN INTERVAL, WE CAN USE THE SECOND DERIVATIVE TO DETERMINE WHETHER THE CURVE IS CONCAVE UP OR CONCAVE DOWN ON AN INTERVAL.

A curve is said to be **concave up** on an interval (a, b) if at each point on the interval the curve is above its tangent at the point (Figure 10.9(a) on the next page). If the curve is below all its tangents on a given interval, it is **concave down** on the interval (Figure 10.9(b)).

Looking at Figure 10.9(a), we see that the *slopes* of the tangent lines increase over the interval where the graph is concave up. Because $f'(x)$ gives the slopes of those tangents, it follows that $f'(x)$ is increasing over the interval where $f(x)$ is concave up. However, we know that $f'(x)$ is increasing when its derivative, $f''(x)$, is positive. That is, if the second derivative is positive, the curve is concave up.

Similarly, if the second derivative of a function is negative over an interval, the slopes of the tangents to the graph decrease over that interval. This happens when the tangent lines are above the graph, as in Figure 10.9(b), so the graph must be concave down on this interval.

Thus we see that the second derivative can be used to determine the concavity of a curve.

Figure 10.9

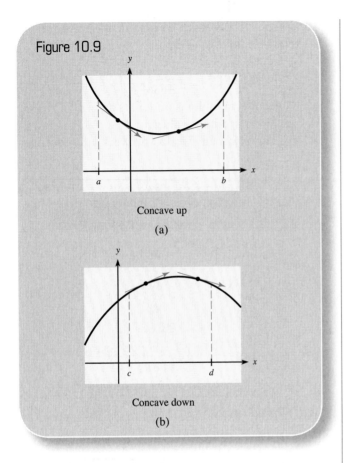

Concave up

(a)

Concave down

(b)

Concave Up and Concave Down

Assume that the first and second derivatives of function f exist.

The function f is **concave up** on an interval I if $f''(x) > 0$ on I, and **concave up** at the point $(a, f(a))$ if $f''(a) > 0$.

The function f is **concave down** on an interval I if $f''(x) < 0$ on I, and **concave down** at the point $(a, f(a))$ if $f''(a) < 0$.

EXAMPLE 10.4 Concavity at a Point

Is the graph of $f(x) = x^3 - 4x^2 + 3$ concave up or down at the point
(a) $(1, 0)$?　　(b) $(2, -5)$?

SOLUTION

(a) We must find $f''(x)$ before we can answer this question.

$$f'(x) = 3x^2 - 8x \quad f''(x) = 6x - 8$$

Then $f''(1) = 6(1) - 8 = -2$, so the graph is concave down at $(1, 0)$.

(b) Because $f''(2) = 6(2) - 8 = 4$, the graph is concave up at $(2, -5)$. The graph of $f(x) = x^3 - 4x^2 + 3$ is shown in Figure 10.10(a).

Points of Inflection

Looking at the graph of $y = x^3 - 4x^2 + 3$ (Figure 10.10(a)), we see that the curve is concave down on the left and concave up on the right. Thus it has changed from concave down to concave up, and from Example 10.4 we would expect the concavity to change somewhere between $x = 1$ and $x = 2$. Figure 10.10(b) shows

Figure 10.10

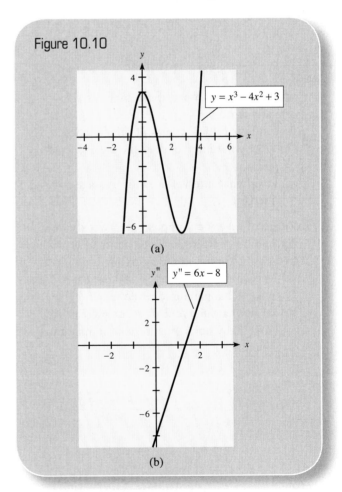

$y = x^3 - 4x^2 + 3$

(a)

$y'' = 6x - 8$

(b)

the graph of $y'' = f''(x) = 6x - 8$, and we can see that $y'' = 0$ when $x = \frac{4}{3}$ and that $y'' < 0$ for $x < \frac{4}{3}$ and $y'' > 0$ for $x > \frac{4}{3}$. Thus the second derivative changes sign at $x = \frac{4}{3}$, so the concavity of the graph of $y = f(x)$ changes at $x = \frac{4}{3}$, $y = -\frac{47}{27}$. The point where concavity changes is called a **point of inflection**.

> **Point of Inflection:** A point (x_0, y_0) on the graph of a function f is called a **point of inflection** if the curve is concave up on one side of the point and concave down on the other side. The second derivative at this point, $f''(x_0)$, will be 0 or undefined.

In general, we can find points of inflection and information about concavity as follows.

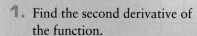

Procedure

To find the point(s) of inflection of a curve and intervals where it is concave up and where it is concave down:

1. Find the second derivative of the function.

2. Set the second derivative equal to 0, and solve for x. Potential points of inflection occur at these values of x or at values of x where $f(x)$ is defined and $f''(x)$ is undefined.

3. Find the potential points of inflection.

4. If the second derivative has opposite signs on the two sides of one of these values of x, a point of inflection occurs.

   The curve is concave up where $f''(x) > 0$ and concave down where $f''(x) < 0$.

   The changes in the sign of $f''(x)$ correspond to changes in concavity and occur at points of inflection.

EXAMPLE 10.5 Finding Points of Inflection and Concavity

→ Find the points of inflection and concavity of the graph of $y = \dfrac{x^4}{2} - x^3 + 5$.

SOLUTION

We use the steps outlined in the preceding procedure.
1. $y' = f'(x) = 2x^3 - 3x^2$
   $y'' = f''(x) = 6x^2 - 6x$
2. $0 = 6x^2 - 6x = 6x(x - 1)$ has solutions $x = 0$, $x = 1$. $f''(x)$ is defined everywhere.
3. $(0, 5)$ and $(1, \frac{9}{2})$ are potential points of inflection.
4. A **sign diagram for** $f''(x)$ is

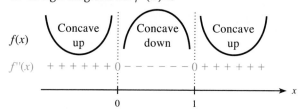

$(0, 5)$ and $(1, \frac{9}{2})$ are points of inflection.
See the graph in Figure 10.11.

The graph of $y = \frac{1}{2}x^4 - x^3 + 5$ is shown in Figure 10.11. Note the point of inflection at $(0, 5)$ is a horizontal point of inflection because $f'(x)$ is also 0 at $x = 0$.

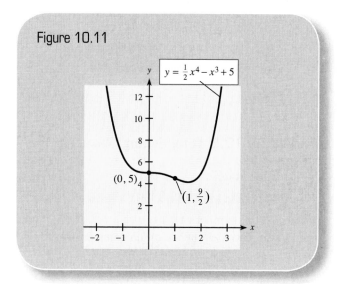

Figure 10.11

Suppose that a real estate developer wishes to remove pollution from a small lake so that she can sell lakefront homes on a "crystal clear" lake. The graph in Figure 10.12 on the following page shows the relation between dollars spent on cleaning the lake and the purity of the water. The point of inflection on the graph

Shutterstock (both)

is called the **point of diminishing returns** on her investment because it is where the *rate* of return on her investment changes from increasing to decreasing.

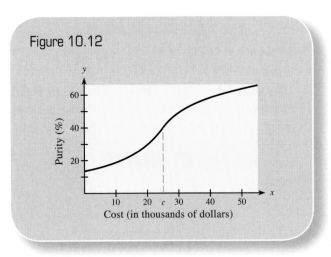

Figure 10.12

EXAMPLE 10.6 Diminishing Returns

Suppose the annual profit for a store (in thousands of dollars) is given by

$$P(x) = -0.2x^3 + 3x^2 + 6$$

where x is the number of years past 2008. If this model is accurate, find the point of diminishing returns for the profit.

SOLUTION

The point of diminishing returns occurs at the point of inflection. Thus we seek the point where the graph of this function changes from concave up to concave down, if such a point exists.

$$P'(x) = -0.6x^2 + 6x$$

$$P''(x) = -1.2x + 6$$

$P''(x) = 0$ when $0 = -1.2x + 6$ or   when $x = 5$

Thus $x = 5$ is a possible point of inflection. We test $P''(x)$ to the left and the right of $x = 5$.

$$P''(4) = 1.2 > 0 \Rightarrow \text{concave up to the left of } x = 5$$

$$P''(6) = -1.2 < 0 \Rightarrow \text{concave down to the right of } x = 5$$

Thus $(5, 56)$ is the point of inflection for the graph, and the point of diminishing returns for the profit is when $x = 5$ (in the year 2013) and is $P(5) = 56$ thousand dollars. Figure 10.13 shows the graphs of $P(x)$ and $P'(x)$. At $x = 5$, we see that the point of diminishing returns on the graph of $P(x)$ corresponds to the maximum point of the graph of $P'(x)$.

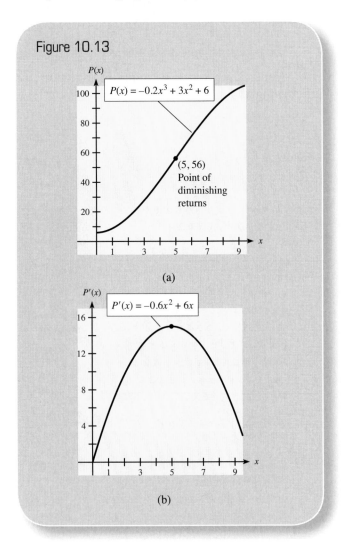

Figure 10.13

Second-Derivative Test

We can use information about points of inflection and concavity to help sketch graphs. For example, if we know that the curve is concave up at a critical point where $f'(x) = 0$, then the point must be a relative minimum because the tangent to the curve is horizontal at the critical point, and only a point at the bottom of

Figure 10.14

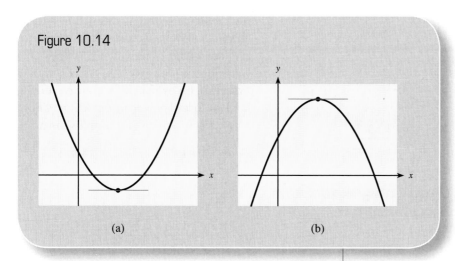

(a)   (b)

a "concave up" curve could have a horizontal tangent (see Figure 10.14(a)).

On the other hand, if the curve is concave down at a critical point where $f'(x) = 0$, then the point is a relative maximum (see Figure 10.14(b)).

Thus we can use the **second-derivative test** to determine whether a critical point where $f'(x) = 0$ is a relative maximum or minimum.

Procedure

To find relative maxima and minima of a function:

1. Find the critical values of the function.

2. Substitute the critical values into $f(x)$ to find the critical points.

3. Evaluate $f''(x)$ at each critical value for which $f'(x) = 0$.
   (a) If $f''(x_0) < 0$, a relative maximum occurs at x_0.
   (b) If $f''(x_0) > 0$, a relative minimum occurs at x_0.
   (c) If $f''(x_0) = 0$, or $f''(x_0)$ is undefined, the second-derivative test fails; use the first-derivative test.

Second-Derivative Test

EXAMPLE 10.7 Second-Derivative Test

Find the relative maxima and minima of
$$y = f(x) = \tfrac{1}{3}x^3 - x^2 - 3x + 2$$

SOLUTION

Follow the procedure outlined in the box above.

1. $f'(x) = x^2 - 2x - 3$
$0 = (x - 3)(x + 1)$ has solutions $x = -1$ and $x = 3$. No values of x make $f'(x)$ undefined.

2. $f(-1) = \tfrac{11}{3}$ $f(3) = -7$
The critical points are $(-1, \tfrac{11}{3})$ and $(3, -7)$.

3. $f''(x) = 2x - 2$
$f''(-1) = 2(-1) - 2 = -4 < 0$, so $(-1, \tfrac{11}{3})$ is a relative maximum point.
$f''(3) = 2(3) - 2 = 4 > 0$, so $(3, -7)$ is a relative minimum point. (The graph is shown in Figure 10.15.)

Figure 10.15

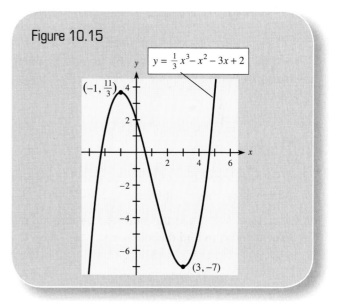

$$y = \tfrac{1}{3}x^3 - x^2 - 3x + 2$$

Because the second-derivative test (just shown) and the first-derivative test (in Section 10.1) are both methods for classifying critical values, let's compare their advantages and disadvantages.

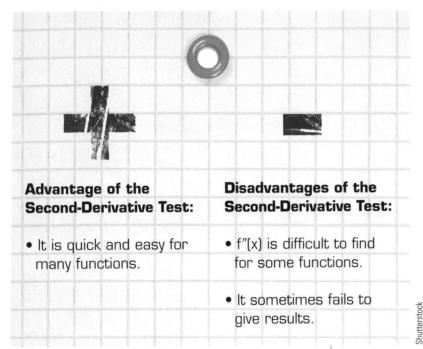

Advantage of the Second-Derivative Test:

- It is quick and easy for many functions.

Disadvantages of the Second-Derivative Test:

- $f''(x)$ is difficult to find for some functions.

- It sometimes fails to give results.

The relationships among $f(x)$, $f'(x)$, and $f''(x)$ can be summarized in the following table.

$f(x)$	Concave Upward	Concave Downward	Point of Inflection	
$f'(x)$	increasing	decreasing	maximum	minimum
$f''(x)$	positive $(+)$	negative $(-)$	$(+)$ to $(-)$	$(-)$ to $(+)$

See the Tech Card for details on using a graphing calculator or Excel to explore the relationships among f, f', and f''.

10.2 Exercises

1. Determine whether $f(x) = x^3 - 3x^2 + 1$ is concave up or concave down at (a) $x = -2$ and (b) $x = 3$.

In Problems 2 and 3, use the indicated x-values on the graph of $y = f(x)$ to find the following.

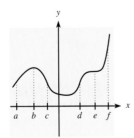

2. Find intervals where $f''(x) > 0$.

3. Find the x-coordinates of three points of inflection.

In Problems 4 and 5, a function and its graph are given. Use the second derivative to determine intervals on which the function is concave up, to determine intervals on which it is concave down, and to locate points of inflection. Check these results against the graph shown.

4. $f(x) = x^3 - 6x^2 + 5x + 6$

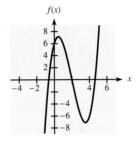

5. $f(x) = \frac{1}{4}x^4 + \frac{1}{2}x^3 - 3x^2 + 3$

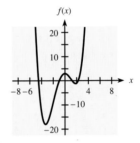

6. Let $y = \frac{1}{3}x^3 - 2x^2 + 3x + 2$. Find the relative maxima, relative minima, and points of inflection. Then sketch the graph of the function.

7. A function and its first and second derivatives are given. Use these to find critical values, relative maxima, relative minima, and points of inflection; sketch the graph of the function.

$$y = x^{1/3}(x - 4)$$
$$y' = \frac{4(x - 1)}{3x^{2/3}}$$
$$y'' = \frac{4(x + 2)}{9x^{5/3}}$$

8. Use the graph shown in the figure and for each part identify one point from A through I that satisfies the given conditions.
   (a) $f'(x) > 0$ and $f''(x) > 0$
   (b) $f'(x) < 0$ and $f''(x) < 0$
   (c) $f'(x) = 0$ and $f''(x) > 0$
   (d) $f'(x) > 0$ and $f''(x) = 0$

(e) $f'(x) = 0$ and $f''(x) = 0$

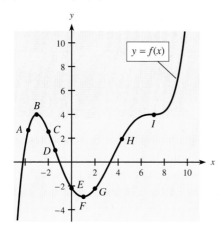

Applications

9. *Advertising and sales* The figure shows the daily sales volume S as a function of time t since an ad campaign began.

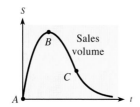

(a) Which of A, B, and C is the point of inflection for the graph?

(b) On which side of C is $d^2S/dt^2 > 0$?

(c) Does the *rate of change* of sales volume attain its minimum at C?

10. *Production* Suppose that the total number of units produced by a worker in t hours of an 8-hour shift can be modeled by the production function $P(t)$:

$$P(t) = 27t + 12t^2 - t^3$$

(a) Find the number of hours before production is maximized.

(b) Find the number of hours before the rate of production is maximized. That is, find the point of diminishing returns.

11. *Advertising and sales—diminishing returns* Suppose that a company's daily sales volume attributed to an advertising campaign is given by

$$S(t) = \frac{3}{t + 3} - \frac{18}{(t + 3)^2} + 1$$

(a) Find how long it will be before sales volume is maximized.

(b) Find how long it will be before the rate of change of sales volume is minimized. That is, find the point of diminishing returns.

10.3 Optimization in Business and Economics

MOST COMPANIES ARE INTERESTED IN OBTAINING THE GREATEST POSSIBLE REVENUE OR PROFIT. SIMILARLY, MANUFACTURERS OF PRODUCTS ARE CONCERNED ABOUT PRODUCING THEIR PRODUCTS FOR THE LOWEST POSSIBLE AVERAGE COST PER UNIT. THEREFORE, RATHER THAN JUST FINDING THE RELATIVE MAXIMA OR RELATIVE MINIMA OF A FUNCTION, WE WILL CONSIDER WHERE THE ABSOLUTE MAXIMUM OR ABSOLUTE MINIMUM OF A FUNCTION OCCURS IN A GIVEN INTERVAL.

Absolute Extrema

As their name implies, **absolute extrema** are the functional values that are the largest or smallest values over the entire domain of the function (or over the interval of interest).

Absolute Extrema:

The value $f(a)$ is the **absolute maximum** of f if $f(a) \geq f(x)$ for all x in the domain of f (or over the interval of interest).

The value $f(b)$ is the **absolute minimum** of f if $f(b) \leq f(x)$ for all x in the domain of f (or over the interval of interest).

IN THE MANAGEMENT, LIFE, AND SOCIAL SCIENCES, A LIMITED DOMAIN OCCURS VERY OFTEN, BECAUSE MANY QUANTITIES ARE REQUIRED TO BE **POSITIVE**, OR AT LEAST NONNEGATIVE.

Let us begin by considering the graph of $y = (x - 1)^2$ shown in Figure 10.16(a). This graph has a relative minimum at $(1, 0)$. Note that the relative minimum is the lowest point on the graph. In this case, the point $(1, 0)$ is an **absolute minimum point,** and 0 is the absolute minimum for the function. Similarly, when there is a point that is the highest point on the graph over the domain of the function, we call the point an **absolute maximum point** of the graph of the function.

In Figure 10.16(a), we see that there is no relative maximum. However, if the domain of the function is restricted to the interval $[\frac{1}{2}, 2]$, then we get the graph shown in Figure 10.16(b). In this case, there is an absolute maximum of 1 at the point $(2, 1)$ and the absolute minimum of 0 is still at $(1, 0)$.

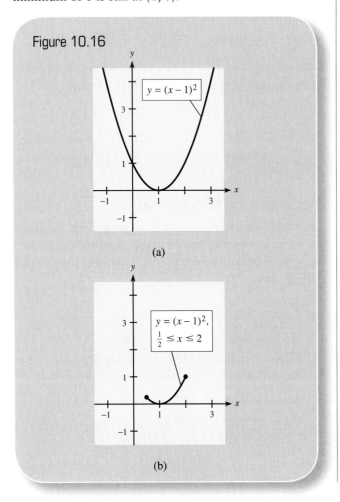

Figure 10.16

$y = (x - 1)^2$

(a)

$y = (x - 1)^2,$
$\frac{1}{2} \leq x \leq 2$

(b)

As the preceding discussion indicates, if the domain of a continuous function is limited to a closed interval, the absolute maximum or minimum may occur at an endpoint of the domain. In testing functions with limited domains for absolute maxima and minima, we must compare the function values at the endpoints of the domain with the function values at the critical values found by taking derivatives.

Maximizing Revenue

Because the marginal revenue is the first derivative of the total revenue, it should be obvious that the total revenue function will have a critical point at the point where the marginal revenue equals 0. It is important to verify where the maximum value occurs. For example, say the total revenue in dollars for a firm is given by

$$R(x) = 8000x - 40x^2 - x^3$$

where x is the number of units sold per day. If only 50 units can be sold per day, the owner needs to find the number of units that must be sold to maximize revenue. To do this, she must first realize that this revenue function is limited to x in the interval $[0, 50]$. Thus, the maximum revenue will occur at a critical value in this interval or at an endpoint. $R'(x) = 8000 - 80x - 3x^2$, so we must solve $8000 - 80x - 3x^2 = 0$ for x.

$$(40 - x)(200 + 3x) = 0$$

$$40 - x = 0 \quad 200 + 3x = 0$$

so

$$x = 40 \quad \text{or} \quad x = -\frac{200}{3}$$

We reject the negative value of x and then use either the second-derivative test or the first-derivative test with a sign diagram.

The tests in the table on page 255 show that a relative maximum occurs at $x = 40$, giving revenue $R(40) = \$192,000$. Checking the endpoints $x = 0$ and $x = 50$ gives $R(0) = \$0$ and $R(50) = \$175,000$. Thus $R = \$192,000$ at $x = 40$ is the (absolute) maximum revenue.

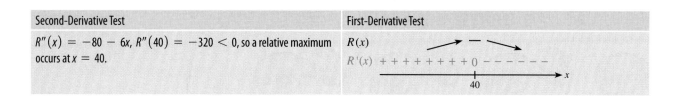

Second-Derivative Test	First-Derivative Test
$R''(x) = -80 - 6x$, $R''(40) = -320 < 0$, so a relative maximum occurs at $x = 40$.	

EXAMPLE 10.8 Revenue

A travel agency will plan tours for groups of 25 or larger. If the group contains exactly 25 people, the price is $300 per person. However, the price per person is reduced by $10 for each additional person above the 25. What size group will produce the largest revenue for the agency?

SOLUTION

The total revenue is

$$R = (\text{number of people})(\text{price per person})$$

The following table shows how the revenue is changed by increases in the size of the group.

Number in Group	Price per Person	Revenue
25	300	7500
25 + 1	300 − 10	7540
25 + 2	300 − 20	7560
⋮	⋮	⋮
25 + x	300 − 10x	$(25 + x)(300 - 10x)$

Thus when x is the number of people added to the 25, the total revenue will be a function of x,

$$R = R(x) = (25 + x)(300 - 10x)$$

or

$$R(x) = 7500 + 50x - 10x^2$$

This is a quadratic function, so its graph is a parabola that is concave down. A maximum will occur at its vertex, where $R'(x) = 0$.

$R'(x) = 50 - 20x$, and the solution to $0 = 50 - 20x$ is $x = 2.5$. Thus adding 2.5 people to the group should maximize the total revenue. But we cannot add half a person, so we will test the total revenue function for 27 people and 28 people. This will determine the optimal group size because $R(x)$ is concave down for all x.

For $x = 2$ (giving 27 people) we get $R(2) = 7500 + 50(2) - 10(2)^2 = 7560$. For $x = 3$ (giving 28 people) we get $R(3) = 7500 + 50(3) - 10(3)^2 = 7560$. Note that both 27 and 28 people give the same total revenue and that this revenue is greater than the revenue for 25 people. Thus the revenue is maximized at either 27 or 28 people in the group. (See Figure 10.17.)

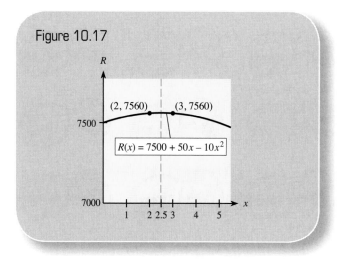

Figure 10.17

Minimizing Average Cost

Because the total cost function is always increasing for $x \geq 0$, the number of units that will make the total cost a minimum is always $x = 0$ units, which gives an absolute minimum. However, it is more useful to find the number of units that will make the average cost per unit a minimum.

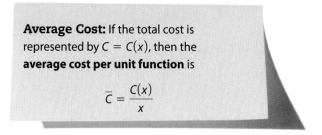

Average Cost: If the total cost is represented by $C = C(x)$, then the **average cost per unit function** is

$$\overline{C} = \frac{C(x)}{x}$$

We can use derivatives to find the minimum of the average cost function, as the following example shows.

EXAMPLE 10.9 Average Cost

If the total cost function for a commodity is given by $C = \frac{1}{4}x^2 + 4x + 100$ dollars, where x represents the number of units produced, producing how many units will result in a minimum *average cost* per unit? Find the minimum average cost per unit.

20 UNITS @ $14

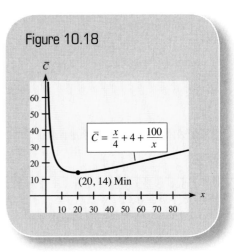

Figure 10.18

$$\bar{C} = \frac{x}{4} + 4 + \frac{100}{x}$$

(20, 14) Min

Shutterstock

SOLUTION

The average cost function is given by

$$\bar{C} = \frac{\frac{1}{4}x^2 + 4x + 100}{x} = \frac{1}{4}x + 4 + \frac{100}{x}$$

Then

$$\bar{C}' = \bar{C}'(x) = \frac{1}{4} - \frac{100}{x^2}$$

Setting $\bar{C}' = 0$ gives

$$0 = \frac{1}{4} - \frac{100}{x^2}$$

$$0 = x^2 - 400, \quad \text{or} \quad x = \pm 20$$

Because the quantity produced must be positive, 20 units should minimize the average cost per unit. We show that it is an absolute minimum by using the second derivative.

$$\bar{C}''(x) = \frac{200}{x^3} \quad \text{so} \quad \bar{C}''(x) > 0 \quad \text{when } x > 0$$

Thus the minimum average cost per unit occurs if 20 units are produced. The graph of the average cost per unit is shown in Figure 10.18. The minimum average cost per unit is $\bar{C}(20) = \$14$.

Maximizing Profit

We have defined the marginal profit function as the derivative of the profit function. That is,

$$\overline{MP} = P'(x)$$

In this chapter we have seen how to use the derivative to find maxima and minima for various functions. Now we can apply those same techniques in the context of **profit maximization.** We can use marginal profit to maximize profit functions.

EXAMPLE 10.10 Profit

Suppose that the production capacity for a certain commodity cannot exceed 30. If the total profit for this commodity is

$$P(x) = 4x^3 - 210x^2 + 3600x - 200 \text{ dollars}$$

where x is the number of units sold, find the number of items that will maximize profit.

SOLUTION

The restrictions on capacity mean that $P(x)$ is restricted by $0 \leq x \leq 30$. The marginal profit function is

$$P'(x) = 12x^2 - 420x + 3600$$

Setting $P'(x)$ equal to 0, we get

$$0 = 12(x - 15)(x - 20)$$

so $P'(x) = 0$ at $x = 15$ *and* $x = 20$. A sign diagram for $P'(x)$ tests these critical values.

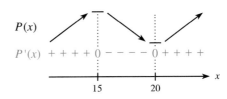

Thus, at (15, 20,050) the total profit function has a *relative* maximum, but we must check the endpoints (0 and 30) before deciding whether it is the absolute maximum.

$$P(0) = -200 \quad \text{and} \quad P(30) = 26,800$$

Thus the absolute maximum profit is $26,800, and it occurs at the endpoint, $x = 30$. Figure 10.19 shows the graph of the profit function.

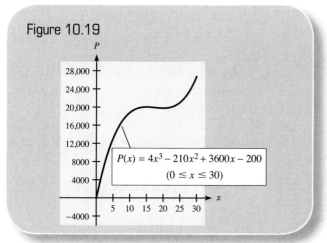

Figure 10.19

$$P(x) = 4x^3 - 210x^2 + 3600x - 200$$
$$(0 \le x \le 30)$$

In a **monopolistic market,** the seller who has a monopoly controls the supply of a product and can force the price higher by limiting supply.

For example, suppose that in a monopoly market, the price of a product in dollars is related to the number of units x demanded daily by $p = 168 - 0.2x$, and the daily average cost for this product is

$$\overline{C} = 120 + x \text{ dollars}$$

Then the total revenue function is

$$R(x) = px = (168 - 0.2x)x = 168x - 0.2x^2$$

and the total cost function is

$$C(x) = \overline{C} \cdot x = (120 + x)x = 120x + x^2$$

Thus the profit function is

$$P(x) = R(x) - C(x) = 168x - 0.2x^2 - (120x + x^2)$$
$$= 48x - 1.2x^2$$

From $P(x)$ we can use $P'(x)$ to find the maximum profit.

In a **competitive market,** each firm is so small that its actions in the market cannot affect the price of the product. The price of the product is determined in the market by the intersection of the market demand curve (from all consumers) and market supply curve (from all firms that supply this product). The firm can sell as little or as much as it desires at the market equilibrium price.

MONOPOLY

One big player controls supply and dictates pricing.

COMPETITION

Several players are so small that individual actions cannot affect product price, so supply and demand dictate price.

Therefore, a firm in a competitive market has a total revenue function given by $R(x) = px$, where p is the market equilibrium price for the product and x is the quantity sold.

As we have seen, graphing calculators and Excel can be used to locate maxima and minima for graphing. They can also be used to locate absolute maxima and absolute minima in applied problems. See the Tech Card for details.

10.3 Exercises

1. Find the absolute maxima and minima for $f(x) = x^3 - 2x^2 - 4x + 2$ on $[-1, 3]$.

Maximizing Revenue

2. (a) If the total revenue function for a radio is $R = 36x - 0.01x^2$, then sale of how many units, x, will maximize the total revenue in dollars? Find the maximum revenue.
   (b) Find the maximum revenue if production is limited to at most 1500 radios.

3. If the total revenue function for a computer is $R(x) = 2000x - 20x^2 - x^3$, find the level of sales, x, that maximizes revenue and find the maximum revenue in dollars.

4. A cable TV company has 4000 customers paying $55 each month. If each $1 reduction in price attracts 100 new customers, find the price that yields maximum revenue. Find the maximum revenue.

Minimizing Average Cost

5. If the total cost function for a lamp is $C(x) = 250 + 33x + 0.1x^2$ dollars, producing how many units, x, will result in a minimum average cost per unit? Find the minimum average cost.

6. If the total cost function for a product is $C(x) = 100(0.02x + 4)^3$ dollars, where x represents the number of hundreds of units produced, producing how many units will minimize average cost? Find the minimum average cost.

Maximizing Profit

7. If the profit function for a product is $P(x) = 5600x + 85x^2 - x^3 - 200{,}000$ dollars, selling how many items, x, will produce a maximum profit? Find the maximum profit.

8. A firm can produce only 1000 units per month. The monthly total cost is given by $C(x) = 300 + 200x$ dollars, where x is the number produced. If the total revenue is given by $R(x) = 250x - \frac{1}{100}x^2$ dollars, how many items, x, should the firm produce for maximum profit? Find the maximum profit.

9. The figure shows the graph of a quadratic revenue function and a linear cost function.
   (a) At which of the four x-values shown is the distance between the revenue and the cost greatest?
   (b) At which of the four x-values shown is the profit largest?
   (c) At which of the four x-values shown is the slope of the tangent to the revenue curve equal to the slope of the cost line?
   (d) What is the relationship between marginal cost and marginal revenue when profit is at its maximum value?

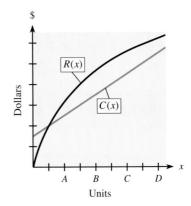

10. A firm has monthly average costs, in dollars, given by

$$\overline{C} = \frac{45{,}000}{x} + 100 + x$$

where x is the number of units produced per month. The firm can sell its product in a competitive market for $1600 per unit. If production is limited to 600 units per month, find the number of units that gives maximum profit, and find the maximum profit.

Miscellaneous Applications

11. **Dow Jones Industrial Average** The figure shows the Dow Jones Industrial Average for all of 2001, the year of the terrorist attacks on New York City and Washington, D.C.
    (a) Approximate when during 2001 the Dow reached its absolute maximum for that year.
    (b) When do you think the Dow reached its absolute minimum for this period? What happened to trigger this?

Dow Jones Industrial Average

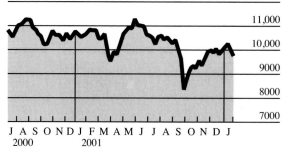

J A S O N D J F M A M J J A S O N D J
2000   2001

SOURCE: *Wall Street Journal*, January 17, 2002. Copyright © 2002 by Dow Jones & Co. Reprinted by permission of Dow Jones & Co. via Copyright Clearance Center.

Need more practice?
Find more here: cengagebrain.com

10.4 Applications of Maxima and Minima

IF WE HAVE FUNCTIONS THAT MODEL COST, REVENUE, OR POPULATION GROWTH, WE CAN APPLY THE METHODS OF THIS CHAPTER TO FIND THE MAXIMA AND MINIMA OF THOSE FUNCTIONS.

EXAMPLE 10.11 Company Growth

Suppose that a new company begins production in 2010 with eight employees and the growth of the company over the next 10 years is predicted by

$$N = N(t) = 8\left(1 + \frac{160t}{t^2 + 16}\right), \quad 0 \le t \le 10$$

where N is the number of employees t years after 2010.

Determine in what year the number of employees will be maximized and the maximum number of employees.

SOLUTION

This function will have a relative maximum when $N'(t) = 0$.

$$N'(t) = 8\left[\frac{(t^2 + 16)(160) - (160t)(2t)}{(t^2 + 16)^2}\right]$$

$$= 8\left[\frac{160t^2 + 2560 - 320t^2}{(t^2 + 16)^2}\right]$$

$$= 8\left[\frac{2560 - 160t^2}{(t^2 + 16)^2}\right]$$

Because $N'(t) = 0$ when its numerator is 0 (note that this denominator is never 0), we must solve

$$2560 - 160t^2 = 0$$
$$160(4 + t)(4 - t) = 0$$

so

$$t = -4 \quad \text{or} \quad t = 4$$

We are interested only in positive t-values, so we evaluate $N'(t)$ on either side of $t = 4$.

$$N'(0) = 8\left[\frac{2560}{256}\right] > 0$$
$$N'(10) = 8\left[\frac{-13{,}440}{(116)^2}\right] < 0 \quad \Biggr\} \Rightarrow \text{relative maximum}$$

The relative maximum is $N(4) = 8\left(1 + \dfrac{640}{32}\right) = 168$

At $t = 0$, the number of employees is $N(0) = 8$, and it increases to $N(4) = 168$. After $t = 4$ (in 2014), $N(t)$ decreases to $N(10) = 118$ (approximately), so $N(4) = 168$ is the maximum number of employees. Figure 10.20 verifies these conclusions.

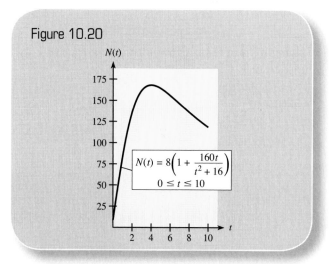

Figure 10.20

Sometimes we must develop the function we need from the statement of the problem. In this case, it is important to understand what is to be maximized or minimized and to express that quantity as a function of *one* variable.

EXAMPLE 10.12 Minimizing Cost

A farmer needs to enclose a rectangular pasture containing 1,600,000 square feet. Suppose that along the road adjoining his property he wants to use a more expensive fence and that he needs no fence on one side perpendicular to the road because a river bounds his property on that side. If the fence costs $15 per foot along the road and $10 per foot along the two remaining sides that must be fenced, what dimensions of his rectangular field will minimize his cost? (See Figure 10.21.)

Figure 10.21

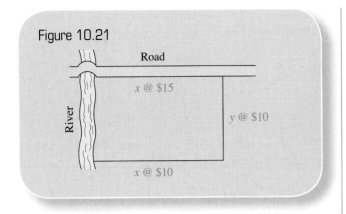

SOLUTION

In Figure 10.21, x represents the length of the pasture along the road (and parallel to the road) and y represents the width. The cost function for the fence used is

$$C = 15x + 10y + 10x = 25x + 10y$$

We cannot use a derivative to find where C is minimized unless we write C as a function of x or y only. Because the area of the rectangular field must be 1,600,000 square feet, we have

$$A = xy = 1,600,000$$

Solving for y in terms of x and substituting give C as a function of x.

$$y = \frac{1,600,000}{x}$$

$$C = 25x + 10\left(\frac{1,600,000}{x}\right) = 25x + \frac{16,000,000}{x}$$

The derivative of C with respect to x is

$$C'(x) = 25 - \frac{16,000,000}{x^2}$$

and we find the relative minimum of C as follows:

$$0 = 25 - \frac{16,000,000}{x^2}$$

$$0 = 25x^2 - 16,000,000$$

$$25x^2 = 16,000,000$$

$$x^2 = 640,000 \Rightarrow x = \pm\, 800$$

We use $x = 800$ feet

Testing to see whether $x = 800$ gives the minimum cost, we find

$$C''(x) = \frac{32,000,000}{x^3}$$

$C''(x) > 0$ for $x > 0$, so $C(x)$ is concave up for all positive x. Thus $x = 800$ gives the absolute minimum, and $C(800) = 40,000$ is the minimum cost. The other dimension of the rectangular field is $y = 1,600,000 / 800 = 2000$ feet.

We next consider **inventory cost models,** in which x items are produced in each production run and items are removed from inventory at a fixed constant rate. Because items are removed at a constant rate, the average number stored at any time is $x/2$. Also, when $x = 0$, new items must be added to inventory from a production run. Thus the number of units in storage changes with time and is illustrated in Figure 10.22. In these models there are costs associated with both production and storage, but lowering one of these costs means increasing the other. To see how inventory cost models work, consider the following example.

EXAMPLE 10.13
Inventory Cost Model

→ Suppose that a company needs 1,000,000 items during a year and that preparation costs are $800 for each production run. Suppose further that it costs the company $6 to produce each item and $1 to store an item for up

x = 800 ft @ $15/ft = $12,000 for the road side

Shutterstock

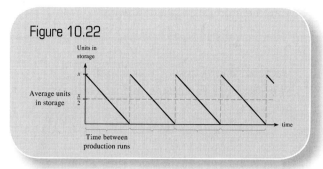

Figure 10.22

to a year. If each production run consists of x items, find x so that the total costs of production and storage are minimized.

SOLUTION

The total production costs are given by

$$\begin{pmatrix} \text{No. of} \\ \text{runs} \end{pmatrix} \begin{pmatrix} \text{cost} \\ \text{per run} \end{pmatrix} + \begin{pmatrix} \text{no. of} \\ \text{items} \end{pmatrix} \begin{pmatrix} \text{cost} \\ \text{per item} \end{pmatrix}$$

$$= \left(\frac{1,000,000}{x} \right)(\$800) + (1,000,000)(\$6)$$

The total storage costs are

$$\begin{pmatrix} \text{Average} \\ \text{no. stored} \end{pmatrix} \begin{pmatrix} \text{storage cost} \\ \text{per item} \end{pmatrix} = \left(\frac{x}{2} \right)(\$1)$$

Thus the total costs of production and storage are

$$C = \left(\frac{1,000,000}{x} \right)(800) + 6,000,000 + \frac{x}{2}$$

We wish to find x so that C is minimized.

$$C' = \frac{-800,000,000}{x^2} + \frac{1}{2}$$

If $x > 0$, critical values occur when $C' = 0$.

$$0 = \frac{-800,000,000}{x^2} + \frac{1}{2}$$

$$\frac{800,000,000}{x^2} = \frac{1}{2}$$

$$1,600,000,000 = x^2$$

$$x = \pm 40,000$$

Because x must be positive, we test $x = 40,000$ with the second derivative.

$$C''(x) = \frac{1,600,000,000}{x^3}, \quad \text{so} \quad C''(40,000) > 0$$

Note that $x = 40,000$ yields an absolute minimum value of C, because $C'' > 0$ for all $x > 0$. That is, production runs of 40,000 items yield minimum total costs for production and storage.

Problems of the type studied in this section can also be solved (or approximated) with a graphing calculator or Excel. See the Tech Card for details.

10.4 Exercises

1. **Productivity** Analysis of daily output of a factory during an 8-hour shift shows that the hourly number of units y produced after t hours of production is

$$y = 70t + \tfrac{1}{2}t^2 - t^3, \quad 0 \le t \le 8$$

   (a) After how many hours will the hourly number of units be maximized?
   (b) What is the maximum hourly output?

2. **Consumer expenditure** Suppose that the demand x (in units) for a product is $x = 10,000 - 100p$, where p dollars is the market price per unit. Then the consumer expenditure for the product is

$$E = px = 10,000p - 100p^2$$

   For what market price will expenditure be greatest?

3. **Advertising and sales** An inferior product with a large advertising budget sells well when it is introduced, but sales fall as people discontinue use of the product. Suppose that the weekly sales S are given by

$$S = \frac{200t}{(t+1)^2}, \quad t \ge 0$$

   where S is in millions of dollars and t is in weeks. After how many weeks will sales be maximized?

4. **Minimum cost** A rectangular field with one side along a river is to be fenced. Suppose that no fence is needed along the river, the fence on the side opposite the river costs $20 per foot, and the fence on the other sides costs $5 per foot. If the field must contain 45,000 square feet, what dimensions will minimize costs?

5. **Optimization at a fixed cost** A rectangular area is to be enclosed and divided into thirds. The family has $800 to spend for the fencing material. The outside fence costs $10 per running foot installed, and the dividers cost $20 per running foot installed. What are the dimensions that will maximize the area enclosed? (The answer contains a fraction.)

6. **Inventory cost model** Suppose that a company needs 1,500,000 items during a year and that preparation for each production run costs $600. Suppose also that it costs $15 to produce each item and $2 per year to store an item. Use the inventory cost model to find the number of items in each

production run so that the total costs of production and storage are minimized.

7. *Inventory cost model* A company needs 150,000 items per year. It costs the company $360 to prepare a production run of these items and $7 to produce each item. If it also costs the company $0.75 per year for each item stored, find the number of items that should be produced in each run so that total costs of production and storage are minimized.

8. *Revenue* The owner of an orange grove must decide when to pick one variety of oranges. She can sell them for $8 a bushel if she sells them now, with each tree yielding an average of 5 bushels. The yield increases by half a bushel per week for the next 5 weeks, but the price per bushel decreases by $0.50 per bushel each week. When should the oranges be picked for maximum return?

Need more practice?
Find more here: cengagebrain.com

10.5 Rational Functions: More Curve Sketching

IN THIS SECTION, WE CONSIDER HOW TO USE INFORMATION ABOUT ASYMPTOTES ALONG WITH THE FIRST AND SECOND DERIVATIVES, AND WE PRESENT A UNIFIED APPROACH TO CURVE SKETCHING.

Asymptotes

Recall the following definition of vertical asymptote presented in Chapter 9.

Vertical Asymptote: The line $x = x_0$ is a **vertical asymptote** of the graph of $y = f(x)$ if the values of $f(x)$ approach $+\infty$ or $-\infty$ as x approaches x_0 (from the left or the right).

We can, therefore, show the following.

Vertical Asymptote of a Rational Function: The graph of the rational function

$$h(x) = \frac{f(x)}{g(x)}$$

has a vertical asymptote at $x = c$ if $g(c) = 0$ and $f(c) \neq 0$.

Because a horizontal asymptote tells us the behavior of the values of the function (y-coordinates) when x increases or decreases without bound, we use limits at infinity to determine the existence of horizontal asymptotes.

Horizontal Asymptote: The graph of a rational function $y = f(x)$ will have a **horizontal asymptote** at $y = b$, for a constant b, if

$$\lim_{x \to +\infty} f(x) = b \quad \text{or} \quad \lim_{x \to -\infty} f(x) = b$$

Otherwise, the graph has no horizontal asymptote.

We can use limits at infinity to prove the statements regarding horizontal asymptotes of the graphs of rational functions on the next page.

EXAMPLE 10.14 Vertical and Horizontal Asymptotes

➡ Find any vertical and horizontal asymptotes for

(a) $f(x) = \dfrac{2x - 1}{x + 2}$ (b) $f(x) = \dfrac{x^2 + 3}{1 - x}$

SOLUTION

(a) The denominator of this function is 0 at $x = -2$, and because this value does not make the numerator 0, there is a vertical asymptote at $x = -2$. Because the function is rational, with the degree of the numerator equal to that of the denominator and with the ratio of the leading coefficients equal to 2, the graph of the function has a horizontal asymptote at $y = 2$. The graph is shown in Figure 10.23(a).

(b) At $x = 1$, the denominator of $f(x)$ is 0 and the numerator is not, so a vertical asymptote occurs at $x = 1$. The function is rational with the degree of the numerator greater than that of the denominator, so there is no horizontal asymptote. The graph is shown in Figure 10.23(b).

Figure 10.23

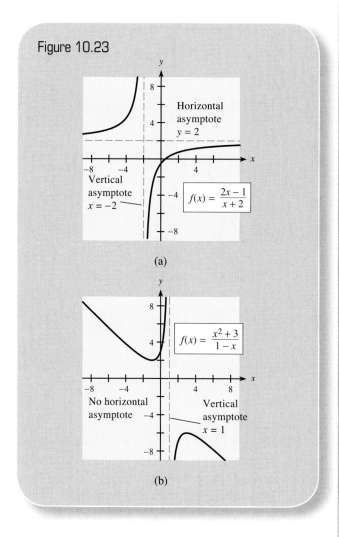

(a)

(b)

More Curve Sketching

We now extend our first- and second-derivative techniques of curve sketching to include functions that have asymptotes.

In general, the "Graphing Guidelines" steps shown to the right are helpful when we sketch the graph of a function.

EXAMPLE 10.15 Graphing with Asymptotes

Sketch the graph of the function $f(x) = \dfrac{x^2}{(x + 1)^2}$.

Horizontal Asymptotes of Rational Functions

Consider the rational function
$$y = \frac{f(x)}{g(x)} = \frac{a_n x^n + \cdots + a_1 x + a_0}{b_m x^m + \cdots + b_1 x + b_0}.$$

1. If $n < m$ (that is, if the degree of the numerator is less than that of the denominator), a horizontal asymptote occurs at $y = 0$ (the x-axis).

2. If $n = m$ (that is, if the degree of the numerator equals that of the denominator), a horizontal asymptote occurs at $y = \dfrac{a_n}{b_m}$ (the ratio of the leading coefficients).

3. If $n > m$ (that is, if the degree of the numerator is greater than that of the denominator), there is no horizontal asymptote.

GRAPHING   GUIDELINES

1. Determine the domain of the function. The domain may be restricted by the nature of the problem or by the equation.

2. Look for vertical asymptotes, especially if the function is a rational function.

3. Look for horizontal asymptotes, especially if the function is a rational function.

4. Find the relative maxima and minima by using the first-derivative test or the second-derivative test.

5. Use the second derivative to find the points of inflection if this derivative is easily found.

6. Use other information (intercepts, for example) and plot additional points to complete the sketch of the graph.

SOLUTION

1. The domain is the set of all real numbers except $x = -1$.

2. Because $x = -1$ makes the denominator 0 and does not make the numerator 0, there is a vertical asymptote at $x = -1$.

3. Because $\dfrac{x^2}{(x+1)^2} = \dfrac{x^2}{x^2 + 2x + 1}$ the function is rational with the degree of the numerator equal to that of the denominator and with the ratio of the leading coefficients equal to 1. Hence, the graph of the function has a horizontal asymptote at $y = 1$.

4. To find any maxima and minima, we first find $f'(x)$.

$$f'(x) = \frac{(x+1)^2(2x) - x^2[2(x+1)]}{(x+1)^4}$$

$$= \frac{2x(x+1)[(x+1) - x]}{(x+1)^4}$$

$$= \frac{2x}{(x+1)^3}$$

Thus $f'(x) = 0$ when $x = 0$ (and $y = 0$), and $f'(x)$ is undefined at $x = -1$ (where the vertical asymptote occurs). Testing $f'(x)$ on either side of $x = 0$ and $x = -1$ gives the following sign diagram. The sign diagram for f' shows that the critical point $(0, 0)$ is a relative minimum and shows how the graph approaches

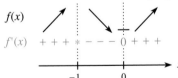

*$x = -1$ is a vertical asymptote.

the vertical asymptote at $x = -1$.

5. To see how the graph approaches the horizontal asymptote, we check $f(x)$ for large values of $|x|$

$$f(-100) = \frac{(-100)^2}{(-99)^2} = \frac{10{,}000}{9{,}801} > 1,$$

$$f(100) = \frac{100^2}{101^2} = \frac{10{,}000}{10{,}201} < 1$$

Thus the graph has the characteristics shown in

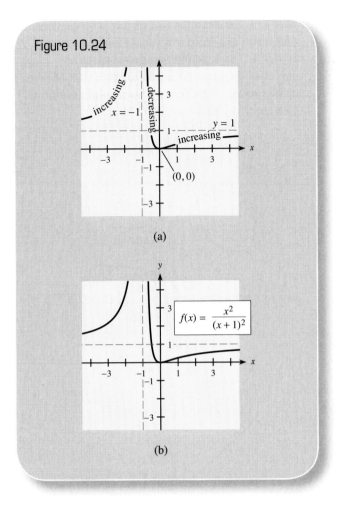

Figure 10.24

(a)

(b)

Figure 10.24(a). The graph is shown in Figure 10.24(b).

When we wish to learn about a function $f(x)$ or sketch its graph, it is important to understand what information we obtain from $f(x)$, from $f'(x)$, and from $f''(x)$. The following summary may be helpful.

Summary

Source	Information Provided
$f(x)$	*y*-coordinates; horizontal asymptotes, vertical asymptotes; domain restrictions
$f'(x)$	Increasing $[f'(x) > 0]$; decreasing $[f'(x) < 0]$; critical points $[f'(x) = 0$ or $f'(x)$ undefined$]$; sign-diagram tests for maxima and minima
$f''(x)$	Concave up $[f''(x) > 0]$; concave down $[f''(x) < 0]$; possible points of inflection $[f''(x) = 0$ or $f''(x)$ undefined$]$; sign-diagram tests for points of inflection; second-derivative test for maxima and minima

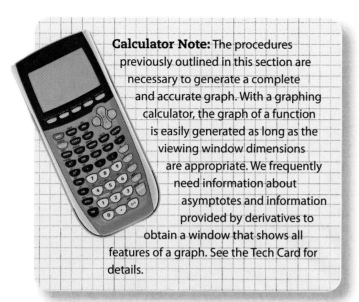

Calculator Note: The procedures previously outlined in this section are necessary to generate a complete and accurate graph. With a graphing calculator, the graph of a function is easily generated as long as the viewing window dimensions are appropriate. We frequently need information about asymptotes and information provided by derivatives to obtain a window that shows all features of a graph. See the Tech Card for details.

10.5 Exercises

1. Use the given graph of $f(x) = \dfrac{x-4}{x-2}$ to find each of the following, if they exist. Then confirm your results analytically.

   (a) vertical asymptotes
   (b) $\displaystyle\lim_{x\to\infty} f(x)$
   (c) $\displaystyle\lim_{x\to-\infty} f(x)$
   (d) horizontal asymptotes

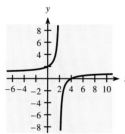

2. Find any horizontal and vertical asymptotes for the function $y = \dfrac{x+1}{x^2-4}$.

In Problems 3 and 4, for each function, find any horizontal and vertical asymptotes, and use information from the first derivative to sketch the graph.

3. $f(x) = \dfrac{2x+2}{x-3}$

4. $y = \dfrac{x^2+4}{x}$

In Problems 5 and 6, a function and its first and second derivatives are given. Use these to find any horizontal and vertical asymptotes, critical points, relative maxima, relative minima, and points of inflection. Then sketch the graph of each function.

5. $y = \dfrac{x}{(x-1)^2}$

   $y' = -\dfrac{x+1}{(x-1)^3}$

   $y'' = \dfrac{2x+4}{(x-1)^4}$

6. $f(x) = \dfrac{9(x-2)^{2/3}}{x^2}$

   $f'(x) = \dfrac{12(3-x)}{x^3(x-2)^{1/3}}$

   $f''(x) = \dfrac{4(7x^2-42x+54)}{x^4(x-2)^{4/3}}$

7. Given $f(x) = \dfrac{9x}{17-4x}$ and its graph as shown:

   (a) Use the graph to estimate the locations of any horizontal or vertical asymptotes.
   (b) Use the function to determine precisely the locations of any asymptotes.

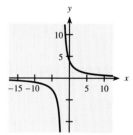

Applications

8. **Cost-benefit** The percent p of particulate pollution that can be removed from the smokestacks of an industrial plant by spending C dollars is given by

$$p = \dfrac{100C}{7300 + C}$$

   (a) Find any C-values at which the rate of change of p with respect to C does not exist. Make sure that these make sense in the problem.
   (b) Find C-values for which p is increasing.
   (c) If there is a horizontal asymptote, find it.
   (d) Can 100% of the pollution be removed?

9. **Productivity** The figure is a typical graph of worker productivity per hour P as a function of time t on the job.
   (a) What is the horizontal asymptote?
   (b) What is $\lim\limits_{x \to \infty} P(t)$?
   (c) What is the horizontal asymptote for $P'(t)$?
   (d) What is $\lim\limits_{x \to \infty} P'(t)$?

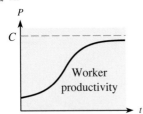

10. **Female workers in the workforce** For selected years from 1970 to 2006, the following table shows the percent of total U.S. workers who were female.

Year	% Female	Year	% Female
1970	30.0	1995	40.2
1975	31.9	2000	41.2
1980	35.3	2005	41.4
1985	37.9	2006	41.5
1991	39.2		

Source: U.S. Census Bureau

Assume these data can be modeled with the function

$$p(t) = \frac{84.7t + 1458}{1.63t + 48.8}$$

where $p(t)$ is the percent of the U.S. workforce that is female and t is the number of years past 1970.
(a) Find $\lim\limits_{t \to \infty} p(t)$.
(b) Interpret your answer to part (a).
(c) Does $p(t)$ have any vertical asymptotes within its domain $t \geq 0$?
(d) Whenever $p(t) < 0$ or $p(t) \geq 100$, the model would be inappropriate. Determine whether the model is ever inappropriate for $t \geq 0$.

....

Need more practice?
Find more here: cengagebrain.com

Chapter Exercises

In Problems 1–3, find all critical points and determine whether they are relative maxima, relative minima, or horizontal points of inflection.

1. $p = q^2 - 4q - 5$

2. $f(x) = 1 - 3x + 3x^2 - x^3$

3. $f(x) = \dfrac{3x}{x^2 + 1}$

In Problems 4–8:
   (a) Find all critical values, including those at which $f'(x)$ is undefined.
   (b) Find the relative maxima and minima, if any exist.
   (c) Find the horizontal points of inflection, if any exist.
   (d) Sketch the graph.

4. $y = x^3 + x^2 - x - 1$

5. $f(x) = 4x^3 - x^4$

6. $y = 5x^7 - 7x^5 - 1$

7. $y = x^{2/3} - 1$

8. $y = x^{2/3}(x - 4)^2$

9. Is the graph of $y = x^4 - 3x^3 + 2x - 1$ concave up or concave down at $x = 2$?

10. Find intervals on which the graph of $y = x^4 - 2x^3 - 12x^2 + 6$ is concave up and intervals on which it is concave down, and find points of inflection.

11. Find the relative maxima, relative minima, and points of inflection of the graph of $y = x^3 - 3x^2 - 9x + 10$.

In Problems 12 and 13, find any relative maxima, relative minima, and points of inflection, and sketch each graph.

12. $y = x^3 - 12x$

13. $y = 2 + 5x^3 - 3x^5$

14. Given $R = 280x - x^2$, find the absolute maximum and minimum for R when (a) $0 \leq x \leq 200$ and (b) $0 \leq x \leq 100$.

15. Given $y = 6400x - 18x^2 - \dfrac{x^3}{3}$, find the absolute maximum and minimum for y when (a) $0 \le x \le 50$ and (b) $0 \le x \le 100$.

16. Use the graph to find the following items.
   (a) vertical asymptotes
   (b) horizontal asymptotes
   (c) $\lim\limits_{x \to +\infty} f(x)$
   (d) $\lim\limits_{x \to -\infty} f(x)$

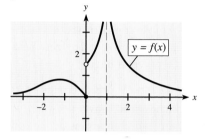

In Problems 17 and 18, find any horizontal asymptotes and any vertical asymptotes.

17. $y = \dfrac{3x + 2}{2x - 4}$ 18. $y = \dfrac{x^2}{1 - x^2}$

In Problems 19–21:
   (a) Find any horizontal and vertical asymptotes.
   (b) Find any relative maxima and minima.
   (c) Sketch each graph.

19. $y = \dfrac{3x}{x + 2}$ 20. $y = \dfrac{8(x - 2)}{x^2}$

21. $y = \dfrac{x^2}{x - 1}$

22. A function and its graph are given.
   (a) Use the graph to determine (estimate) x-values where $f'(x) > 0$, where $f'(x) < 0$, and where $f'(x) = 0$.
   (b) Use the graph to determine x-values where $f''(x) > 0$, where $f''(x) < 0$, and where $f''(x) = 0$.
   (c) Check your conclusions in part (a) by finding $f'(x)$ and graphing it with a graphing utility.
   (d) Check your conclusions in part (b) by finding $f''(x)$ and graphing it with a graphing utility.

$$f(x) = x^3 - 4x^2 + 4x$$

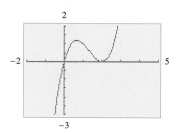

23. (a) Use the graph of $f'(x)$ to determine (estimate) where the graph of $f(x)$ is increasing, where it is decreasing, and where it has relative extrema.
   (b) Use the graph of $f'(x)$ to determine where $f''(x) > 0$, where $f''(x) < 0$, and where $f''(x) = 0$.
   (c) Verify that $f(x)$ has $f'(x)$ as its derivative, and graph $f(x)$ to check your conclusions in part (a).
   (d) Calculate $f''(x)$ and graph it to check your conclusions in part (b).

$$f'(x) = x^2 + 4x - 5 \quad \left(\text{for } f(x) = \dfrac{x^3}{3} + 2x^2 - 5x\right)$$

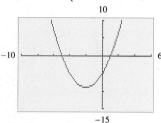

24. The graph shown is $f''(x)$.
   (a) Use the graph to determine (estimate) where the graph of $f(x)$ is concave up, where it is concave down, and where it has points of inflection.
   (b) Verify that $f(x)$ has $f''(x)$ as its second derivative, and graph $f(x)$ to check your conclusions in part (a).

$$f''(x) = 4 - x \quad \left(\text{for } f(x) = 2x^2 - \dfrac{x^3}{6}\right)$$

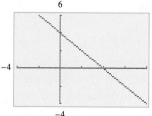

Applications

In Problems 25–28, cost, revenue, and profit are in dollars and x is the number of units.

25. **Cost** Suppose the total cost function for a product is

$$C(x) = 3x^2 + 15x + 75$$

How many units will minimize the average cost? Find the minimum average cost.

26. **Revenue** Suppose the total revenue function for a product is given by

$$R(x) = 32x - 0.01x^2$$

   (a) How many units will maximize the total revenue? Find the maximum revenue.
   (b) If production is limited to 1500 units, how many units will maximize the total revenue? Find the maximum revenue.

27. **Profit** Suppose the profit function for a product is

$$P(x) = 1080x + 9.6x^2 - 0.1x^3 - 50,000$$

   Find the maximum profit.

28. **Profit** How many units (x) will maximize profit if $R(x) = 46x - 0.01x^2$ and $C(x) = 0.05x^2 + 10x + 1100$?

29. **Marginal profit** The figure shows the graph of a marginal profit function for a company. At what level of sales will profit be maximized? Explain.

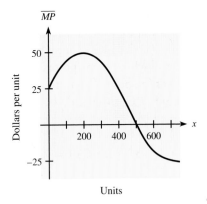

30. **Revenue** MMR II Extreme Bike Shop sells 54 basic-style mountain bikes per month at a price of $385 each. Market research indicates that MMR II could sell 10 more of these bikes if the price were $25 lower. At what selling price will MMR II maximize the revenue from these bikes?

31. **Profit** If in Problem 30 the mountain bikes cost the shop $200 each, at what selling price will MMR II's profit be a maximum?

32. **Profit** Suppose that for a product in a competitive market, the demand function is $p = 1200 - 2x$ and the supply function is $p = 200 + 2x$, where x is the number of units and p is in dollars. A firm's average cost function for this product is

$$\overline{C}(x) = \frac{12,000}{x} + 50 + x$$

Find the maximum profit. (*Hint:* First find the equilibrium price.)

33. **Profit** The monthly demand function for x units of a product sold at $\$p$ per unit by a monopoly is $p = 800 - x$ and its average cost is $\overline{C} = 200 + x$.
   (a) Determine the quantity that will maximize profit.
   (b) Find the selling price at the optimal quantity.

34. **Product design** A playpen manufacturer wants to make a rectangular enclosure with maximum play area. To remain competitive, he wants the perimeter of the base to be only 16 feet. What dimensions should the playpen have?

35. **Printing design** A printed page is to contain 56 square inches and have a $\frac{3}{4}$-inch margin at the bottom and 1-inch margins at the top and on both sides. Find the dimensions that minimize the size of the page (and hence the costs for paper).

36. **Drug sensitivity** The reaction R to an injection of a drug is related to the dose x, in milligrams, according to

$$R(x) = x^2\left(500 - \frac{x}{3}\right)$$

The sensitivity to the drug is defined by dR/dx. Find the dose that maximizes sensitivity.

37. **Federal tax per capita** For the years from 1970 to 2005, the federal tax per capita T (in dollars) can be modeled by the function

$$T(x) = -0.1384x^3 + 9.660x^2 + 27.17x + 1005$$

where x is the number of years past 1970. (*Source:* Internal Revenue Service)
   (a) When does the rate of change of federal tax per capita reach its maximum?
   (b) On a graph of $T(x)$, what feature occurs at the x-value found in part (a)?

38. **Inventory cost model** A company needs to produce 288,000 items per year. Production costs are $1500 to prepare for a production run and $30 for each item produced. Inventory costs are $1.50 per year for each item stored. Find the number of items that should be produced in each run so that the total costs of production and storage are minimum.

39. **Market share** Suppose a company's percent share of the market (actual and projected) for a new product t quarters after its introduction is given by

$$M(t) = \frac{3.8t^2 + 3}{0.1t^2 + 1}$$

(a) Find the company's market share when the product is introduced.

(b) Find any horizontal asymptote of the graph of $M(t)$, and write a sentence that explains the meaning of this asymptote.

Derivatives
Continued

11.1 Derivatives of Logarithmic Functions

IN THIS CHAPTER WE WILL DEVELOP DERIVATIVE FORMULAS FOR LOGARITHMIC AND EXPONENTIAL FUNCTIONS, FOCUSING PRIMARILY ON BASE e EXPONENTIALS AND LOGARITHMS. WE WILL APPLY LOGARITHMIC AND EXPONENTIAL FUNCTIONS AND USE THEIR DERIVATIVES TO SOLVE MAXIMIZATION AND MINIMIZATION PROBLEMS IN THE MANAGEMENT AND LIFE SCIENCES.

We will also develop methods for finding the derivative of one variable with respect to another even when the first variable is not a function of the other. This method is called **implicit differentiation.** We will use implicit differentiation with respect to time to solve problems involving rates of change of two or more variables. These problems are called **related-rates** problems.

Recall that we defined the logarithmic function $y = \log_a x$ as follows.

Logarithmic Function: For $a > 0$ and $a \neq 1$, the **logarithmic function**

$$y = \log_a x \quad \text{(logarithmic form)}$$

has domain $x > 0$, base a, and is defined by

$$a^y = x \quad \text{(exponential form)}$$

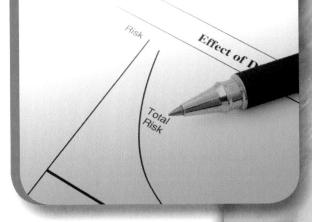

The a is called the **base** in both $\log_a x = y$ and $a^y = x$, and y is the *logarithm* in $\log_a x = y$ and the *exponent* in $a^y = x$. Thus **a logarithm is an exponent.** Although logarithmic functions can have any base a, where $a > 0$ and $a \neq 1$, most problems in calculus and many of the applications to the management, life, and social sciences involve logarithms with base e, called **natural logarithms.**

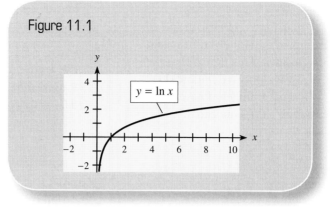

Figure 11.1

$y = \ln x$

In this section we'll see why this base is so important. Recall that the natural logarithmic function ($y = \log_e x$) is written $y = \ln x$; see Figure 11.1 for the graph.

Derivative of $y = \ln x$

From Figure 11.1 we see that for $x > 0$, the graph of $y = \ln x$ is always increasing, so the slope of the tangent line to any point must be positive. This means that the derivative of $y = \ln x$ is always positive. Figure 11.2 shows the graph of $y = \ln x$ with tangent lines drawn at several points and with their slopes indicated.

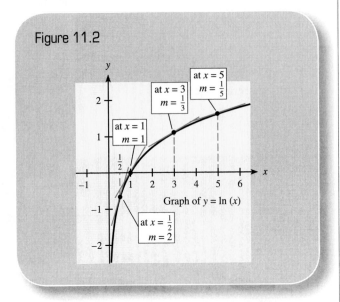

Figure 11.2

Note in Figure 11.2 that at each point where a tangent line is drawn, the slope of the tangent line is the reciprocal of the x-coordinate. In fact, this is true for every point on $y = \ln x$, so the slope of the tangent at any point is given by $1/x$. Thus we have the following:

Derivative of $y = \ln x$:

If $y = \ln x$, then $\dfrac{dy}{dx} = \dfrac{1}{x}$.

This formula for the derivative of $y = \ln x$ is proved in the Enrichment material on CourseMate for MATH APPS at cengagebrain.com.

EXAMPLE 11.1 Derivatives Involving Logarithms

(a) If $y = x^3 + 3 \ln x$, find dy/dx.
(b) If $y = x^2 \ln x$, find y'.

SOLUTION

(a) $\dfrac{dy}{dx} = 3x^2 + 3\left(\dfrac{1}{x}\right) = 3x^2 + \dfrac{3}{x}$

(b) By the Product Rule,

$$y' = x^2 \cdot \dfrac{1}{x} + (\ln x)(2x) = x + 2x \ln x$$

We can use the Chain Rule to find the formula for the derivative of $y = \ln u$, where $u = f(x)$.

Derivatives of Natural Logarithmic Functions: If $y = \ln u$, where u is a differentiable function of x, then

$$\dfrac{dy}{dx} = \dfrac{1}{u} \cdot \dfrac{du}{dx}$$

EXAMPLE 11.2 Derivative $y = \ln (u)$

Find the derivative for each of the functions.
(a) $f(x) = \ln (x^4 - 3x + 7)$
(b) $f(x) = \frac{1}{3} \ln (2x^6 - 3x + 2)$
(c) $g(x) = \dfrac{\ln (2x + 1)}{2x + 1}$

SOLUTION

(a) $f'(x) = \dfrac{1}{x^4 - 3x + 7}(4x^3 - 3) = \dfrac{4x^3 - 3}{x^4 - 3x + 7}$

(b) $f'(x)$ is $\frac{1}{3}$ of the derivative of $\ln (2x^6 - 3x + 2)$.

$$f'(x) = \dfrac{1}{3} \cdot \dfrac{1}{2x^6 - 3x + 2}(12x^5 - 3)$$

(c) We begin with the Quotient Rule.

$$g'(x) = \dfrac{(2x + 1)\dfrac{1}{2x + 1}(2) - [\ln (2x + 1)]2}{(2x + 1)^2}$$

$$= \dfrac{2 - 2 \ln (2x + 1)}{(2x + 1)^2}$$

Using Properties of Logarithms

A logarithmic function of products, quotients, or powers, such as $y = \ln [x(x^5 - 2)^{10}]$, can be rewritten with properties of logarithms so that finding the derivative is much easier. The properties of logarithms, which were introduced in Section 5.2, are stated for natural logarithms in the Properties of Logarithms box on the next page.

EXAMPLE 11.3 Logarithm Properties and Derivatives

Find the derivative of $f(x) = \ln \left(\dfrac{\sqrt[3]{3x + 5}}{x^2 + 11}\right)^4$.

SOLUTION

We begin by using logarithm properties.

$$f(x) = \ln \left(\dfrac{\sqrt[3]{3x + 5}}{x^2 + 11}\right)^4 = 4 \ln \left(\dfrac{\sqrt[3]{3x + 5}}{x^2 + 11}\right) \quad \text{Property V}$$

$$f(x) = 4[\tfrac{1}{3} \ln (3x + 5) - \ln (x^2 + 11)] \quad \text{Properties IV and V}$$

Properties of Logarithms

Let M, N, and p be real numbers with $M > 0$ and $N > 0$.

Natural Logarithms
I. $\ln(e^x) = x$
II. $e^{\ln x} = x$
III. $\ln(MN) = \ln M + \ln N$
IV. $\ln\left(\dfrac{M}{N}\right) = \ln M - \ln N$
V. $\ln(M^p) = p \ln M$

We now take the derivative.

$$f'(x) = 4\left[\frac{1}{3} \cdot \frac{1}{3x+5} \cdot 3 - \frac{1}{x^2+11} \cdot 2x\right]$$

$$= 4\left[\frac{1}{3x+5} - \frac{2x}{x^2+11}\right] = \frac{4}{3x+5} - \frac{8x}{x^2+11}$$

EXAMPLE 11.4 Cost

Suppose the cost function for x skateboards is given by

$$C(x) = 18{,}250 + 615 \ln(4x + 10)$$

where $C(x)$ is in dollars. Find the marginal cost when 100 units are produced, and explain what it means.

SOLUTION

Marginal cost is given by $C'(x)$.

$$\overline{MC} = C'(x) = 615\left(\frac{1}{4x+10}\right)(4) = \frac{2460}{4x+10}$$

$$\overline{MC}(100) = \frac{2460}{4(100)+10} = \frac{2460}{410} = 6$$

When 100 units are produced, the marginal cost is 6. This means that the approximate cost of producing the 101st skateboard is \$6.

See the Tech Card for details on approximating derivatives, finding critical values, and locating optimal values for functions involving logarithms.

Derivative of $y = \log_a(x)$

So far we have found derivatives of natural logarithmic functions. If we have a logarithmic function with a base other than e, then we can use the **change-of-base formula.**

Change-of-Base Formula:
To express a logarithm base a as a natural logarithm, use

$$\log_a x = \frac{\ln x}{\ln a}$$

COST

$$C(x) = 18{,}250 + 615 \ln(4x + 10)$$

$$\overline{MC} = C'(x) = 615\left(\frac{1}{4x+10}\right)(4) = \frac{2460}{4x+10}$$

100

When 100 units are produced, the marginal cost is 6. This means that the approximate cost of producing the 101st skateboard is \$6.

\$6

We can apply this change-of-base formula to find the derivative of a logarithm with any base, as the following example illustrates.

EXAMPLE 11.5 Derivative of $y = \log_a(u)$

➡️ If $y = \log_4(x^3 + 1)$, find dy/dx.

SOLUTION

By using the change-of-base formula, we have

$$y = \log_4(x^3 + 1) = \frac{\ln(x^3 + 1)}{\ln 4} = \frac{1}{\ln 4} \cdot \ln(x^3 + 1)$$

Thus

$$\frac{dy}{dx} = \frac{1}{\ln 4} \cdot \frac{1}{x^3 + 1} \cdot 3x^2 = \frac{3x^2}{(x^3 + 1)\ln 4}$$

11.1 Exercises

In Problems 1–3, find the derivative of each function.

1. $f(x) = 4 \ln x$

2. $f(x) = \ln(4x + 9)$

3. $y = \ln(2x^2 - x) + 3x$

4. Find the derivative of the function in part (a). Then find the derivative of the function in part (b) or show that the function in part (b) is the same function as that in part (a).
   (a) $y = \ln x - \ln(x - 1)$
   (b) $y = \ln \dfrac{x}{x - 1}$

5. Find $\dfrac{dy}{dx}$ if $y = \ln(x^3\sqrt{x + 1})$.

In Problems 6–8, find y'.

6. $y = \dfrac{\ln x}{x}$

7. $y = [\ln(x^4 + 3)]^2$

8. $y = \log_6(x^4 - 4x^3 + 1)$

Applications

9. **Marginal cost** Suppose that the total cost (in dollars) for a product is given by

   $$C(x) = 1500 + 200 \ln(2x + 1)$$

   where x is the number of units produced.
   (a) Find the marginal cost function.

(b) Find the marginal cost when 200 units are produced, and interpret your result.

10. **Marginal revenue** The total revenue, in dollars, from the sale of x units of a product is given by

    $$R(x) = \frac{2500x}{\ln(10x + 10)}$$

    (a) Find the marginal revenue function.
    (b) Find the marginal revenue when 100 units are sold, and interpret your result.

11. **Demand** The demand function for a product is given by $p = \dfrac{4000}{\ln(x + 10)}$ where p is the price per unit in dollars when x units are demanded.

    (a) Find the rate of change of price with respect to the number of units sold when 40 units are sold.
    (b) Find the rate of change of price with respect to the number of units sold when 90 units are sold.
    (c) Find the second derivative to see whether the rate at which the price is changing at 40 units is increasing or decreasing.

> **Need more practice?**
> Find more here: **cengagebrain.com**

11.2 Derivatives of Exponential Functions

JUST AS BASE e LOGARITHMS ARE MOST CONVENIENT, SO ARE BASE e EXPONENTIALS. THUS, WE'LL START WITH EXAMINING DERIVATIVES OF $Y = e^X$.

Figure 11.3(a) shows the graph of $F(x) = e^x$ and Figure 11.3(b) shows the same graph with tangent lines drawn to several points.

Note in Figure 11.3(b) that when $x < 0$, tangent lines have slopes near 0, much like the y-coordinates

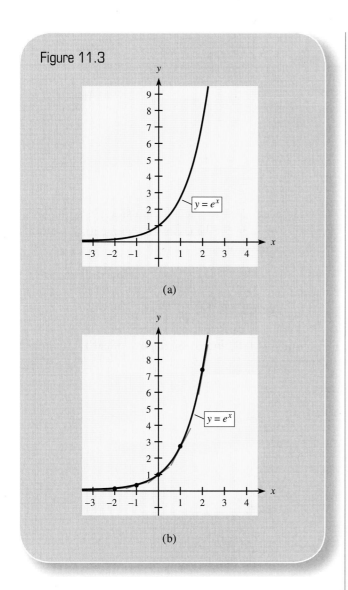

Figure 11.3

(a)

(b)

of $f(x) = e^x$. Furthermore, as x increases to $x = 0$ and for $x > 0$, the slopes of the tangents increase just as the values of the function do. This suggests that the function that gives the slope of the tangent to the graph of $f(x) = e^x$ (that is, the derivative) is similar to the function itself. In fact, the derivative of $f(x) = e^x$ is exactly the function itself (see the Enrichment material on CourseMate for MATH APPS at cengagebrain.com for the derivation).

> **Derivative of $y = e^x$**
>
> If $y = e^x$, then $\dfrac{dy}{dx} = e^x$.

EXAMPLE 11.6 Derivative of an Exponential Function

➡️ If $y = 3e^x + 4x - 11$, find $\dfrac{dy}{dx}$.

SOLUTION

$$\frac{dy}{dx} = 3e^x + 4$$

Derivative of $y = e^u$

As with logarithmic functions, the Chain Rule permits us to expand our derivative formulas.

> **Derivatives of $y = e^u$:**
>
> If $y = e^u$, where u is a differentiable function of x, then
>
> $$\frac{dy}{dx} = e^u \cdot \frac{du}{dx}$$

EXAMPLE 11.7 Derivative of $y = e^u$

➡️ If $f(x) = e^{4x^3}$, find $f(x)$.

SOLUTION

$$f(x) = e^{4x^3} \cdot (12x^2) = 12x^2 e^{4x^3}$$

Notice that to find u' for $u = w/e^{3w}$, we must begin with the Quotient Rule.

EXAMPLE 11.8 Future Value

➡️ When \$100 is invested at 8% compounded continuously, the amount that accrues after t years, which is called the future value, is $S(t) = 100e^{0.08t}$. At what rate is the money in this account growing at the end of 1 year?

SOLUTION

The rate of growth of the money is given by

$$S'(t) = 100e^{0.08t}(0.08) = 8e^{0.08t}$$

The rate of growth of the money at the end of 1 year is

$$S'(1) = 8e^{0.08} = 8.666$$

Thus the future value will change by about \$8.67 during the next year.

EXAMPLE 11.9 Revenue

➡️ North Forty, Inc. is a manufacturer of wilderness camping equipment. The revenue function for its best-selling tent, the Sierra, can be modeled by the function

$$R(x) = 250xe^{(1-0.01x)}$$

where $R(x)$ is the revenue in thousands of dollars from the sale of x thousand Sierra tents. Find the marginal revenue when 75,000 tents are sold, and explain what it means.

Shutterstock

SOLUTION

The marginal revenue function is $\overline{MR} = R'(x)$, and to find this derivative we begin with the Product Rule.

$$\overline{MR} = 250x[e^{(1-0.01x)} \cdot (-0.01)] + e^{(1-0.01x)}(250)$$

$$\overline{MR} = 250e^{(1-0.01x)}(1 - 0.01x)$$

To find the marginal revenue when 75,000 tents are sold, we use

$$\overline{MR}(75) - 250e^{(1-0.75)}(1 - 0.75) = 80.25$$

This means that the sale of one (thousand) more Sierra tents will yield approximately \$80.25 (thousand) in additional revenue.

Derivative of $y = a^u$

In a manner similar to that used to find the derivative of $y = e^x$, we can develop a formula for the derivative of $y = a^x$ for any base $a > 0$ and $a \neq 1$.

> **Derivative of $y = a^u$:**
>
> If $y = a^x$ with $a > 0$, $a \neq 1$, then
>
> $$\frac{dy}{dx} = a^x \ln a$$
>
> If $y = a^u$, with $a > 0$, $a \neq 1$, where u is a differentiable function of x, then
>
> $$\frac{dy}{dx} = a^u \frac{du}{dx} \ln a$$

EXAMPLE 11.10 Derivatives of $y = a^u$

➡️ (a) If $y = 4^x$, find dy/dx.
(b) If $y = 5^{x^2+x}$, find y'.

SOLUTION

(a) $\dfrac{dy}{dx} = 4^x \ln 4$

(b) $y' = 5^{x^2+x}(2x + 1) \ln 5$

See the Tech Card for details on using a graphing calculator to study the behavior of exponential functions and their derivatives.

....

11.2 Exercises

In Problems 1–8, find the derivative of each function.

1. $y = 5e^x - x$

2. $y = 6e^{3x^2}$

3. $y = 2e^{(x^2+1)^3}$

4. $s = t^2e^t$

5. $y = \ln(e^{4x} + 2)$

6. $y = \dfrac{1 + e^{5x}}{e^{3x}}$

7. $y = (e^{3x} + 4)^{10}$

8. $y = 4^{x^2}$

Applications

9. *Future value* If \$P is invested for n years at 10% compounded continuously, the future value that results after n years is given by the function

$$S = Pe^{0.1n}$$

(a) At what rate is the future value growing at any time (for any nonnegative n)?
(b) At what rate is the future value growing after 1 year?
(c) Is the rate of growth of the future value after 1 year greater than 10%? Why?

10. *Marginal cost* Suppose that the total cost in dollars of producing x units of a product is given by

$$C(x) = 10,000 + 20xe^{x/600}$$

Find the marginal cost when 600 units are produced.

11. **Drugs in a bloodstream** The percent concentration y of a certain drug in the bloodstream at any time t (in hours) is given by

$$y = 100(1 - e^{-0.462t})$$

(a) What function gives the instantaneous rate of change of the concentration of the drug in the bloodstream?

(b) Find the rate of change of the concentration after 1 hour. Give your answer to three decimal places.

12. **National health care** Using data from the U.S. Department of Health and Human Services, the national health expenditure H can be modeled by

$$H = 13e^{0.0908t}$$

where t is the number of years past 1960 and H is in billions of dollars. If this model is accurate, at what rate will health care expenditures change in 2012?

13. **Spread of disease** Suppose that the spread of a disease through the student body at an isolated college campus can be modeled by

$$y = \frac{10,000}{1 + 9999e^{-0.99t}}$$

where y is the total number affected at time t (in days). Find the rate of change of y.

Need more practice?
Find more here: **cengagebrain.com**

11.3 Implicit Differentiation

UP TO THIS POINT, FUNCTIONS INVOLVING X AND Y HAVE BEEN WRITTEN IN THE FORM $Y = F(X)$, DEFINING Y AS AN *EXPLICIT FUNCTION* OF X. HOWEVER, NOT ALL EQUATIONS INVOLVING X AND Y CAN BE WRITTEN IN THE FORM $Y = F(X)$, AND WE NEED A NEW TECHNIQUE FOR TAKING THEIR DERIVATIVES.

For example, solving $y^2 = x$ for y gives $y = \pm\sqrt{x}$ so that y is not a function of x. We can write $y = \sqrt{x}$ and $y = -\sqrt{x}$, but then finding the derivative $\frac{dy}{dx}$ at a point on the graph of $y^2 = x$ would require determining which of these functions applies before taking the derivative. Alternatively, we can *imply* that y is a function of x and use a technique called **implicit differentiation**.

EXAMPLE 11.11 Implicit Differentiation

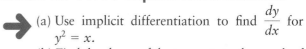

(a) Use implicit differentiation to find $\frac{dy}{dx}$ for $y^2 = x$.

(b) Find the slopes of the tangents to the graph of $y^2 = x$ at the points $(4, 2)$ and $(4, -2)$.

SOLUTION

(a) First take the derivative of both sides of the equation with respect to x.

$$\frac{d}{dx}(y^2) = \frac{d}{dx}(x)$$

Because y is an implied, or implicit, function of x, we can think of $y^2 = x$ as meaning $[u(x)]^2 = x$ for some function u. We use the Chain Rule to take the derivative of y^2 in the same way we would for $[u(x)]^2$. Thus

$$\frac{d}{dx}y^2 = \frac{d}{dx}(x) \text{ gives } 2y\frac{dy}{dx} = 1$$

Solving for $\frac{dy}{dx}$ gives $\frac{dy}{dx} = \frac{1}{2y}$.

(b) To find the slopes of the tangents at the points $(4, 2)$ and $(4, -2)$, we use the coordinates of the points to evaluate $\frac{dy}{dx}$ at those points.

$$\frac{dy}{dx}\bigg|_{(4, 2)} = \frac{1}{2(2)} = \frac{1}{4} \text{ and } \frac{dy}{dx}\bigg|_{(4, -2)} = \frac{1}{2(-2)} = \frac{-1}{4}$$

Figure 11.4 on the following page shows the graph of $y^2 = x$ with tangent lines drawn at $(4, 2)$ and $(4, -2)$.

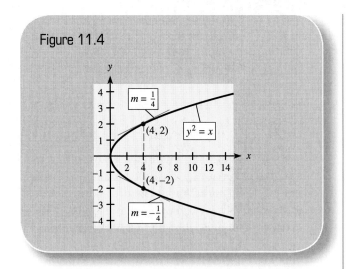

Figure 11.4

Let us now compare the results obtained in Example 11.11 with the results from the derivatives of the two functions $y = \sqrt{x}$ and $y = -\sqrt{x}$, which give the upper and lower braches of $y^2 = x$. The derivatives of these two functions are as follows:

For $y = x^{1/2}$, then $\dfrac{dy}{dx} = \dfrac{1}{2}x^{-1/2} = \dfrac{1}{2\sqrt{x}}$

For $y = -x^{1/2}$, then $\dfrac{dy}{dx} = -\dfrac{1}{2}x^{-1/2} = \dfrac{-1}{2\sqrt{x}}$

We cannot find the slope of a tangent line at $x = 4$ unless we also know the y-coordinate and hence which function and which derivative to use.

At $(4, 2)$ use $y = \sqrt{x}$ so $\dfrac{dy}{dx}\Big|_{(4, 2)} = \dfrac{1}{2\sqrt{4}} = \dfrac{1}{4}$

At $(4, -2)$ use $y = -\sqrt{x}$ so $\dfrac{dy}{dx}\Big|_{(4, -2)} = \dfrac{-1}{2\sqrt{4}} = \dfrac{-1}{4}$

These are the same results that we obtained directly after implicit differentiation.

EXAMPLE 11.12 Implicit Differentiation

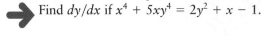 Find dy/dx if $x^4 + 5xy^4 = 2y^2 + x - 1$.

SOLUTION

By viewing $5xy^4$ as the product $(5x)(y^4)$ and differentiating implicitly, we get

$$\frac{d}{dx}(x^4) + \frac{d}{dx}(5xy^4) = \frac{d}{dx}(2y^2) + \frac{d}{dx}(x) - \frac{d}{dx}(1)$$

$$4x^3 + \left[5x\left(4y^3\frac{dy}{dx}\right) + y^4(5)\right] = 4y\frac{dy}{dx} + 1$$

To solve for $\dfrac{dy}{dx}$ we first rewrite the equation with terms containing $\dfrac{dy}{dx}$ on one side and the other terms on the other side.

$$20xy^3\frac{dy}{dx} - 4y\frac{dy}{dx} = 1 - 4x^3 - 5y^4$$

Now $\dfrac{dy}{dx}$ is a factor of one side. To complete the solution, we factor out $\dfrac{dy}{dx}$ and divide both sides by its coefficient.

$$(20xy^3 - 4y)\frac{dy}{dx} = 1 - 4x^3 - 5y^4$$

$$\frac{dy}{dx} = \frac{1 - 4x^3 - 5y^4}{20xy^3 - 4y}$$

EXAMPLE 11.13 Production

Suppose that a company's weekly production output is $384,000 and that this output is related to hours of labor x and dollars of capital investment y by

$$384,000 = 30x^{1/3}y^{2/3}$$

(This relationship is an example of a Cobb-Douglas production function, studied in more detail in Chapter 14.) Find and interpret the rate of change of capital investment with respect to labor hours when labor hours are 512 and capital investment is $64,000.

SOLUTION

The desired rate of change is given by the value of dy/dx when $x = 512$ and $y = 64,000$. Taking the derivative implicitly gives

$$\frac{d}{dx}(384,000) = \frac{d}{dx}(30x^{1/3}y^{2/3})$$

$$0 = 30x^{1/3}\left(\frac{2}{3}y^{-1/3}\frac{dy}{dx}\right) + y^{2/3}(10x^{-2/3})$$

$$0 = \frac{20x^{1/3}}{y^{1/3}} \cdot \frac{dy}{dx} + \frac{10y^{2/3}}{x^{2/3}}$$

$$\frac{-20x^{1/3}}{y^{1/3}} \cdot \frac{dy}{dx} = \frac{10y^{2/3}}{x^{2/3}}$$

Multiplying both sides by $\dfrac{-y^{1/3}}{20x^{1/3}}$ gives

$$\frac{dy}{dx} = \left(\frac{10y^{2/3}}{x^{2/3}}\right)\left(\frac{-y^{1/3}}{20x^{1/3}}\right) = \frac{-y}{2x}$$

When $x = 512$ and $y = 64{,}000$, we obtain

$$\frac{dy}{dx} = \frac{-64{,}000}{2(512)} = -62.5$$

This means that when labor hours are 512 and capital investment is $64,000, if labor hours change by 1 hour, then capital investment could decrease by about $62.50.

EXAMPLE 11.14 Implicit Derivatives and Logarithms and Exponentials

 Find dy/dx if
(a) $\ln xy = 6$
(b) $4x^2 + e^{xy} = 6y$

SOLUTION

(a) Using the properties of logarithms, we have

$$\ln x + \ln y = 6$$

which leads to the implicit derivative

$$\frac{1}{x} + \frac{1}{y}\frac{dy}{dx} = 0$$

Solving gives

$$\frac{1}{y}\frac{dy}{dx} = -\frac{1}{x} \quad \text{or} \quad \frac{dy}{dx} = -\frac{y}{x}$$

(b) We take the derivative of both sides and obtain

$$8x + e^{xy} \cdot \frac{d}{dx}(xy) = 6\frac{dy}{dx}$$

$$8x + e^{xy}\left(x\frac{dy}{dx} + y\right) = 6\frac{dy}{dx}$$

$$8x + xe^{xy}\frac{dy}{dx} + ye^{xy} = 6\frac{dy}{dx}$$

$$8x + ye^{xy} = 6\frac{dy}{dx} - xe^{xy}\frac{dy}{dx}$$

$$\frac{8x + ye^{xy}}{6 - xe^{xy}} = \frac{dy}{dx}$$

11.3 Exercises

1. Find dy/dx for $xy^2 = 8$ at $(2, 2)$. Use implicit differentiation; do not solve for y first.

2. If $x^2 + 2y^2 - 4 = 0$, find dy/dx.

3. If $3x^5 - 5y^3 = 5x^2 + 3y^5$, find dy/dx.

4. If $x^4 + 2x^3y^2 = x - y^3$, find dy/dx.

5. Find the slope of the tangent to the curve $x^2 + 4x + y^2 + 2y - 4 = 0$ at $(1, -1)$.

6. Write the equation of the line tangent to the curve $4x^2 + 3y^2 - 4y - 3 = 0$ at $(-1, 1)$.

7. If $\ln x = y^2$, find dy/dx.

8. Find the slope of the tangent to the curve $y^2 \ln x + x^2y = 3$ at $(1, 3)$.

9. If $ye^x - y = 3$, find dy/dx.

10. Write the equation of the line tangent to the curve $xe^y = 2y + 3$ at $(3, 0)$.

Applications

11. *Advertising and sales* Suppose that a company's sales volume y (in thousands of units) is related to its advertising expenditures x (in thousands of dollars) according to

$$xy - 20x + 10y = 0$$

Find the rate of change of sales volume with respect to advertising expenditures when $x = 10$ (thousand dollars).

12. *Production* Suppose that a company can produce 12,000 units when the number of hours of skilled labor y and unskilled labor x satisfy

$$384 = (x + 1)^{3/4}(y + 2)^{1/3}$$

Find the rate of change of skilled-labor hours with respect to unskilled-labor hours when $x = 255$ and $y = 214$. This can be used to approximate the change in skilled-labor hours required to maintain the same production level when unskilled-labor hours are increased by 1 hour.

13. *Demand* The demand function for q units of a product at p per unit is given by

$$p(q + 1)^2 = 200{,}000$$

Find the rate of change of quantity with respect to price when $p = \$80$. Interpret this result.

Need more practice?
Find more here: cengagebrain.com

11.4 Related Rates

WE HAVE SEEN THAT THE DERIVATIVE REP-
RESENTS THE INSTANTANEOUS RATE OF
CHANGE OF ONE VARIABLE WITH RESPECT
TO ANOTHER. WHEN THE DERIVATIVE IS
TAKEN WITH RESPECT TO TIME, IT REPRE-
SENTS THE RATE AT WHICH THAT VARIABLE
IS CHANGING WITH RESPECT TO TIME (OR
THE VELOCITY).

For example, if distance x is measured in miles and time t in hours, then dx/dt is measured in miles per hour and indicates how fast x is changing. Similarly, if V represents the volume (in cubic feet) of water in a swimming pool and t is time (in minutes), then dV/dt is measured in cubic feet per minute (ft³/min) and might measure the rate at which the pool is being filled with water or being emptied.

Sometimes, two (or more) quantities that depend on time are also related to each other. For example, the height of a tree h (in feet) is related to the radius r (in inches) of its trunk, and this relationship can be modeled by

$$h = kr^{2/3}$$

where k is a constant.* Of course, both h and r are also related to time, so the rates of change dh/dt and dr/dt are related to each other. Thus they are called related rates.

The specific relationship between dh/dt and dr/dt can be found by differentiating $h = kr^{2/3}$ implicitly with respect to time t.

Related rates use
**IMPLICIT
DERIVATIVES**
with respect to time *t*.

*T. McMahon, "Size and Shape in Biology," *Science* 179 (1979): 1201.

EXAMPLE 11.15 Tree Height and Trunk Radius

→ Suppose that for a certain type of tree, the height of the tree (in feet) is related to the radius of its trunk (in inches) by

$$h = 15r^{2/3}$$

Suppose that the rate of change of r is $\frac{3}{4}$ inch per year. Find how fast the height is changing when the radius is 8 inches.

SOLUTION

To find how the rates dh/dt and dr/dt are related, we differentiate $h = 15r^{2/3}$ implicitly with respect to time t.

$$\frac{dh}{dt} = 10r^{-1/3}\frac{dr}{dt}$$

Using $r = 8$ inches and $dr/dt = \frac{3}{4}$ inch per year gives

$$\frac{dh}{dt} = 10(8)^{-1/3}(3/4) = \frac{15}{4} = 3\frac{3}{4} \text{ feet per year}$$

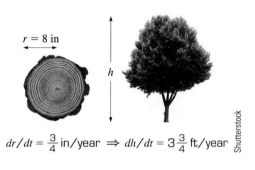

$dr/dt = \frac{3}{4}$ in/year $\Rightarrow$ $dh/dt = 3\frac{3}{4}$ ft/year

Shutterstock

Percent Rates of Change

The work in Example 11.15 shows how to obtain related rates, but the different units (feet per year and inches per year) may be somewhat difficult to interpret. For this reason, many applications in the life sciences deal with **percent rates of change.** The percent rate of change of a quantity is the rate of change of the quantity divided by the quantity.

EXAMPLE 11.16 Blood Flow

→ Poiseuille's law expresses the flow of blood F as a function of the radius r of the vessel according to

$$F = kr^4$$

where k is a constant. When the radius of a blood vessel is restricted, such as by cholesterol deposits, drugs can be administered that will increase the radius of the blood vessel (and hence the blood flow). Find the percent rate of change of the flow of blood that corresponds to the percent rate of change of the radius of a blood vessel caused by the drug.

SOLUTION

We seek the percent rate of change of flow, $(dF/dt)/F$, that results from a given percent rate of change of the radius $(dr/dt)/r$. We first find the related rates of change by differentiating

$$F = kr^4$$

implicitly with respect to time.

$$\frac{dF}{dt} = k\left(4r^3\frac{dr}{dt}\right)$$

Then the percent rate of change of flow can be found by dividing both sides of the equation by F.

$$\frac{\frac{dF}{dt}}{F} = \frac{4kr^3\frac{dr}{dt}}{F}$$

If we replace F on the right side of the equation with kr^4 and reduce, we get

$$\frac{\frac{dF}{dt}}{F} = \frac{4kr^3\frac{dr}{dt}}{kr^4} = 4\left(\frac{\frac{dr}{dt}}{r}\right)$$

Thus we see that the percent rate of change of the flow of blood is 4 times the corresponding percent rate of change of the radius of the blood vessel. This means that a drug that would cause a 12% increase in the radius of a blood vessel at a certain time would produce a corresponding 48% increase in blood flow through that vessel at that time.

Solving Related-Rates Problems

In the preceding examples, the equation relating the time-dependent variables has been given. For some problems, the original equation relating the variables must first be developed from the statement of the problem. These problems can be solved with the aid of the following procedure.

Procedure

To solve related-rates problems:

1. Use geometric and/or physical conditions to write an equation that relates the time-dependent variables.

2. Substitute into the equation values or relationships that are true *at all times*.

3. Differentiate both sides of the equation implicitly with respect to time. This equation is valid for all times.

4. Substitute the values that are known at the instant specified, and solve the equation.

5. Solve for the specified quantity at the given time.

Solving Related Rates Problems

EXAMPLE 11.17 Related Rates

Sand falls at a rate of 5 ft³/min on a conical pile, with the diameter always equal to the height of the pile. At what rate is the height increasing when the pile is 10 ft high?

SOLUTION

Using the Procedure for Solving Related Rates Problems, we have the following.

1. The conical pile has its volume given by

$$V = \frac{1}{3}\pi r^2 h$$

2. The radius $r = \frac{1}{2}h$ at all times, so

$$V = \frac{1}{3}\pi\left[\frac{1}{4}h^2\right]h = \frac{\pi}{12}h^3$$

3. $\dfrac{dV}{dt} = \dfrac{\pi}{12}\left[3h^2\dfrac{dh}{dt}\right] = \dfrac{\pi}{4}h^2\dfrac{dh}{dt}$

4. $\dfrac{dV}{dt} = 5$ at all times, so when $h = 10$,

$$5 = \frac{\pi}{4}(10^2)\frac{dh}{dt}$$

5. $\dfrac{dh}{dt} = \dfrac{20}{100\pi} = \dfrac{1}{5\pi}$ (ft/min) is the rate of change of height.

Note that you should *not* substitute numerical values for any quantity that varies with time until after the derivative is taken. If values are substituted before the derivative is taken, that quantity will have the constant value resulting from the substitution and hence will have a derivative equal to zero.

EXAMPLE 11.18 Hot Air Balloon

→ A hot air balloon has a velocity of 50 ft/min and is flying at a constant height of 500 ft. An observer on the ground is watching the balloon approach. How fast is the distance between the balloon and the observer changing when the balloon is 1000 ft from the observer?

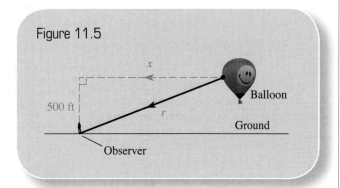

Figure 11.5

SOLUTION

If we let r be the distance between the balloon and the observer and let x be the horizontal distance from the balloon to a point directly above the observer, then we see that these quantities are related by the equation

$$x^2 + 500^2 = r^2 \quad \text{(See Figure 11.5.)}$$

Because the distance x is decreasing, we know that dx/dt must be negative. Thus we are given that $dx/dt = -50$ at all times, and we need to find dr/dt when $r = 1000$. Taking the derivative with respect to t of both sides of the equation $x^2 + 500^2 = r^2$ gives

$$2x \frac{dx}{dt} + 0 = 2r \frac{dr}{dt}$$

Using $dx/dt = -50$ and $r = 1000$, we get

$$2x(-50) = 2000 \frac{dr}{dt}$$

$$\frac{dr}{dt} = \frac{-100x}{2000} = \frac{-x}{20}$$

Using $r = 1000$ in $x^2 + 500^2 = r^2$ gives $x^2 = 750{,}000$. Thus $x = 500\sqrt{3}$, and

$$\frac{dr}{dt} = \frac{-500\sqrt{3}}{20} = -25\sqrt{3} \approx -43.3 \text{ ft/min}$$

The distance is decreasing at 43.3 ft/min.

11.4 Exercises

1. Let $y = x^3 - 3x$. Find dy/dt given that $x = 2$ and $dx/dt = 4$.

2. Assume that x and y are differentiable functions of t and $x^2 + y^2 = 169$. Find dx/dt given that $x = 5, y = 12$, and $dy/dt = 2$.

3. If $x^2 + y^2 = z^2$, find dy/dt when $x = 3, y = 4$, $dx/dt = 10$, and $dz/dt = 2$.

4. The radius of a circle is increasing at a rate of 2 ft/min. At what rate is its area changing when the radius is 3 ft? (Recall that for a circle, $A = \pi r^2$.)

5. The volume of a cube is increasing at a rate of 64 in³/sec. At what rate is the length of each edge of the cube changing when the edges are 6 in long? (Recall that for a cube, $V = x^3$.)

Applications

6. **Profit** Suppose that the daily profit (in dollars) from the production and sale of x units of a product is given by

$$P = 180x - \frac{x^2}{1000} - 2000$$

At what rate per day is the profit changing when the number of units produced and sold is 100 and is increasing at a rate of 10 units per day?

7. **Capital investment and production** Suppose that for a particular product, the number of units x produced per month depends on the number of thousands of dollars y invested, with $x = 30y + 20y^2$. At what rate will production increase if \$10,000 is invested and if the investment capital is increasing at a rate of \$1000 per month?

8. **Allometric relationships—crabs** For fiddler crabs, data gathered by Thompson* show that the

---

*d'Arcy Thompson, *On Growth and Form* (Cambridge, England: Cambridge University Press, 1961).

allometric relationship between the weight C of the claw and the weight W of the body is given by

$$C = 0.11W^{1.54}$$

Find the percent rate of change of the claw weight in terms of the percent rate of change of the body weight for fiddler crabs.

9. *Flight* A plane is flying at a constant altitude of 1 mile and a speed of 300 mph. If it is flying toward an observer on the ground, how fast is the plane approaching the observer when it is 5 miles from the observer?

10. *Distance* Two cars are approaching an intersection on roads that are perpendicular to each other. Car A is north of the intersection and traveling south at 40 mph. Car B is east of the intersection and traveling west at 55 mph. How fast is the distance between the cars changing when car A is 15 miles from the intersection and car B is 8 miles from the intersection?

....

Need more practice?
Find more here: cengagebrain.com

11.5 Applications in Business and Economics

LET'S SEE HOW THE CONCEPTS WE'VE EXPLORED IN THE CHAPTER THUS FAR APPLY TO THE WORLD OF BUSINESS AND ECONOMICS, SPECIFICALLY, IN DETERMINING ELASTICITY OF DEMAND AND MAXIMUM TAX REVENUE.

Elasticity of Demand

We know from the law of demand that consumers will respond to changes in prices; if prices increase, the quantities demanded will decrease. But the degree of responsiveness of the consumers to price changes will vary widely for different products. For example, a price increase in insulin will not greatly decrease the demand for it by diabetics, but a price increase in clothes may cause consumers to buy less and wear their old clothes longer. When the response to price changes is considerable, we say the demand is *elastic*. When price changes cause relatively small changes in demand for a product, the demand is said to be *inelastic* for that product.

Economists measure the **point elasticity of demand** as follows.

> **Elasticity:** The **elasticity of demand** at the point (q_A, p_A) is
>
> $$\eta = -\frac{p}{q} \cdot \frac{dq}{dp}\bigg|_{(q_A, p_A)}$$

EXAMPLE 11.19 Elasticity

Find the elasticity of the demand function $p + 5q = 100$ when (a) the price is \$40, (b) the price is \$60, and (c) the price is \$50.

SOLUTION

Solving the demand function for q gives $q = 20 - \frac{1}{5}p$. Then $dq/dp = -\frac{1}{5}$ and

$$\eta = -\frac{p}{q}\left(-\frac{1}{5}\right)$$

(a) When $p = 40$, $q = 12$ and $\eta = -\frac{p}{q}\left(-\frac{1}{5}\right)\bigg|_{(12, 40)}$
$$= -\frac{40}{12}\left(-\frac{1}{5}\right) = \frac{2}{3}.$$

(b) When $p = 60$, $q = 8$ and $\eta = -\frac{p}{q}\left(-\frac{1}{5}\right)\bigg|_{(8, 60)}$
$$= -\frac{60}{8}\left(-\frac{1}{5}\right) = \frac{3}{2}.$$

(c) When $p = 50$, $q = 10$ and $\eta = -\frac{p}{q}\left(-\frac{1}{5}\right)\bigg|_{(10, 50)}$
$$= -\frac{50}{10}\left(-\frac{1}{5}\right) = 1.$$

Note that in Example 11.19 the demand equation was $p + 5q = 100$, so the demand "curve" is a straight line, with slope $m = -5$. But the elasticity was $\eta = \frac{2}{3}$ at $(12, 40)$, $\eta = \frac{3}{2}$ at $(8, 60)$, and $\eta = 1$ at $(10, 50)$. This illustrates that the elasticity of demand may be different at different points on the demand curve, even though the slope of the demand "curve" is constant. (See Figure 11.6 on the next page.)

Shutterstock

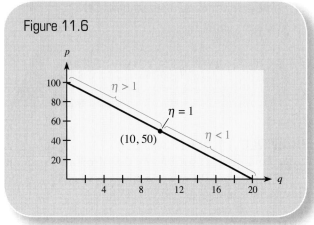

Figure 11.6

Economists use η to measure how responsive demand is to price at different points on the demand curve for a product.

ELASTICITY OF DEMAND

- If $\eta > 1$, the demand is **elastic,** and the percent decrease in demand is greater than the corresponding percent increase in price.
- If $\eta < 1$, the demand is **inelastic,** and the percent decrease in demand is less than the corresponding percent increase in price.
- If $\eta = 1$, the demand is **unitary elastic,** and the percent decrease in demand is approximately equal to the corresponding percent increase in price.

Elasticity and Revenue

Elasticity is related to revenue in a special way and summarized as follows.

ELASTICITY AND REVENUE

The rate of change of revenue R with respect to price p is related to elasticity in the following way.

- Elastic ($\eta > 1$) means $\dfrac{dR}{dp} < 0$

  Hence if price increases, revenue decreases, and if price decreases, revenue increases.

- Inelastic ($\eta < 1$) means $\dfrac{dR}{dp} > 0$

  Hence if price increases, revenue increases, and if price decreases, revenue decreases.

- Unitary elastic ($\eta = 1$) means $\dfrac{dR}{dp} = 0$

  Hence an increase or decrease in price will not change revenue. Revenue is optimized at this point.

See the Enrichment material on CourseMate for MATH APPS at cengagebrain.com for the development of these relationships.

EXAMPLE 11.20 Elasticity and Revenue

The demand for a product is given by

$$p = 10\sqrt{100 - q}, \quad 0 \le q \le 100$$

(a) Find the point at which demand is of unitary elasticity, and find intervals in which the demand is inelastic and in which it is elastic.
(b) Find q where revenue is increasing, where it is decreasing, and where it is maximized.

SOLUTION

The elasticity is

$$\eta = \frac{-p}{q} \cdot \frac{dq}{dp} = -\frac{10\sqrt{100 - q}}{q} \cdot \frac{dq}{dp}$$

Finding dq/dp implicitly from $p = 10\sqrt{(100 - q)}$, we have

$$1 = 10\left[\frac{1}{2}(100 - q)^{-1/2}\left(-\frac{dq}{dp}\right)\right] = \frac{-5}{\sqrt{100 - q}} \cdot \frac{dq}{dp}$$

so

$$\frac{dq}{dp} = \frac{-\sqrt{100 - q}}{5}$$

Thus

$$\eta = -\frac{10\sqrt{100 - q}}{q}\left[\frac{-\sqrt{100 - q}}{5}\right] = \frac{200 - 2q}{q}$$

(a) Unitary elasticity occurs where $\eta = 1$.

$$1 = \frac{200 - 2q}{q}$$
$$q = 200 - 2q$$
$$3q = 200$$
$$q = 66\frac{2}{3}$$

so unitary elasticity occurs when $66\frac{2}{3}$ units are sold, at a price of $57.74. For values of q between 0 and $66\frac{2}{3}$, $\eta > 1$ and demand is elastic. For values of q between $66\frac{2}{3}$ and 100, $\eta < 1$ and demand is inelastic.

(b) When q increases over $0 < q < 66\frac{2}{3}$, p decreases, so $\eta > 1$ means R increases. Similarly, when q increases over $66\frac{2}{3} < q < 100$, p decreases, so $\eta < 1$ means R decreases. Revenue is maximized where $\eta = 1$, at $q = 66\frac{2}{3}$ and $p = 57.74$. See Figure 11.7.

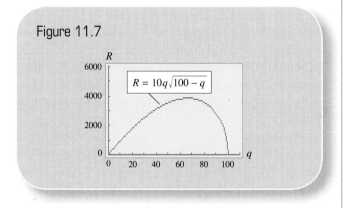

Figure 11.7

$$R = 10q\sqrt{100 - q}$$

Taxation in a Competitive Market

Many taxes imposed by governments are "hidden." That is, the tax is levied on goods produced, and the producers must pay the tax. Of course, the tax becomes a cost to the producers, and they pass that cost on to the consumer in the form of higher prices for goods.

Suppose the government imposes a tax of t dollars on each unit produced and sold by producers. If we are in pure competition in which the consumers' demand depends only on price, the *demand function* will not change. The tax will change the supply function, of course, because at each level of output q, the firm will want to charge a price $p + t$ per unit, where p is the original price per unit and t is the tax per unit.

The graphs of the market demand function, the original market supply function, and the market supply function after taxes are shown in Figure 11.8. Because the tax added to each item is constant, the graph of the new supply function is t units above the original supply function. If $p = f(q)$ defines the original supply function, then $p = f(q) + t$ defines the supply function after taxation.

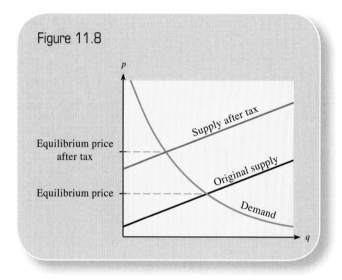

Figure 11.8

Note in this case that after the taxes are imposed, *no* items are supplied at the price that was the equilibrium price before taxation. After the taxes are imposed, the consumers simply have to pay more for the product. Because taxation does not change the demand curve, the quantity purchased at market equilibrium will be less than it was before taxation. Thus governments planning taxes should recognize that they will not collect taxes on the original equilibrium quantity. They will collect on the *new* equilibrium quantity, a quantity reduced by their taxation. Thus a large tax on each item may reduce the quantity demanded at the new market equilibrium so much that very little revenue results from the tax!

If the tax revenue is represented by $T = tq$, where t is the tax per unit and q is the equilibrium quantity of the supply and demand functions after taxation, we can use the following procedure for maximizing the total tax revenue in a competitive market.

Shutterstock

Procedure

To find the tax per item (under pure competition) that will maximize total tax revenue:

1. Write the supply function after taxation.

2. Set the demand function and the new supply function equal, and solve for t in terms of q.

3. Form the total tax revenue function, $T = tq$, by multiplying the expression for t by q, and then take its derivative with respect to q.

4. Set $T' = 0$, and solve for q. This is the q that should maximize T. Use the second-derivative test to verify it.

5. Substitute the value of q in the equation for t (in Step 2). This is the value of t that will maximize T.

Maximizing Total Tax Revenue

EXAMPLE 11.21 Maximizing Total Tax Revenue

→ Suppose the demand and supply functions are given by $p = 600 - q$ and $p = 200 + \frac{1}{3}q$, respectively, where $\$p$ is the price per unit and q is the number of units. Find the tax t that will maximize the total tax revenue T.

SOLUTION

Following the steps in the Procedure for Maximizing Total Tax Revenue, we have the following:

1. $p = 200 + \dfrac{1}{3}q + t$

2. $600 - q = 200 + \dfrac{1}{3}q + t$

   $400 - \dfrac{4}{3}q = t$

3. $T = tq = 400q - \dfrac{4}{3}q^2$

   $T'(q) = \dfrac{dT}{dq} = 400 - \dfrac{8}{3}q$

4. $0 = 400 - \dfrac{8}{3}q$

   $q = 150$

   $T''(q) = -\dfrac{8}{3}$. Thus T is maximized at $q = 150$.

5. $t = 400 - \dfrac{4}{3}(150) = 200$

A tax of \$200 per item will maximize the total tax revenue. The total tax revenue for the period would be $\$200 \cdot (150) = \$30{,}000$.

Note that in the example just given, if a tax of \$300 were imposed, market equilibrium would occur at $q = 75$, and the total tax revenue the government would receive would be

$$(\$300)(75) = \$22{,}500$$

Thus, with a tax of \$300 per unit rather than \$200, the government's tax revenue would be \$22,500 rather than \$30,000. In addition, with a \$300 tax, suppliers would sell only 75 units rather than 150, and consumers would pay $p = 200 + 25 + 300 = \$525$ per item, which is \$75 more than the price with a \$200 tax. Thus everyone would suffer if the tax rate were raised to \$300.

An infamous example of a tax increase that resulted in decreased tax revenue and economic disaster is the "luxury tax" enacted in 1991. This was a 10% excise tax on the sale of expensive jewelry, furs, airplanes, certain expensive boats, and automobiles costing more than \$30,000. The Congressional Joint Tax Committee had estimated that the luxury tax would raise \$6 million from airplanes alone, but it raised only \$53,000 while it destroyed the small-airplane market (one company lost \$130 million and 480 jobs in a single year). It also capsized the boat market. The luxury tax was repealed at the end of 1993 (except for automobiles).[*]

11.5 Exercises

In Problems 1 and 2, p is in dollars and q is the number of units.

1. (a) Find the elasticity of the demand function $p^2 + 2p + q = 49$ at $p = 6$.
   (b) How will a price increase affect total revenue?

*Fortune, September 6, 1993; Motor Trend, December 1993.

2. Suppose that the demand for a product is given by $pq + p + 100q = 50{,}000$.
   (a) Find the elasticity when $p = \$401$.
   (b) Tell what type of elasticity this is.
   (c) How would a price increase affect revenue?

In Problems 3 and 4, p is the price per unit in dollars and q is the number of units.

3. If the demand and supply functions for a product are $p = 800 - 2q$ and $p = 100 + 0.5q$, respectively, find the tax per unit t that will maximize the tax revenue T.

4. Suppose the weekly demand for a product is given by $p + 2q = 840$ and the weekly supply before taxation is given by $p = 0.02q^2 + 0.55q + 7.4$. Find the tax per item that produces maximum tax revenue. Find the tax revenue.

··

Need more practice?
Find more here: cengagebrain.com

Chapter Exercises

In Problems 1–10, find the indicated derivative.

1. If $y = e^{3x^2 - x}$, find dy/dx.

2. If $p = \ln\left(\dfrac{q}{q^2 - 1}\right)$, find $\dfrac{dp}{dq}$.

3. If $y = xe^{x^2}$, find dy/dx.

4. If $f(x) = 5e^{2x} - 40e^{-0.1x} + 11$, find $f'(x)$.

5. If $g(x) = (2e^{3x + 1} - 5)^3$, find $g'(x)$.

6. If $y = \ln(3x^4 + 7x^2 - 12)$, find dy/dx.

7. If $s = \dfrac{3}{4}\ln(x^{12} - 2x^4 + 5)$, find ds/dx.

8. If $y = 3^{3x - 4}$, find dy/dx.

9. If $y = 1 + \log_8(x^{10})$, find dy/dx.

10. If $y = \dfrac{\ln x}{x}$, find $\dfrac{dy}{dx}$.

11. Write the equation of the line tangent to $y = 4e^{x^3}$ at $x = 1$.

12. Write the equation of the line tangent to $y = x \ln x$ at $x = 1$.

In Problems 13–17, find the indicated derivative.

13. If $y \ln x = 5y^2 + 11$, find dy/dx.

14. Find dy/dx for $e^{xy} = y$.

15. Find dy/dx for $x^2 + 3y^2 + 2x - 3y + 2 = 0$.

16. Find dy/dx for $3x^2 + 2x^3y^2 - y^5 = 7$.

17. Find the second derivative y'' if $x^2 + y^2 = 1$.

18. Find the slope of the tangent to the curve $x^2 + 4x - 3y^2 + 6 = 0$ at $(3, 3)$.

19. Suppose $3x^2 - 2y^3 = 10y$, where x and y are differentiable functions of t. If $dx/dt = 2$, find dy/dt when $x = 10$ and $y = 5$.

20. A right triangle with legs of lengths x and y has its area given by

$$A = \frac{1}{2}xy$$

If the rate of change of x is 2 units per minute and the rate of change of y is 5 units per minute, find the rate of change of the area when $x = 4$ and $y = 1$.

Applications

21. **Deforestation** One of the major causes of rain forest deforestation is agricultural and residential development. The number of hectares destroyed in a particular year t can be modeled by

$$y = -3.91435 + 2.62196 \ln t$$

where $t = 0$ represents 1950.
   (a) Find $y'(t)$.
   (b) Find and interpret $y(50)$ and $y'(50)$.

22. **Nonmarital childbearing** The percent of live births to unmarried mothers for the years 1970–2007 can be modeled by the function

$$y = -33.410 + 18.035 \ln x$$

where x is the number of years past 1960.
   (a) What does the model predict the rate of change of this percent to be in 2015?
   (b) Is the percent of live births to unmarried mothers predicted to increase or decrease after 2015?

23. **Compound interest** If the future value of $1000 invested for n years at 12%, compounded continuously, is given by

$$S = 1000e^{0.12n} \text{ dollars}$$

find the rate at which the future value is growing after 1 year.

24. **Marginal cost** The average cost of producing x units of a product is $\overline{C} = 600e^{x/600}$ dollars per unit. What is the marginal cost when 600 units are produced?

25. **Inflation** The impact of inflation on a $20,000 pension can be measured by the purchasing power P of $20,000 after t years. For an inflation rate of 5% per year, compounded annually, P is given by

$$P = 20{,}000e^{-0.0495t}$$

At what rate is purchasing power changing when $t = 10$? (*Source: Viewpoints, VALIC*)

26. **Evaporation** A spherical droplet of water evaporates at a rate of 1 mm³/min. Find the rate of change of the radius when the droplet has a radius of 2.5 mm.

27. **Worker safety** A sign is being lowered over the side of a building at the rate of 2 ft/min. A worker handling a guide line is 7 ft away from a spot directly below the sign. How fast is the worker taking in the guide line at the instant the sign is 25 ft from the worker's hands? See the figure.

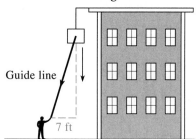

28. **Environment** Suppose that in a study of water birds, the relationship between the area A of wetlands (in square miles) and the number of different species S of birds found in the area was determined to be

$$S = kA^{1/3}$$

where k is constant. Find the percent rate of change of the number of species in terms of the percent rate of change of the area.

29. **Taxes** Can increasing the tax per unit sold actually lead to a decrease in tax revenues?

30. **Taxes** Suppose the demand and supply functions for a product are

$$p = 2800 - 8q - \frac{q^2}{3} \quad \text{and} \quad p = 400 + 2q$$

respectively, where p is in dollars and q is the number of units. Find the tax per unit t that will maximize the tax revenue T, and find the maximum tax revenue.

31. **Taxes** Suppose the supply and demand functions for a product are

$$p = 40 + 20q \quad \text{and} \quad p = \frac{5000}{q + 1}$$

respectively, where p is in dollars and q is the number of units. Find the tax t that maximizes the tax revenue T, and find the maximum tax revenue.

32. **Elasticity** Suppose the demand for a product is given by

$$p^2(2q + 1) = 10{,}000$$

where p is in dollars and q is the number of units. Find the elasticity of demand when $p = \$20$.

33. **Elasticity** Suppose the weekly demand function for a product is given by

$$p = 100e^{-0.1q}$$

where p is the price in dollars and q is the number of tons demanded. What is the elasticity of demand when the price is $36.79 and the quantity demanded is 10?

34. **Revenue** A product has the demand function

$$p = 100 - 0.5q$$

where p is in dollars and q is the number of units.
(a) Find the elasticity $\eta(q)$ as a function of q, and graph the function

$$f(q) = \eta(q)$$

(b) Find where $f(q) = 1$, which gives the quantity for which the product has unitary elasticity.
(c) The revenue function for this product is

$$R(q) = pq = (100 - 0.5q)q$$

Graph $R(q)$ and find the q-value for which the maximum revenue occurs.
(d) What is the relationship between elasticity and maximum revenue?

Need more practice?
Find more here: cengagebrain.com

TURN TO
A TRUSTED
RESOURCE

With the Student Solutions Manual, now available online, you will have all the learning resources you need for your course in one convenient place, at no additional charge!

Solutions to all odd-numbered exercises, organized by chapter, are easily found online.

Simply visit **login.cengagebrain.com** to access this trusted resource.

Indefinite
Integrals

When we know the derivative of a function, it is often useful to determine the function itself. For example, accountants can use linear regression to translate information about marginal cost into a linear equation defining (approximately) the marginal cost function and then use the process of **antidifferentiation** (or **integration**) as part of finding the (approximate) total cost function. We can also use integration to find total revenue functions from marginal revenue functions, to optimize profit from information about marginal cost and marginal revenue, and to find national consumption from information about marginal propensity to consume.

Integration can also be used in the social and life sciences to predict growth or decay from expressions giving rates of change. For example, we can find equations for population size from the rate of change of growth, we can write equations for the number of radioactive atoms remaining in a substance if we know the rate of decay of the substance, and we can determine the volume of blood flow from information about rate of flow.

objectives

12.1 The Indefinite Integral

12.2 The Power Rule

12.3 Integrals Involving Exponential and Logarithmic Functions

12.4 The Indefinite Integral in Business and Economics

12.5 Differential Equations

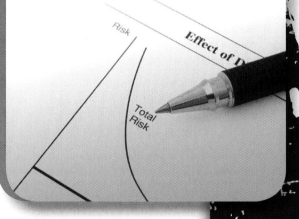

12.1 The Indefinite Integral

WE HAVE STUDIED PROCEDURES FOR AND APPLICATIONS OF FINDING DERIVATIVES OF A GIVEN FUNCTION. WE NOW TURN OUR ATTENTION TO REVERSING THIS PROCESS OF DIFFERENTIATION. WHEN WE KNOW THE

DERIVATIVE OF A FUNCTION, THE PROCESS OF FINDING THE FUNCTION ITSELF IS CALLED **ANTIDIFFERENTIATION.**

For example, if the derivative of a function is $2x$, we know that the function could be $f(x) = x^2$ because $\frac{d}{dx}(x^2) = 2x$. But the function could also be $f(x) = x^2 + 4$ because $\frac{d}{dx}(x^2 + 4) = 2x$. It is clear that any function of the form $f(x) = x^2 + C$, where C is an arbitrary constant, will have $f'(x) = 2x$ as its derivative. Thus we say that the **general anti-derivative** of $f'(x) = 2x$ is $f(x) = x^2 + C$, where C is an arbitrary constant.

Shutterstock

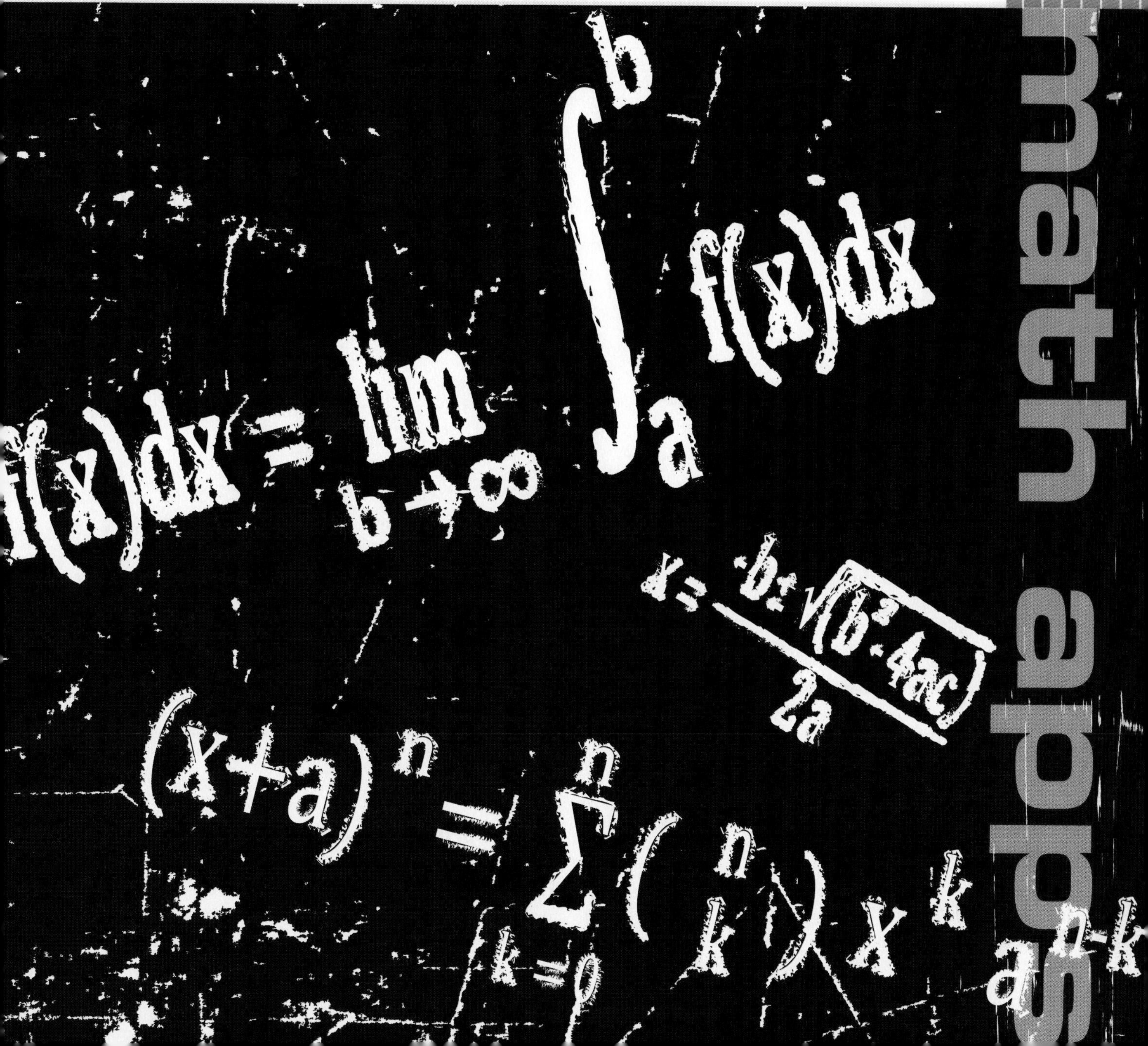

math apps

The process of finding an antiderivative is also called **integration**. The function that results when integration takes place is called an **indefinite integral** or, more simply, an **integral**. We can denote the indefinite integral (that is, the general antiderivative) of a function $f(x)$ by $\int f(x)dx$. Thus we can write $\int 2x\, dx$ to indicate the general antiderivative of the function $f(x) = 2x$. The expression is read as "the integral of $2x$ with respect to x." In this case, $2x$ is called the **integrand**. The **integral sign**, $\int$, indicates the process of integration, and the dx indicates that the integral is to be taken with respect to x. Because the antiderivative of $2x$ is $x^2 + C$, we can write

$$\int 2x\, dx = x^2 + C$$

EXAMPLE 12.1 Integration

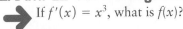 If $f'(x) = x^3$, what is $f(x)$?

SOLUTION

We know that $\dfrac{d}{dx}(x^4) = 4x^3$, so the derivative of $f(x) = \frac{1}{4}x^4$ is $f'(x) = x^3$. Thus

$$f(x) = \int f'(x)\, dx = \int x^3 dx = \tfrac{1}{4}x^4 + C$$

Powers of x Formula

It is easily seen that

$$\int x^4 dx = \frac{x^5}{5} + C \text{ because } \frac{d}{dx}\left(\frac{x^5}{5} + C\right) = x^4$$

$$\int x^5 dx = \frac{x^6}{6} + C \text{ because } \frac{d}{dx}\left(\frac{x^6}{6} + C\right) = x^5$$

In general, we have the following.

Powers of x Formula:
$$\int x^n dx = \frac{x^{n+1}}{n+1} + C \quad (\text{for } n \neq -1)$$

In the Powers of x Formula, we see that $n \neq -1$ is essential, because if $n = -1$, then the denominator $n + 1 = 0$. We will discuss the case when $n = -1$ later. (Can you think what function has $1/x$ as its derivative?)

In addition, we can see that this Powers of x Formula applies for any $n \neq -1$ by noting that

$$\frac{d}{dx}\left(\frac{x^{n+1}}{n+1} + C\right) = \frac{d}{dx}\left(\frac{1}{n+1}x^{n+1} + C\right) = \frac{n+1}{n+1}x^n = x^n$$

EXAMPLE 12.2 Powers of x Formula

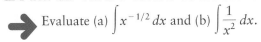 Evaluate (a) $\int x^{-1/2}\, dx$ and (b) $\int \dfrac{1}{x^2}\, dx$.

SOLUTION

(a) Using the formula, we get

$$\int x^{-1/2}\, dx = \frac{x^{(-1/2)+1}}{(-1/2)+1} + C = \frac{x^{1/2}}{1/2} + C = 2x^{1/2} + C$$

We can check by noting that the derivative of $2x^{1/2} + C$ is $x^{-1/2}$. ✔

(b) We write the power of x in the numerator so that the integral has the form in the Powers of x Formula.

$$\int \frac{1}{x^2}\, dx = \int x^{-2}\, dx = \frac{x^{-2+1}}{-2+1} + C = \frac{x^{-1}}{-1} + C = \frac{-1}{x} + C$$

Note that the indefinite integral in Example 12.2(a) is a function (actually a number of functions, one for each value of C). Graphs of several members of this family of functions are shown in Figure 12.1. Note that at any given x-value, the tangent line to each curve has the same slope, indicating that all family members have the same derivative. These graphs also can be generated with a graphing calculator. See the Tech Card for details.

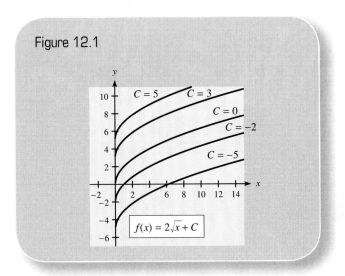

Figure 12.1

$f(x) = 2\sqrt{x} + C$

Other Formulas and Properties

Other formulas will be useful in evaluating integrals. The following chart shows how some new integration formulas result from differentiation formulas.

Integration Formulas

Derivative	Resulting Integral
$\dfrac{d}{dx}(x) = 1$	$\displaystyle\int 1\,dx = \int dx = x + C$
$\dfrac{d}{dx}[c \cdot u(x)] = c \cdot \dfrac{d}{dx}u(x)$	$\displaystyle\int c\,u(x)\,dx = c\int u(x)\,dx$
$\dfrac{d}{dx}[u(x) \pm v(x)] = \dfrac{d}{dx}u(x) \pm \dfrac{d}{dx}v(x)$	$\displaystyle\int [u(x) \pm v(x)]\,dx = \int u(x)\,dx \pm \int v(x)\,dx$

EXAMPLE 12.3 Using Integration Formulas

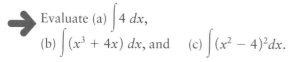

Evaluate (a) $\displaystyle\int 4\,dx$, (b) $\displaystyle\int (x^3 + 4x)\,dx$, and (c) $\displaystyle\int (x^2 - 4)^2\,dx$.

SOLUTION

(a) $\displaystyle\int 4\,dx = 4\int dx = 4(x + C_1) = 4x + C$

(Because C_1 is an unknown constant, we can write $4C_1$ as the unknown constant C.)

(b) $\displaystyle\int (x^3 + 4x)\,dx = \int x^3\,dx + \int 4x\,dx$

$$= \left(\frac{x^4}{4} + C_1\right) + \left(4 \cdot \frac{x^2}{2} + C_2\right)$$

$$= \frac{x^4}{4} + 2x^2 + C_1 + C_2$$

$$= \frac{x^4}{4} + 2x^2 + C$$

Note that we need only one constant because the sum of C_1 and C_2 is just a new constant.

(c) We expand $(x^2 - 4)^2$ so that the integrand is in a form that fits the basic integration formulas.

$$\int (x^2 - 4)^2\,dx = \int (x^4 - 8x^2 + 16)\,dx = \frac{x^5}{5} - \frac{8x^3}{3} + 16x + C$$

EXAMPLE 12.4 Revenue

Sales records at Jarus Technologies show that the rate of change of the revenue (that is, the marginal revenue) in dollars per unit for a motherboard is $\underline{MR} = 300 - 0.2x$, where x represents the quantity sold. Find the total revenue function for the product. Then find the total revenue from the sale of 1000 motherboards.

SOLUTION

We know that the marginal revenue can be found by differentiating the total revenue function. That is,

$$R'(x) = 300 - 0.2x$$

Thus integrating the marginal revenue function gives the total revenue function.

$$R(x) = \int (300 - 0.2x)\,dx = 300x - 0.1x^2 + K*$$

We can use the fact that there is no revenue when no units are sold to evaluate K. Setting $x = 0$ and $R = 0$ gives $0 = 300(0) - 0.1(0)^2 + K$, so $K = 0$. Thus the total revenue function is

$$R(x) = 300x - 0.1x^2$$

*Here we are using K rather than C to represent the constant of integration to avoid confusion between the constant C and the cost function $C = C(x)$.

The total revenue from the sale of 1000 motherboards is

$$R(1000) = 300(1000) - 0.1(1000^2) = \$200,000$$

The numerical integration feature of a graphing calculator can be used to check the integral of a function. See the Tech Card for details.

12.1 Exercises

1. If $f'(x) = x^6$, what is $f(x)$?

In Problems 2–8, evaluate each integral. Check your answers by differentiating.

2. $\int 8x^5 \, dx$

3. $\int (3 - x^{3/2}) \, dx$

4. $\int (x^4 - 9x^2 + 3) \, dx$

5. $\int (2 + 2\sqrt{x}) \, dx$

6. $\int \dfrac{5}{x^4} \, dx$

7. $\int \left(x^3 - 4 + \dfrac{5}{x^6} \right) dx$

8. $\int \left(x^9 - \dfrac{1}{x^3} + \dfrac{2}{\sqrt[3]{x}} \right) dx$

9. Given $\int \dfrac{x + 1}{x^3} \, dx$, use algebra to rewrite the integrand and then integrate and simplify.

10. Given $F(x) = x^3 - 3x^2 + C$ and its graph as shown, write the indefinite integral that gives the function.

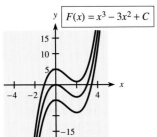

Applications

11. **Revenue** If the marginal revenue (in dollars per unit) for a month is given by $\overline{MR} = -0.3x + 450$,

what is the total revenue from the production and sale of 50 units?

12. **Average cost** The DeWitt Company has found that the rate of change of its average cost for a product is

$$\overline{C}'(x) = \frac{1}{4} - \frac{100}{x^2}$$

where x is the number of units and cost is in dollars. The average cost of producing 20 units is $40.00.
(a) Find the average cost function for the product.
(b) Find the average cost of 100 units of the product.

13. **National health care** National expenditures for health care H (in billions of dollars) have risen dramatically in the past 20 years. By using U.S. Department of Health and Human Services, Center for Medicare and Medical Services data (actual and projected) from 2000 to 2015, the rate of change of expenditures can be modeled by

$$\frac{dH}{dt} = 10.066t + 100.5$$

billions of dollars per year, where $t = 0$ represents the year 2000.
(a) Find the function that models the national expenditures for health care if expenditures in 2009 were $2689 billion.
(b) Use the model from part (a) to predict the national health care expenditures for 2015.

14. **U.S. population** With U.S. Census Bureau data (actual and projected) for selected years from 1960 to 2050, the rate of change of U.S. population P can be modeled by

$$\frac{dP}{dt} = -0.0002187t^2 + 0.0276t + 1.98$$

million people per year, where $t = 0$ represents the year 1960.
(a) In what year does this rate of change reach its maximum?
(b) In 1960, the U.S. population was 181 million. Use this to find a model for $P(t)$.
(c) For 2025, the Census Bureau's predicted U.S. population is 348 million. What does the model predict?

Need more practice?
Find more here: cengagebrain.com

12.2 The Power Rule

OUR GOAL IN THIS SECTION IS TO EXTEND THE POWERS OF *X* FORMULA,

$$\int x^n \, dx = \frac{x^{n+1}}{n+1} + C, \, n \neq -1$$

TO POWERS OF A FUNCTION OF *X*. IN ORDER TO DO THIS, WE MUST UNDERSTAND THE SYMBOL *DX*.

Recall that the derivative of $y = f(x)$ with respect to x can be denoted by dy/dx. As we will see, there are advantages to using dy and dx as separate quantities whose ratio dy/dx equals $f'(x)$.

> **Differentials:** If $y = f(x)$ is a differentiable function with derivative $dy/dx = f'(x)$, then the **differential of x** is dx, and the **differential of y** is dy, where
>
> $$dy = f'(x) \, dx$$

Although differentials are useful in certain approximation problems, we are interested in the differential notation at this time.

EXAMPLE 12.5 Differential

Find (a) dy if $y = x^3 - 4x^2 + 5$ and (b) du if $u = 5x^4 + 11$.

SOLUTION

(a) $dy = f'(x) \, dx = (3x^2 - 8x) \, dx$

(b) If the dependent variable in a function is u, then $du = u'(x) \, dx$.

$$du = u'(x) \, dx = 20x^3 \, dx$$

The Power Rule

In terms of our goal of extending the Powers of x Formula, it turns out that if x is replaced by a function of x, then dx should be replaced by the differential of that function.

> **Power Rule for Integration:**
>
> $$\int [u(x)]^n \cdot u'(x) \, dx = \frac{[u(x)]^{n+1}}{n+1} + C, \, n \neq -1$$

Using the fact that

$$du = u'(x) \, dx \quad \text{or} \quad du = u' \, dx$$

we can write the Power Rule in the following alternative form.

> **Power Rule (Alternative Form):**
> If $u = u(x)$, then
>
> $$\int u^n \, du = \frac{u^{n+1}}{n+1} + C, \, n \neq -1$$

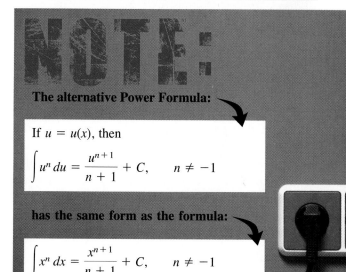

The alternative Power Formula:

If $u = u(x)$, then

$$\int u^n \, du = \frac{u^{n+1}}{n+1} + C, \quad n \neq -1$$

has the same form as the formula:

$$\int x^n \, dx = \frac{x^{n+1}}{n+1} + C, \quad n \neq -1$$

with the *function u substituted for x and du substituted for dx.*

EXAMPLE 12.6 Power Rule

Evaluate $\int (3x^2 + 4)^5 \cdot 6x \, dx$.

SOLUTION

To use the Power Rule, we must be sure that we have the function $u(x)$, its derivative $u'(x)$, and n. If $u = 3x^2 + 4$ and $n = 5$, then $u' = 6x$. All required parts are present, so the integral is of the form

$$\int (3x^2 + 4)^5 \, 6x \, dx = \int u^5 \cdot u' \, dx = \int u^5 \, du$$

$$= \frac{u^6}{6} + C = \frac{(3x^2 + 4)^6}{6} + C$$

We can check the integration by noting that the derivative of

$$\frac{(3x^2 + 4)^6}{6} + C \text{ is } (3x^2 + 4)^5 \cdot 6x$$

EXAMPLE 12.7 Power Rule

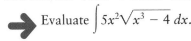 Evaluate $\int 5x^2 \sqrt{x^3 - 4}\, dx$.

SOLUTION

If we let $u = x^3 - 4$, then $u' = 3x^2$. Thus we need the factor 3, rather than 5, in the integrand. If we first reorder the factors and then multiply by the constant factor 3 (and divide it out), we have

$$\int \sqrt{x^3 - 4} \cdot 5x^2\, dx = \int \sqrt{x^3 - 4} \cdot \frac{5}{3}(3x^2)\, dx$$

$$= \frac{5}{3}\int (x^3 - 4)^{1/2} \cdot 3x^2\, dx$$

This integral is of the form $\frac{5}{3}\int u^{1/2} \cdot u'\, dx$, resulting in

$$\frac{5}{3} \cdot \frac{u^{3/2}}{3/2} + C = \frac{5}{3} \cdot \frac{(x^3 - 4)^{3/2}}{3/2} + C = \frac{10}{9}(x^3 - 4)^{3/2} + C$$

Note that we can factor a constant outside the integral sign to obtain the integrand in the form we seek, but if the integral requires the introduction of a variable to obtain the form $u^n \cdot u'\, dx$, we *cannot* use this form and must try something else.

EXAMPLE 12.8 Power Rule Fails

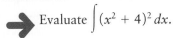 Evaluate $\int (x^2 + 4)^2\, dx$.

SOLUTION

If we let $u = x^2 + 4$, then $u' = 2x$. Because we would have to introduce a variable to get u' in the integral, we cannot solve this problem by using the Power Rule. We must find another method. We can evaluate this integral by squaring and then integrating term by term.

$$\int (x^2 + 4)^2\, dx = \int (x^4 + 8x^2 + 16)\, dx$$

$$= \frac{x^5}{5} + \frac{8x^3}{3} + 16x + C$$

Note that if we tried to introduce the factor $2x$ into the integral in Example 12.8, we would get

$$\int (x^2 + 4)^2\, dx = \int (x^2 + 4)^2 \cdot \frac{1}{2x}(2x)\, dx$$

Although it is tempting to factor $1/2x$ outside the integral and use the Power Rule, this leads to an "answer" that does not check.

EXAMPLE 12.9 Power Rule

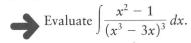 Evaluate $\int \frac{x^2 - 1}{(x^3 - 3x)^3}\, dx$.

SOLUTION

This integral can be treated as $\int u^{-3}\, u'\, dx$ if we let $u = x^3 - 3x$.

$$\int \frac{x^2 - 1}{(x^3 - 3x)^3}\, dx = \int (x^3 - 3x)^{-3}(x^2 - 1)\, dx$$

Then $u' = 3x^2 - 3$ and we can multiply and divide by 3 to get the form we need.

$$= \int (x^3 - 3x)^{-3} \cdot \frac{1}{3} \cdot 3(x^2 - 1)\, dx$$

$$= \frac{1}{3}\int (x^3 - 3x)^{-3}(3x^2 - 3)\, dx$$

$$= \frac{1}{3}\left[\frac{(x^3 - 3x)^{-2}}{-2} \right] + C$$

$$= \frac{-1}{6(x^3 - 3x)^2} + C$$

Recall that graphing calculators can be used to check integrals and to graph some members of the family of functions given by an indefinite integral. See the Tech Card for details.

EXAMPLE 12.10 Revenue

Suppose that the marginal revenue for a product is given by

$$\overline{MR} = \frac{600}{\sqrt{3x + 1}} + 2$$

Find the total revenue function.

SOLUTION

$$R(x) = \int \overline{MR}\, dx = \int \left[\frac{600}{(3x + 1)^{1/2}} + 2 \right] dx$$

$$= \int 600(3x + 1)^{-1/2}\, dx + \int 2\, dx$$

$$= 600\left(\frac{1}{3}\right)\int (3x + 1)^{-1/2}(3\, dx) + 2\int dx$$

$$= 200\frac{(3x + 1)^{1/2}}{1/2} + 2x + K$$

$$= 400\sqrt{3x + 1} + 2x + K$$

We know that $R(0) = 0$, so we have

$$0 = 400\sqrt{1} + 0 + K \text{ or } K = -400$$

Thus the total revenue function is

$$R(x) = 400\sqrt{3x + 1} + 2x - 400$$

Note in Example 12.10 that even though $R(0) = 0$, the constant of integration K was *not* 0. This is because $x = 0$ does not necessarily mean that $u(x)$ will also be 0.

Note that there are no integration formulas that correspond to "reversing" the derivative formulas for a product or for a quotient. This means that functions that may be easy to differentiate can be quite difficult (or even impossible) to integrate. Hence, in general, integration is a more difficult process than differentiation.

12.2 Exercises

1. Given $u = 2x^5 + 9$, find the differential du.

In Problems 2–8, evaluate each integral. Check your results by differentiation.

2. $\displaystyle\int (5x^3 + 11)^4 \, 15x^2 \, dx$

3. $\displaystyle\int 4x^3(7x^4 + 12)^3 \, dx$

4. $\displaystyle\int 8x^5(4x^6 + 15)^{-3} \, dx$

5. $\displaystyle\int 2(x^3 - 1)(x^4 - 4x + 3)^{-5} \, dx$

6. $\displaystyle\int 7x^3\sqrt{x^4 + 6} \, dx$

7. $\displaystyle\int (x^3 + 1)^2(3x \, dx)$

8. $\displaystyle\int \frac{3x^4 \, dx}{(2x^5 - 5)^4}$

Applications

9. **Revenue** Suppose that the marginal revenue for a product is given by

$$\overline{MR} = \frac{-30}{(2x + 1)^2} + 30$$

where x is the number of units and revenue is in dollars. Find the total revenue.

10. **Data entry speed** The rate of change in data entry speed of a typical student is $ds/dx = 5(x + 1)^{-1/2}$, where x is the number of lessons the student has had and s is in entries per minute.
   (a) Find the data entry speed as a function of the number of lessons if a typical student can complete 10 entries per minute with no lessons ($x = 0$).
   (b) How many entries per minute can a typical student complete after 24 lessons?

11. **Film attendance** An excellent film with a very small advertising budget must depend largely on word-of-mouth advertising. In this case, the rate at which weekly attendance might grow can be given by

$$\frac{dA}{dt} = \frac{-100}{(t + 10)^2} + \frac{2000}{(t + 10)^3}$$

where t is the time in weeks since release and A is attendance in millions.
   (a) Find the function that describes weekly attendance at this film.
   (b) Find the attendance at this film in the tenth week.

Need more practice?
Find more here: cengagebrain.com

12.3 Integrals Involving Exponential and Logarithmic Functions

IN THIS SECTION, WE CONSIDER INTEGRATION FORMULAS THAT RESULT IN NATURAL LOGARITHMS AND FORMULAS FOR INTEGRATING EXPONENTIALS.

Integrals Involving Exponential Functions

We know that

$$\frac{d}{dx}(e^x) = e^x \quad \text{and} \quad \frac{d}{dx}(e^u) = e^u \cdot u'$$

The corresponding integrals are given by the following.

> **Exponential Formula:** If u is a function of x,
>
> $$\int e^u \cdot u' \, dx = \int e^u \, du = e^u + C$$
>
> In particular, $\displaystyle\int e^x \, dx = e^x + C$.

EXAMPLE 12.11 Integrals of Exponentials

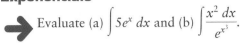

Evaluate (a) $\int 5e^x \, dx$ and (b) $\int \dfrac{x^2 \, dx}{e^{x^3}}$.

SOLUTION

(a) $\int 5e^x \, dx = 5\int e^x \, dx = 5e^x + C$

(b) In order to use $\int e^u \cdot u' \, dx$, we write the exponential in the numerator. Thus

$$\int \frac{x^2 \, dx}{e^{x^3}} = \int e^{-x^3}(x^2 \, dx)$$

This is *almost* of the form $\int e^u \cdot u' \, dx$. Letting $u = -x^3$ gives $u' = -3x^2$. Thus

$$\int e^{-x^3}(x^2 \, dx) = -\frac{1}{3}\int e^{-x^3}(-3x^2 \, dx)$$

$$= -\frac{1}{3}e^{-x^3} + C = \frac{-1}{3e^{x^3}} + C$$

EXAMPLE 12.12 Real Estate Inflation

As the housing market pulls out of the collapse of 2008, suppose the rate of change of the value of a home in a resort area that cost \$200,000 in 2010 is predicted to be modeled by

$$\frac{dV}{dt} = 15.4e^{0.077t}$$

where V is the market value of the home in thousands of dollars and t is the time in years since 2010.

(a) Find the function that expresses the value V in terms of t.

(b) Find the predicted value in 2020 (after 10 years).

SOLUTION

(a) $V = \int \dfrac{dV}{dt}\, dt = \int 15.4e^{0.077t}\, dt$

$$= 15.4\int e^{0.077t}\left(\frac{1}{0.077}\right)(0.077\, dt)$$

$$V = 15.4\left(\frac{1}{0.077}\right)\int e^{0.077t}(0.077\, dt) = 200e^{0.077t} + C$$

Using $V = 200$ (thousand) when $t = 0$, we have $200 = 200 + C$, so $0 = C$.

Thus we have the value as a function of time given by $V = 200e^{0.077t}$.

(b) The predicted value after 10 years is found by using $t = 10$.

$$V = 200e^{0.077(10)} = 200e^{0.77} \approx 431.95$$

Thus, in 2020, the predicted value of the home is \$431,950.

Integrals Involving Logarithmic Functions

Recall that the Power Rule for integrals applies only if $n \neq -1$. That is,

$$\int u^n u' \, dx = \frac{u^{n+1}}{n+1} + C \text{ if } n \neq -1$$

The following formula applies when $n = -1$.

Logarithmic Formula: If u is a function of x, then

$$\int u^{-1} u' \, dx = \int \frac{u'}{u}\, dx = \int \frac{1}{u}\, du = \ln|u| + C$$

In particular, $\int \dfrac{1}{x}\, dx = \ln|x| + C$.

EXAMPLE 12.13 Integrals Resulting in Logarithmic Functions

Evaluate $\int \dfrac{4}{4x + 8}\, dx$.

SOLUTION

This integral is of the form

$$\int \frac{u'}{u}\, dx = \ln|u| + C$$

with $u = 4x + 8$ and $u' = 4$. Thus

$$\int \frac{4}{4x + 8}\, dx = \ln|4x + 8| + C$$

EXAMPLE 12.14 Integral of *du/u*

Evaluate $\int \dfrac{x - 3}{x^2 - 6x + 1}\, dx$.

SOLUTION

This integral is of the form $\int (u'/u)\, dx$, *almost*. If we let $u = x^2 - 6x + 1$, then $u' = 2x - 6$. If we multiply (and divide) the numerator by 2, we get

$$\int \frac{x - 3}{x^2 - 6x + 1}\, dx = \frac{1}{2}\int \frac{2(x - 3)}{x^2 - 6x + 1}\, dx$$

$$= \frac{1}{2}\int \frac{2x - 6}{x^2 - 6x + 1}\, dx$$

$$= \frac{1}{2}\int \frac{u'}{u}\, dx = \frac{1}{2}\ln |u| + C$$

$$= \frac{1}{2}\ln |x^2 - 6x + 1| + C$$

EXAMPLE 12.15 Population Growth

Because the world contains only about 10 billion acres of arable land, world population is limited. Suppose that world population is limited to 40 billion people and that the rate of population growth per year is given by

$$\frac{dP}{dt} = k(40 - P)$$

where P is the population in billions at time t and k is a positive constant. Then the relationship between the year and the population during that year is given by the integral

$$t = \frac{1}{k}\int \frac{1}{40 - P}\, dP$$

where $40 - P > 0$ because 40 billion is the population's upper limit.

Shutterstock

(a) Evaluate this integral to find the relationship.
(b) Use properties of logarithms and exponential functions to write P as a function of t.

SOLUTION

(a) $t = \dfrac{1}{k}\int \dfrac{1}{40 - P}\, dP = -\dfrac{1}{k}\int \dfrac{-dP}{40 - P}$

$\quad = -\dfrac{1}{k}\ln |40 - P| + C_1$

(b) Because $40 - P > 0$ means $|40 - P| = 40 - P$,

it follows that $t = \dfrac{1}{k}ln\,(40 - P) + C_1$.

Solving this equation for P requires converting to exponential form.

$$-k(t - C_1) = \ln\,(40 - P)$$

$$e^{C_1 k - kt} = 40 - P$$

$$e^{C_1 k} \cdot e^{-kt} = 40 - P$$

Because $e^{C_1 k}$ is an unknown constant, we replace it with C and solve for P.

$$P = 40 - Ce^{-kt}$$

....

12.3 Exercises

In Problems 1–8, evaluate each integral.

1. $\displaystyle\int 3e^{3x}\, dx$

2. $\displaystyle\int 840e^{-0.7x}\, dx$

3. $\displaystyle\int x^3 e^{3x^4}\, dx$

4. $\displaystyle\int \frac{3}{e^{2x}}\, dx$

5. $\displaystyle\int \frac{3x^2}{x^3 + 4}\, dx$

6. $\displaystyle\int \frac{dz}{4z + 1}$

7. $\displaystyle\int \frac{6x^3}{2x^4 + 1}\, dx$

8. $\displaystyle\int \frac{z^2 + 1}{z^3 + 3z + 17}\, dz$

9. Given $F(x) = x + \ln |x| + C$ and its graph as shown, find the function $f(x)$ such that $F(x) = \int f(x)\, dx$.

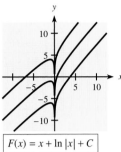

$$F(x) = x + \ln |x| + C$$

Applications

10. **Revenue** Suppose the marginal revenue from the sale of x units of a product is $\overline{MR} = R'(x) = 6e^{0.01x}$. What is the revenue in dollars from the sale of 100 units of the product?

11. **Memorization** The rate of vocabulary memorization of the average student in a foreign language course is given by

$$\frac{dv}{dt} = \frac{40}{t + 1}$$

where t is the number of continuous hours of study, $0 < t \le 4$, and v is the number of words. How many words would the average student memorize in 3 hours?

12. **Blood pressure in the aorta** The rate at which blood pressure decreases in the aorta of a normal adult after a heartbeat is

$$\frac{dp}{dt} = -46.645e^{-0.491t}$$

where t is time in seconds.
(a) What function describes the blood pressure in the aorta if $p = 95$ when $t = 0$?
(b) What is the blood pressure 0.1 second after a heartbeat?

Need more practice?
Find more here: cengagebrain.com

12.4 The Indefinite Integral in Business and Economics

IN THIS SECTION, WE'LL USE INTEGRATION TO DERIVE TOTAL COST AND PROFIT FUNCTIONS FROM THE MARGINAL COST AND MARGINAL REVENUE FUNCTIONS, AS WELL AS TO FIND THE MARGINAL PROPENSITY TO CONSUME AND THE NATIONAL CONSUMPTION FUNCTION.

Total Cost and Profit

We know that the marginal cost for a commodity is the derivative of the total cost function—that is, $\overline{MC} = C'(x)$, where $C(x)$ is the total cost function. Thus if we have the marginal cost function, we can integrate (or "reverse" the process of differentiation) to find the total cost. That is, $C(x) = \int \overline{MC}\, dx$. We cannot, however, determine the total cost function from the marginal cost unless additional information is available to help us determine the value of the constant of integration.

EXAMPLE 12.16 Total Cost

Suppose the marginal cost function for a month for a certain product is $\overline{MC} = 3x + 50$, where x is the number of units and cost is in dollars. If the fixed costs related to the product amount to $100 per month, find the total cost function for the month.

SOLUTION
The total cost function is

$$C(x) = \int (3x + 50)\, dx = \frac{3x^2}{2} + 50x + K$$

The constant of integration K is found by using the fact that $C(0) = FC = 100$. Thus

$$3(0)^2 + 50(0) + K = 100, \text{ so } K = 100$$

and the total cost for the month is given by

$$C(x) = \frac{3x^2}{2} + 50x + 100$$

EXAMPLE 12.17 Cost

Suppose monthly records show that the rate of change of the cost (that is, the marginal cost) for a luxury handbag is $\overline{MC} = 3(2x + 25)^{1/2}$, where x is the number of units and cost is in dollars. If the fixed costs for the month are \$11,125, what would be the total cost of producing 300 handbags per month?

SOLUTION

We can integrate the marginal cost to find the total cost function.

$$C(x) = \int \overline{MC}\, dx = \int 3(2x + 25)^{1/2}\, dx$$

$$= 3 \cdot \left(\frac{1}{2}\right)\int (2x + 25)^{1/2}(2\, dx)$$

$$= \left(\frac{3}{2}\right)\frac{(2x + 25)^{3/2}}{3/2} + K$$

$$= (2x + 25)^{3/2} + K$$

We can find K by using the fact that fixed costs are \$11,125.

$$C(0) = 11,125 = (25)^{3/2} + K$$

Thus $K = 11,000$ and the total cost function is

$$C(x) = (2x + 25)^{3/2} + 11,000$$

It can be shown that the profit is usually maximized when $\overline{MR} = \overline{MC}$.

Shutterstock (both)

and the cost of producing 300 luxury handbags per month is

$$C(300) = (625)^{3/2} + 11,000 = 26,625 \text{ (dollars)}$$

EXAMPLE 12.18 Maximum Profit

Given that $\overline{MR} = 200 - 4x$, $\overline{MC} = 50 + 2x$, and the total cost of producing 10 Wagbats is \$700, at what level should the Wagbat firm hold production in order to maximize the profits?

SOLUTION

Setting $\overline{MR} = \overline{MC}$, we can solve for the production level that maximizes profit.

$$200 - 4x = 50 + 2x$$
$$150 = 6x$$
$$25 = x$$

The level of production that should optimize profit is 25 units. To see whether 25 units maximizes profits or minimizes the losses (in the short run), we must find the total revenue and total cost functions.

$$R(x) = \int (200 - 4x)\, dx = 200x - 2x^2 + K$$

$$= 200x - 2x^2, \text{ because } R(0) = 0 \text{ means } K = 0$$

$$C(x) = \int (50 + 2x)\, dx = 50x + x^2 + K$$

We find K by noting that $C(x) = 700$ when $x = 10$.

$$700 = 50(10) + (10)^2 + K \text{ so } K = 100.$$

Thus the cost is given by

$$C = C(x) = 50x + x^2 + 100.$$

At $x = 25$, $R = R(25) = 200(25) - 2(25)^2 = \$3750$ and $C = C(25) = 50(25) + (25)^2 + 100 = \$1975$.

We see that the total revenue is greater than the total cost, so production should be held at 25 units, which results in a maximum profit.

National Consumption and Savings

If C represents national consumption (in billions of dollars), then a **national consumption function** has the form $C = f(y)$, where y is disposable national income (also in billions of dollars). The **marginal propensity to consume** is the derivative of the national consumption function with respect to y, or $dC/dy = f'(y)$. For example, suppose that

$$C = f(y) = 0.8y + 6$$

is a national consumption function; then the marginal propensity to consume is

$$f'(y) = 0.8.$$

If we know the marginal propensity to consume, we can integrate with respect to y to find national consumption:

$$C = \int f'(y)\, dy = f(y) + K$$

We can find the unique national consumption function if we have additional information to help us determine the value of K, the constant of integration.

EXAMPLE 12.19 National Consumption

If consumption is \$6 billion when disposable income is 0, and if the marginal propensity to consume is $dC/dy = 0.3 + 0.4/\sqrt{y}$ (in billions of dollars), find the national consumption function.

SOLUTION

$$C = \int \left(\frac{dC}{dy}\right) dy = \int \left(0.3 + \frac{0.4}{\sqrt{y}}\right) dy = \int (0.3 + 0.4y^{-1/2})\, dy$$

$$= 0.3y + 0.4\frac{y^{1/2}}{1/2} + K = 0.3y + 0.8y^{1/2} + K$$

If $C = 6$ when $y = 0$, then $6 = 0.3(0) + 0.8\sqrt{0} + K$. Thus the constant of integration is $K = 6$, and the consumption function is

$$C = 0.3y + 0.8\sqrt{y} + 6 \quad \text{(billions of dollars)}$$

If S represents national savings, we can assume that the disposable national income is given by $y = C + S$, or $S = y - C$. Then the **marginal propensity to save** is $dS/dy = 1 - dC/dy$.

EXAMPLE 12.20 Consumption and Savings

If the consumption is \$9 billion when income is 0, and if the marginal propensity to save is 0.25, find the consumption function.

SOLUTION

If $dS/dy = 0.25$, then $0.25 = 1 - dC/dy$, or $dC/dy = 0.75$. Thus

$$C = \int 0.75\, dy = 0.75y + K$$

If $C = 9$ when $y = 0$, then $9 = 0.75(0) + K$, or $K = 9$. Then the consumption function is $C = 0.75y + 9$ (in billions of dollars).

12.4 Exercises

Total Cost and Profit

In Problems 1–5, cost, revenue, and profit are in dollars and x is the number of units.

1. If the monthly marginal cost for a product is $\overline{MC} = 2x + 100$, with fixed costs amounting to \$200, find the total cost function for the month.

2. If the marginal cost for a product is $\overline{MC} = 4x + 2$, and the production of 10 units results in a total cost of \$300, find the total cost function.

3. A firm knows that its marginal cost for a product is $\overline{MC} = 3x + 20$, that its marginal revenue is $\overline{MR} = 44 - 5x$, and that the cost of production and sale of 80 units is \$11,400.
   (a) Find the optimal level of production.
   (b) Find the profit function.
   (c) Find the profit or loss at the optimal level.

4. Suppose that the marginal revenue for a product is $\overline{MR} = 900$ and the marginal cost is $\overline{MC} = 30\sqrt{x + 4}$, with a fixed cost of \$1000.
   (a) Find the profit or loss from the production and sale of 5 units.
   (b) How many units will result in a maximum profit?

5. The average cost of a product changes at the rate

$$\overline{C}'(x) = -6x^{-2} + 1/6$$

   and the average cost of 6 units is \$10.00.
   (a) Find the average cost function.
   (b) Find the average cost of 12 units.

National Consumption and Savings

6. If consumption is \$7 billion when disposable income is 0, and if the marginal propensity to consume is 0.80, find the national consumption function (in billions of dollars).

7. If consumption is \$8 billion when income is 0, and if the marginal propensity to consume is

$$\frac{dC}{dy} = 0.3 + \frac{0.2}{\sqrt{y}} \text{ (in billions of dollars)}$$

find the national consumption function.

8. Suppose that the marginal propensity to consume is

$$\frac{dC}{dy} = 0.7 - e^{-2y} \text{ (in billions of dollars)}$$

and that consumption is \$5.65 billion when disposable income is 0. Find the national consumption function.

9. Suppose that the marginal propensity to save is

$$\frac{dS}{dy} = 0.15 \text{ (in billions of dollars)}$$

and that consumption is \$5.15 billion when disposable income is 0. Find the national consumption function.

....

Need more practice?
Find more here: cengagebrain.com

12.5 Differential Equations

RECALL THAT WE INTRODUCED THE DERIVATIVE AS AN INSTANTANEOUS RATE OF CHANGE AND DENOTED THE INSTANTANEOUS RATE OF CHANGE OF Y WITH RESPECT TO TIME AS DY/DT. FOR MANY GROWTH OR DECAY PROCESSES, SUCH AS CARBON-14 DECAY, THE RATE OF CHANGE OF THE AMOUNT OF A SUBSTANCE WITH RESPECT TO TIME IS PROPORTIONAL TO THE AMOUNT PRESENT. THIS CAN BE REPRESENTED BY THE EQUATION

$$\frac{dy}{dt} = ky \ (k = \text{constant})$$

An equation of this type, where y is an unknown function of x or t, is called a **differential equation** because it contains derivatives (or differentials). In this section, we restrict ourselves to differential equations where the highest derivative present in the equation is the first derivative. These differential equations are called **first-order differential equations.** Examples are

$$f'(x) = \frac{1}{x+1}, \ \frac{dy}{dt} = 2t, \text{ and } x \, dy = (y+1) \, dx$$

Solution of Differential Equations

The solution to a differential equation is a function [say $y = f(x)$] that, when used in the differential equation, results in an identity. For example, to show that $y = 4e^{-5t}$ is a solution to $dy/dt + 5y = 0$, we must show that substituting $y = 4e^{-5t}$ into the equation $dy/dt + 5y = 0$ results in an identity:

$$\frac{d}{dt}(4e^{-5t}) + 5(4e^{-5t}) = 0$$
$$-20e^{-5t} + 20e^{-5t} = 0$$
$$0 = 0$$

Thus $y = 4e^{-5t}$ is a solution.

Now that we know what it means for a function to be a solution to a differential equation, let us consider how to find solutions.

The most elementary differential equations are of the form

$$\frac{dy}{dx} = f(x)$$

where $f(x)$ is a continuous function. These equations are elementary to solve because the solutions are found by integration:

$$y = \int f(x) \, dx$$

EXAMPLE 12.21 Solving a Differential Equation

Find the solution of

$$f'(x) = \frac{1}{x+1}$$

SOLUTION
The solution is

$$f(x) = \int f'(x) \, dx = \int \frac{1}{x+1} dx = \ln|x+1| + C$$

The solution in Example 12.21, $f(x) =$ ln $|x + 1| + C$, is called the **general solution** because every solution to the equation has this form, and different values of C give different **particular solutions.** Figure 12.2 shows the graphs of several members of the family of solutions to this differential equation. (We cannot, of course, show all of them.)

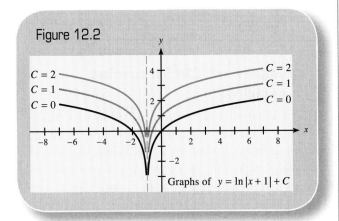

Figure 12.2

Graphs of $y = \ln|x + 1| + C$

We can find a particular solution to a differential equation when we know that the solution must satisfy additional conditions, such as **initial conditions** or **boundary conditions.** For instance, to find the particular solution to

$$f'(x) = \frac{1}{x + 1} \text{ with the condition that } f(-2) = 2$$

we use $f(-2) = 2$ in the general solution, $f(x) = \ln|x + 1| + C$.

$$2 = f(-2) = \ln|-2 + 1| + C$$
$$2 = \ln|-1| + C \text{ so } C = 2$$

Thus the particular solution is

$$f(x) = \ln|x + 1| + 2$$

and is shown in Figure 12.2 with $C = 2$.

See the Tech Card for details on finding the particular solution to a differential equation.

Separable Differential Equations

Just as we can find the differential of both sides of an equation, we can find the solution to a differential equation of the form

$$G(y) \, dy = f(x) \, dx$$

by integrating both sides.

It is frequently necessary to change the form of a differential equation before it can be solved by integrating both sides.

Separable Differential Equations:
When a differential equation can be equivalently expressed in the form

$$g(y) \, dy = f(x) \, dx$$

we say that the equation is **separable.**
The solution of a separable differential equation is obtained by integrating both sides of the equation after the variables have been separated.

EXAMPLE 12.22 Separable Differential Equation

 Solve the differential equation

$$(x^2y + x^2) \, dy = x^3 \, dx$$

SOLUTION

To write the equation in separable form, we first factor x^2 from the left side and divide both sides by it.

$$x^2(y + 1) \, dy = x^3 \, dx$$
$$(y + 1) \, dy = \frac{x^3}{x^2} \, dx$$

The equation is now separated, so we integrate both sides.

$$\int (y + 1) \, dy = \int x \, dx$$
$$\frac{y^2}{2} + y + C_1 = \frac{x^2}{2} + C_2$$

This equation, as well as the equation

$$y^2 + 2y - x^2 = C, \text{ where } C = 2(C_2 - C_1)$$

gives the solution implicitly.

Note that we need not write both C_1 and C_2 when we integrate, because it is always possible to combine the two constants into one.

EXAMPLE 12.23 Separable Differential Equation

 Solve the differential equation

$$\frac{dy}{dt} = ky \quad (k = \text{constant})$$

SOLUTION

To solve the equation, we write it in separated form and integrate both sides:

$$\frac{dy}{y} = k\,dt \Rightarrow \int \frac{dy}{y} = \int k\,dt \Rightarrow \ln|y| = kt + C_1$$

Assuming that $y > 0$ and writing $\ln|y| = kt + C_1$ in exponential form gives

$$y = e^{kt + C_1}$$

$$y = e^{kt} \cdot e^{C_1} = Ce^{kt}, \text{ where } C = e^{C_1}$$

$$y = Ce^{kt}$$

The case of $y < 0$ is covered by values of $C < 0$.

In many applied problems that can be modeled with differential equations, we know conditions that allow us to obtain a particular solution.

Applications of Differential Equations

We now consider two applications that can be modeled by differential equations. These are radioactive decay and one-container mixture problems (as a model for drugs in an organ).

EXAMPLE 12.24 Carbon-14 Dating

When an organism dies, the rate of change of the amount of carbon-14 present is proportional to the amount present and is represented by the differential equation

$$\frac{dy}{dt} = ky$$

Shutterstock (both)

where y is the amount present, k is a constant, and t is time in years. If we denote the initial amount of carbon-14 in an organism as y_0, then $y = y_0$ represents the amount present at time $t = 0$ (when the organism dies). Suppose that archaeologists discover a fossil that contains 1% of the initial amount of carbon-14. Find the age of the fossil. (Recall that the half-life of carbon-14 is 5730 years.)

SOLUTION

We must find a particular solution to

$$\frac{dy}{dt} = ky$$

subject to the fact that when $t = 0$, $y = y_0$, and we must determine the value of k on the basis of the half-life of carbon-14 ($t = 5730$ years, $y = \frac{1}{2}y_0$ units.) From Example 12.23, we know that the general solution to the differential equation $dy/dt = ky$ is $y = Ce^{kt}$. Using $y = y_0$ when $t = 0$, we obtain $y_0 = C$, so the equation becomes $y = y_0 e^{kt}$. Using $t = 5730$ and $y = \frac{1}{2}y_0$ in this equation gives

$$\frac{1}{2}y_0 = y_0 e^{5730k} \text{ or } 0.5 = e^{5730k}$$

Rewriting this equation in logarithmic form and then solving for k, we get

$$\ln(0.5) = 5730k$$

$$-0.69315 = 5730k$$

$$-0.00012097 = k$$

Thus the equation we seek is

$$y = y_0 e^{-0.00012097t}$$

Using the fact that $y = 0.01y_0$ when the fossil was discovered, we can find its age t by solving

$$0.01y_0 = y_0 e^{-0.00012097t} \text{ or } 0.01 = e^{-0.00012097t}$$

Rewriting this in logarithmic form and then solving give

$$\ln(0.01) = -0.00012097t$$

$$-4.6051702 = -0.00012097t$$

$$38,069 \approx t$$

Thus the fossil is approximately 38,069 years old.

The differential equation $dy/dt = ky$, which describes the decay of radioactive substances in Example 12.24, also models the rate of growth of an investment that is compounded continuously and the rate of decay of purchasing power due to inflation.

Another application of differential equations comes from a group of applications called *one-container mixture problems*. In problems of this type, there is a substance whose amount in a container is changing with time, and the goal is to determine the amount of the substance at any time t. The differential equations that model these problems are of the following form:

$$\begin{bmatrix} \text{Rate of change} \\ \text{of the amount} \\ \text{of the substance} \end{bmatrix} =$$

$$\begin{bmatrix} \text{Rate at which} \\ \text{the substance} \\ \text{enters the container} \end{bmatrix} - \begin{bmatrix} \text{Rate at which} \\ \text{the substance} \\ \text{leaves the container} \end{bmatrix}$$

We consider this application as it applies to the amount of a drug in an organ.

EXAMPLE 12.25 Drug in an Organ

A liquid carries a drug into an organ of volume 300 cc at a rate of 5 cc/s, and the liquid leaves the organ at the same rate. If the concentration of the drug in the entering liquid is 0.1 g/cc, and if x represents the amount of drug in the organ at any time t, then using the fact that the rate of change of the amount of the drug in the organ, dx/dt, equals the rate at which the drug enters minus the rate at which it leaves, we have

$$\frac{dx}{dt} = \left(\frac{5 \text{ cc}}{s}\right)\left(\frac{0.1 \text{ g}}{\text{cc}}\right) - \left(\frac{5 \text{ cc}}{s}\right)\left(\frac{x \text{ g}}{300 \text{ cc}}\right)$$

or

$$\frac{dx}{dt} = 0.5 - \frac{x}{60} = \frac{30}{60} - \frac{x}{60} = \frac{30 - x}{60}, \text{ in g/s}$$

Find the amount of the drug in the organ as a function of time t.

SOLUTION

Multiplying both sides of the equation $\dfrac{dx}{dt} = \dfrac{30 - x}{60}$ by $\dfrac{dt}{(30 - x)}$ gives

$$\frac{dx}{30 - x} = \frac{1}{60} dt$$

The equation is now separated, so we can integrate both sides.

$$\int \frac{dx}{30 - x} = \int \frac{1}{60} dt$$

$$-\ln(30 - x) = \frac{1}{60}t + C_1 \ (30 - x > 0)$$

$$\ln(30 - x) = -\frac{1}{60}t - C_1$$

One-Container Mixture Problem

| $\begin{bmatrix} \text{Rate of change} \\ \text{of the amount} \\ \text{of the substance} \end{bmatrix}$ | $=$ | $\begin{bmatrix} \text{Rate at which} \\ \text{the substance} \\ \text{enters the container} \end{bmatrix}$ | $-$ | $\begin{bmatrix} \text{Rate at which} \\ \text{the substance} \\ \text{leaves the container} \end{bmatrix}$ |

Shutterstock (both)

Rewriting this in exponential form gives

$$30 - x = e^{-t/60 - C_1} = e^{-t/60} \cdot e^{-C_1}$$

Letting $C = e^{-C_1}$ yields

$$30 - x = Ce^{-t/60}$$

so

$$x = 30 - Ce^{-t/60}$$

and we have the desired function.

12.5 Exercises

1. Show that $y = x^2$ is a solution to $4y - 2xy' = 0$.

2. Use integration to find the general solution to the differential equation $3y^2\, dy = (2x - 1)\, dx$.

3. Find the particular solution to $dy = \left(\dfrac{1}{x} - x\right) dx$; $y(1) = 0$

In Problems 4 and 5, find the general solution to the given differential equation.

4. $dx = x^3 y\, dy$

5. $(x + 1)\dfrac{dy}{dx} = y$

In Problems 6 and 7, find the particular solution to each differential equation.

6. $2y^2\, dx = 3x^2\, dy$, when $x = 2$, $y = -1$

7. $x^2 e^{2y}\, dy = (x^3 + 1)\, dx$, when $x = 1$, $y = 0$

Applications

8. *Investing* When the interest on an investment is compounded continuously, the investment grows at a rate that is proportional to the amount in the account, so that if the amount present is P, then

$$\frac{dP}{dt} = kP$$

where P is in dollars, t is in years, and k is a constant. If \$100,000 is invested (when $t = 0$) and the amount in the account after 15 years is \$211,700, find the function that gives the value of the investment as a function of t. What is the interest rate on this investment?

9. *Sales and pricing* Suppose that in a certain company, the relationship between the price per unit p of its product and the weekly sales volume y, in thousands of dollars, is given by

$$\frac{dy}{dp} = -\frac{2}{5}\left(\frac{y}{p + 8}\right)$$

Solve this differential equation if $y = 8$ when $p = \$24$.

10. *Half-life* A breeder reactor converts uranium-238 into an isotope of plutonium-239 at a rate proportional to the amount present at any time. After 10 years, 0.03% of the radioactivity has dissipated (that is, 0.9997 of the initial amount remains). Suppose that initially there is 100 pounds of this substance. Find the half-life.

11. *Drug in an organ* Suppose that a liquid carries a drug with concentration 0.1 g/cc into a 200-cc organ at a rate of 5 cc/s and leaves the organ at the same rate. If initially there is 10 g of the drug in the organ, find the amount of drug in the organ as a function of time t.

12. *Impact of inflation* The impact of a 5% inflation rate on an \$80,000-per-year pension can be severe. If P represents the purchasing power (in dollars) of an \$80,000 pension, then the effect of a 5% inflation rate can be modeled by the differential equation

$$\frac{dP}{dt} = -0.05P, \quad P(0) = 80{,}000$$

where t is in years.
(a) Find the particular solution to this differential equation.
(b) Find the purchasing power after 15 years.

Need more practice?
Find more here: cengagebrain.com

Chapter Exercises

In Problems 1–23, evaluate each integral.

1. $\displaystyle\int x^6\, dx$

2. $\displaystyle\int x^{1/2}\, dx$

3. $\displaystyle\int (12x^3 - 3x^2 + 4x + 5)\, dx$

4. $\displaystyle\int 7(x^2 - 1)^2\, dx$

5. $\displaystyle\int 7x\,(x^2 - 1)^2\, dx$

6. $\displaystyle\int (x^3 - 3x^2)^5(x^2 - 2x)\, dx$

7. $\displaystyle\int (x^3 + 4)^2 3x\, dx$

8. $\displaystyle\int 5x^2(3x^3 + 7)^6\, dx$

9. $\displaystyle\int \frac{x^2}{x^3 + 1}\, dx$

10. $\int \dfrac{x^2}{(x^3 + 1)^2}\, dx$

11. $\int \dfrac{x^2\, dx}{\sqrt[3]{x^3 - 4}}$

12. $\int \dfrac{x^2\, dx}{x^3 - 4}$

13. $\int \dfrac{x^3 + 1}{x^2}\, dx$

14. $\int y^2 e^{y^3}\, dy$

15. $\int (3x - 1)^{12}\, dx$

16. $\int \dfrac{3x^2}{2x^3 - 7}\, dx$

17. $\int \dfrac{5\, dx}{e^{4x}}$

18. $\int (x^3 - e^{3x})\, dx$

19. $\int x e^{1 + x^2}\, dx$

20. $\int \dfrac{6x^7}{(5x^8 + 7)^3}\, dx$

21. $\int \dfrac{7x^3}{\sqrt{1 - x^4}}\, dx$

22. $\int \left(\dfrac{e^{2x}}{2} + \dfrac{2}{e^{2x}} \right) dx$

23. $\int \left[x - \dfrac{1}{(x + 1)^2} \right] dx$

In Problems 24–28, find the general solution to each differential equation.

24. $\dfrac{dy}{dt} = 4.6 e^{-0.05t}$

25. $dy = (64 + 76x - 36x^2)\, dx$

26. $\dfrac{dy}{dx} = \dfrac{4x}{y - 3}$

27. $\dfrac{dy}{dx} = \dfrac{x}{e^y}$

28. $\dfrac{dy}{dt} = \dfrac{4y}{t}$

29. Find the particular solution to
$$y' = \dfrac{x^2}{y + 1},\ y(0) = 4$$

Applications

30. **Revenue** If the marginal revenue for a month for a product is $\overline{MR} = 0.06x + 12$ dollars per unit, find the total revenue from the sale of $x = 800$ units of the product.

31. **Productivity** Suppose that the rate of change of production of the average worker at a factory is given by
$$\dfrac{dp}{dt} = 27 + 24t - 3t^2,\ 0 \le t \le 8$$
where p is the number of units the worker produces in t hours. How many units will the average worker produce in an 8-hour shift? (Assume that $p = 0$ when $t = 0$.)

32. **Oxygen levels in water** The rate of change of the oxygen level (in mmol/l) per month in a body of water after an oil spill is given by
$$P'(t) = 400 \left[\dfrac{5}{(t + 5)^2} - \dfrac{50}{(t + 5)^3} \right]$$
where t is the number of months after the spill. What function gives the oxygen level P at any time t if $P = 400$ mmol/l when $t = 0$?

33. **Bacterial growth** A population of bacteria grows at the rate
$$\dfrac{dp}{dt} = \dfrac{100{,}000}{(t + 100)^2}$$
where p is the population and t is time. If the population is 1000 when $t = 1$, write the equation that gives the size of the population at any time t.

34. **Market share** The rate of change of the market share (as a percent) a firm expects for a new product is
$$\dfrac{dy}{dt} = 2.4 e^{-0.04t}$$
where t is the number of months after the product is introduced.
    (a) Write the equation that gives the expected market share y at any time t. (Note that $y = 0$ when $t = 0$.)

(b) What market share does the firm expect after 1 year?

35. **Revenue** If the marginal revenue for a product is $\overline{MR} = \dfrac{800}{x+2}$, find the total revenue function.

36. **Cost** The marginal cost for a product is $\overline{MC} = 6x + 4$ dollars per unit, and the cost of producing 100 items is $31,400.
    (a) Find the fixed costs.
    (b) Find the total cost function.

37. **Profit** Suppose a product has a daily marginal revenue $\overline{MR} = 46$ and a daily marginal cost $\overline{MC} = 30 + \frac{1}{5}x$, both in dollars per unit. If the daily fixed cost is $200, how many units will give maximum profit and what is the maximum profit?

38. **National consumption** If consumption is $8.5 billion when disposable income is 0, and if the marginal propensity to consume is
    $$\frac{dC}{dy} = \frac{1}{\sqrt{2y+16}} + 0.6 \text{ (in billions of dollars)}$$
    find the national consumption function.

39. **National consumption** Suppose that the marginal propensity to save is
    $$\frac{dS}{dy} = 0.2 - 0.1e^{-2y} \text{ (in billions of dollars)}$$
    and consumption is $7.8 billion when disposable income is 0. Find the national consumption function.

40. **Allometric growth** For many species of fish, the length L and weight W of a fish are related by
    $$\frac{dW}{dL} = \frac{3W}{L}$$
    The general solution to this differential equation expresses the allometric relationship between the length and weight of a fish. Find the general solution.

41. **Investment** When the interest on an investment is compounded continuously, the investment grows at a rate that is proportional to the amount in the account, so that if the amount present is P, then
    $$\frac{dP}{dt} = kP$$
    where P is in dollars, t is in years, and k is a constant.
    (a) Solve this differential equation to find the relationship.
    (b) Use properties of logarithms and exponential functions to write P as a function of t.
    (c) If $50,000 is invested (when $t = 0$) and the amount in the account after 10 years is $135,914, find the function that gives the value of the investment as a function of t.
    (d) In part (c), what does the value of k represent?

42. **Fossil dating** Radioactive beryllium is sometimes used to date fossils found in deep-sea sediment. The amount of radioactive material x satisfies
    $$\frac{dx}{dt} = kx$$
    Suppose that 10 units of beryllium are present in a living organism and that the half-life of beryllium is 4.6 million years. Find the age of a fossil if 20% of the original radioactivity is present when the fossil is discovered.

43. **Drug in an organ** Suppose that a liquid carries a drug into a 120-cc organ at a rate of 4 cc/s and leaves the organ at the same rate. If initially there is no drug in the organ and if the concentration of drug in the liquid is 3 g/cc, find the amount of drug in the organ as a function of time.

> **Need more practice?**
> **Find more here: cengagebrain.com**

chapter 13

Definite Integrals:
Techniques of Integration

In this chapter we define the definite integral and discuss a theorem and techniques that are useful in evaluating or approximating it. We will also see how it can be used in many interesting applications, such as consumer's and producer's surplus and total value, present value, and future value of continuous income streams. Improper integrals can be used to find the capital value of a continuous income stream.

13.1 The Definite Integral: The Fundamental Theorem of Calculus

ONE WAY TO FIND THE ACCUMULATED PRODUCTION (SUCH AS THE PRODUCTION OF ORE FROM A MINE) OVER A PERIOD OF TIME IS TO GRAPH THE RATE OF PRODUCTION AS A FUNCTION OF TIME AND FIND THE AREA UNDER THE RESULTING CURVE OVER A SPECIFIED TIME INTERVAL. FOR EXAMPLE, IF A COAL MINE PRODUCES AT A RATE OF 30 TONS PER DAY, THE PRODUCTION OVER 10 DAYS ($30 \cdot 10 = 300$) COULD BE REPRESENTED BY THE AREA UNDER THE LINE $Y = 30$ BETWEEN $X = 0$ AND $X = 10$ (SEE FIGURE 13.1).

Using area to determine the accumulated production is very useful when the rate-of-production function varies at different points in time. For example, if the rate of production (in tons per day) is represented by

$$y = 100e^{-0.1x}$$

where x represents the number of days, then the area under the curve (and above the x-axis) from $x = 0$ to $x = 10$ represents the total production over the 10-day period (see Figure 13.2(a) on the following page).

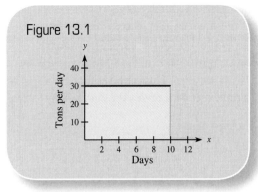

Figure 13.1

Estimating the Area under a Curve

To estimate the accumulated production for the previous example, we approximate the area under the graph of the production rate function. We can find a rough approximation of the area under this curve by fitting two rectangles to the curve as shown in Figure 13.2(b).

Shutterstock

The area of the first rectangle is $5 \cdot 100 = 500$ square units, and the area of the second rectangle is $(10 - 5)[100e^{-0.1(5)}] \approx 5(60.65) = 303.25$ square units, so this rough approximation is 803.25 square units or 803.25 tons of ore. This approximation is clearly larger than the exact area under the curve. Why?

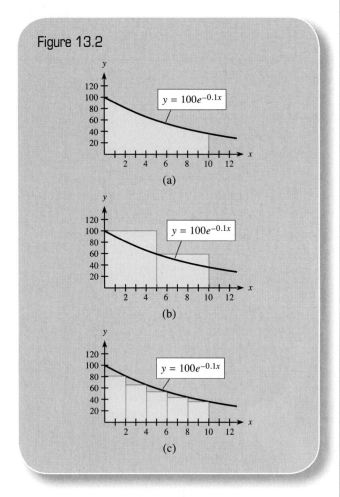

Figure 13.2

(a)

(b)

(c)

EXAMPLE 13.1 Ore Production

Find another, more accurate approximation of the tons of ore produced by approximating the area under the curve in Figure 13.2(a). Fit five rectangles with equal bases inside the area under the curve $y = 100e^{-0.1x}$, and use them to approximate the area under the curve from $x = 0$ to $x = 10$ (see Figure 13.2(c)).

SOLUTION

Each of the five rectangles has base 2, and the height of each rectangle is the value of the function at the right-hand endpoint of the interval forming its base. Thus the areas of the rectangles are as follows.

Rectangle	Base	Height	Area = Base × Height
1	2	$100e^{-0.1(2)} \approx 81.87$	$2(81.87) = 163.74$
2	2	$100e^{-0.1(4)} \approx 67.03$	$2(67.03) = 134.06$
3	2	$100e^{-0.1(6)} \approx 54.88$	$2(54.88) = 109.76$
4	2	$100e^{-0.1(8)} \approx 44.93$	$2(44.93) = 89.86$
5	2	$100e^{-0.1(10)} \approx 36.79$	$2(36.79) = 73.58$

The area under the curve is approximately equal to

$$163.74 + 134.06 + 109.76 + 89.86 + 73.58 = 571$$

so approximately 571 tons of ore are produced in the 10-day period. The area is actually 632.12, to two decimal places (or 632.12 tons of ore), so the approximation 571 is smaller than the actual area but is much better than the one we obtained with just two rectangles. In general, if we use bases of equal width, the approximations of the area under a curve approach the exact area as the widths of the bases approaches zero. (See the Enrichment material on CourseMate for MATH APPS at cengagebrain.com for more about the area under a curve. See the Tech Card for details on using Excel to approximate the area under a curve.)

In fact, we can use subintervals that are not of equal width, and we can use any point within each subinterval, denoted by x_i^*, to determine the height of each rectangle. For the ith rectangle, we denote the width as Δx_i and the height as $f(x_i^*)$, so its area is $f(x_i^*)\Delta x_i$. Then the sum of the areas of these rectangles is $S = \sum f(x_i^*)\Delta x_i$ and this sum approximates the exact area under the curve.

Increasing the number of subintervals and making sure that every interval becomes smaller will in the long run improve the approximation. Thus for any subdivision of $[a, b]$ and any x_i^*, the exact area is given by $A = \lim_{\max \Delta x_i \to 0} \sum f(x_i^*)\Delta x_i$, provided that this limit exists.

In addition to giving the exact area, this limit has other important applications and is called the **definite integral** of $f(x)$ over interval $[a, b]$.

Definite Integral

If f is a function on the interval $[a, b]$, then, for any subdivision of $[a, b]$ and any choice of x_i^* in the ith subinterval, the *definite integral* of f from a to b is

$$\int_b^a f(x)\,dx = \lim_{\max \Delta x_i \to 0} \sum f(x_i^*)\Delta x_i$$

If f is a continuous function, and $\Delta x_i \to 0$, then the limit exists and we say that f is integrable on $[a, b]$.

Note that for some intervals, values of f may be negative. In this case, the product $f(x_i^*)\Delta x_i$ will be negative and can be thought of geometrically as a "signed area." (Remember that area is a positive number.) Thus a definite integral can be thought of geometrically as the sum of signed areas, just as a derivative can be thought of geometrically as the slope of a tangent line. In the case where $f(x)$ is positive for all x from a to b, the definite integral equals the area between the graph of $y = f(x)$ and the x-axis.

Fundamental Theorem of Calculus

The obvious question is how this definite integral is related to the indefinite integral (the antiderivative) discussed in Chapter 12. The connection between these two concepts is the most important result in calculus, because it connects derivatives, indefinite integrals, and definite integrals.

> **Fundamental Theorem of Calculus**
>
> Let f be a continuous function on the closed interval $[a, b]$; then the definite integral of f exists on this interval, and
>
> $$\int_a^b f(x)\, dx = F(b) - F(a)$$
>
> where F is any function such that $F'(x) = f(x)$ for all x in $[a, b]$.

Stated differently, this theorem says that if the function F is an indefinite integral of a function f that is continuous on the interval $[a, b]$, then

$$\int_a^b f(x)\, dx = F(b) - F(a)$$

Thus, we apply the Fundamental Theorem of Calculus by using the following two steps.

1. Integration of $f(x)$: $\quad \displaystyle\int_a^b f(x)\, dx = F(x)\Big|_a^b$

2. Evaluation of $F(x)$: $\quad F(x)\Big|_a^b = F(b) - F(a)$

EXAMPLE 13.2 Definite Integral

➡ Evaluate $\displaystyle\int_2^4 (x^3 + 4)\, dx$.

SOLUTION

1. $\displaystyle\int_2^4 (x^3 + 4)\, dx = \dfrac{x^4}{4} + 4x + C\ \Big|_2^4$

2. $\left[\dfrac{(4)^4}{4} + 4(4) + C\right] - \left[\dfrac{(2)^4}{4} + 4(2) + C\right]$

$= (64 + 16 + C) - (4 + 8 + C)$

$= 68 \quad$ Note that the Cs subtract out

Note that the Fundamental Theorem states that F can be *any* indefinite integral of f, so we need not add the constant of integration to the integral.

EXAMPLE 13.3 Fundamental Theorem

➡ Evaluate $\displaystyle\int_1^3 (3x^2 + 6x)\, dx$.

SOLUTION

$\displaystyle\int_1^3 (3x^2 + 6x)\, dx = x^3 + 3x^2\ \Big|_1^3$

$= (3^3 + 3 \cdot 3^2) - (1^3 + 3 \cdot 1^2)$

$= 54 - 4 = 50$

Properties

Two properties of definite integrals are

1. $\displaystyle\int_a^b [f(x) \pm g(x)]\, dx = \int_a^b f(x)\, dx \pm \int_a^b g(x)\, dx$

2. $\displaystyle\int_a^b kf(x)\, dx = k\int_a^b f(x)\, dx$, where k is a constant

The following example uses both of these properties.

EXAMPLE 13.4 Definite Integral

➡ Evaluate $\displaystyle\int_3^5 (\sqrt{x^2 - 9} + 2)x\, dx$.

SOLUTION

$\displaystyle\int_3^5 (\sqrt{x^2 - 9} + 2)x\, dx = \int_3^5 \sqrt{x^2 - 9}(x\, dx) + \int_3^5 2x\, dx$

$= \dfrac{1}{2}\int_3^5 (x^2 - 9)^{1/2}(2x\, dx) + 2\int_3^5 x\, dx$

$= \dfrac{1}{2}\left[\dfrac{2}{3}(x^2 - 9)^{3/2}\right]\Big|_3^5 + x^2\ \Big|_3^5$

$= \dfrac{1}{3}[(16)^{3/2} - (0)^{3/2}] + (25 - 9)$

$= \dfrac{64}{3} + 16 = \dfrac{64}{3} + \dfrac{48}{3} = \dfrac{112}{3}$

In the integral $\int_a^b f(x)\,dx$, we call a the *lower limit* and b the *upper limit* of integration. Although we developed the definite integral with the assumption that the lower limit was less than the upper limit, the following properties permit us to evaluate the definite integral even when that is not the case.

3. $\int_a^a f(x)\,dx = 0$

4. If f is integrable on $[a, b]$, then

$$\int_b^a f(x)\,dx = -\int_a^b f(x)\,dx$$

To illustrate these properties, note the following:

$$\int_4^4 x^2\,dx = \frac{x^3}{3}\Big|_4^4 = \frac{4^3}{3} - \frac{4^3}{3} = 0$$

and

$$\int_2^4 3x^2\,dx = x^3\Big|_2^4 = 4^3 - 2^3 = 56$$

$$\int_4^2 3x^2\,dx = x^3\Big|_4^2 = 2^3 - 4^3 = -56$$

Another property of definite integrals is called the additive property.

5. If f is continuous on some interval containing a, b, and c,* then

$$\int_a^c f(x)\,dx + \int_c^b f(x)\,dx = \int_a^b f(x)\,dx$$

Thus by property 5, $\int_2^3 4x\,dx + \int_3^5 4x\,dx = \int_2^5 4x\,dx$.

The Definite Integral and Areas

Let us now return to area problems, to see the relationship between the definite integral and the area under a curve. By the formula for the area of a triangle, the area under the curve (line) $y = x$ from $x = 0$ to $x = 1$ is $\frac{1}{2}$ (see Figure 13.3(a)). Using the definite integral to find the area gives

$$A = \int_0^1 x\,dx = \frac{x^2}{2}\Big|_0^1 = \frac{1}{2} - 0 = \frac{1}{2}$$

However, not every definite integral represents the area between the curve and the x-axis over an interval. For example,

$$\int_0^2 (x - 2)\,dx = \frac{x^2}{2} - 2x\,\Big|_0^2 = (2 - 4) - (0) = -2$$

Figure 13.3

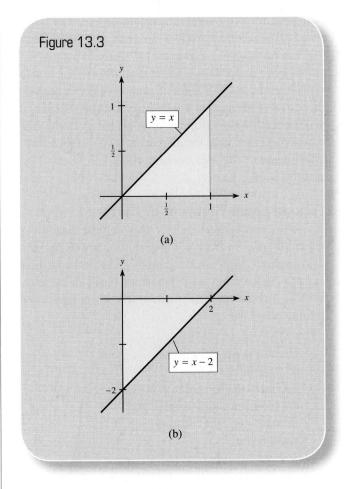

(a)

(b)

Notice that the shaded region in Figure 13.3(b) is a triangle with area 2. Thus the value of the definite integral over *this* interval does not represent the area between the curve and the x-axis.

In general, the definite integral will give the area under the curve and above the x-axis only when $f(x) \geq 0$ for all x in $[a, b]$.

Area under a Curve

If f is a continuous function on $[a, b]$ and $f(x) \geq 0$ on $[a, b]$, then the exact area between $y = f(x)$ and the x-axis from $x = a$ to $x = b$ is given by

$$\text{Area (shaded)} = \int_a^b f(x)\,dx$$

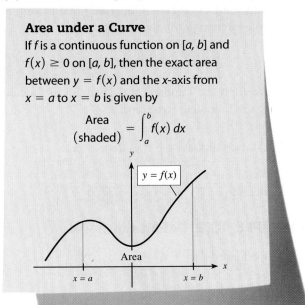

Note also that if $f(x) \le 0$ for all x in $[a, b]$, then

$$\int_a^b f(x) \, dx = -(\text{Area between } f(x) \text{ and the } x\text{-axis})$$

If the rate of growth of some function with respect to time t is $f'(t)$, then the total growth of the function during the period from $t = 0$ to $t = k$ can be found by evaluating the definite integral

$$\int_0^k f'(t) \, dt = f(t) \Big|_0^k = f(k) - f(0)$$

For nonnegative rates of growth, this definite integral (and thus growth) is the same as the area under the graph of $f'(t)$ from $t = 0$ to $t = k$.

EXAMPLE 13.5 Income Stream

Suppose that money flows continuously into a slot machine at a casino and grows at a rate given by

$$A'(t) = 100e^{0.1t}$$

where t is time in hours and $0 \le t \le 10$. Find the total amount that accumulates in the machine during the 10-hour period, if no money is paid out.

SOLUTION

The total amount is given by

$$A = \int_0^{10} 100e^{0.1t} \, dt = \frac{100}{0.1} \int_0^{10} e^{0.1t}(0.1) \, dt$$

$$= 1000e^{0.1t} \Big|_0^{10}$$

$$= 1000e - 1000$$

$$\approx 1718.28 \ (\text{dollars})$$

In Section 8.4, we stated that the total area under the normal curve is 1 and that the area under the curve from value x_1 to value x_2 represents the probability that a score chosen at random will lie between x_1 and x_2.

The normal distribution is an example of a **continuous probability distribution** because the values of the random variable are considered over intervals rather than at discrete values. The preceding statements relating probability and area under the graph apply to other continuous probability distributions determined by **probability density functions**. In fact, if x is a continuous random

variable with probability density function $f(x)$, then the probability that x is between a and b is

$$\Pr(a \le x \le b) = \int_a^b f(x) \, dx$$

See the Tech Card for details on using a graphing calculator to approximate both the area under the graph of a function and the definite integral of a function.

13.1 Exercises

In Problems 1–7, evaluate each definite integral.

1. $\displaystyle\int_2^4 x^3 \, dx$

2. $\displaystyle\int_0^5 4\sqrt[3]{x^2} \, dx$

3. $\displaystyle\int_2^4 (x^2 + 2)^3 x \, dx$

4. $\displaystyle\int_0^4 \sqrt{4x + 9} \, dx$

5. $\displaystyle\int_1^3 \frac{3}{y^2} \, dy$

6. $\displaystyle\int_0^1 e^{3x} \, dx$

7. $\displaystyle\int_1^e \frac{4}{z} \, dz$

8. (a) Write the integral that describes the area of the shaded region and (b) find the area.

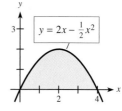

9. Find the area between the curve $y = -x^2 + 3x - 2$ and the x-axis from $x = 1$ to $x = 2$.

10. If $\displaystyle\int_1^2 (2x - x^2) \, dx = \frac{2}{3}$ and $\displaystyle\int_2^4 (2x - x^2) \, dx = -\frac{20}{3}$, what does $\displaystyle\int_1^4 (x^2 - 2x) \, dx$ equal?

11. Evaluate $\displaystyle\int_4^4 \sqrt{x^2 - 2}\, dx$.

Applications

12. **Depreciation** The rate of depreciation of a building is given by $D'(t) = 3000(20 - t)$ dollars per year, $0 \le t \le 20$; see the figure.
    (a) Use the graph to find the total depreciation of the building over the first 10 years ($t = 0$ to $t = 10$).
    (b) Use the definite integral to find the total depreciation over the first 10 years.

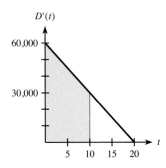

13. **Sales and advertising** A store finds that its sales revenue changes at a rate given by

    $$S'(t) = -30t^2 + 360t \quad \text{dollars per day}$$

    where t is the number of days after an advertising campaign ends and $0 \le t \le 30$.
    (a) Find the total sales for the first week after the campaign ends ($t = 0$ to $t = 7$).
    (b) Find the total sales for the second week after the campaign ends ($t = 7$ to $t = 14$).

14. **Total income** The income from an oil change service chain can be considered as flowing continuously at an annual rate given by

    $$f(t) = 10{,}000e^{0.02t} \quad \text{(dollars/ year)}$$

    Find the total income for this chain over the first 2 years (from $t = 0$ to $t = 2$).

15. **Telecommunications revenue** Total market revenue for worldwide telecommunications (in billions of dollars per year) can be modeled by

    $$R(t) = 72.11t + 81.16$$

    where $t = 0$ represents 1985. Evaluate $\displaystyle\int_{15}^{25} R(t)\, dt$ and tell what it represents. (*Source:* International Telecommunications Union)

16. **Customer service** The duration t (in minutes) of customer service calls received by a certain company is given by the probability density function

    $$f(t) = 0.3e^{-0.3t}, \ t \ge 0$$

    Find the probability that a call selected at random lasts
    (a) 3 minutes or less.
    (b) between 5 and 10 minutes.

> **Need more practice?**
> Find more here: **cengagebrain.com**

13.2 Area between Two Curves

WE HAVE USED THE DEFINITE INTEGRAL TO FIND THE AREA BETWEEN A CURVE AND THE *X*-AXIS. IN THIS SECTION WE WILL USE DEFINITE INTEGRALS TO FIND THE AREA BETWEEN TWO CURVES.

Suppose that the graphs of both $y = f(x)$ and $y = g(x)$ lie above the x-axis and that the graph of $y = f(x)$ lies above $y = g(x)$ throughout the interval from $x = a$ to $x = b$; that is, $f(x) \ge g(x)$ on $[a, b]$.

Then Figures 13.4(a) and 13.4(b) show the areas under $y = f(x)$ and $y = g(x)$. Figure 13.4(c) shows how the difference of these two areas can be used to find the area of the region between the graphs of $y = f(x)$ and $y = g(x)$. That is,

$$\text{Area between the curves} = \int_a^b f(x)\, dx - \int_a^b g(x)\, dx$$

Although Figure 13.4(c) shows the graphs of both $y = f(x)$ and $y = g(x)$ lying above the x-axis, this difference of their integrals will always give the area between their graphs if both functions are continuous and if $f(x) \ge g(x)$ on the interval $[a, b]$. Using the fact that

$$\int_a^b f(x)\, dx - \int_a^b g(x)\, dx = \int_a^b [f(x) - g(x)]\, dx$$

we have the following result.

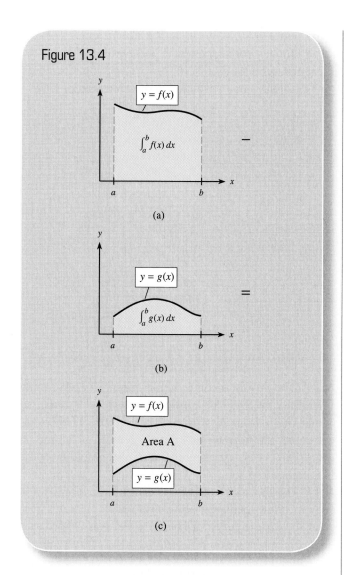

Figure 13.4

$$y = f(x)$$

$$\int_a^b f(x)\, dx$$

(a)

$$-$$

$$y = g(x)$$

$$\int_a^b g(x)\, dx$$

(b)

$$=$$

$$y = f(x)$$

Area A

$$y = g(x)$$

(c)

Area between Two Curves

If f and g are continuous functions on $[a, b]$ and if $f(x) \geq g(x)$ on $[a, b]$, then the area of the region bounded by $y = f(x)$, $y = g(x)$, $x = a$, and $x = b$ is

$$A = \int_a^b [f(x) - g(x)]\, dx$$

EXAMPLE 13.6 Area between Two Curves

Find the area of the region bounded by $y = x^2 + 4$, $y = x$, $x = 0$, and $x = 3$.

SOLUTION

We first sketch the graphs of the functions. The graph of the region is shown in Figure 13.5. Because $y = x^2 + 4$

lies above $y = x$ in the interval from $x = 0$ to $x = 3$, the area is

$$A = \int (\text{top curve} - \text{bottom curve})\, dx$$

$$A = \int_0^3 [(x^2 + 4) - x]\, dx = \frac{x^3}{3} + 4x - \frac{x^2}{2} \Big|_0^3$$

$$= \left(9 + 12 - \frac{9}{2} \right) - (0 + 0 - 0)$$

$$= 16\tfrac{1}{2} \text{ square units}$$

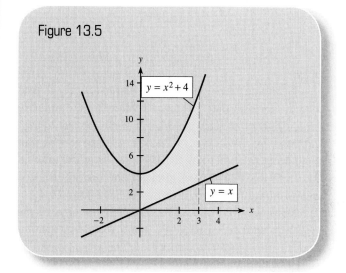

Figure 13.5

$$y = x^2 + 4$$

$$y = x$$

We are sometimes asked to find the area enclosed by two curves. In that case, we find the points of intersection of the curves to determine a and b. Some graphs enclose two or more regions because they have *more* than two points of intersection. Let's see how this works.

EXAMPLE 13.7 A Region with Two Sections

Find the area of the region enclosed by the graphs of

$$y = f(x) = x^3 - x^2$$

and

$$y = g(x) = 2x$$

SOLUTION

To find the points of intersection of the graphs, we set the y-values equal and solve for x.

$$x^3 - x^2 = 2x$$

$$x^3 - x^2 - 2x = 0$$

$$x(x - 2)(x + 1) = 0$$

$$x = 0, \; x = 2, \; x = -1$$

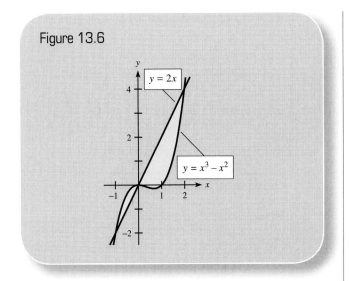

Figure 13.6

Graphing these functions between $x = -1$ and $x = 2$, we see that for any x-value in the interval $(-1, 0)$, $f(x) \geq g(x)$, so $f(x) \geq g(x)$ for the region enclosed by the curves from $x = -1$ to $x = 0$. But evaluating the functions for any x-value in the interval $(0, 2)$ shows that $f(x) \leq g(x)$ for the region enclosed by the curves from $x = 0$ to $x = 2$. See Figure 13.6.

Thus we need one integral to find the area of the region from $x = -1$ to $x = 0$ and a second integral to find the area from $x = 0$ to $x = 2$. The area is found by summing these two integrals.

$$A = \int_{-1}^{0} [(x^3 - x^2) - (2x)]\, dx + \int_{0}^{2} [(2x) - (x^3 - x^2)]\, dx$$

$$= \int_{-1}^{0} (x^3 - x^2 - 2x)\, dx + \int_{0}^{2} (2x - x^3 + x^2)\, dx$$

$$= \left(\frac{x^4}{4} - \frac{x^3}{3} - x^2\right)\Big|_{-1}^{0} + \left(x^2 - \frac{x^4}{4} + \frac{x^3}{3}\right)\Big|_{0}^{2}$$

$$= \left[(0) - \left(\frac{1}{4} - \frac{-1}{3} - 1\right)\right] + \left[\left(4 - \frac{16}{4} + \frac{8}{3}\right) - (0)\right] = \frac{37}{12}$$

The area between the curves is $\frac{37}{12}$ square units.

See the Tech Card for details on ways to use a graphing calculator to find the area between two curves.

Determining the area between two curves has practical applications as well, for example when analyzing a country's income distribution. In economics, the Lorenz curve is used to represent the inequality of income distribution among different groups in the population of a country. The curve is constructed by plotting the cumulative percent of families at or below a given income level and the cumulative percent of total personal income received by these families. For example,

the point (0.80, 0.50) is on the Lorenz curve because the families with incomes in the bottom 80% of the country received 50.0% of the total income. The graph of a Lorenz curve $y = L(x)$ is shown in Figure 13.7.

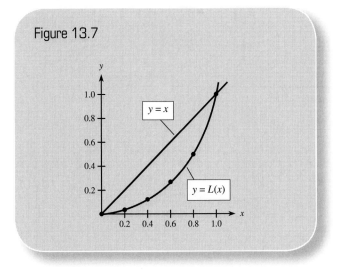

Figure 13.7

Equality of income would result if each family received an equal proportion of the total income, so that the bottom 20% would receive 20% of the total income, the bottom 40% would receive 40%, and so on. The Lorenz curve representing this would have the equation $y = x$.

The inequality of income distribution is measured by the **Gini coefficient** of income, which measures how far the Lorenz curve falls below $y = x$. It is defined as

$$\frac{\text{Area between } y = x \text{ and } y = L(x)}{\text{Area below } y = x} = \frac{\int_{0}^{1} [x - L(x)]\, dx}{1/2}$$

$$= 2\int_{0}^{1} [x - L(x)]\, dx$$

The function $y = L(x) = 2.266x^4 - 3.175x^3 + 2.056x^2 - 0.148x$ models the following 2006 income distribution data.

U.S. Income Distribution for 2006
(Points on the Lorenz Curve)

x, Cumulative Proportion of Families Below Income Level	$y = L(x)$, Cumulative Proportion of Total Income
0	0
0.20	0.034
0.40	0.120
0.60	0.265
0.80	0.498
1	1

Source: U.S. Bureau of the Census

We can use this $L(x)$ to find the Gini coefficient of income for 2006:

$$2\int_0^1 [x - L(x)]\,dx$$

$$= 2\int_0^1 [x - (2.266x^4 - 3.175x^3 + 2.056x^2 - 0.148x)]\,dx$$

$$= 2[-0.453x^5 + 0.794x^4 - 0.685x^3 + 0.574x^2]_0^1$$

$$\approx 0.460$$

After finding the Gini coefficient of income, we can analyze data from individual years in various ways. For example, we can determine during which year in the series the distribution of income is more nearly equal. Absolute equality of income occurs when the Gini coefficient of income is 0; smaller coefficients indicate more nearly equal incomes. Thus, if the Census Bureau Gini coefficient of income for 1991 is 0.428, that means the distribution of income was more nearly equal in 1991 than in 2006.

Average Value

The average value of a continuous function $y = f(x)$ over the interval $[a, b]$ is

$$\text{Average value} = \frac{1}{b-a}\int_a^b f(x)\,dx$$

For the development of this formula, see the Enrichment material on CourseMate for MATH APPS at cengagebrain.com.

EXAMPLE 13.8 Average Cost

Suppose that the total cost in dollars for x table lamps is given by $C(x) = 400 + x + 0.3x^2$.
(a) What is the average value of $C(x)$ for 10 to 20 units?
(b) Find the average cost per unit if 40 lamps are produced.

SOLUTION

(a) The average value of $C(x)$ from $x = 10$ to $x = 20$ is

$$\frac{1}{20 - 10}\int_{10}^{20} (400 + x + 0.3x^2)\,dx$$

$$= \frac{1}{10}\left(400x + \frac{x^2}{2} + 0.1x^3\right)\Big|_{10}^{20}$$

$$= \frac{1}{10}[(8000 + 200 + 800) - (4000 + 50 + 100)]$$

$$= 485 \text{ (dollars)}$$

Thus the total cost averages $485 when between 10 and 20 lamps are produced.

(b) The average cost per unit if 40 units are produced is the average cost function evaluated at $x = 40$.

The average cost function is

$$\overline{C}(x) = \frac{C(x)}{x} = \frac{400}{x} + 1 + 0.3x$$

Thus the *average cost per unit* if 40 units are produced is

$$\overline{C}(40) = \frac{400}{40} + 1 + 0.3(40)$$

$$= 23 \text{ (dollars per unit (i.e., lamp))}$$

$23/unit ($23/lamp)

Shutterstock

13.2 Exercises

In Problems 1 and 2, for each shaded region (a) form the integral that represents the area of the shaded region and (b) find the area of the region.

1.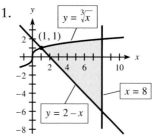

2.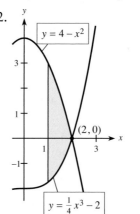

In Problems 3 and 4, for each shaded region (a) find the points of intersection of the curves, (b) form the integral that represents the area of the shaded region, and (c) find the area of the shaded region.

3.

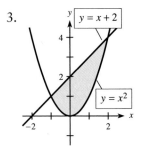

4.

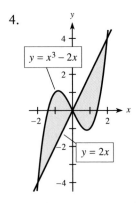

In Problems 5–7, equations are given whose graphs enclose a region. In each problem, find the area of the region.

5. $f(x) = x^2 + 2$; $g(x) = -x^2$; $x = 0$; $x = 2$

6. $y = \frac{1}{2}x^2$; $y = x^2 - 2x$

7. $f(x) = x^3$; $g(x) = x^2 + 2x$

In Problems 8–10, find the average value of each function over the given interval.

8. $f(x) = 9 - x^2$ over $[0, 3]$

9. $f(x) = x^3 - x$ over $[-1, 1]$

10. $f(x) = \frac{1}{2}x^3 + 1$ over $[-2, 0]$

Applications

11. **Sales and advertising** The number of daily sales of a product was found to be given by

$$S = 100xe^{-x^2} + 100$$

x days after the start of an advertising campaign for this product.
(a) Find the average daily sales during the first 20 days of the campaign—that is, from $x = 0$ to $x = 20$.

(b) If no new advertising campaign is begun, what is the average number of sales per day for the next 10 days (from $x = 20$ to $x = 30$)?

12. **Tax burden** The U.S. federal tax burden per capita (in dollars) for selected years since 1960 can be modeled by

$$T(t) = 257.8e^{0.067t}$$

where $t = 0$ represents 1950. (*Source:* Internal Revenue Service) If the model remains valid, find the predicted average federal tax burden per capita from 2000 to 2010.

13. **Income distribution** U.S. political analysts tend to believe that the tax policies of Republican administrations tend to favor the wealthy and those of Democratic administrations tend to favor the poor. The Lorenz curves for income distribution in 1988 (Reagan's final year) and 2000 (Clinton's final year) are given. Find the Gini coefficient of income for each year and determine in which administration's final year the income distribution was more nearly equal. Does this support the conventional wisdom regarding the tax policies of Republicans versus Democrats?

$$1988: y = x^{2.3521} \quad 2000: y = x^{2.4870}$$

Need more practice?
Find more here: cengagebrain.com

13.3 Definite Integrals in Business and Economics

THE DEFINITE INTEGRAL CAN BE USED IN A NUMBER OF APPLICATIONS IN BUSINESS AND ECONOMICS. FOR A CONTINUOUS INCOME STREAM OVER A FIXED NUMBER OF YEARS, THE DEFINITE INTEGRAL CAN BE USED TO FIND THE STREAM'S TOTAL INCOME, PRESENT VALUE, AND FUTURE VALUE.

DEFINITE INTEGRALS ALSO CAN BE USED TO DETERMINE THE SAVINGS REALIZED IN THE MARKETPLACE BY SOME CONSUMERS (CALLED CONSUMER'S SURPLUS) AND SOME PRODUCERS (CALLED PRODUCER'S SURPLUS).

Continuous Income Streams

An oil company's profits depend on the amount of oil that can be pumped from a well. Thus we can consider a pump at an oil field as producing a **continuous stream of income** for the owner. Because both the pump and the oil field "wear out" with time, the continuous stream of income is a function of time. Suppose $f(t)$ is the (annual) *rate* of flow of income from this pump; then we can find the total income from the rate of income by using integration. In particular, the total income for k years is given by

$$\text{Total income} = \int_0^k f(t)\, dt$$

Now ———————————→ 10 Years From Now

$2,594,000 = Total Income

SOLUTION

$$\text{Total income} = \int_0^{10} f(t)\, dt = \int_0^{10} 600e^{-0.2t}\, dt$$

$$= \frac{600}{-0.2} e^{-0.2t} \Big|_0^{10}$$

$$\approx 2594 \quad (\text{to the nearest integer})$$

Thus the total income is approximately $2,594,000.

In addition to the total income from a continuous income stream, the **present value** of the stream is also important. The present value is the value today of a continuous income stream that will be providing income in the future. The present value is useful in deciding when to replace machinery or what new equipment to select.

> **Present Value of a Continuous Income Stream**
> If $f(t)$ is the rate of continuous income flow earning interest at rate r, compounded continuously, then the present value of the continuous income stream is
> $$\text{Present value} = \int_0^k f(t)e^{-rt}\, dt$$
> where $t = 0$ to $t = k$ is the time interval.

EXAMPLE 13.9 Oil Revenue

A small oil company considers the continuous pumping of oil from a well as a continuous income stream with its annual rate of flow at time t given by

$$f(t) = 600e^{-0.2t}$$

in thousands of dollars per year. Find an estimate of the total income from this well over the next 10 years.

EXAMPLE 13.10 Present Value

Suppose that the oil company in Example 13.9 is planning to sell one of its wells because of its remote location. Suppose further that the company wants to use the present value of this well over the next 10 years to help establish its selling price. If the company determines that the annual rate of flow is

$$f(t) = 600e^{-0.2(t+5)}$$

in thousands of dollars per year, and if money is worth 10%, compounded continuously, find this present value.

SOLUTION

$$\text{Present value} = \int_0^{10} f(t)e^{-rt}\, dt$$

$$= \int_0^{10} 600e^{-0.2(t+5)}e^{-0.1t}\, dt$$

$$= \int_0^{10} 600e^{-0.3t-1}\, dt$$

If $u = -0.3t - 1$, then $u' = -0.3$ and we get

$$\frac{1}{-0.3}\int 600e^{-0.3t-1}(-0.3\, dt) = \frac{600}{-0.3}e^{-0.3t-1}\bigg|_0^{10}$$

$$= -2000(e^{-4} - e^{-1})$$

$$\approx 699 \quad \text{(to the nearest integer)}$$

Thus the present value is \$699,000.

Recall that the future value of a continuously compounded investment at rate r after k years is Pe^{rk}, where P is the amount invested (or the present value). Thus, for a continuous income stream, the future value is found as follows.

> **Future Value of a Continuous Income Stream**
>
> If $f(t)$ is the rate of continuous income flow for k years earning interest at rate r, compounded continuously, then the future value of the continuous income stream is
>
> $$FV = e^{rk}\int_0^k f(t)e^{-rt}\, dt$$

EXAMPLE 13.11 Future Value

If the rate of flow of income from an asset is $1000e^{0.02t}$, in millions of dollars per year, and if the income is invested at 6% compounded continuously, find the future value of the asset 4 years from now.

SOLUTION

The future value is given by

$$FV = e^{rk}\int_0^k f(t)e^{-rt}\, dt$$

$$= e^{(0.06)4}\int_0^4 1000e^{0.02t}e^{-0.06t}\, dt = e^{0.24}\int_0^4 1000e^{-0.04t}\, dt$$

$$= e^{0.24}\big[-25,000e^{-0.04t}\big]_0^4 = -25,000e^{0.24}\big[e^{-0.16} - 1\big]$$

$$\approx 4699.05 \quad \text{(millions of dollars)}$$

Consumer's Surplus

Suppose that the demand for a product is given by $p = f(x)$ and that the supply of the product is described by $p = g(x)$. The price p_1 where the graphs of these functions intersect is the **equilibrium price** (see Figure 13.8(a)). As the demand curve shows, some consumers (but not all) would be willing to pay more than \$$p_1$ for the product.

For example, some consumers would be willing to buy x_3 units if the price were \$$p_3$. Those consumers willing to pay more than \$$p_1$ are benefiting from the lower price. The total gain for all those consumers willing to pay more than \$$p_1$ is called the **consumer's surplus**, and

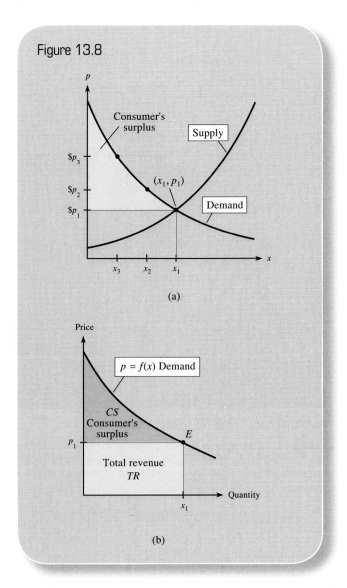

Figure 13.8

under proper assumptions the area of the shaded region in Figure 13.8(a) represents this consumer's surplus.

Looking at Figure 13.8(b), we see that if the demand curve has equation $p = f(x)$, the consumer's surplus is given by the area between $f(x)$ and the x-axis from 0 to x_1, *minus* the area of the rectangle denoted *TR*:

$$CS = \int_0^{x_1} f(x)\, dx - p_1 x_1$$

Note that with equilibrium price p_1 and equilibrium quantity x_1, the product $p_1 x_1$ is the area of the rectangle that represents the total dollars spent by consumers and received as revenue by producers (see Figure 13.8(b)).

EXAMPLE 13.12 Consumer's Surplus

→ The demand function for x units of a product is $p = 1020/(x + 1)$ dollars. If the equilibrium price is $20, what is the consumer's surplus?

SOLUTION

We must first find the quantity that will be purchased at this price. Letting $p = 20$ and solving for x, we get

$$20 = \frac{1020}{x + 1}$$

$$20(x + 1) = 1020$$

$$20x + 20 = 1020 \quad \text{or} \quad 20x = 1000$$

$$x = 50$$

Thus the equilibrium point is (50, 20). The consumer's surplus is given by

$$
\begin{aligned}
CS &= \int_0^{x_1} f(x)\, dx - p_1 x_1 = \int_0^{50} \frac{1020}{x + 1}\, dx - 20 \cdot 50 \\
&= 1020 \ln |x + 1| \Big|_0^{50} - 1000 \\
&= 1020(\ln 50 - \ln 1) - 1000 \\
&\approx 1020(3.912 - 0) - 1000 \\
&= 3990.24 - 1000 \\
&= 2990.24
\end{aligned}
$$

The consumer's surplus is $2990.24.

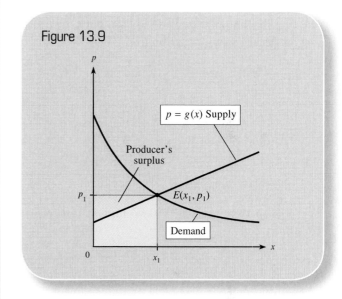

Figure 13.9

Producer's Surplus

When a product is sold at the equilibrium price, some producers will also benefit, for they would have sold the product at a lower price. The area between the line $p = p_1$ and the supply curve (from $x = 0$ to $x = x_1$) gives the producer's surplus (see Figure 13.9).

If the supply function is $p = g(x)$, the **producer's surplus** is given by the area between the graph of $p = g(x)$ and the x-axis from 0 to x_1 *subtracted from* $p_1 x_1$, the area of the rectangle shown in Figure 13.9.

$$PS = p_1 x_1 - \int_0^{x_1} g(x)\, dx$$

Note that $p_1 x_1$ represents the total revenue at the equilibrium point.

MONEY ON THE TABLE

Another way to understand consumer surplus is through the popular concept of "money on the table." Imagine you have $300 to spend on an iPod. When you get to the Apple store, the model you want costs only $250. You're delighted to spend only $250, but you **would** have spent all your money if the iPod had been priced at $300. In this example, Apple left money—$50 to be exact—on the table (or in your pocket). That's a consumer's surplus.

GRAVY

Now think of the iPod example in another way. Suppose Apple meets all its profit targets at a price of $200 per iPod. The fact that you pay $250 means that there is a producer surplus. Apple would sell that iPod at the $200 price point, but because you were willing to part with $250, the company receives an extra $50—or gravy—for its bottom line. That is, a producer's surplus.

Shutterstock (both)

EXAMPLE 13.13 Producer's Surplus

➡️ Suppose that the supply function for x million units of a product is $p = x^2 + x$ dollars per unit. If the equilibrium price is $20 per unit, what is the producer's surplus?

SOLUTION

Because $p = 20$, we can find the equilibrium quantity x as follows:

$$20 = x^2 + x$$
$$0 = x^2 + x - 20$$
$$0 = (x + 5)(x - 4)$$
$$x = -5, x = 4$$

The equilibrium point is $x = 4$ million units, $p = \$20$. The producer's surplus is given by

$$PS = 20 \cdot 4 - \int_0^4 (x^2 + x)\, dx$$
$$= 80 - \left(\frac{x^3}{3} + \frac{x^2}{2}\right)\Big|_0^4$$
$$= 80 - \left(\frac{64}{3} + 8\right)$$
$$\approx 50.67$$

The producer's surplus is $50.67 million (see Figure 13.10).

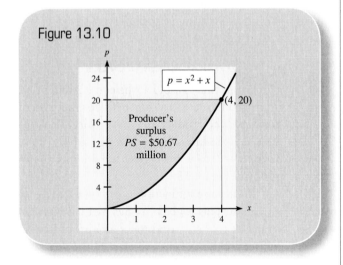

Figure 13.10

13.3 Exercises

Continuous Income Streams

1. Find the total income over the next 10 years from a continuous income stream that has an annual rate of flow at time t given by $f(t) = 12{,}000e^{0.01t}$ (dollars per year)

2. Suppose that a steel company views the production of its continuous caster as a continuous income stream with a monthly rate of flow at time t given by

$$f(t) = 24{,}000e^{0.03t} \quad \text{(dollars per month)}$$

Find the total income from this caster in the first year.

3. A small brewery considers the output of its bottling machine as a continuous income stream with an annual rate of flow at time t given by

$$f(t) = 80e^{-0.1t}$$

in thousands of dollars per year. Find the income from this stream for the next 10 years.

4. A franchise models the profit from its store as a continuous income stream with a monthly rate of flow at time t given by

$$f(t) = 3000e^{0.004t} \quad \text{(dollars per month)}$$

When a new store opens, its manager is judged against the model, with special emphasis on the second half of the first year. Find the total profit for the second 6-month period ($t = 6$ to $t = 12$).

5. The income from an established chain of Laundromats is a continuous stream with its annual rate of flow at time t given by $f(t) = 630{,}000$ (dollars per year). If money is worth 7% compounded continuously, find the present value and future value of this chain over the next 5 years.

Consumer's Surplus

In Problems 6–8, p is in dollars and x is the number of units.

6. The demand function for a product is $p = 34 - x^2$. If the equilibrium price is $9 per unit, what is the consumer's surplus?

7. The demand function for a product is $p = 200/(x + 2)$. If the equilibrium quantity is 8 units, what is the consumer's surplus?

8. The demand function for a certain product is $p = 81 - x^2$ and the supply function is $p = x^2 + 4x + 11$. Find the equilibrium point and the consumer's surplus there.

Producer's Surplus

In Problems 9–11, p is in dollars and x is the number of units.

9. Suppose that the supply function for a good is $p = 4x^2 + 2x + 2$. If the equilibrium price is \$422 per unit, what is the producer's surplus there?

10. If the supply function for a commodity is $p = 10e^{x/3}$, what is the producer's surplus when 15 units are sold?

11. Find the producer's surplus for a product if its demand function is $p = 81 - x^2$ and its supply function is $p = x^2 + 4x + 11$.

....

Need more practice?
Find more here: cengagebrain.com

13.4 Using Tables of Integrals

THE FORMULAS IN TABLE 13.1, AND OTHERS LISTED IN OTHER RESOURCES, SUCH AS THE CHEMICAL RUBBER COMPANY'S *STANDARD MATHEMATICAL TABLES,* EXTEND THE NUMBER OF INTEGRALS THAT CAN BE EVALUATED. USING THE FORMULAS IS NOT QUITE AS EASY AS IT MAY SOUND BECAUSE FINDING THE CORRECT FORMULA AND USING IT PROPERLY MAY PRESENT PROBLEMS. THE EXAMPLES IN THIS SECTION ILLUSTRATE HOW SOME OF THESE FORMULAS ARE USED.

EXAMPLE 13.14 Using Formulas

Evaluate $\displaystyle\int \frac{dx}{\sqrt{x^2 + 4}}$.

SOLUTION

We must find a formula in Table 13.1 that is of the same form as this integral. We see that formula 8 has the desired form, *if* we let $u = x$ and $a = 2$. Thus

$$\int \frac{dx}{\sqrt{x^2 + 4}} = \ln\left|x + \sqrt{x^2 + 4}\right| + C$$

TABLE 13.1 Integration Formulas

1. $\displaystyle\int u^n\,du = \frac{u^{n+1}}{n+1} + C,\ \text{for } n \ne -1$

2. $\displaystyle\int \frac{du}{u} = \int u^{-1}\,du = \ln|u| + C$

3. $\displaystyle\int a^u\,du = a^u \log_a e + C = \frac{a^u}{\ln a} + C$

4. $\displaystyle\int e^u\,du = e^u + C$

5. $\displaystyle\int \frac{du}{a^2 - u^2} = \frac{1}{2a} \ln\left|\frac{a+u}{a-u}\right| + C$

6. $\displaystyle\int \sqrt{u^2 + a^2}\,du = \frac{1}{2}\left(u\sqrt{u^2+a^2} + a^2 \ln\left|u + \sqrt{u^2+a^2}\right|\right) + C$

7. $\displaystyle\int \sqrt{u^2 - a^2}\,du = \frac{1}{2}\left(u\sqrt{u^2-a^2} - a^2 \ln\left|u + \sqrt{u^2-a^2}\right|\right) + C$

8. $\displaystyle\int \frac{du}{\sqrt{u^2 + a^2}} = \ln\left|u + \sqrt{u^2+a^2}\right| + C$

9. $\displaystyle\int \frac{du}{u\sqrt{a^2 - u^2}} = -\frac{1}{a} \ln\left|\frac{a + \sqrt{a^2-u^2}}{u}\right| + C$

10. $\displaystyle\int \frac{du}{\sqrt{u^2 - a^2}} = \ln\left|u + \sqrt{u^2-a^2}\right| + C$

11. $\displaystyle\int \frac{du}{u\sqrt{a^2 + u^2}} = -\frac{1}{a} \ln\left|\frac{a + \sqrt{a^2+u^2}}{u}\right| + C$

12. $\displaystyle\int \frac{u\,du}{au+b} = \frac{u}{a} - \frac{b}{a^2} \ln|au+b| + C$

13. $\displaystyle\int \frac{du}{u(au+b)} = \frac{1}{b} \ln\left|\frac{u}{au+b}\right| + C$

14. $\displaystyle\int \ln u\,du = u(\ln u - 1) + C$

15. $\displaystyle\int \frac{u\,du}{(au+b)^2} = \frac{1}{a^2}\left(\ln|au+b| + \frac{b}{au+b}\right) + C$

16. $\displaystyle\int u\sqrt{au+b}\,du = \frac{2(3au - 2b)(au+b)^{3/2}}{15a^2} + C$

EXAMPLE 13.15 Fitting a Formula

Evaluate $\displaystyle\int_1^2 \frac{dx}{x^2 + 2x}$.

SOLUTION

There does not appear to be any formula that has exactly the same form as our integral. But if we rewrite our integral as

$$\int_1^2 \frac{dx}{x(x+2)}$$

we see that formula 13 will work. Letting $u = x$, $a = 1$, and $b = 2$, we get

$$\int_1^2 \frac{dx}{x(x+2)} = \frac{1}{2} \ln \left| \frac{x}{x+2} \right| \Big|_1^2 = \frac{1}{2} \ln \left| \frac{2}{4} \right| - \frac{1}{2} \ln \left| \frac{1}{3} \right|$$

$$= \frac{1}{2} \left(\ln \frac{1}{2} - \ln \frac{1}{3} \right)$$

$$= \frac{1}{2} \ln \frac{3}{2} \approx 0.2027$$

Although the formulas in Table 13.1 are given in terms of the variable u, they may be used with any variable.

EXAMPLE 13.16 Using Formulas

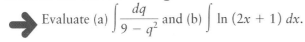 Evaluate (a) $\int \frac{dq}{9 - q^2}$ and (b) $\int \ln (2x + 1)\, dx$.

SOLUTION

(a) The formula that applies in this case is formula 5, with $a = 3$ and $u = q$. Then

$$\int \frac{dq}{9 - q^2} = \frac{1}{2 \cdot 3} \ln \left| \frac{3 + q}{3 - q} \right| + C = \frac{1}{6} \ln \left| \frac{3 + q}{3 - q} \right| + C$$

(b) This integral has the form of formula 14, with $u = 2x + 1$. But if $u = 2x + 1$, du must be represented by the differential of $2x + 1$ (that is, $2\,dx$). Thus

$$\int \ln (2x + 1)\, dx = \frac{1}{2} \int \ln (2x + 1)(2\, dx)$$

$$= \frac{1}{2} \int \ln (u)\, du = \frac{1}{2} u[\ln (u) - 1] + C$$

$$= \frac{1}{2} (2x + 1)[\ln (2x + 1) - 1] + C$$

Remember that the formulas given in Table 13.1 represent only a very small sample of all possible integration formulas. Additional formulas may be found in books of mathematical tables.

Now that we've considered several integration formulas in theory, let's examine how we might use them in practice. With data from 1990 through 2007, the rate of change of total market revenue for worldwide telecommunications (in billions of dollars per year) can be modeled by

$$\frac{dR}{dt} = 37.988(1.101^t)$$

where $t = 0$ represents 1990 and $R = \$920$ billion in 1999. (*Source*: International Telecommunications Union)

This information allows us to find the function that models the total market revenue for worldwide telecommunications as well as the predicted total market revenue for a given year.

To find market revenue, or $R(t)$, we integrate dR/dt by using formula 3, with $a = 1.101$ and $u = t$.

$$R(t) = \int 37.988(1.101^t)\, dt = 37.988 \int (1.101^t)\, dt$$

$$= 37.988 \frac{(1.101^t)}{\ln (1.101)} + C$$

$$R(t) \approx 394.808(1.101^t) + C$$

We use the fact that $R = \$920$ billion in 1999 (when $t = 9$) to find the value of C.

$$920 = 394.808(1.101^9) + C \Rightarrow$$
$$920 = 938.6 + C \Rightarrow C = -18.6$$

Thus $R(t) = 394.808(1.101^t) - 18.6$. We can then find the predicted revenue for a given year (say, in 2015, when $t = 25$) as follows:

$$R(25) = 394.808(1.101^{25}) - 18.6 \approx 4357 \text{ billion dollars}$$

$4357 billion

13.4 Exercises

Evaluate the integrals in Problems 1–8. Identify the formula used.

1. $\displaystyle\int \frac{dx}{16 - x^2}$

2. $\displaystyle\int_1^4 \frac{dx}{x\sqrt{9 + x^2}}$

3. $\displaystyle\int \ln w \, dw$

4. $\displaystyle\int_0^2 \frac{q \, dq}{6q + 9}$

5. $\displaystyle\int \frac{dv}{v(3v + 8)}$

6. $\displaystyle\int_5^7 \sqrt{x^2 - 25} \, dx$

7. $\displaystyle\int w\sqrt{4w + 5} \, dw$

8. $\displaystyle\int \frac{3x \, dx}{(2x - 5)^2}$

Applications

9. **Producer's surplus** If the supply function for x units of a commodity is $p = 40 + 100 \ln (x + 1)^2$ dollars, what is the producer's surplus at $x = 20$?

10. **Cost** (a) If the marginal cost for x units of a good is $\overline{MC} = \sqrt{x^2 + 9}$ (dollars per unit) and if the fixed cost is \$300, what is the total cost function of the good? (b) What is the total cost of producing 4 units of this good?

11. **Income streams** Suppose that when a new oil well is opened, its production is viewed as a continuous income stream with monthly rate of flow

$$f(t) = 10 \ln (t + 1) - 0.1t$$

where t is time in months and $f(t)$ is in thousands of dollars per month. Find the total income over the next 10 years (120 months).

> **Need more practice?**
> Find more here: **cengagebrain.com**

13.5 | Integration by Parts

INTEGRATION BY PARTS IS AN INTEGRATION TECHNIQUE THAT USES A FORMULA THAT FOLLOWS FROM THE PRODUCT RULE FOR DERIVATIVES (ACTUALLY DIFFERENTIALS).

Using the product rule to differentiate uv gives

$$\frac{d}{dx}(uv) = u\frac{dv}{dx} + v\frac{du}{dx} \quad \text{so} \quad d(uv) = u \, dv + v \, du$$

Rearranging the differential form and integrating both sides give the following.

$$u \, dv = d(uv) - v \, du$$

$$\int u \, dv = \int d(uv) - \int v \, dv$$

$$\int u \, dv = uv - \int v \, dv$$

> **Integration by Parts Formula**
>
> $$\int u \, dv = uv - \int v \, du$$

Integration by parts is useful if the integral we seek to evaluate can be treated as the product of one function, u, and the differential dv of a second function, so that the two integrals $\int dv$ and $\int v \, du$ can be found. Let us consider an example using this method.

EXAMPLE 13.17 Integration by Parts

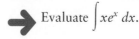 Evaluate $\displaystyle\int xe^x \, dx$.

SOLUTION

We cannot evaluate this integral using methods we have learned. But we can "split" the integrand into two parts, setting one part equal to u and the other part equal to dv. This "split" must be done in such a way that $\int dv$ and $\int v \, du$ can be evaluated. Letting $u = x$ and letting $dv = e^x \, dx$ are possible choices. If we make these choices, we have

$$u = x \qquad dv = e^x \, dx$$
$$du = 1 \, dx \quad v = \int e^x \, dx = e^x$$

Then

$$\int xe^x \, dx = uv - \int v \, du$$
$$= xe^x - \int e^x \, dx$$
$$= xe^x - e^x + C$$

We see that choosing $u = x$ and $dv = e^x \, dx$ worked in evaluating $\int xe^x \, dx$ in Example 13.17. If we had chosen $u = e^x$ and $dv = x \, dx$, the results would not have been so successful.

How can we select u and dv to make integration by parts work? As a general guideline, we do the following.

First, identify the types of functions occurring in the problem in the order

Logarithm, Polynomial (or Power of x), Radical, Exponential*

Thus, in Example 13.17, we had x and e^x, a polynomial and an exponential.

Second, we choose u to equal the function whose type occurs first on the list; hence in Example 13.17 we chose $u = x$. Then dv equals the rest of the integrand (and always includes dx) so that $u\, dv$ equals the original integrand. A helpful way to remember the order of the function types that help us choose u is the sentence

"Lazy People Rarely Excel."

in which the first letters, LPRE, coordinate with the order and types of functions. Consider the following examples.

EXAMPLE 13.18 Integration by Parts

 Evaluate $\int x \ln x\, dx$.

SOLUTION

The integral contains a logarithm ($\ln x$) and a polynomial (x). Thus, let $u = \ln x$ and $dv = x\, dx$. Then

$$du = \frac{1}{x}\, dx \quad \text{and} \quad v = \frac{x^2}{2}$$

so

$$\int x \ln x\, dx = u \cdot v - \int v\, du$$
$$= (\ln x)\frac{x^2}{2} - \int \frac{x^2}{2} \cdot \frac{1}{x}\, dx$$
$$= \frac{x^2}{2} \ln x - \int \frac{x}{2}\, dx$$
$$= \frac{x^2}{2} \ln x - \frac{x^2}{4} + C$$

*This order is related to the ease with which the function types can be integrated.

Note that letting $dv = \ln x\, dx$ is contrary to our guidelines and would lead to great difficulty in evaluating $\int dv$ and $\int v\, du$.

Sometimes it is necessary to repeat integration by parts to complete the evaluation. When this occurs, at each use of integration by parts it is important to choose u and dv consistently according to our guidelines.

EXAMPLE 13.19 Repeated Integration by Parts

 Evaluate $\int x^2 e^{2x}\, dx$.

SOLUTION

Let $u = x^2$ and $dv = e^{2x}\, dx$, so $du = 2x\, dx$ and $v = \frac{1}{2}e^{2x}$. Then

$$\int x^2 e^{2x}\, dx = \frac{1}{2}x^2 e^{2x} - \int x e^{2x}\, dx$$

We cannot evaluate $\int x e^{2x}\, dx$ directly, but this new integral is simpler than the original, and a second integration by parts will be successful. Letting $u = x$ and $dv = e^{2x}\, dx$ gives $du = dx$ and $v = \frac{1}{2}e^{2x}$. Thus

$$\int x^2 e^{2x}\, dx = \frac{1}{2}x^2 e^{2x} - \left(\frac{1}{2}x e^{2x} - \int \frac{1}{2}e^{2x}\, dx\right)$$
$$= \frac{1}{2}x^2 e^{2x} - \frac{1}{2}x e^{2x} + \frac{1}{4}e^{2x} + C$$
$$= \frac{1}{4}e^{2x}(2x^2 - 2x + 1) + C$$

The most obvious choices for u and dv are not always the correct ones. Integration by parts may still involve some trial and error. Furthermore, some functions that arise in practical problems cannot be integrated with any formula. However, a definite integral of such a function can be approximated by using numerical integration methods (see the Enrichment material).

13.5 Exercises

In Problems 1–6, use integration by parts to evaluate the integral.

1. $\int x e^{2x}\, dx$

2. $\int x^2 \ln x\, dx$

3. $\int_4^6 q\sqrt{q - 4}\, dq$

4. $\int \dfrac{\ln x}{x^2}\, dx$

5. $\int_{1}^{e} \ln x\, dx$

6. $\int x^2 e^{-x}\, dx$ (Note this may require integration by parts more than once.)

Applications

7. **Producer's surplus** If the supply function for x units of a commodity is $p = 30 + 50 \ln (2x + 1)^2$ dollars, what is the producer's surplus at $x = 30$?

8. **Present value** Suppose that a machine's production can be considered as a continuous income stream with annual rate of flow at time t given by

$$f(t) = 10{,}000 - 500t \text{ (dollars per year)}$$

If money is worth 10%, compounded continuously, find the present value of the machine over the next 5 years.

9. **Income distribution** Suppose the Lorenz curve for the distribution of income of a certain country is given by

$$y = xe^{x-1}$$

Find the Gini coefficient of income.

Need more practice?
Find more here: cengagebrain.com

13.6 Improper Integrals and Their Applications

SOME APPLICATIONS OF CALCULUS TO STATISTICS OR BUSINESS INVOLVE DEFINITE INTEGRALS OVER INTERVALS OF INFINITE LENGTH (IMPROPER INTEGRALS). THE AREA OF A REGION THAT EXTENDS INFINITELY TO THE LEFT OR RIGHT ALONG THE *X*-AXIS (SEE FIGURE 13.11) COULD BE DESCRIBED BY AN **IMPROPER INTEGRAL**.

Figure 13.11

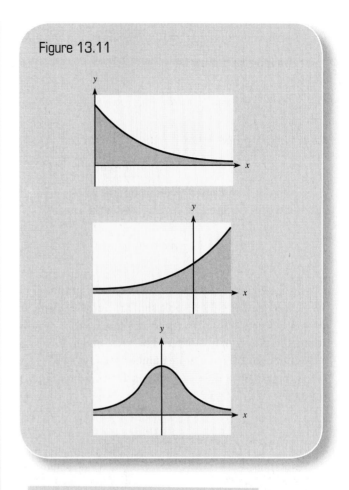

Improper Integral

$$\int_{a}^{\infty} f(x)\, dx = \lim_{b \to \infty} \int_{a}^{b} f(x)\, dx$$

If the limit defining the improper integral is a unique finite number, we say that the integral *converges*; otherwise, we say that the integral *diverges*.

EXAMPLE 13.20 Improper Integrals

Evaluate the following improper integrals, if they converge.

(a) $\displaystyle\int_{1}^{\infty} \dfrac{1}{x^3}\, dx$ (b) $\displaystyle\int_{1}^{\infty} \dfrac{1}{x}\, dx$

SOLUTION

(a) $\displaystyle\int_{1}^{\infty} \dfrac{1}{x^3}\, dx = \lim_{b \to \infty} \int_{1}^{b} \dfrac{1}{x^3}\, dx = \lim_{b \to \infty} \left[\dfrac{x^{-2}}{-2} \right]_{1}^{b}$

$$= \lim_{b \to \infty} \left[\dfrac{-1}{2b^2} - \left(\dfrac{-1}{2(1)^2} \right) \right]$$

$$= \lim_{b \to \infty} \left(\dfrac{-1}{2b^2} + \dfrac{1}{2} \right)$$

Now, as $b \to \infty$, $\frac{-1}{2b^2} \to 0$, so the limit and the integral converge to $\frac{1}{2}$. That is,

$$\int_1^\infty \frac{1}{x^3} \, dx = \frac{1}{2}$$

(b) $\int_1^\infty \frac{1}{x} \, dx = \lim_{b \to \infty} \int_1^b \frac{1}{x} \, dx = \lim_{b \to \infty} \left[\ln |x| \right]_1^b$

$$= \lim_{b \to \infty} (\ln b - \ln 1)$$

Now $\ln b$ increases without bound as $b \to \infty$, so the limit and the integral diverge. We write this as

$$\int_1^\infty \frac{1}{x} \, dx = \infty$$

As Figure 13.12 shows, the graphs of $y = 1/x^2$ and $y = 1/x$ look similar, but the graph of $y = 1/x^2$ gets "close" to the x-axis much more rapidly than the graph of $y = 1/x$. The area under $y = 1/x$ does not converge to a finite number because as $x \to \infty$ the graph of $1/x$ does not approach the x-axis rapidly enough.

Figure 13.12

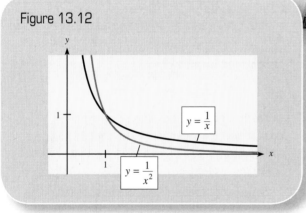

Now that we've examined improper integrals in the abstract, let's consider how they can be applied to economic problems. Suppose that an organization wants to establish a trust fund that will provide a perpetual continuous income stream with an annual rate of flow at time t given by $f(t) = 10,000$ dollars per year. If the interest rate remains at 10% compounded continuously, find the *capital value* of the fund (i.e., the present value needed to establish the fund).

The formula for the capital value of a continuous income stream is given by

$$\int_0^\infty f(t) e^{-rt} \, dt$$

where $f(t)$ is the annual rate of flow at time t, and r is the annual interest rate, compounded continuously.

$$\int_0^\infty 10{,}000 e^{-0.10t} \, dt = \lim_{b \to \infty} \int_0^b 10{,}000 e^{-0.10t} \, dt$$

$$= \lim_{b \to \infty} \left[-100{,}000 e^{-0.10t} \right]_0^b$$

$$= \lim_{b \to \infty} \left(\frac{-100{,}000}{e^{0.10b}} + 100{,}000 \right)$$

$$= 100{,}000$$

Thus the capital value of the fund is $100,000.

Another term for a fund like the one on our example is a **perpetuity**. Usually the rate of flow of a perpetuity is a constant. If the rate of flow is a constant A, it can be shown that the capital value is given by A/r.

PERPETUITY

Two additional improper integrals that involve infinite limits are defined next.

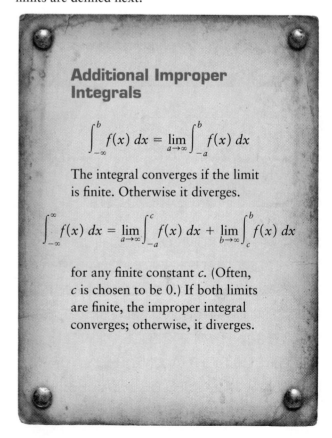

Additional Improper Integrals

$$\int_{-\infty}^b f(x) \, dx = \lim_{a \to \infty} \int_{-a}^b f(x) \, dx$$

The integral converges if the limit is finite. Otherwise it diverges.

$$\int_{-\infty}^\infty f(x) \, dx = \lim_{a \to \infty} \int_{-a}^c f(x) \, dx + \lim_{b \to \infty} \int_c^b f(x) \, dx$$

for any finite constant c. (Often, c is chosen to be 0.) If both limits are finite, the improper integral converges; otherwise, it diverges.

EXAMPLE 13.21 Improper Integrals

 Evaluate the following integrals.

(a) $\int_{-\infty}^{4} e^{3x}\, dx$ (b) $\int_{-\infty}^{\infty} \frac{x^3}{(x^4 + 3)^2}\, dx$

SOLUTION

(a) $\displaystyle\int_{-\infty}^{4} e^{3x}\, dx = \lim_{a \to \infty} \int_{-a}^{4} e^{3x}\, dx$

$= \displaystyle\lim_{a \to \infty} \left[\left(\frac{1}{3} \right) e^{3x} \right]_{-a}^{4}$

$= \displaystyle\lim_{a \to \infty} \left[\left(\frac{1}{3} \right) e^{12} - \left(\frac{1}{3} \right) e^{-3a} \right]$

$= \displaystyle\lim_{a \to \infty} \left[\left(\frac{1}{3} \right) e^{12} - \left(\frac{1}{3} \right)\left(\frac{1}{e^{3a}} \right) \right]$

$= \dfrac{1}{3} e^{12}$ (because $1/e^{3a} \to 0$ as $a \to \infty$)

(b) $\displaystyle\int_{-\infty}^{\infty} \frac{x^3}{(x^4 + 3)^2}\, dx$

$= \displaystyle\lim_{a \to \infty} \int_{-a}^{0} \frac{x^3}{(x^4 + 3)^2}\, dx + \lim_{b \to \infty} \int_{0}^{b} \frac{x^3}{(x^4 + 3)^2}\, dx$

$= \displaystyle\lim_{a \to \infty} \left[\frac{1}{4} \frac{(x^4 + 3)^{-1}}{-1} \right]_{-a}^{0} + \lim_{b \to \infty} \left[\frac{1}{4} \frac{(x^4 + 3)^{-1}}{-1} \right]_{0}^{b}$

$= \displaystyle\lim_{a \to \infty} \left[\frac{-1}{4} \left(\frac{1}{3} - \frac{1}{a^4 + 3} \right) \right] + \lim_{b \to \infty} \left[\frac{-1}{4} \left(\frac{1}{b^4 + 3} - \frac{1}{3} \right) \right]$

$= \left[-\dfrac{1}{12} + 0 \right] + \left[0 + \dfrac{1}{12} \right] = 0$

$\left(\text{since } \displaystyle\lim_{a \to \infty} \frac{1}{a^4 + 3} = 0 \text{ and } \lim_{b \to \infty} \frac{1}{b^4 + 3} = 0 \right)$

13.6 Exercises

In Problems 1–5, evaluate the improper integrals that converge.

1. $\displaystyle\int_{1}^{\infty} \frac{dx}{x^6}$

2. $\displaystyle\int_{1}^{\infty} \frac{dt}{t^{3/2}}$

3. $\displaystyle\int_{-\infty}^{-1} \frac{10}{x^2}\, dx$

4. $\displaystyle\int_{-\infty}^{-1} \frac{6}{x}\, dx$

5. $\displaystyle\int_{-\infty}^{\infty} x^3 e^{-x^4}\, dx$

6. For what value of c does $\displaystyle\int_{0}^{\infty} \frac{c}{e^{0.5t}}\, dt = 1$?

7. Find the area, if it exists, of the region under the graph of $y = f(x) = \dfrac{x}{e^{x^2}}$ and to the right of $x = 1$.

Application

8. **Capital value** Suppose that a business provides a continuous income stream with an annual rate of flow at time t given by $f(t) = 120 e^{0.04t}$ in thousands of dollars per year. If the interest rate is 9% compounded continuously, find the capital value of the business.

> **Need more practice?**
> Find more here: cengagebrain.com

Chapter Exercises

1. Use a definite integral to find the area under the graph of $y = 3x^2$ from $x = 0$ to $x = 1$.

In Problems 2–13, evaluate each integral.

2. $\displaystyle\int_{1}^{4} 4\sqrt{x^3}\, dx$

3. $\displaystyle\int_{-3}^{2} (x^3 - 3x^2 + 4x + 2)\, dx$

4. $\displaystyle\int_{0}^{5} (x^3 + 4x)\, dx$

5. $\displaystyle\int_{-1}^{3} (3x + 4)^{-2}\, dx$

6. $\displaystyle\int_{-3}^{-1} (x + 1)\, dx$

7. $\displaystyle\int_{2}^{3} \frac{x^2}{2x^3 - 7}\, dx$

8. $\displaystyle\int_{-1}^{2} (x^2 + x)\, dx$

9. $\displaystyle\int_{1}^{4} \left(\frac{1}{x} + \sqrt{x} \right) dx$

10. $\int_0^2 5x^2(6x^3 + 1)^{1/2}\, dx$

11. $\int_0^1 \dfrac{x}{x^2 + 1}\, dx$

12. $\int_0^1 e^{-2x}\, dx$

13. $\int_0^1 xe^{x^2}\, dx$

In Problems 14–16, find the area between the curves.

14. $y = x^2 - 3x + 2$ and $y = x^2 + 4$ from $x = 0$ to $x = 5$

15. $y = x^2$ and $y = 4x + 5$

16. $y = x^3 - 1$ and $y = x - 1$

In Problems 17–20, evaluate the integrals, using the formulas in Table 13.1.

17. $\int \sqrt{x^2 - 4}\, dx$

18. $\int_0^1 3^x\, dx$

19. $\int x \ln x^2\, dx$

20. $\int \dfrac{dx}{x(3x + 2)}$

In Problems 21–24, use integration by parts to evaluate.

21. $\int x^5 \ln x\, dx$

22. $\int x^2 e^{-2x}\, dx$

23. $\int \dfrac{x\, dx}{\sqrt{x + 5}}$

24. $\int_1^e \ln x\, dx$

In Problems 25–28, evaluate the improper integrals.

25. $\int_1^\infty \dfrac{1}{x + 1}\, dx$

26. $\int_{-\infty}^{-1} \dfrac{200}{x^3}\, dx$

27. $\int_0^\infty 5e^{-3x}\, dx$

28. $\int_{-\infty}^0 \dfrac{x}{(x^2 + 1)^2}\, dx$

Applications

29. **Maintenance** Maintenance costs for buildings increase as the buildings age. If the rate of increase in maintenance costs for a building is

$$M'(t) = \dfrac{14{,}000}{\sqrt{t + 16}}$$

where M is in dollars and t is time in years, $0 \le t \le 15$, find the total maintenance cost for the first 9 years ($t = 0$ to $t = 9$).

30. **Quality control** Suppose the probability density function for the life expectancy of a "disposable" telephone is

$$f(x) = \begin{cases} 1.4e^{-1.4x} & x \ge 0 \\ 0 & x < 0 \end{cases}$$

Find the probability that the telephone lasts 2 years or less.

31. **Savings** The future value of $1000 invested in a savings account at 10%, compounded continuously, is $S = 1000e^{0.1t}$, where t is in years. Find the average amount in the savings account during the first 5 years.

32. **Income streams** Suppose the total income in dollars from a video machine is given by

$$I = 50e^{0.2t},\ 0 \le t \le 4,\ t \text{ in hours}$$

Find the average income over this 4-hour period.

33. **Income distribution** In 1969, after the "Great Society" initiatives of the Johnson administration, the Lorenz curve for the U.S. income distribution was $L(x) = x^{2.1936}$. In 2000, after the stock market's historic 10-year growth, the Lorenz curve for the U.S. income distribution was $L(x) = x^{2.4870}$. Find the Gini coefficient of income for both years, and determine in which year income was more equally distributed.

34. **Consumer's surplus** The demand function for a product under pure competition is $p = \sqrt{64 - 4x}$, and the supply function is $p = x - 1$, where x is the number of units and p is in dollars.
    (a) Find the market equilibrium.
    (b) Find the consumer's surplus at market equilibrium.

35. *Producer's surplus* Find the producer's surplus at market equilibrium for Problem 34.

36. *Income streams* Find the total income over the next 10 years from a continuous income stream that has an annual flow rate at time t given by $f(t) = 125e^{0.05t}$ in thousands of dollars per year.

37. *Income streams* Suppose that a machine's production is considered a continuous income stream with an annual rate of flow at time t given by $f(t) = 150e^{-0.2t}$ in thousands of dollars per year. Money is worth 8%, compounded continuously.
    (a) Find the present value of the machine's production over the next 5 years.
    (b) Find the future value of the production 5 years from now.

38. *Average cost* Suppose the cost function for x units of a product is given by $C(x) = \sqrt{40{,}000 + x^2}$ dollars. Find the average cost over the first 150 units.

39. *Producer's surplus* Suppose the supply function for x units of a certain lamp is given by

$$p = 0.02x + 50.01 - \frac{10}{\sqrt{x^2 + 1}}$$

where p is in dollars. Find the producer's surplus if the equilibrium price is $70 and the equilibrium quantity is 1000.

40. *Income streams* Suppose the present value of a continuous income stream over the next 5 years is given by

$$P = 9000 \int_0^5 te^{-0.08t}\, dt,\ P \text{ in dollars, } t \text{ in years}$$

Find the present value.

41. *Cost* If the marginal cost for x units of a product is $\overline{MC} = 3 + 60(x + 1) \ln (x + 1)$ dollars per unit and if the fixed cost is $2000, find the total cost function.

42. *Capital value* Find the capital value of a business if its income is considered a continuous income stream with annual rate of flow given by

$$f(t) = 120e^{0.03t}$$

in thousands of dollars per year, and the current interest rate is 6% compounded continuously.

Functions of
Two or More Variables

In this chapter we will extend our study to functions of two or more variables. We will use these functions to solve problems in the management, life, and social sciences. In particular, we will discuss joint cost functions, utility functions that describe the customer satisfaction derived from the consumption of two products, Cobb-Douglas production functions, and wind chill temperatures as a function of air temperature and wind speed.

We will use derivatives with respect to one of two variables (called **partial derivatives**) to find marginal cost, marginal productivity, marginal utility, marginal demand, and other rates of change. We will use partial derivatives to find maxima and minima of functions of two variables, and we will use Lagrange multipliers to optimize functions of two variables subject to a condition that constrains these variables. These skills are used to maximize profit, production, and utility and to minimize cost subject to constraints.

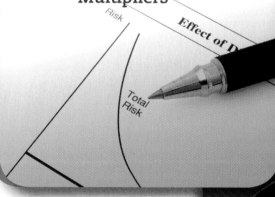

objectives

14.1 Functions of Two or More Variables

14.2 Partial Differentiation

14.3 Functions of Two Variables in Business and Economics

14.4 Maxima and Minima

14.5 Constrained Optimization and Lagrange Multipliers

14.1 Functions of Two or More Variables

WE WRITE $Z = F(X, Y)$ TO STATE THAT Z IS A FUNCTION OF BOTH X AND Y. THE VARIABLES X AND Y ARE CALLED THE **INDEPENDENT VARIABLES** AND Z IS CALLED THE **DEPENDENT VARIABLE.** THUS THE FUNCTION F ASSOCIATES WITH EACH PAIR OF POSSIBLE VALUES FOR THE INDEPENDENT VARIABLES (X AND Y) EXACTLY ONE VALUE OF THE DEPENDENT VARIABLE (Z).

The equation $z = x^2 - xy$ defines z as a function of x and y. We can denote this by writing $z = f(x, y) = x^2 - xy$. The domain of the function is the set of all ordered pairs (of real numbers), and the range is the set of all real numbers.

EXAMPLE 14.1 Domain

 Give the domain of the function

$$g(x, y) = \frac{x^2 - 3y}{x - y}$$

SOLUTION

The domain of the function is the set of ordered pairs that do not give a 0 denominator. That is, the domain is the set of all ordered pairs where the first and second elements are not equal (that is, where $x \neq y$).

We graph the function $z = f(x, y)$ by using three dimensions. We can construct a three-dimensional coordinate space by drawing three mutually perpendicular axes, as in Figure 14.1. By setting up scales of measurement along the three axes from the origin of each axis, we can determine the three coordinates (x, y, z) for any point P. The point shown in Figure 14.1 is $+2$ units in the x-direction, $+3$ units in the y-direction, and $+4$ units in the z-direction, so the coordinates of the point are $(2, 3, 4)$.

The pairs of axes determine the three **coordinate planes**; the xy-plane, the yz-plane, and the xz-plane. The planes divide the space into eight **octants**. The point $P(2, 3, 4)$ is in the first octant.

If we are given a function $z = f(x, y)$, we can find the z-value corresponding to $x = a$ and $y = b$ by evaluating $f(a, b)$.

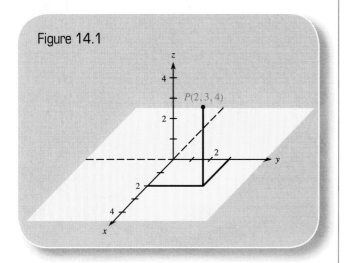

Figure 14.1

EXAMPLE 14.2 Function Values

If $z = f(x, y) = x^2 - 4xy + xy^3$, find the following.

(a) $f(1, 2)$ (b) $f(-1, 3)$

SOLUTION

(a) $f(1, 2) = 1^2 - 4(1)(2) + (1)(2)^3 = 1$
(b) $f(-1, 3) = (-1)^2 - 4(-1)(3) + (-1)(3)^3$
$\qquad = -14$

Let's consider the practical application of these concepts. Suppose a small furniture company's cost (in dollars) to manufacture 1 unit of several different all-wood items is given by $C(x, y) = 5 + 5x + 22y$ where x represents the number of board-feet of material used and y represents the number of work-hours of labor required for assembly and finishing. A certain bookcase uses 20 board-feet of material and requires

2.5 work-hours for assembly and finishing. The cost of manufacturing the specific bookcase would be $C(20, 2.5) = 5 + 5(20) + 22(2.5) = 160$ dollars.

C(20, 2.5) = 160 dollars

For a given function $z = f(x, y)$, we can construct a table of values by assigning values to x and y and finding the corresponding values of z. To each pair of values for x and y there corresponds a unique value of z, and thus a unique point in a three-dimensional coordinate system. From a table of values such as this, a finite number of points can be plotted. All points that satisfy the equation form a "surface" in space. Because z is a function of x and y, lines parallel to the z-axis will intersect such a surface in at most one point. The graph of the equation $z = 4 - x^2 - y^2$ is the surface shown in Figure 14.2(a). The portion of the surface above the xy-plane resembles a bullet and is called a **paraboloid**. Some points on the surface are given in Table 14.1.

TABLE 14.1

x	y	z
−2	0	0
0	−2	0
−1	0	3
0	−1	3
−1	−1	2
0	0	4
1	0	3
0	1	3
1	1	2
2	0	0
0	2	0

Figure 14.2

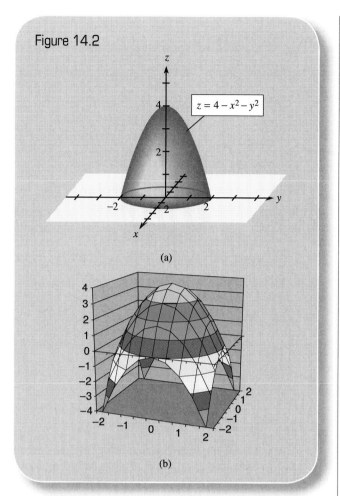

(a)

(b)

Figure 14.3

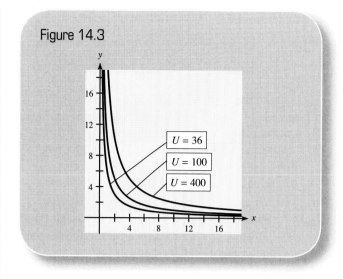

See the Tech Card for details on using Excel to create three-dimensional plots of surfaces such as the plot of $z = 4 - x^2 - y^2$ in Figure 14.1(b).

The properties of functions of one variable can be extended to functions of two variables. The precise definition of continuity for functions of two variables is technical and may be found in more advanced books. We will limit our study to functions that are continuous and have continuous derivatives in the domain of interest to us. We may think of continuous functions as functions whose graphs consist of surfaces without "holes" or "breaks" in them.

Let the function $U = f(x, y)$ represent the **utility** (that is, satisfaction) derived by a consumer from the consumption of two goods, X and Y, where x and y represent the amounts of X and Y, respectively. Because we will assume that the utility function is continuous, a given level of utility can be derived from an infinite number of combinations of x and y. The graph of all points (x, y) that give the same utility is called an **indifference curve**. A set of indifference curves corresponding to different levels of utility is called an **indifference map** (see Figure 14.3).

EXAMPLE 14.3 Utility

Suppose that the utility function for two goods, X and Y, is $U = x^2y^2$ and a consumer purchases 10 units of X and 2 units of Y.
(a) What is the level of utility U for these two products?
(b) If the consumer purchases 5 units of X, how many units of Y must be purchased to retain the same level of utility?
(c) Graph the indifference curve for this level of utility.
(d) Graph the indifference curves for this utility function if $U = 100$ and if $U = 36$.

SOLUTION

(a) If $x = 10$ and $y = 2$ satisfy the utility function, then the level of utility is $U = 10^2 \cdot 2^2 = 400$.
(b) If x is 5, y must satisfy $400 = 5^2y^2$, so $y = \pm 4$, and 4 units of Y must be purchased.
(c) The indifference curve for $U = 400$ is $400 = x^2y^2$. The graph $U = 400$ for positive x and y is shown in Figure 14.3.
(d) The indifference map in Figure 14.3 also contains these indifference curves.

A graphing utility can be used to graph each indifference curve in the indifference map shown in Figure 14.3. To graph the indifference curve for a given value of U, we must recognize that y will be positive and solve for y to express it as a function of x. Try it for $U = 100$. Does your graph agree with the one shown in Figure 14.3?

Sometimes functions of two variables are studied by fixing a value for one variable and graphing the resulting function of a single variable. We'll do this in Section 14.3 with production functions.

EXAMPLE 14.4 Production

Suppose a company has the Cobb-Douglas production function

$$Q = 4K^{0.4}L^{0.6}$$

where Q is thousands of dollars of production value, K is hundreds of dollars of capital investment per week, and L is work-hours of labor per week.

(a) If current capital investment is $72,900 per week and work-hours are 3072 per week, find the current weekly production value.

(b) If weekly capital investment is increased to $97,200 and new employees are hired so that there are 4096 total weekly work-hours, find the percent increase in the production value.

SOLUTION

(a) Capital investment of $72,900 means that $K = 729$. We use this value and $L = 3072$ in the production function.

$$Q = 4(729)^{0.4}(3072)^{0.6} = 6912$$

Thus the weekly production value is $6,912,000.

(b) In this case we use $K = 972$ and $L = 4096$.

$$Q = 4(972)^{0.4}(4096)^{0.6} = 9216$$

This is an increase in production value of $9216 - 6912 = 2304$, which is equivalent to a weekly increase of

$$\frac{2304}{6912} = 0.33\tfrac{1}{3} = 33\tfrac{1}{3}\%$$

14.1 Exercises

In Problems 1 and 2, give the domain of each function.

1. $z = \dfrac{4x - 3}{y}$

2. $z = \dfrac{4x^3y - x}{2x - y}$

In Problems 3 and 4, evaluate the functions at the given values of the independent variables.

3. $z = x^3 + 4xy + y^2$; $x = 1$, $y = -1$

4. $C(x_1, x_2) = 600 + 4x_1 + 6x_2$; $x_1 = 400$, $x_2 = 50$

In Problems 5–7, evaluate each function as indicated.

5. $q_1(p_1, p_2) = \dfrac{p_1 + 4p_2}{p_1 - p_2}$; find $q_1(40, 35)$.

6. $z(x, y) = xe^{x+y}$; find $z(3, -3)$.

7. $w(x, y, z) = \dfrac{x^2 + 4yz}{xyz}$; find $w(1, 3, 1)$.

Applications

8. **Investment** The future value S of an investment earning 6% compounded continuously is a function of the principal P and the length of time t that the principal has been invested. It is given by

$$S = f(P, t) = Pe^{0.06t}$$

Find $f(2000, 20)$, and interpret your answer.

9. **Wilson's lot size formula** In economics, the most economical quantity Q of goods (TVs, dresses, gallons of paint, etc.) for a store to order is given by Wilson's lot size formula

$$Q = f(K, M, h) = \sqrt{2KM/h}$$

where K is the cost of placing the order, M is the number of items sold per week, and h is the weekly holding cost for each item (the cost of storage space, utilities, taxes, security, etc.). Find $f(200, 625, 1)$ and interpret your answer.

10. **Temperature-humidity models** There are different models for measuring the effects of high temperature and humidity. Two of these are the Summer Simmer Index (S) and the Apparent Temperature (A),[*] and they are given by

*W. Bosch and L. G. Cobb, "Temperature-Humidity Indices," UMAP Unit 691, *The UMAP Journal*, 10, no. 3 (Fall 1989): 237–256.

Shutterstock

$$S = 1.98T - 1.09(1 - H)(T - 58) - 56.9$$
$$A = 2.70 + 0.885T - 78.7H + 1.20TH$$

where T is the air temperature (in degrees Fahrenheit) and H is the relative humidity (expressed as a decimal). At the Dallas–Fort Worth Airport, the average daily temperatures and humidities for July are

Maximum: 97.8°F with 44% humidity
Minimum: 74.7°F with 80% humidity**

Calculate the Summer Simmer Index S and the Apparent Temperature A for both the average daily maximum and the average daily minimum temperature.

11. **Production** Suppose that the number of units of a good produced, z, is given by $z = 20xy$, where x is the number of machines working properly and y is the average number of work-hours per machine. Find the production for a week in which
(a) 12 machines are working properly and the average number of work-hours per machine is 30.
(b) 10 machines are working properly and the average number of work-hours per machine is 25.

Need more practice?
Find more here: cengagebrain.com

14.2 Partial Differentiation

IN THIS SECTION WE FIND PARTIAL DERIVATIVES OF FUNCTIONS OF TWO OR MORE VARIABLES AND USE THESE DERIVATIVES TO FIND RATES OF CHANGE AND SLOPES OF TANGENTS TO SURFACES. WE WILL ALSO FIND SECOND- AND HIGHER-ORDER PARTIAL DERIVATIVES OF FUNCTIONS OF TWO VARIABLES.

**James Ruffner and Frank Bair (eds.), *Weather of U.S. Cities*, Gale Research Co., Detroit, MI, 1987.

First-Order Partial Derivatives

If $z = f(x, y)$, we find the partial derivative of z with respect to x (denoted $\partial z / \partial x$) by treating the variable y as a constant and taking the derivative of $z = f(x, y)$ with respect to x. We can also take the derivative of z with respect to y by holding the variable x constant and taking the derivative of $z = f(x, y)$ with respect to y. We denote this derivative as $\partial z / \partial y$. Note that dz/dx represents the derivative of a function of one variable, x, and that $\partial z / \partial x$ represents the partial derivative of a function of two or more variables. Notations used to represent the partial derivative of $z = f(x, y)$ with respect to x are

$$\frac{\partial z}{\partial x}, \frac{\partial f}{\partial x}, \frac{\partial}{\partial x} f(x, y), f_x(x, y), f_x, \text{ and } z_x$$

and notations used to represent the partial derivative of $z = f(x, y)$ with respect to y are

$$\frac{\partial z}{\partial y}, \frac{\partial f}{\partial y}, \frac{\partial}{\partial y} f(x, y), f_y(x, y), f_y, \text{ and } z_y$$

EXAMPLE 14.5 Partial Derivatives

➡ If $z = 4x^2 + 5x^2y^2 + 6y^3 - 7$, find $\partial z/\partial x$ and $\partial z/\partial y$.

SOLUTION

To find $\partial z / \partial x$, we hold y constant so that the term $6y^3$ is constant, and its derivative is 0; the term $5x^2y^2$ is viewed as the constant $(5y^2)$ times (x^2), so its derivative is $(5y^2)(2x) = 10xy^2$. Thus

$$\frac{\partial z}{\partial x} = 8x + 10xy^2$$

Similarly for $\partial z/\partial y$, the term $4x^2$ is constant and the partial derivative of $5x^2y^2$ is the constant $5x^2$ times the derivative of y^2, so its derivative is $(5x^2)(2y) = 10x^2y$. Thus

$$\frac{\partial z}{\partial y} = 10x^2y + 18y^2$$

EXAMPLE 14.6 Power Rule

➡ If $f(x, y) = (x^2 - y^2)^2$, find the following.
(a) f_x
(b) f_y

SOLUTION

(a) $f_x = 2(x^2 - y^2)2x = 4x^3 - 4xy^2$
(b) $f_y = 2(x^2 - y^2)(-2y) = -4x^2y + 4y^3$

We may evaluate partial derivatives by substituting values for x and y in the same way we did with derivatives of functions of one variable. For example, if $\partial z / \partial x = 2x - xy$, the value of the partial derivative with respect to x at $x = 2$, $y = 3$ is

$$\frac{\partial z}{\partial x}\bigg|_{(2,\,3)} = 2(2) - 2(3) = -2$$

Other notations used to denote evaluation of partial derivatives with respect to x at (a, b) are

$$\frac{\partial}{\partial x} f(a, b) \text{ and } f_x(a, b)$$

EXAMPLE 14.7 Partial Derivative at a Point

Find the partial derivative of $f(x, y) = x^2 + 3xy + 4$ with respect to x at the point $(1, 2, 11)$.

SOLUTION

$$f_x(x, y) = 2x + 3y$$
$$f_x(1, 2) = 2(1) + 3(2) = 8$$

EXAMPLE 14.8 Marginal Sales

Suppose that a company's sales are related to its television advertising by

$$s = 20{,}000 + 10nt + 20n^2$$

where n is the number of commercials per day and t is the length of the commercials in seconds. Find the partial derivative of s with respect to n, and use the result to find the instantaneous rate of change of sales with respect to the number of commercials per day, if the company is currently running ten 30-second commercials per day.

+1 AD = +700 SALES

Shutterstock

SOLUTION

The partial derivative of s with respect to n is $\partial s/\partial n = 10t + 40n$. At $n = 10$ and $t = 30$, the rate of change in sales is

$$\frac{\partial s}{\partial n}\bigg|_{\substack{n=10 \\ t=30}} = 10(30) + 40(10) = 700$$

Thus, increasing the number of 30-second commercials by 1 would result in approximately 700 additional sales. This is the marginal sales with respect to the number of commercials per day at $n = 10$, $t = 30$.

Just as the derivative of $f(x)$ represents the slope of the tangent to $y = f(x)$, the partial derivative of $z = f(x, y)$ with respect to x represents the slope of the tangent to the surface $z = f(x, y)$ in the positive x-direction. Similarly, the partial derivative with respect to y gives the slope of the tangent to the surface $z = f(x, y)$ in the positive y-direction.

EXAMPLE 14.9 Slopes of Tangents

Let $z = 4x^3 - 4e^x + 4y^2$ and let P be the point $(0, 2, 12)$. Find the slope of the tangent to z at the point P in the positive direction of (a) the x-axis and (b) the y-axis.

SOLUTION

(a) The slope of z at P in the positive x-direction is given by $\dfrac{\partial z}{\partial x}$, evaluated at P.

$$\frac{\partial z}{\partial x} = 12x^2 - 4e^x \text{ and } \frac{\partial z}{\partial x}\bigg|_{(0,\,2)} = 12(0)^2 - 4e^0 = -4$$

This tells us that z *decreases* approximately 4 units for an increase of 1 unit in x at this point.

(b) The slope of z at P in the positive y-direction is given by $\dfrac{\partial z}{\partial y}$, evaluated at P.

$$\frac{\partial z}{\partial y} = 8y \text{ and } \frac{\partial z}{\partial y}\bigg|_{(0,\,2)} = 8(2) = 16$$

Thus, at the point $P(0, 2, 12)$, the function *increases* approximately 16 units in the z-value for a unit increase in y.

Up to this point, we have considered derivatives of functions of two variables. We can easily extend the concept to functions of three or more variables.

EXAMPLE 14.10 Functions of Four Variables

If $u = f(w, x, y, z) = 3x^2y + w^3 - 4xyz$, find the following.

(a) $\dfrac{\partial u}{\partial w}$ (b) $\dfrac{\partial u}{\partial x}$ (c) $\dfrac{\partial u}{\partial y}$ (d) $\dfrac{\partial u}{\partial z}$

SOLUTION

(a) $\dfrac{\partial u}{\partial w} = 3w^2$ (b) $\dfrac{\partial u}{\partial x} = 6xy - 4yz$

(c) $\dfrac{\partial u}{\partial y} = 3x^2 - 4xz$ (d) $\dfrac{\partial u}{\partial z} = -4xy$

Higher-Order Partial Derivatives

Just as we have taken derivatives of derivatives to obtain higher-order derivatives of functions of one variable, we may also take partial derivatives of partial derivatives to obtain higher-order partial derivatives of a function of more than one variable. If $z = f(x, y)$, then the partial derivative functions z_x and z_y are called *first partials*. Partial derivatives of z_x and z_y are called *second partials*, so $z = f(x, y)$ has *four* **second partial derivatives**. The notations for these second partial derivatives follow.

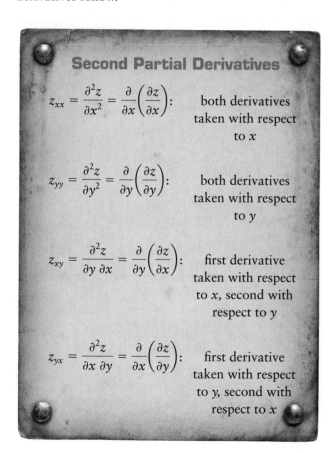

Second Partial Derivatives

$z_{xx} = \dfrac{\partial^2 z}{\partial x^2} = \dfrac{\partial}{\partial x}\left(\dfrac{\partial z}{\partial x}\right)$:   both derivatives taken with respect to x

$z_{yy} = \dfrac{\partial^2 z}{\partial y^2} = \dfrac{\partial}{\partial y}\left(\dfrac{\partial z}{\partial y}\right)$:   both derivatives taken with respect to y

$z_{xy} = \dfrac{\partial^2 z}{\partial y\, \partial x} = \dfrac{\partial}{\partial y}\left(\dfrac{\partial z}{\partial x}\right)$:   first derivative taken with respect to x, second with respect to y

$z_{yx} = \dfrac{\partial^2 z}{\partial x\, \partial y} = \dfrac{\partial}{\partial x}\left(\dfrac{\partial z}{\partial y}\right)$:   first derivative taken with respect to y, second with respect to x

EXAMPLE 14.11 Second Partial Derivatives

If $z = x^3 y - 3xy^2 + 4$, find each of the second partial derivatives of the function.

SOLUTION

Because $z_x = 3x^2 y - 3y^2$ and $z_y = x^3 - 6xy$,

$$z_{xx} = \frac{\partial}{\partial x}\left(3x^2 y - 3y^2\right) = 6xy$$

$$z_{xy} = \frac{\partial}{\partial y}\left(3x^2 y - 3y^2\right) = 3x^2 - 6y$$

$$z_{yy} = \frac{\partial}{\partial y}\left(x^3 - 6xy\right) = -6x$$

$$z_{yx} = \frac{\partial}{\partial x}\left(x^3 - 6xy\right) = 3x^2 - 6y$$

Note that z_{xy} and z_{yx} are equal for the function in Example 14.11. This will always occur if the derivatives of the function are continuous.

$$z_{xy} = z_{yx}$$

If the second partial derivatives z_{xy} and z_{yx} of a function $z = f(x, y)$ are continuous at a point, they are equal there.

We can find partial derivatives of orders higher than the second. For example, we can find the third-order partial derivatives z_{xyx} and z_{xyy} for the function in Example 14.11 from the second derivative $z_{xy} = 3x^2 - 6y$.

$$z_{xyx} = 6x$$

$$z_{xyy} = -6$$

EXAMPLE 14.12 Third Partial Derivatives

If $y = 4y \ln x + e^{xy}$, find z_{xyy}.

SOLUTION

$$z_x = 4y \cdot \frac{1}{x} + e^{xy} \cdot y$$

$$z_{xy} = 4 \cdot \frac{1}{x} \cdot 1 + e^{xy} \cdot 1 + y \cdot e^{xy} \cdot x = \frac{4}{x} + e^{xy} + xye^{xy}$$

$$z_{xyy} = 0 + e^{xy} \cdot x + xy \cdot e^{xy} \cdot x + e^{xy} \cdot x = x^2 y e^{xy} + 2x e^{xy}$$

14.2 Exercises

1. If $z = x^3 + 4x^2 y + 6y^2$, find z_x and z_y.

2. If $f(x, y) = (x^3 + 2y^2)^3$, find $\dfrac{\partial f}{\partial x}$ and $\dfrac{\partial f}{\partial y}$.

3. If $C(x, y) = 600 - 4xy + 10x^2 y$, find $\dfrac{\partial C}{\partial x}$ and $\dfrac{\partial C}{\partial y}$.

4. If $Q(s, t) = \dfrac{2s - 3t}{s^2 + t^2}$, find $\dfrac{\partial Q}{\partial s}$ and $\dfrac{\partial Q}{\partial t}$.

5. If $z = e^{2x} + y \ln x$, find z_x and z_y.

6. Find the slope of the tangent in the positive x-direction to the surface $z = 5x^3 - 4xy$ at the point $(1, 2, -3)$.

7. Find the slope of the tangent in the positive y-direction to the surface $z = 3x + 2y - 7e^{xy}$ at $(3, 0, 2)$.

8. If $C(x_1, x_2, x_3) = 4x_1^2 + 5x_1x_2 + 6x_2^2 + x_3$, find the following.
   (a) $\dfrac{\partial C}{\partial x_1}$ (b) $\dfrac{\partial C}{\partial x_2}$ (c) $\dfrac{\partial C}{\partial x_3}$

9. If $z = x^2 + 4x - 5y^3$, find the following.
   (a) z_{xx} (b) z_{xy} (c) z_{yx} (d) z_{yy}

10. If $f(x, y) = y^2 - \ln xy$, find the following.
    (a) $\dfrac{\partial^2 f}{\partial x^2}$ (b) $\dfrac{\partial^2 f}{\partial y \, \partial x}$
    (c) $\dfrac{\partial^2 f}{\partial x \, \partial y}$ (d) $\dfrac{\partial^2 f}{\partial y^2}$

Applications

11. *Mortgage* When a homeowner has a 25-year variable-rate mortgage loan, the monthly payment R is a function of the amount of the loan A and the current interest rate i (as a percent); that is, $R = f(A, i)$. Interpret each of the following.
    (a) $f(100{,}000, 8) = 1289$
    (b) $\dfrac{\partial f}{\partial i}(100{,}000, 8) = 62.51$

12. *Pesticide* Suppose that the number of thousands of insects killed by two brands of pesticide is given by
    $$f(x, y) = 10{,}000 - 6500e^{-0.01x} - 3500e^{-0.02y}$$
    where x is the number of liters of brand 1 and y is the number of liters of brand 2. What is the rate of change of insect deaths with respect to the number of liters of brand 1 if 100 liters of each brand are currently being used? What does this mean?

13. *Utility* If $U = f(x, y)$ is the utility function for goods X and Y, the *marginal utility* of X is $\partial U/\partial x$ and the *marginal utility* of Y is $\partial U/\partial y$. If $U = x^2y^2$, find the marginal utility of
    (a) X.   (b) Y.

Need more practice?
Find more here: cengagebrain.com

14.3 Functions of Two Variables in Business and Economics

FUNCTIONS OF TWO VARIABLES ARE COMMON IN BUSINESS AND ECONOMICS. IN THIS SECTION, WE'LL EXAMINE HOW THEY APPLY TO JOINT AND MARGINAL COST AND TO PRODUCTION AND DEMAND FUNCTIONS.

Joint Cost and Marginal Cost

Suppose that a firm produces two products using the same inputs in different proportions. In such a case the **joint cost function** is of the form $C = Q(x, y)$, where x represents the quantity of product X and y represents the quantity of product Y. Then $\partial C/\partial x$ is the **marginal cost** of the first product and $\partial C/\partial y$ is the marginal cost of the second product.

EXAMPLE 14.13 Joint Cost

 If the joint cost function for two products is
$$C = Q(x, y) = 50 + x^2 + 8xy + y^3$$
where x represents the quantity of product X and y represents the quantity of product Y, find the marginal cost with respect to the number of units of
(a) Product X (b) Product Y
(c) Product X at $(5, 3)$ (d) Product Y at $(5, 3)$

SOLUTION

(a) The marginal cost with respect to the number of units of product X is $\partial C/\partial x = 2x + 8y$.

(b) The marginal cost with respect to the number of units of product Y is $\dfrac{\partial C}{\partial y} = 8x + 3y^2$.

(c) $\left.\dfrac{\partial C}{\partial x}\right|_{(5, 3)} = 2(5) + 8(3) = 34$

   Thus if 5 units of product X and 3 units of product Y are produced, the total cost will increase approximately \$34 for a unit increase in product X if y is held constant.

(d) $\left.\dfrac{\partial C}{\partial y}\right|_{(5, 3)} = 8(5) + 3(3)^2 = 67$

   Thus if 5 units of product X and 3 units of product Y are produced, the total cost will increase approximately $67 for a unit increase in product Y if x is held constant.

Production Functions

An important problem in economics concerns how the factors necessary for production determine the output of a product. For example, the output of a product depends on available labor, land, capital, material, and machines. If the amount of output z of a product depends on the amounts of two inputs x and y, then the quantity z is given by the **production function** $z = f(x, y)$.

If $z = f(x, y)$ is a production function, $\partial z / \partial x$ represents the rate of change in the output z with respect to input x while input y remains constant. This partial derivative is called the **marginal productivity of x**. The partial derivative $\partial z / \partial y$ is the **marginal productivity of y** and measures the rate of change of z with respect to input y.

Marginal productivity (for either input) will be positive over a wide range of inputs, but it increases at a decreasing rate, and it may eventually reach a point where it no longer increases and begins to decrease.

EXAMPLE 14.14 Production

If a production function is given by $z = 5x^{1/2}y^{1/4}$, find the marginal productivity of
(a) x. (b) y.

SOLUTION

(a) $\dfrac{\partial z}{\partial x} = \dfrac{5}{2}x^{-1/2}y^{1/4}$ (b) $\dfrac{\partial z}{\partial y} = \dfrac{5}{4}x^{1/2}y^{-3/4}$

Note that the marginal productivity of x is positive for all values of x but that it decreases as x gets larger (because of the negative exponent). The same is true for the marginal productivity of y.

Demand Functions

Suppose that two products are sold at prices p_1 and p_2 (both in dollars) in a competitive market consisting of a fixed number of consumers with given tastes and incomes. Then the amount of each *one* of the products demanded by the consumers is dependent on the prices of *both* products on the market. If q_1 represents the demand for the number of units of the first product, then $q_1 = f(p_1, p_2)$ is the **demand function** for that product. The graph of such a function is called a **demand surface**.

EXAMPLE 14.15 Demand

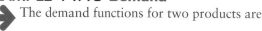

The demand functions for two products are

$$q_1 = 50 - 5p_1 - 2p_2$$
$$q_2 = 100 - 3p_1 - 8p_2$$

where q_1 and q_2 are the numbers of units and p_1 and p_2 are in dollars.
(a) What is the demand for each of the products if the price of the first is $p_1 = \$5$ and the price of the second is $p_2 = \$8$?
(b) Find a pair of prices p_1 and p_2 such that the demands for product 1 and product 2 are equal.

SOLUTION

(a) $q_1 = 50 - 5(5) - 2(8) = 9$
$q_2 = 100 - 3(5) - 8(8) = 21$
Thus if these are the prices, the demand for product 2 is higher than the demand for product 1.
(b) We want q_1 to equal q_2. Setting $q_1 = q_2$, we see that

$$50 - 5p_1 - 2p_2 = 100 - 3p_1 - 8p_2$$
$$6p_2 - 50 = 2p_1$$
$$p_1 = 3p_2 - 25$$

Hence, any pair of positive values that satisfies this equation will make the demands equal. Letting $p_2 = 10$, we see that $p_1 = 5$ will satisfy the equation. Thus the prices $p_1 = 5$ and $p_2 = 10$ will make the demands equal. The prices $p_1 = 2$ and $p_2 = 9$ will also make the demands equal. Many pairs of values (that is, all those satisfying $p_1 = 3p_2 - 25$) will equalize the demands.

If the demand functions for a pair of related products, product 1 and product 2, are $q_1 = f(p_1, p_2)$ and $q_2 = g(p_1, p_2)$, respectively, then the partial derivatives of q_1 and q_2 are called **marginal demand functions**.

$\dfrac{\partial q_1}{\partial p_1}$ is the marginal demand of q_1 with respect to p_1.

$\dfrac{\partial q_1}{\partial p_2}$ is the marginal demand of q_1 with respect to p_2.

$\dfrac{\partial q_2}{\partial p_1}$ is the marginal demand of q_2 with respect to p_1.

$\dfrac{\partial q_2}{\partial p_2}$ is the marginal demand of q_2 with respect to p_2.

For typical demand functions, if the price of product 2 is fixed, the demand for product 1 will decrease as its price p_1 increases. In this case the marginal demand of q_1 with respect to p_1 will be negative; that is, $\partial q_1 / \partial p_1 < 0$. Similarly, $\partial q_2 / \partial p_2 < 0$.

But what about $\partial q_2 / \partial p_1$ and $\partial q_1 / \partial p_2$? If $\partial q_2 / \partial p_1$ and $\partial q_1 / \partial p_2$ are both positive, the two products are **competitive** because an increase in price p_1 will result in an increase in demand for product 2 (q_2) if the price p_2 is held constant, and an increase in price p_2 will increase the demand for product 1 (q_1) if p_1 is held constant. Stated more simply, an increase in the price of one of the two products will result in an increased demand for the other, so the products are in competition. For example, an increase in the price of a Japanese automobile will result in an increase in demand for an American automobile if the price of the American automobile is held constant.

If $\partial q_2 / \partial p_1$ and $\partial q_1 / \partial p_2$ are both negative, the products are **complementary** because an increase in the price of one product will cause a decrease in demand for the other product if the price of the second product doesn't change. Under these conditions, a *decrease* in the price of product 1 will result in an *increase* in the demand for product 2, and a decrease in the price of product 2 will result in an increase in the demand for product 1. For example, a decrease in the price of gasoline will result in an increase in the demand for large automobiles.

Shutterstock

If the signs of $\partial q_2 / \partial p_1$ and $\partial q_1 / \partial p_2$ are different, the products are neither competitive nor complementary. This situation rarely occurs but is possible.

EXAMPLE 14.16 Demand

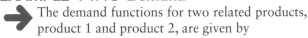 The demand functions for two related products, product 1 and product 2, are given by

$$q_1 = 400 - 5p_1 + 6p_2 \quad q_2 = 250 + 4p_1 - 5p_2$$

(a) Determine the four marginal demands.
(b) Are product 1 and product 2 complementary or competitive?

SOLUTION

(a) $\dfrac{\partial q_1}{\partial p_1} = -5 \quad \dfrac{\partial q_2}{\partial p_2} = -5 \quad \dfrac{\partial q_1}{\partial p_2} = 6 \quad \dfrac{\partial q_2}{\partial p_1} = 4$

(b) Because $\partial q_1 / \partial p_2$ and $\partial q_2 / \partial p_1$ are positive, products 1 and 2 are competitive.

14.3 Exercises

Joint Cost and Marginal Cost

1. The cost (in dollars) of manufacturing one item is given by

$$C(x, y) = 30 + 3x + 5y$$

where x is the cost of 1 hour of labor and y is the cost of 1 pound of material. If the hourly cost of labor is $20, and the material costs $3 per pound, what is the cost of manufacturing one of these items?

2. The total cost of producing 1 unit of a product is

$$C(x, y) = 30 + 2x + 4y + \frac{xy}{50} \text{ dollars}$$

where x is the cost per pound of raw materials and y is the cost per hour of labor.
(a) If labor costs are held constant, at what rate will the total cost increase for each increase of $1 per pound in material cost?
(b) If material costs are held constant, at what rate will the total cost increase for each $1 per hour increase in labor costs?

3. The joint cost (in dollars) for two products is given by

$$C(x, y) = 30 + x^2 + 3y + 2xy$$

where x represents the quantity of product X produced and y represents the quantity of product Y produced.
(a) Find and interpret the marginal cost with respect to x if 8 units of product X and 10 units of product Y are produced.

(b) Find and interpret the marginal cost with respect to y if 8 units of product X and 10 units of product Y are produced.

Production Functions

4. Suppose that the production function for a product is $z = \sqrt{4xy}$, where x represents the number of work-hours per month and y is the number of available machines. Determine the marginal productivity of
   (a) x.   (b) y.

5. Suppose that the number of crates of an agricultural product is given by

$$z = \frac{11xy - 0.0002x^2 - 5y}{0.03x + 3y}$$

where x is the number of hours of labor and y is the number of acres of the crop. Find and interpret the marginal productivity of the number of hours of labor (x) when $x = 300$ and $y = 500$.

Demand Functions

In Problems 6 and 7, prices p_1 and p_2 are in dollars and q_1 and q_2 are numbers of units.

6. The demand functions for two products are given by

$$q_1 = 300 - 8p_1 - 4p_2$$
$$q_2 = 400 - 5p_1 - 10p_2$$

Find the demand for each of the products if the price of the first is $p_1 = \$10$ and the price of the second is $p_2 = \$8$.

7. Find a pair of prices p_1 and p_2 such that the demands for the two products in Problem 6 will be equal.

8. Assume the demand functions for q_A and q_B units of two related products, A and B, are as follows:

$$\begin{cases} q_A = 400 - 3p_A - 2p_B \\ q_B = 250 - 5p_A - 6p_B \end{cases}$$

Complete parts (a)–(e). Assume p_A and p_B are in dollars.
   (a) Find the marginal demand of q_A with respect to p_A.
   (b) Find the marginal demand of q_A with respect to p_B.
   (c) Find the marginal demand of q_B with respect to p_B.
   (d) Find the marginal demand of q_B with respect to p_A.
   (e) Are the two goods competitive or complementary?

....

Need more practice?
Find more here: cengagebrain.com

14.4 Maxima and Minima

IN OUR STUDY OF DIFFERENTIABLE FUNCTIONS OF ONE VARIABLE, WE SAW THAT FOR A RELATIVE MAXIMUM OR MINIMUM TO OCCUR AT A POINT, THE DERIVATIVE HAD TO EQUAL ZERO AT THAT POINT. THE FUNCTION $z = f(x, y)$ DESCRIBES A SURFACE IN THREE DIMENSIONS. IF ALL PARTIAL DERIVATIVES OF $f(x, y)$ EXIST, THEN THE FIRST PARTIAL DERIVATIVES MUST BOTH EQUAL ZERO AT A POINT FOR THE SURFACE TO HAVE A RELATIVE MAXIMUM AT THAT POINT (SEE FIGURE 14.4A ON THE FOLLOWING PAGE) OR A MINIMUM AT THAT POINT (SEE FIGURE 14.4B). THUS THOSE POINTS WHERE *BOTH* $\partial z/\partial x = 0$ AND $\partial z/\partial y = 0$ ARE CALLED **CRITICAL POINTS** FOR THE SURFACE.

How can we determine whether a critical point is a relative maximum, a relative minimum, or neither of these? Finding that $\partial^2 z/\partial x^2 < 0$ and $\partial^2 z/\partial y^2 < 0$ is not enough to tell us that we have a relative maximum. The "second-derivative" test we must use involves the values of the second partial derivatives and the value of D at the critical point (a, b), where D is defined as follows:

$$D = \frac{\partial^2 z}{\partial x^2} \cdot \frac{\partial^2 z}{\partial y^2} - \left(\frac{\partial^2 z}{\partial x \partial y} \right)^2$$

We shall state, without proof, the result that determines

Figure 14.4

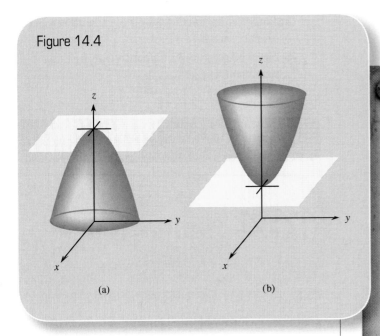

(a)   (b)

whether there is a relative maximum, a relative minimum, or neither at the critical point (a, b).

Test for Maxima and Minima

Let $z = f(x, y)$ be a function for which both

$$\frac{\partial z}{\partial x} = 0 \text{ and } \frac{\partial z}{\partial y} = 0 \text{ at a point } (a, b)$$

and suppose that all second partial derivatives are continuous there. Evaluate

$$D = \frac{\partial^2 z}{\partial x^2} \cdot \frac{\partial^2 z}{\partial y^2} - \left(\frac{\partial^2 z}{\partial x \partial y}\right)^2$$

at the critical point (a, b), and conclude the following:

(a) If $D > 0$ and $\partial^2 z / \partial x^2 > 0$ at (a, b), then a relative minimum occurs at (a, b). In this case, $\partial^2 z / \partial y^2 > 0$ at (a, b) also.

(b) If $D > 0$ and $\partial^2 z / \partial x^2 < 0$ at (a, b), then a relative maximum occurs at (a, b). In this case, $\partial^2 z / \partial y^2 < 0$ at (a, b) also.

(c) If $D < 0$ at (a, b), there is neither a relative maximum nor a relative minimum at (a, b).

(d) If $D = 0$ at (a, b), the test fails; investigate the function near the point.

We can test for relative maxima and minima by using the following procedure.

Maxima and Minima of $z = f(x, y)$

Procedure

To find relative maxima and minima of $z = f(x, y)$:

1. Find $\partial z / \partial x$ and $\partial z / \partial y$.

2. Find the point(s) that satisfy *both* $\partial z / \partial x = 0$ and $\partial z / \partial y = 0$. These are the critical points.

3. Find all second partial derivatives.

4. Evaluate D at each critical point.

5. Use the test for maxima and minima to determine whether relative maxima or minima occur.

EXAMPLE 14.17 Relative Minima

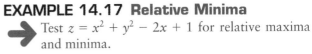
Test $z = x^2 + y^2 - 2x + 1$ for relative maxima and minima.

SOLUTION

1. $\dfrac{\partial z}{\partial x} = 2x - 2; \quad \dfrac{\partial z}{\partial y} = 2y$

2. $\dfrac{\partial z}{\partial x} = 0$ if $x = 1$; $\dfrac{\partial z}{\partial y} = 0$ if $y = 0$

    Both are 0 if $x = 1$ *and* $y = 0$, so the critical point is $(1, 0, 0)$.

3. $\dfrac{\partial^2 z}{\partial x^2} = 2; \quad \dfrac{\partial^2 z}{\partial y^2} = 2; \quad \dfrac{\partial^2 z}{\partial x \partial y} = \dfrac{\partial^2 z}{\partial y \, \partial x} = 0$

4. At $(1, 0)$, $D = 2 \cdot 2 - 0^2 = 4$.

5. $D > 0$, $\partial^2 z / \partial x^2 > 0$, and $\partial^2 z / \partial y^2 > 0$. A relative minimum occurs at $(1, 0)$. (See Figure 14.5.)

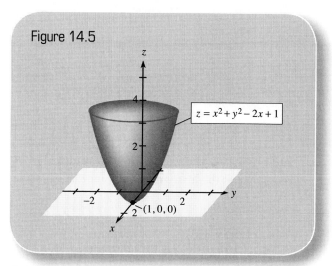

Figure 14.5

$z = x^2 + y^2 - 2x + 1$

$(1, 0, 0)$

EXAMPLE 14.18 Saddle Points

Test $z = y^2 - x^2$ for relative maxima and minima.

SOLUTION

1. $\dfrac{\partial z}{\partial x} = -2x$; $\dfrac{\partial z}{\partial y} = 2y$

2. $\dfrac{\partial z}{\partial x} = 0$ if $x = 0$; $\dfrac{\partial z}{\partial y} = 0$ if $y = 0$

Thus both equal 0 if $x = 0$, $y = 0$. The critical point is $(0, 0, 0)$.

3. $\dfrac{\partial^2 z}{\partial x^2} = -2$; $\dfrac{\partial^2 z}{\partial y^2} = 2$; $\dfrac{\partial^2 z}{\partial x\, \partial y} = \dfrac{\partial^2 z}{\partial y\, \partial x} = 0$

4. $D = (-2)(2) - 0 = -4$

5. $D < 0$, so the critical point is neither a relative maximum nor a relative minimum. As Figure 14.6 shows, the surface formed has the shape of a saddle. For this reason, critical points that are neither relative maxima nor relative minima are called **saddle points**.

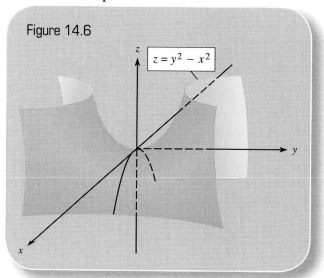

Figure 14.6

$z = y^2 - x^2$

EXAMPLE 14.19 Maximum Profit

Adele Lighting manufactures 20-inch lamps and 31-inch lamps. Suppose that x is the number of thousands of 20-inch lamps and that the demand for these is given by $p_1 = 50 - x$, where p_1 is in dollars. Similarly, suppose that y is the number of thousands of 31-inch lamps and that the demand for these is given by $p_2 = 60 - 2y$, where p_2 is also in dollars. Adele Lighting's joint cost function for these lamps is $C = 2xy$ (in thousands of dollars). Therefore, Adele Lighting's profit (in thousands of dollars) is a function of the two variables x and y. Determine Adele's maximum profit.

Shutterstock

SOLUTION

The profit function is $P(x, y) = p_1 x + p_2 y - C(x, y)$. Thus,

$$P(x, y) = (50 - x)x + (60 - 2y)y - 2xy$$
$$= 50x - x^2 + 60y - 2y^2 - 2xy$$

gives the profit in thousands of dollars. To maximize the profit, we proceed as follows.

$$P_x = 50 - 2x - 2y \text{ and } P_y = 60 - 4y - 2x$$

Solving $P_x = 0$ and $P_y = 0$ simultaneously, we have

$$\begin{cases} 0 = 50 - 2x - 2y \\ 0 = 60 - 2x - 4y \end{cases}$$

Subtraction gives $0 = -10 + 2y$, so $y = 5$. With $y = 5$, the equation $0 = 50 - 2x - 2y$ becomes $0 = 40 - 2x$, so $x = 20$. Now

$$P_{xx} = -2, \ P_{yy} = -4, \text{ and } P_{xy} = -2, \text{ and}$$
$$D = (P_{xx})(P_{yy}) - (P_{xy})^2 = (-2)(-4) - (-2)^2 = 4$$

Because $P_{xx} < 0$, $P_{yy} < 0$, and $D > 0$, the values $x = 20$ and $y = 5$ yield maximum profit. Therefore,

ADELE LIGHTING'S MAXIMUM PROFIT IS $650,000 WHEN THE COMPANY SELLS 20,000 OF THE 20-INCH LAMPS AT $30 EACH AND 5000 OF THE 31-INCH LAMPS AT $50 EACH.

when $x = 20$ and $y = 5$, $p_1 = 30$, $p_2 = 50$, and the maximum profit is

$$P(20, 5) = 600 + 250 - 200 = 650$$

That is, Adele Lighting's maximum profit is $650,000 when the company sells 20,000 of the 20-inch lamps at $30 each and 5000 of the 31-inch lamps at $50 each.

An important application of these max-min techniques is in the development of linear regression formulas (see the Enrichment material on CourseMate for MATH APPS at cengagebrain.com).

14.4 Exercises

In Problems 1–5, test for relative maxima and minima.

1. $z = x^2 + y^2 + 4$

2. $z = x^2 - y^2 + 4x - 6y + 11$

3. $z = x^2 + y^2 - 2x + 4y + 5$

4. $z = 24 - x^2 + xy - y^2 + 36y$

5. $z = x^3 + y^3 - 6xy$

Applications

6. **Profit** Suppose that the profit in dollars from the sale of Kisses and Kreams is given by

$$P(x, y) = 10x + 6.4y - 0.001x^2 - 0.025y^2$$

where x is the number of pounds of Kisses and y is the number of pounds of Kreams. Selling how many pounds of Kisses and Kreams will maximize profit? What is the maximum profit?

7. **Profit** Suppose that a manufacturer produces two brands of a product, brand 1 and brand 2. Suppose the demand for brand 1 is $x = 70 - p_1$ thousand units and the demand for brand 2 is $y = 80 - p_2$ thousand units, where p_1 and p_2 are prices in dollars. If the joint cost function is $C = xy$, in thousands of dollars, how many of each brand should be produced to maximize profit? What is the maximum profit?

8. **Manufacturing** Find the values for each of the dimensions of an open-top box of length x, width y, and height $500,000/(xy)$ (in inches) such that the box requires the least amount of material to make.

> **Need more practice?**
> Find more here: **cengagebrain.com**

14.5 Constrained Optimization and Lagrange Multipliers

WE CAN OBTAIN MAXIMA AND MINIMA FOR A FUNCTION $Z = F(X, Y)$ SUBJECT TO THE CONSTRAINT $G(X, Y) = 0$ BY USING THE METHOD OF LAGRANGE MULTIPLIERS, NAMED FOR THE FAMOUS EIGHTEENTH-CENTURY MATHEMATICIAN JOSEPH LOUIS LAGRANGE. LAGRANGE MULTIPLIERS CAN BE USED WITH FUNCTIONS OF TWO OR MORE VARIABLES WHEN THE CONSTRAINTS ARE GIVEN BY AN EQUATION.

In order to find the critical values of a function $f(x, y)$ subject to the constraint $g(x, y) = 0$, we will use the new variable λ to form the **objective function**

$$F(x, y, \lambda) = f(x, y) + \lambda g(x, y)$$

It can be shown that the critical values of $F(x, y, \lambda)$ will satisfy the constraint $g(x, y) = 0$ and will also be critical points of $f(x, y)$. Thus we need only find the critical points of $F(x, y, \lambda)$ to find the required critical points.

To find the critical points of $F(x, y, \lambda)$, we must find the points that make all the partial derivatives equal to 0. That is, the points must satisfy

$$\partial F/\partial x = 0, \quad \partial F/\partial y = 0, \quad \text{and} \quad \partial F/\partial \lambda = 0$$

Because $F(x, y, \lambda) = f(x, y) + \lambda g(x, y)$, these equations may be written as

$$\frac{\partial f}{\partial x} + \lambda \frac{\partial g}{\partial x} = 0$$

$$\frac{\partial f}{\partial y} + \lambda \frac{\partial g}{\partial y} = 0$$

$$g(x, y) = 0$$

Finding the values of x and y that satisfy these three equations simultaneously gives the critical values.

This method will not tell us whether the critical points correspond to maxima or minima, but this can be determined either from the physical setting for the problem or by testing according to a procedure similar to that used for unconstrained maxima and minima. The following examples illustrate the use of Lagrange multipliers.

EXAMPLE 14.20 Maxima Subject to Constraints

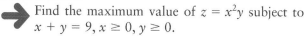 Find the maximum value of $z = x^2 y$ subject to $x + y = 9$, $x \geq 0$, $y \geq 0$.

SOLUTION

The function to be maximized is $f(x, y) = x^2 y$. The constraint is $g(x, y) = 0$, where $g(x, y) = x + y - 9$. The objective function is

$$F(x, y, \lambda) = f(x, y) + \lambda g(x, y)$$

or

$$F(x, y, \lambda) = x^2 y + \lambda(x + y - 9)$$

Thus

$$\frac{\partial F}{\partial x} = 2xy + \lambda(1) = 0, \text{ or } 2xy + \lambda = 0$$

$$\frac{\partial F}{\partial y} = x^2 + \lambda(1) = 0, \text{ or } x^2 + \lambda = 0$$

$$\frac{\partial F}{\partial \lambda} = 0 + 1(x + y - 9) = 0, \text{ or } x + y - 9 = 0$$

Solving the first two equations for λ and substituting gives

$$\lambda = -2xy$$
$$\lambda = -x^2$$
$$2xy = x^2$$
$$2xy - x^2 = 0$$
$$x(2y - x) = 0$$

so

$$x = 0 \text{ or } x = 2y$$

Because $x = 0$ could not make $z = x^2 y$ a maximum, we substitute $x = 2y$ into $x + y - 9 = 0$.

$$2y + y = 9$$
$$y = 3$$
$$x = 6$$

Thus the function $z = x^2 y$ is maximized at 108 when $x = 6$, $y = 3$, if the constraint is $x + y = 9$. Testing values near $x = 6$, $y = 3$, and satisfying the constraint shows that the function is maximized there. (Try $x = 5.5$, $y = 3.5$; $x = 7$, $y = 2$; and so on.)

We can also use Lagrange multipliers to find the maxima and minima of functions of three (or more) variables, subject to two (or more) constraints. The method involves using two multipliers, one for each constraint, to form an objective function $F = f + \lambda g_1 + \mu g_2$. We leave further discussion for more advanced courses.

We can easily extend the method to functions of three or more variables, as the following example shows.

EXAMPLE 14.21 Minima of Function of Three Variables

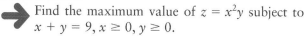 Find the minimum value of the function $w = x + y^2 + z^2$, subject to the constraint $x + y + z = 1$.

SOLUTION

The function to be minimized is $f(x, y, z) = x + y^2 + z^2$. The constraint is $g(x, y, z) = 0$, where $g(x, y, z) = x + y + z - 1$. The objective function is

$$F(x, y, z, \lambda) = f(x, y, z) + \lambda g(x, y, z)$$

or

$$F(x, y, z, \lambda) = x + y^2 + z^2 + \lambda(x + y + z - 1)$$

Then

$$\frac{\partial F}{\partial x} = 1 + \lambda = 0$$

$$\frac{\partial F}{\partial y} = 2y + \lambda = 0$$

$$\frac{\partial F}{\partial z} = 2z + \lambda = 0$$

$$\frac{\partial F}{\partial \lambda} = x + y + z - 1 = 0$$

Solving the first three equations simultaneously gives

$$\lambda = -1, \ y = \frac{1}{2}, \ z = \frac{1}{2}$$

Substituting these values in the fourth equation (which is the constraint), we get $x + \frac{1}{2} + \frac{1}{2} - 1 = 0$, so $x = 0$, $y = \frac{1}{2}$, $z = \frac{1}{2}$. These give $w = \frac{1}{2}$, which is the minimum value of the function, because other values of x, y, and z that satisfy $x + y + z = 1$ give larger values of w.

EXAMPLE 14.22 Production

Suppose that the Cobb-Douglas production function for a certain manufacturer gives the number of units of production z according to

$$z = f(x, y) = 100x^{4/5}y^{1/5}$$

where x is the number of units of labor and y is the number of units of capital. Suppose further that labor costs \$160 per unit, capital costs \$200 per unit, and the total cost for capital and labor is limited to \$100,000, so that production is constrained by

$$160x + 200y = 100{,}000$$

Find the number of units of labor and the number of units of capital that maximize production.

$160/UNIT

$200/UNIT

Shutterstock

SOLUTION

The objective function is

$$F(x, y, \lambda) = 100x^{4/5}y^{1/5} + \lambda(160x + 200y - 100{,}000)$$

$$\frac{\partial F}{\partial x} = 80x^{-1/5}y^{1/5} + 160\lambda, \quad \frac{\partial F}{\partial y} = 20x^{4/5}y^{-4/5} + 200\lambda$$

$$\frac{\partial F}{\partial \lambda} = 160x + 200y - 100{,}000$$

Setting these partial derivatives equal to 0 and solving for λ gives

$$\lambda = \frac{-80x^{-1/5}y^{1/5}}{160} = \frac{-20x^{4/5}y^{-4/5}}{200} \quad \text{so} \quad \frac{y^{1/5}}{2x^{1/5}} = \frac{x^{4/5}}{10y^{4/5}}$$

This means $5y = x$. Using this in $\frac{\partial F}{\partial \lambda} = 0$ gives

$$160(5y) + 200y - 100{,}000 = 0$$

$$1000y = 100{,}000$$

$$y = 100$$

$$x = 5y = 500$$

Thus production is maximized at $z = 100(500)^{4/5}(100)^{1/5} \approx 36{,}239$ when $x = 500$ (units of labor) and $y = 100$ (units of capital). See Figure 14.7.

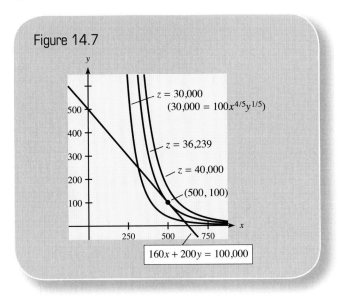

Figure 14.7

In problems of this type, economists call the value of $-\lambda$ the **marginal productivity of money**. In this case,

$$-\lambda = \frac{y^{1/5}}{2x^{1/5}} = \frac{(100)^{0.2}}{2(500)^{0.2}} \approx 0.362$$

This means that each additional dollar spent on production results in approximately 0.362 additional unit produced.

Finally, Figure 14.7 shows the graph of the constraint, together with some production function curves that correspond to different production levels.

The Excel tool "Solver" can be used to find maxima and minima of functions subject to constraints. See the Tech Card for details.

14.5 Exercises

1. Find the minimum value of $z = x^2 + y^2$ subject to the constraint $x + y = 6$.

2. Find the minimum value of $z = 3x^2 + 5y^2 - 2xy$ subject to the constraint $x + y = 5$.

3. Find the maximum value of $z = x^2 y$ subject to the constraint $x + y = 6$, $x \geq 0$, $y \geq 0$.

4. Find the maximum value of $z = 2xy - 2x^2 - 4y^2$ subject to the constraint $x + 2y = 8$.

5. Find the minimum value of $w = x^2 + y^2 + z^2$ subject to the constraint $x + y + z = 3$.

Applications

6. **Utility** Suppose that the utility function for two commodities is given by $U = xy^2$ and that the budget constraint is $3x + 6y = 18$. What values of x and y will maximize utility?

7. **Production** A company has the Cobb-Douglas production function

$$z = 400x^{0.6}y^{0.4}$$

where x is the number of units of labor, y is the number of units of capital, and z is the units of production. Suppose labor costs $150 per unit, capital costs $100 per unit, and the total cost of labor and capital is limited to $100,000.
(a) Find the number of units of labor and the number of units of capital that maximize production.
(b) Find the marginal productivity of money and interpret it.

8. **Cost** A firm has two plants, X and Y. Suppose that the cost of producing x units at plant X is $x^2 + 1200$ dollars and the cost of producing y units of the same product at plant Y is given by $3y^2 + 800$ dollars. If the firm has an order for 1200 units, how many should it produce at each plant to fill this order and minimize the cost of production?

9. **Revenue** On the basis of past experience a company has determined that its sales revenue (in dollars) is related to its advertising according to the formula $s = 20x + y^2 + 4xy$, where x is the amount spent on radio advertising and y is the amount spent on television advertising. If the company plans to spend $30,000 on these two means of advertising, how much should it spend on each method to maximize its sales revenue?

10. **Manufacturing** Find the dimensions (in centimeters) of the box with square base, open top, and volume 500,000 cubic centimeters that requires the least materials.

> **Need more practice?**
> Find more here: cengagebrain.com

Chapter Exercises

1. What is the domain of $z = \dfrac{3}{2x - y}$?

2. What is the domain of $z = \dfrac{3x + 2\sqrt{y}}{x^2 + y^2}$?

3. If $w(x, y, z) = x^2 - 3yz$, find $w(2, 3, 1)$.

4. If $Q(K, L) = 70K^{2/3}L^{1/3}$, find $Q(64,000, 512)$.

5. Find $\dfrac{\partial z}{\partial x}$ if $z = 5x^3 + 6xy + y^2$.

6. Find $\dfrac{\partial z}{\partial y}$ if $z = 12x^5 - 14x^3y^3 + 6y^4 - 1$.

In Problems 7–12, find z_x and z_y.

7. $z = 4x^2y^3 + \dfrac{x}{y}$

8. $z = \sqrt{x^2 + 2y^2}$

9. $z = (xy + 1)^{-2}$ 10. $z = e^{x^2y^3}$

11. $z = e^{xy} + y \ln x$

12. $z = e^{\ln xy}$

13. Find the partial derivative of $f(x, y) = 4x^3 - 5xy^2 + y^3$ with respect to x at the point $(1, 2, -8)$.

14. Find the slope of the tangent in the x-direction to the surface $z = 5x^4 - 3xy^2 + y^2$ at $(1, 2, -3)$.

In Problems 15–18, find the second partials.

  (a) z_{xx} (b) z_{yy} (c) z_{xy} (d) z_{yx}

15. $z = x^2y - 3xy$

16. $z = 3x^3y^4 - \dfrac{x^2}{y^2}$

17. $z = x^2e^{y^2}$

18. $z = \ln{(xy + 1)}$

19. Test $z = 16 - x^2 - xy - y^2 + 24y$ for maxima and minima.

20. Test $z = x^3 + y^3 - 12x - 27y$ for maxima and minima.

21. Find the minimum value of $z = 4x^2 + y^2$ subject to the constraint $x + y = 10$.

22. Find the maximum value of $z = x^4y^2$ subject to the constraint $x + y = 9, x \geq 0, y \geq 0$.

Applications

23. **Utility** Suppose that the utility function for two goods X and Y is given by $U = x^2y$.
  (a) Write the equation of the indifference curve for a consumer who purchases 6 units of X and 15 units of Y.
  (b) If the consumer purchases 60 units of Y, how many units of X must be purchased to retain the same level of utility?

24. **Savings plans** The accumulated value A of a monthly savings plan over a 20-year period is a function of the monthly contribution R and the interest rate $r\%$, compounded monthly, according to

$$A = f(R, r) = \dfrac{1200R\left[\left(1 + \dfrac{r}{1200}\right)^{240} - 1\right]}{r}$$

  (a) Find the accumulated value of a plan that contributes $100 per month with interest rate 6%.
  (b) Interpret $f(250, 7.8) \approx 143{,}648$.
  (c) Interpret $\dfrac{\partial A}{\partial r}(250, 7.8) \approx 17{,}770$.
  (d) Find $\dfrac{\partial A}{\partial R}(250, 7.8)$ and interpret the result.

25. **Retirement benefits** The monthly benefit B (in thousands of dollars) from a retirement account

that is invested at 9% compounded monthly is a function of the account value V (also in thousands of dollars) and the number of years t that benefits are paid, and it can be approximated by

$$B = f(V, t) = \dfrac{3V}{400 - 400e^{-0.0897t}}$$

  (a) Find the benefit if the account value is $1,000,000 and the monthly benefits last for 20 years.
  (b) Find and interpret $\dfrac{\partial B}{\partial V}(1000, 20)$.
  (c) Find and interpret $\dfrac{\partial B}{\partial t}(1000, 20)$.

26. **Advertising and sales** The number of units of sales of a product, S, is a function of the dollars spent for advertising, A, and the product's price, p. Suppose $S = f(A, p)$ is the function relating these quantities.
  (a) Explain why $\dfrac{\partial S}{\partial A} > 0$.
  (b) Do you think $\dfrac{\partial S}{\partial p}$ is positive or negative? Explain.

27. **Cost** The joint cost, in dollars, for two products is given by $C(x, y) = x^2\sqrt{y^2 + 13}$. Find the marginal cost with respect to
  (a) x if 20 units of x and 6 units of y are produced.
  (b) y if 20 units of x and 6 units of y are produced.

28. **Production** Suppose that the production function for a company is given by

$$Q = 80K^{1/4}L^{3/4}$$

where Q is the output (in hundreds of units), K is the capital expenditures (in thousands of dollars), and L is the work-hours. Find $\partial Q/\partial K$ and $\partial Q/\partial L$ when expenditures are $625,000 and total work-hours are 4096. Interpret the results.

29. **Marginal demand** The demand functions for two related products, product A and product B, are given by

$$q_A = 400 - 2p_A - 3p_B$$
$$q_B = 300 - 5p_A - 6p_B$$

where p_A and p_B are the respective prices in dollars.
  (a) Find the marginal demand of q_A with respect to p_A.

(b) Find the marginal demand of q_B with respect to p_B.

(c) Are the products complementary or competitive?

30. **Marginal demand** Suppose that the demand functions for two related products, A and B, are given by

$$q_A = 800 - 40p_A - \frac{2}{p_B + 1}$$

$$q_B = 1000 - \frac{10}{p_A + 4} - 30p_B$$

where p_A and p_B are the respective prices in dollars. Determine whether the products are competitive or complementary.

31. **Profit** The weekly profit (in dollars) from the sale of two products is given by $P(x, y) = 40x + 80y - x^2 - y^2$, where x is the number of units of product 1 and y is the number of units of product 2. Selling how much of each product will maximize profit? Find the maximum weekly profit.

32. **Cost** Suppose a company has two separate plants that manufacture the same item. Suppose x is the amount produced at plant I and y is the amount at plant II. If the total cost function for the two plants is

$$C(x, y) = 200 - 12x - 30y + 0.03x^2 + 0.001y^3$$

find the production allocation that minimizes the company's total cost.

33. **Utility** If the utility function for two commodities is $U = x^2y$, and the budget constraint is $4x + 5y = 60$, find the values of x and y that maximize utility.

34. **Production** Suppose a company has the Cobb-Douglas production function

$$z = 300x^{2/3}y^{1/3}$$

where x is the number of units of labor, y is the number of units of capital, and z is the units of production. Suppose labor costs are $50 per unit, capital costs are $50 per unit, and total costs are limited to $75,000.

(a) Find the number of units of labor and the number of units of capital that maximize production.

(b) Find the marginal productivity of money and interpret your result.

(c) Graph the constraint with the production function when $z = 180,000$, $z = 300,000$, and when the z-value is optimal.

Need more practice?
Find more here: cengagebrain.com

prepcard CHAPTER 1
LINEAR EQUATIONS AND FUNCTIONS

Chapter Outline

1.1 Solutions of Linear Equations and Inequalities in One Variable
1.2 Functions
1.3 Linear Functions
1.4 Solutions of Systems of Linear Equations
1.5 Applications of Functions in Business and Economics

Chapter Preview

A wide variety of problems from business, the social sciences, and the life sciences may be solved using equations. Managers and economists use equations and their graphs to study costs, sales, national consumption, or supply and demand. Social scientists may plot demographic data or try to develop equations that predict population growth, voting behavior, or learning and retention rates. Life scientists use equations to model the flow of blood or the conduction of nerve impulses and to test theories or develop new models by using experimental data.

This chapter introduces two important applications that will be expanded and used throughout the text as increased mathematical skills permit: supply and demand as functions of price (market analysis); and total cost, total revenue, and profit as functions of the quantity produced and sold (theory of the firm).

Chapter Applications

1.1 Future value of an investment, voting, normal height for a given age, profit
1.2 Income taxes, stock market, mortgage payments
1.3 Depreciation, U.S. banks, pricing
1.4 Investment mix, medicine concentrations, college enrollment
1.5 Total cost, total revenue, profit, break-even analysis, demand and supply, market equilibrium

Chapter Exercises

SECTION	PROBLEMS
1.1	1–7, 36, 37
1.2	8–18, 38–41
1.3	19–29, 42–44
1.4	30–35, 45, 46
1.5	47–52

Chapter Summary

1.1 Solutions of Linear Equations and Inequalities in One Variable
Objectives
- To solve linear equations in one variable
- To solve applied problems by using linear equations
- To solve linear inequalities in one variable

1.2 Functions
Objectives
- To determine whether a relation is a function
- To state the domains and ranges of certain functions
- To use function notation
- To perform operations with functions
- To find the composite of two functions

1.3 Linear Functions
Objectives
- To find the intercept of graphs
- To graph linear functions
- To find the slope of a line from its graph and from its equation
- To find the rate of change of a linear function
- To graph a line, given its slope and y-intercept or its slope and one point on the line
- To write the equation of a line, given information about its graph

1.4 Solutions of Systems of Linear Equations
Objectives
- To solve systems of linear equations in two variables by graphing
- To solve systems of linear equations by substitution
- To solve systems of linear equations by elimination
- To solve systems of three linear equations in three variables

1.5 Applications of Functions in Business and Economics
Objectives
- To formulate and evaluate total cost, total revenue, and profit functions
- To find marginal cost, revenue, and profit, given linear total cost, total revenue, and profit functions
- To find break-even points
- To evaluate and graph supply and demand functions
- To find market equilibrium

Keywords

1.1 Solutions of Linear Equations and Inequalities in One Variable
Linear equation
Stated problems
Linear inequalities

1.2 Functions
Functions
Domain
Range
Function notation
Functions operations
Composite functions

1.3 Linear Functions
Intercepts
Graph
Slope
Rate of change
Point-slope form
Slope-intercept form
Graphing with slope

1.4 Solutions of Systems of Linear Equations
Solution by graphing
Solution by substitution
Solving by elimination
Systems of three equations in three variables

1.5 Applications of Functions in Business and Economics
Cost, revenue, and profit
Marginal cost
Maginal revenue
Marginal profit
Break-even
Supply and demand
Market equilibrium

Go to login.cengage.com to access additional resources.

prepcard

Chapter Outline

2.1 Quadratic Equations
2.2 Quadratic Functions: Parabolas
2.3 Business Applications of Quadratic Functions
2.4 Special Functions and Their Graphs
2.5 Modeling Data with Graphing Utilities (optional)

Chapter Preview

In this chapter we will discuss quadratic functions and their applications, and we will also discuss other types of functions, including identity, constant, power, absolute value, piecewise defined, and reciprocal functions. Graphs of polynomial and rational functions will also be introduced; they will be studied in detail in Chapter 10.

Chapter Applications

2.1 Profit, flight of a ball, wind and pollution, Corvette acceleration, marijuana use
2.2 Profit, crop yield, projectiles, cost, apartment rental, pension resources
2.3 Supply, demand, market equilibrium, break-even, maximization for revenue and profit
2.4 Postal restrictions, pollution, gross domestic product
2.5 Earnings and gender, national health care
EOC Profit, market equilibrium, break-even points, maximum profit, break-even and profit maximization, photosynthesis, cost-benefit, municipal water costs, modeling

Chapter Exercises

Section	Problems
2.1	1–10, 28
2.2	11–19
2.3	29–32
2.4	20–22, 34–35
2.5	26–27, 36
Tech Card	23–25, 33

Chapter Summary

2.1 Quadratic Equations
Objectives
- To solve quadratic equations with factoring methods
- To solve quadratic equations with the quadratic formula

2.2 Quadratic Functions: Parabolas
Objectives
- To find the vertex of the graph of a quadratic function
- To determine whether a vertex is a maximum point or a minimum point
- To find the zeros of a quadratic function
- To graph quadratic functions

2.3 Business Applications of Quadratic Functions
Objectives
- To graph quadratic supply and demand functions
- To find market equilibrium by using quadratic supply and demand functions
- To find break-even points by using quadratic cost and revenue functions
- To maximize quadratic revenue and profit functions

2.4 Special Functions and Their Graphs
Objectives
- To graph and apply basic functions, including constant and power functions
- To graph and apply polynomial and rational functions
- To graph and apply absolute value and piecewise defined functions

2.5 Modeling Data with Graphing Utilities (optional)
Objectives
- To graph data points in a scatter plot
- To determine the function type that will best model data
- To use a graphing utility to create an equation that models the data
- To graph the data points and model on the same graph

Keywords

2.1 Quadratic Equations
Solution by factoring
Quadratic formula

2.2 Quadratic Functions: Parabolas
Vertex
Optimum point
Zeros
Parabolas

2.3 Business Applications of Quadratic Functions
Supply and demand
Market equilibrium
Break-even
Maximum revenue
Maximum profit

2.4 Special Functions and Their Graphs
Special functions
Polynomial functions
Rational functions
Piecewise defined functions
Absolute value function

2.5 Modeling Data with Graphing Utilities (optional)
Scatterplots
Function types
Creating a model
Testing models

Go to login.cengage.com to access additional resources.

prepcard CHAPTER 3
MATRICES

Chapter Outline

3.1 Operations with Matrices
3.2 Multiplication of Matrices
3.3 Gauss-Jordan Elimination: Solving Systems of Equations
3.4 Inverse of a Square Matrix

Chapter Preview

A wide variety of application problems can be solved using matrices. A matrix (plural: matrices) is a rectangular array of numbers. In addition to storing data in a matrix and making comparisons of data, we can analyze data and make business decisions by defining the operations of addition, subtraction, scalar multiplication, and matrix multiplication. Matrices can be used to solve systems of linear equations, as we will see in this chapter. Matrices and matrix operations are the basis for computer spreadsheets such as Excel, which are used extensively in education, research, and business. Matrices are also useful in linear programming, which is discussed in Chapter 4.

Chapter Applications

3.1 Endangered species, sales, international trade
3.2 Car pricing, oil refineries
3.3 Nutrition, investment, car rental patterns
3.4 Medication, investment, manufacturing
EOC Production, cost, manufacturing, investment, nutrition, transportation, economy models

Chapter Exercises

Section	Problems
3.1	1–4, 24
3.2	5–7, 25– 28
3.3	8–15, 29, 30
3.4	17–22, 31
Tech Card	16, 23

Chapter Summary

3.1 Operations with Matrices
Objectives
- To organize and interpret data stored in matrices
- To add and subtract matrices
- To find the transpose of a matrix
- To multiply a matrix by a scalar (real number)

3.2 Multiplication of Matrices
Objectives
- To multiply two matrices

3.3 Gauss-Jordan Elimination: Solving Systems of Equations
Objectives
- To use matrices to solve systems of linear equations with unique solutions
- To use matrices to solve systems of linear equations with nonunique solutions

3.4 Inverse of a Square Matrix
Objectives
- To find the inverse of a square matrix
- To use inverse matrices to solve systems of linear equations
- To find determinants of certain matrices

Keywords

3.1 Operations with Matrices
Matrices
Sums and differences
Transpose
Scalar multiplication

3.2 Multiplication of Matrices
Multiplying matrices

3.3 Gauss-Jordan Elimination: Solving Systems of Equations
Solving systems with matrices
Nonunique solutions

3.4 Inverse of a Square Matrix
Inverse matrix
Matrix equations
Determinants

Go to login.cengage.com to access additional resources.

prepcard

Chapter Outline

Chapter Preview

Most companies seek to maximize profits subject to the limitations imposed by product demand and available resources (such as raw materials and labor) or to minimize production costs subject to the need to fill customer orders. If the relationships among the various resources, production requirements, costs, and profits are all linear, then these activities may be planned (or programmed) in the best possible (optimal) way by using linear programming. Because linear programming provides the best possible solution to problems involving allocation of limited resources among various activities, its impact has been tremendous.

Chapter Applications

4.1 Management, advertising, manufacturing
4.2 Production scheduling, manufacturing, shipping costs
4.3 Manufacturing, production scheduling
4.4 Production scheduling, manufacturing, production
4.5 Production, water purification, manufacturing
EOC Manufacturing, production, profit, nutrition, cost

Chapter Exercises

Section	Problems
4.1	1–4
4.2	5–14, 32, 33
4.3	15–20, 27, 34–36
4.4	21–24, 28, 37–39
4.5	25, 26, 29, 40–42
Tech Card	30, 31

Chapter Summary

4.1 Linear Inequalities in Two Variables
Objectives
- To graph linear inequalities in two variables
- To solve systems of linear inequalities in two variables

4.2 Linear Programming: Graphical Methods
Objectives
- To use graphical methods to find the optimal value of a linear function subject to constraints

4.3 The Simplex Method: Maximization
Objectives
- To use the simplex method to maximize functions subject to constraints

4.4 The Simplex Method: Duality and Minimization
Objectives
- To formulate the dual for minimization problems
- To solve minimization problems using the simplex method on the dual

4.5 The Simplex Method with Mixed Constraints
Objectives
- To solve maximization problems with mixed constraints
- To solve minimization problems with mixed constraints

Keywords

4.1 Linear Inequalities in Two Variables
Linear inequalities
Systems of linear inequalities

4.2 Linear Programming: Graphical Methods
Solution of linear programming problems

4.3 The Simplex Method: Maximization
Slack variables
Simple matrix
Pivoting
Simplex method
Nonunique solutions

4.4 The Simplex Method: Duality and Minimization
Dual problems
Principle of duality
Minimization

4.5 The Simplex Method with Mixed Constraints
Less than or equal to constraints
Maximizing with mixed constraints
Minimizing with mixed constraints

Go to login.cengage.com to access additional resources.

Chapter Outline

- **5.1** Exponential Functions
- **5.2** Logarithmic Functions and Their Properties
- **5.3** Exponential Equations and Applications

Chapter Preview

In this chapter we study exponential and logarithmic functions, which provide models for many applications that at first seem remote and unrelated. In our study of these functions, we will examine their descriptions, their properties, their graphs, and the special inverse relationship between these two functions. We will see how exponential and logarithmic functions are applied to some of the concerns of social scientists, business managers, and life scientists.

Chapter Applications

- **5.1** Compound interest, product liability
- **5.2** Richter scale, decibel readings, doubling time, life span update
- **5.3** Sales decay, inflation, population growth, health care, demand, total revenue, compound interest
- **EOC** Medicare spending, inflation, consumer credit, poverty threshold, percent of paved roads, sales decay, compound interest

Chapter Exercises

Section	Problems
5.1	1– 3, 34
5.2	4– 27
5.3	30– 33, 39, 40
Tech Card	28, 29, 35– 38

Chapter Summary

5.1 Exponential Functions
Objectives
- To graph exponential functions
- To evaluate exponential functions
- To model with exponential functions

5.2 Logarithmic Functions and Their Properties
Objectives
- To convert equations for logarithmic functions from logarithmic to exponential form, and vice versa
- To evaluate some special logarithms
- To graph logarithmic functions
- To model logarithmic functions
- To use properties of logarithmic functions to simplify expressions involving logarithms
- To use the change-of-base formula

5.3 Exponential Equations and Applications
Objectives
- To solve exponential equations
- To solve exponential growth or decay equations when sufficient data are known
- To solve exponential equations representing demand, supply, total revenue, or total cost when sufficient data are known

Keywords

5.1 Exponential Functions
Graphing exponentials
Evaluating exponential functions
Modeling with exponentials

5.2 Logarithmic Functions and Their Properties
Exponential and logarithmic forms
Evaluating logarithms
Graphing logarithmic functions
Modeling with logarithmic functions
Logarithm properties
Change of base formulas

5.3 Exponential Equations and Applications
Exponential equations
Growth and decay
Cost and revenue
Supply and demand

Go to login.cengage.com to access additional resources.

Chapter Outline

6.1 Simple Interest and Arithmetic Sequences
6.2 Compound Interest and Geometric Sequences
6.3 Future Values of Annuities
6.4 Present Values of Annuities
6.5 Loans and Amortization

Chapter Preview

Regardless of whether or not your career is in business, understanding how interest is computed on investments and loans is important to you as a consumer. The proliferation of personal finance and money management software attests to this importance. The goal of this chapter is to provide some understanding of the methods used to determine the interest and future value (principal plus interest) resulting from savings plans and the methods used in repayment of debts.

Chapter Applications

6.1 Salaries, simple interest investments
6.2 Population growth, profit, compound interest
6.3 Savings plans, sinking funds
6.4 Retirement planning, lottery prizes, court settlements
6.5 Home mortgages, car purchase, loan payments
EOC Finance, tuition, salaries, interest, savings goal, future value, college fund, investments, comparing yields, sinking fund, annuity, construction funding, time to reach a goal, Powerball lottery, annuity payments, amortization, cash value, amortization schedule

Chapter Summary

6.1 Simple Interest and Arithmetic Sequences
Objectives
- To find the future value and the amount of interest for a simple interest loan and an investment
- To find the simple interest rate earned on an investment
- To find the time required for a simple interest investment to reach a goal
- To write a specified number of terms of a sequence
- To find specified terms and sums of specified numbers of terms of arithmetic sequences

6.2 Compound Interest and Geometric Sequences
Objectives
- To find the future value of a compound interest investment and the amount of interest earned when interest is compounded at regular intervals or continuously
- To find the annual percentage yield (APY), or the effective annual interest rate, of money invested at compound interest
- To find the time it takes for an investment to reach a specified amount
- To find specified terms, and sums of specified numbers of terms, of geometric sequences

6.3 Future Values of Annuities
Objectives
- To compute the future values of ordinary annuities and annuities due
- To compute the payments required in order for ordinary annuities and annuities due to have specified future values
- To compute the payment required to establish a sinking fund
- To find how long it will take to reach a savings goal

6.4 Present Values of Annuities
Objectives
- To compute the present values of ordinary annuities, annuities due, and deferred annuities
- To compute the payments for a specified present value for an ordinary annuity, an annuity due, and a deferred annuity
- To find how long an annuity will last

Chapter Exercises

Chapter Summary, *continued*

6.5 Loans and Amortization
Objectives
- To find the regular payments required to amortize a debt
- To find the amount that can be borrowed for a specified payment
- To develop an amortization schedule
- To find the unpaid balance of a loan
- To find the effect of paying an extra amount

Keywords

6.1 Simple Interest and Arithmetic Sequences
Simple interest
Future value
Interest rate
Time to reach goal
Sequences
Arithmetic sequences

6.2 Compound Interest and Geometric Sequences
Periodic compounding
Continuous compounding
Annual percentage yield
Doubling time
Geometric sequences

6.3 Future Value of Annuities
Future values
Ordinary annuity
Annuity due
Periodic payments for annuities
Sinking funds
Reaching a savings goal

6.4 Present Value of Annuities
Present values
Ordinary annuity
Annuity due
Deferred annuity
Payments from annuities
Number of payments

6.5 Loans and Amortization
Amortizing debt
Amortization schedule
Unpaid balance

Go to login.cengage.com to access additional resources.

prepcard CHAPTER 7

Chapter Outline

Chapter Preview

An economist cannot predict exactly how the gross national product will change, a physician cannot determine exactly the cause of lung cancer, and a psychologist cannot determine the exact effect of environment on behavior. But each of these determinations can be made with varying probabilities. Thus, an understanding of the meaning and determination of the probabilities of events occurring is important to success in business, economics, the life sciences, and the social sciences. Probability was initially developed to solve gambling problems, but it is now the basis for solving problems in a wide variety of areas.

Chapter Applications

7.1 Drug use, car maintenance, voting, sales promotion, blood types, education
7.2 Drug use, linguistics, salaries, AIDS cases, education
7.3 Blood types, quality control, lactose intolerance, birth control
7.4 Lactose intolerance, quality control, alcoholism, drinking age
7.5 Binz, not Benz, signaling, molecules, poker, license plates, politics, committees
7.6 Politics, license plates, ATMs, telephones, rewards, quality control, banking, diversity, management
EOC Senior citizens, World Series, United Nations, quality control, color blindness, purchasing, management, utilities, juries, blood types, scheduling, lottery, stocks, income levels

Chapter Exercises

Section	Problems
7.1	1–5, 22, 23
7.2	6–10, 24–26
7.3	11–14, 27, 28
7.4	15–17, 29, 30
7.5	18–21, 31–35
7.6	36–40

Chapter Summary

7.1 Probability and Odds
Objectives
- To compute the probability of the occurrence of an event
- To construct a sample space for a probability experiment
- To compute the odds that an event will occur
- To compute the empirical probability that an event will occur

7.2 Union, Intersections, and Complements of Events
Objectives
- To find the probability of the intersection of two events
- To find the probability of the union of two events
- To find the probability of the complement of an event

7.3 Conditional Probability: The Product Rule
Objectives
- To solve probability problems involving conditional probability
- To compute the probability that two or more dependent events will occur
- To compute the probability that two or more independent events will occur

7.4 Probability Trees and Bayes' Formula
Objectives
- To use probability trees to solve problems
- To use Bayes' formula to solve probability problems

7.5 Counting: Permutations and Combinations
Objectives
- To use the Fundamental Counting Principle and permutations to solve counting problems
- To use combinations to solve counting problems

7.6 Permutations, Combinations, and Probability
Objective
- To use counting techniques to solve probability problems

Keywords

7.1 Probability and Odds
Probability of a single event
Sample spaces
Odds
Empirical probability

7.2 Unions, Intersections, and Complements of Events
Probability of E and F
Mutually exclusive events
Probability of E or F
Complement of an event

7.3 Conditional Probability: The Product Rule
Conditional probability
Product Rules
Dependent events
Independent events

7.4 Probability Trees and Bayes' Formula
Probability trees
Bayes' formula

7.5 Counting: Permutations and Combinations
Fundamental Counting Principle
Factorials
Permutations
Combinations

7.6 Permutations, Combinations, and Probability
Counting and probability

Go to login.cengage.com to access additional resources.

prepcard CHAPTER 8
PROBABILITY AND DATA DESCRIPTION

Chapter Outline

8.1 Binomial Probability Experiments
8.2 Describing Data
8.3 Discrete Probability Distributions
8.4 Normal Probability Distribution

Chapter Preview

In this chapter, we discuss how a set of data can be described with **descriptive statistics,** including mode, median, mean, and standard deviation. Descriptive statistics are used by businesses to summarize data about advertising effectiveness, production costs, and profit. We continue our discussion of probability by considering **binomial probability distributions, discrete probability distributions,** and **normal probability distributions.** Social and behavioral scientists collect data about carefully selected **samples** and use probability distributions to reach conclusions about the populations from which the samples were drawn.

Chapter Applications

8.1 Management, genetics, health care, quality control, testing
8.2 Unemployment rates, educational expenditures, birth weights
8.3 Animal relocation, raffle, gambling, insurance, voting, quality control
8.4 Growth, mileage, blood pressure
EOC Genetics, sampling, disease, cancer testing, fraud, testing, net worth, quality control

Chapter Exercises

Section	Problems
8.1	1–3, 29–31
8.2	4–13, 32–34
8.3	14–22, 35, 36
8.4	23–28, 37–39

Chapter Summary

8.1 Binomial Probability Experiments
Objectives
- To solve probability problems related to binomial experiments

8.2 Describing Data
Objectives
- To set up frequency tables and construct frequency histograms for sets of data
- To find the mode of a set of scores (numbers)
- To find the median of a set of scores
- To find the mean of a set of scores
- To find the range of a set of data
- To find the variance and standard deviation of a set of data

8.3 Discrete Probability Distributions
Objectives
- To identify random variables
- To verify that a table or formula describes a discrete probability distribution
- To compute the mean and expected value of a discrete probability distribution
- To make decisions by using expected value
- To find the variance and standard deviation of a discrete probability distribution
- To find the mean and standard deviation of a binomial distribution
- To expand a binomial to a power, using the binomial formula

8.4 Normal Probability Distribution
Objectives
- To calculate the probability that a random variable following the standard normal distribution has values in a certain interval
- To convert normal distribution values to standard normal values (z-scores)
- To find the probability that normally distributed values lie in a certain interval

Keywords

8.1 Binomial Probability Experiments
Binomial probabilities

8.2 Describing Data
Frequency tables
Frequency histograms
Mode of a set of scores
Median
Mean
Range of a set of data
Variance of a sample
Standard deviation of a sample

8.3 Discrete Probability Distributions
Random variable
Discrete random variable

Mean
Expected value
Expected value decisions
Variance and standard deviation
Mean of a binomial distribution
Standard deviation
Binomial formula

8.4 Normal Probability Distribution
Probability with the standard normal distribution
z-scores
Probability with the normal distribution

Go to login.cengage.com to access additional resources.

prepcard CHAPTER 9
DERIVATIVES

Chapter Outline

Chapter Preview

If a firm receives $30,000 in revenue during a 30-day month, its average revenue per day is $30,000/30 = $1000. This does not necessarily mean the actual revenue was $1000 on any one day, just that the average was $1000 per day. Similarly, if a person drove 50 miles in one hour, the average velocity was 50 miles per hour, but the driver could still have received a speeding ticket for traveling 70 miles per hour.

The smaller the time interval, the nearer the average velocity will be to the instantaneous velocity (the speedometer reading). Similarly, changes in revenue over a smaller number of units can give information about the instantaneous rate of change of revenue. The mathematical bridge from average rates of change to instantaneous rates of change is the **limit.**

This chapter is concerned with *limits* and *rates of change*. We will see that the *derivative* of a function can be used to determine instantaneous rates of change.

Chapter Applications

Chapter Summary

9.1 Limits
Objectives
- To use graphs and numerical tables to find limits of functions, when they exist
- To find limits of polynomial functions
- To find limits of rational functions

9.2 Continuous Functions and Limits at Infinity
Objectives
- To determine whether a function is continuous or discontinuous at a point
- To determine where a function is discontinuous
- To find limits at infinity and horizontal asymptotes

9.3 Rates of Change and the Derivative
Objectives
- To define and find average rates of change
- To define the derivative as a rate of change
- To use the definition of derivative to find derivatives of functions
- To use derivatives to find slopes of tangents to curves

9.4 Derivative Formulas
Objectives
- To find derivatives of powers of x
- To find derivatives of constant functions
- To find derivatives of functions involving constant coefficients
- To find derivatives of sums and differences of functions

9.5 The Product Rule and the Quotient Rule
Objectives
- To use the Product Rule to find the derivatives of certain functions
- To use the Quotient Rule to find the derivatives of certain functions

9.6 The Chain Rule and the Power Rule
Objectives
- To use the Chain Rule to differentiate functions
- To use the Power Rule to differentiate functions

9.7 Using Derivative Formulas
Objectives
- To use derivative formulas separately and in combination with each other

Chapter Exercises

Chapter Summary, *continued*

9.8 Higher-Order Derivatives

Objectives

- To find second derivatives and higher-order derivatives of certain functions

9.9 Marginals and Derivatives

Objectives

- To find the marginal cost and marginal revenue at different levels of production
- To find the marginal profit function, given information about cost and revenue

Keywords

9.1 Limits
Limits
Limits of polynomial functions
Limits of rational functions

9.2 Continuous Functions and Limits at Infinity
Continuity
Points of discontinuity
Limits at infinity, horizontal asymptotes

9.3 Rates of Change and the Derivative
Average rates of changes
Derivative
Finding a derivative
Slopes of tangents

9.4 Derivative Formulas
Derivatives of powers of x
Derivatives of constant functions
Derivative of $y = c f(x)$
Sums and differences rules

9.5 The Product Rule and the Quotient Rule
Product Rule
Quotient Rule

9.6 The Chain Rule and the Power Rule
Chain Rule
Power Rule

9.7 Using Derivative Formulas
Using formulas in combination

9.8 Higher-Order Derivatives
Second derivative
Higher-order derivatives

9.9 Marginals and Derivatives
Marginal revenue
Marginal cost
Marginal profit

Go to login.cengage.com to access additional resources.

prepcard CHAPTER 10
APPLICATIONS OF DERIVATIVES

Chapter Outline

Chapter Preview

The derivative can be used to determine where a function has a "turning point" on its graph, so that we can determine where the graph reaches its highest or lowest point within a particular interval. These points are called the relative maxima and relative minima, respectively, and are useful in sketching the graph of the function. The techniques for finding these points are also useful in solving applied problems, such as finding the maximum profit, the minimum average cost, and the maximum productivity. The second derivative can be used to find points of inflection of the graph of a function and to find the point of diminishing returns in certain applications.

Chapter Applications

10.1 Productivity, production costs, marginal revenue, revenue
10.2 Advertising and sales, production, diminishing returns
10.3 Maximizing revenue, minimizing average cost, maximizing profit, marginal revenue and marginal cost, Dow Jones Industrial Average
10.4 Productivity, consumer expenditure, advertising and sales, minimum cost, optimization at a fixed cost, inventory cost model, revenue
10.5 Cost-benefit, productivity, females in the work force
EOC Cost, revenue, profit, marginal profit, product design, printing design, drug sensitivity, federal tax per capita, inventory cost model, market share

Chapter Exercises

Section	Problems
10.1	1– 8, 27, 29
10.2	9–13
10.3	14, 15, 25, 26, 28, 30–33
10.4	34–36, 38
10.5	16–21, 39
Tech Card	22–24, 37

Chapter Summary

10.1 Relative Maxima and Minima: Curve Sketching
Objectives
- To find relative maxima and minima and horizontal points of inflection of functions
- To sketch graphs of functions by using information about maxima, minima, and horizontal points of inflection

10.2 Concavity: Points of Inflection
Objectives
- To find points of inflection of graphs of functions
- To use the second-derivative test to graph functions

10.3 Optimization in Business and Economics
Objectives
- To find absolute maxima and minima
- To maximize revenue, given the total revenue function
- To minimize the average cost, given the total cost function
- To find the maximum profit from total cost and total revenue functions, or from a profit function

10.4 Applications of Maxima and Minima
Objective
- To apply the procedures for finding maxima and minima to solve problems from the management, life, and social sciences

10.5 Rational Functions: More Curve Sketching
Objectives
- To locate horizontal asymptotes
- To locate vertical asymptotes
- To sketch graphs of functions that have vertical and/or horizontal asymptotes

Keywords

10.1 Relative Maxima and Minima: Curve Sketching
Relative maxima and minima
Critical points
Increasing
Decreasing
Sign diagram for $f'(x)$
Horizontal point of inflection
Graphing functions

10.2 Concavity: Points of Inflection
Concave up
Concave down
Points of inflection
Point of diminishing returns
Second derivative test

10.3 Optimization in Business and Economics
Absolute maxima and minima
Maximizing revenue
Minimizing average cost
Maximizing profit, monopoly market
Maximizing profit, competitive market
Applied maxima and minima

10.4 Applications of Maxima and Minima
Inventory cost models

10.5 Rational Functions: More Curve Sketching
Vertical asymptote at $x = c$
Horizontal asymptote at $y = b$
Curve sketching

Go to login.cengage.com to access additional resources.

prepcard

Chapter Outline

11.1 Derivatives of Logarithmic Functions
11.2 Derivatives of Exponential Functions
11.3 Implicit Differentiation
11.4 Related Rates
11.5 Applications in Business and Economics

Chapter Preview

In this chapter we will develop derivative formulas for logarithmic and exponential functions, focusing primarily on base *e* exponentials and logarithms. We will apply logarithmic and exponential functions and use their derivatives to solve maximization and minimization problems in the management and life sciences.

We will also develop methods for finding the derivative of one variable with respect to another even when the first variable is not a function of the other. This method is called **implicit differentiation.** We will use implicit differentiation with respect to time to solve problems involving rates of change of two or more variables. These problems are called **related-rates** problems.

Chapter Applications

11.1 Marginal cost, marginal revenue, demand
11.2 Future value, marginal cost, drugs in a bloodstream, national health care, spread of disease
11.3 Advertising and sales, production, demand
11.4 Profit, capital investment and production, allometric relationships (crabs), flight, distance
11.5 Elasticity of demand, taxation in a competitive market
EOC Deforestation, nonmarital childbearing, compound interest, marginal cost, inflation, evaporation, worker safety, environment, taxes, elasticity, revenue

Chapter Exercises

Section	Problems
11.1	2, 6, 7, 9, 10, 12, 21, 22
11.2	1, 3–5, 8, 11, 23–25
11.3	13–18
11.4	19, 20, 26–28
11.5	29–33
Tech Card	34

Chapter Summary

11.1 Derivatives of Logarithmic Functions
Objective
- To find derivatives of logarithmic functions

11.2 Derivatives of Exponential Functions
Objective
- To find derivatives of exponential functions

11.3 Implicit Differentiation
Objectives
- To find derivatives by using implicit differentiation
- To find slopes of tangents by using implicit differentiation

11.4 Related Rates
Objective
- To use implicit differentiation to solve problems that involve related rates

11.5 Applications in Business and Economics
Objectives
- To find the elasticity of demand
- To find the tax per unit that will maximize tax revenue

Keywords

11.1 Derivatives of Logarithmic Functions
Logarithmic function
Natural logarithm
Logarithmic properties I–IV
Change of base formula
Derivatives of logarithms

11.2 Derivatives of Exponential Functions
Derivatives of exponentials

11.3 Implicit Differentiation
Implicit derivatives
Slopes of tangents

11.4 Related Rates
Related rates
Percent rates of change

11.5 Applications in Business and Economics
Elasticity of demand
Elastic demand
Inelastic demand
Unitary elastic demand
Taxation in a competitive market

Go to login.cengage.com to access additional resources.

prepcard CHAPTER 12

Chapter Outline

12.1 The Indefinite Integral
12.2 The Power Rule
12.3 Integrals Involving Exponential and Logarithmic Functions
12.4 The Indefinite Integral in Business and Economics
12.5 Differential Equations

Chapter Preview

When we know the derivative of a function, it is often useful to determine the function itself. For example, accountants can use linear regression to translate information about marginal cost into a linear equation defining (approximately) the marginal cost function and then use the process of **antidifferentiation** (or **integration**) as part of finding the (approximate) total cost function. We can also use integration to find total revenue functions from marginal revenue functions, to optimize profit from information about marginal cost and marginal revenue, and to find national consumption from information about marginal propensity to consume.

Integration can also be used in the social and life sciences to predict growth or decay from expressions giving rates of change. For example, we can find equations for population size from the rate of change of growth, we can write equations for the number of radioactive atoms remaining in a substance if we know the rate of the decay of the substance, and we can determine the volume of blood flow from information about rate of flow.

Chapter Applications

12.1 Revenue, average cost, national health care, U.S. population
12.2 Revenue, data-entry speed, film attendance
12.3 Revenue, memorization, blood pressure in the aorta
12.4 Cost, revenue, maximum profit, national consumption and savings
12.5 Investing, sales and pricing, half-life, drug in an organ, impact of inflation
EOC Revenue, productivity, oxygen levels in water, bacterial growth, market share, revenue, cost, profit, national consumption, allometric growth, investment, fossil dating, drug in an organ

Chapter Exercises

Section	Problems
12.1	1–4, 30, 31
12.2	5–8, 10, 11, 15, 20, 21, 23, 32, 33
12.3	9, 12–14, 16–19, 22, 34, 35
12.4	36– 39
12.5	24– 29, 40– 43

Chapter Summary

12.1 The Indefinite Integral
Objective
- To find certain indefinite integrals

12.2 The Power Rule
Objective
- To evaluate integrals of the form $\int u^n\, u'dx = \int u^n\, du$ if $n \neq -1$

12.3 Integrals Involving Exponential and Logarithmic Functions
Objectives
- To evaluate integrals of the form $\int e^u\, u'dx$ or, equivalently, $\int e^u\, du$.
- To evaluate integrals of the form $\int \frac{u'}{u}dx$ or, equivalently, $\int \frac{1}{u}du$.

12.4 The Indefinite Integral in Business and Economics
Objectives
- To use integration to find total cost functions from information involving marginal cost
- To optimize profit, given information regarding marginal cost and marginal revenue
- To use integration to find national consumption functions from information about marginal propensity to consume and marginal propensity to save

12.5 Differential Equations
Objectives
- To show that a function is the solution to a differential equation
- To use integration to find the general solution to a differential equation
- To find particular solutions to differential equations using given conditions
- To solve separable differential equations
- To solve applied problems involving separable differential equations

Keywords

12.1 The Indefinite Integral
Differentials
General antiderivative
Integral
Integration formulas

12.2 The Power Rule
Power Rule

12.3 Integrals Involving Exponential and Logarithmic Functions
Exponential formula
Logarithmic formula

12.4 The Indefinite Integral in Business and Economics
Total cost
Total revenue
Profit
National consumption

12.5 Differential Equations
Differential equations
First order equations and general solutions
Particular solutions
Separable solutions
Radioactive decay
Drug in an organ

Go to login.cengage.com to access additional resources.

Chapter Outline

13.1 The Definite Integral: The Fundamental Theorem of Calculus
13.2 Area between Two Curves
13.3 Definite Integrals in Business and Economics
13.4 Using Tables of Integrals
13.5 Integration by Parts
13.6 Improper Integrals and Their Applications

Chapter Preview

In this chapter we define the definite integral and discuss a theorem and techniques that are useful in evaluating or approximating it. We will also see how it can be used in many interesting applications, such as consumer's and producer's surplus and total value, present value, and future value of continuous income streams. Improper integrals can be used to find the capital value of a continuous income stream.

Chapter Applications

13.1 Depreciation, sales and advertising, total income, telecommunications revenue, customer service
13.2 Sales and advertising, tax burden, income distribution
13.3 Continuous income streams, consumer's surplus, producer's surplus
13.4 Producer's surplus, cost, income streams
13.5 Producer's surplus, present value, income distribution
13.6 Capital value
EOC Maintenance, quality control, savings, income streams, income distribution, consumer's surplus, producer's surplus, average cost, cost, capital value

Section	Problems
13.1	1–13, 29– 30
13.2	14–16, 31– 33
13.3	34–37
13.4	17–20, 38– 39
13.5	21–24, 40–41
13.6	25–28, 42

Chapter Summary

13.1 The Definite Integral: The Fundamental Theorem of Calculus

Objectives
- To evaluate definite integrals using the Fundamental Theorem of Calculus
- To use definite integrals to find the area under a curve

13.2 Area between Two Curves

Objectives
- To find the area between two curves
- To find the average value of a function

13.3 Definite Integrals in Business and Economics

Objectives
- To use definite integrals to find total income, present value, and future value of continuous income streams
- To use definite integrals to find the consumer's surplus
- To use definite integrals to find the producer's surplus

13.4 Using Tables of Integrals

Objective
- To use tables of integrals to evaluate certain integrals

13.5 Integration by Parts

Objective
- To evaluate integrals using the method of integration by parts

13.6 Improper Integrals and Their Applications

Objectives
- To evaluate improper integrals
- To apply improper integrals to continuous income streams and to probability density functions

Keywords

13.1 The Definite Integral: The Fundamental Theorem of Calculus
Definite integral
Fundamental theorem of calculus
Area under a curve

13.2 Area between Two Curves
Area between curves
Average value

13.3 Definite Integrals in Business and Economics
Continuous income stream
Total income
Present value
Future value
Consumer's surplus
Producer's surplus

13.4 Using Tables of Integrals
Integral tables

13.5 Integration by Parts
Integration by parts

13.6 Improper Integrals and Their Applications
Improper integrals
Continuous income streams
Applications

Go to login.cengage.com to access additional resources.

prepcard

Chapter Outline

Chapter Preview

In this chapter we will extend our study to functions of two or more variables. We will use these concepts to solve problems in the management, life, and social sciences. In particular, we will discuss joint cost functions, utility functions that describe the customer satisfaction derived from the consumption of two products, Cobb-Douglas production functions, and wind chill temperatures as a function of air temperature and wind speed.

We will use derivatives with respect to one of two variables (called **partial derivatives**) to find marginal cost, marginal productivity, marginal utility, marginal demand, and other rates of change. We will use partial derivatives to find maxima and minima of functions of two variables, and we will use Lagrange multipliers to optimize functions of two variables subject to a condition that constrains these variables. These skills are used to maximize profit, production, and utility and to minimize cost subject to constraints.

Chapter Applications

14.1 Investment, Wilson's lot size formula, temperature and humidity, production
14.2 Mortgage, pesticide, utility
14.3 Joint cost and marginal cost, production functions, demand functions
14.4 Profit, manufacturing
14.5 Utility, production, cost, revenue, manufacturing
EOC Utility, savings plans, retirement benefits, advertising and sales, cost, production, marginal demand, profit

Chapter Exercises

Section	Problems
14.1	1–4, 23
14.2	5–18, 24–26
14.3	27–30
14.4	19, 20, 31, 32
14.5	21, 22, 33, 34

Chapter Summary

14.1 Functions of Two or More Variables
Objectives
- To find the domain of a function of two or more variables
- To evaluate a function of two or more variables given values for the independent variables

14.2 Partial Differentiation
Objectives
- To find partial derivatives of functions of two or more variables
- To evaluate partial derivatives of functions of two or more variables at given points
- To use partial derivatives to find slopes of tangents to surfaces
- To find and evaluate second- and higher-order partial derivatives of functions of two variables

14.3 Functions of Two Variables in Business and Economics
Objectives
- To evaluate cost functions at given levels of production
- To find marginal costs from total cost and joint cost functions
- To find marginal productivity for given production functions
- To find marginal demand functions from demand functions for a pair of related products

14.4 Maxima and Minima
Objectives
- To find relative maxima, minima, and saddle points of functions of two variables

14.5 Constrained Optimization and Lagrange Multipliers
Objectives
- To find the maximum or minimum value of a function of two or more variables subject to a condition that constrains the variables

Keywords

14.1 Functions of Two or More Variables
Domains
Evaluating functions

14.2 Partial Differentiation
Partial derivatives
Partial derivative at a point
Slope of a tangent
Second partial derivative
Higher-order derivatives

14.3 Functions of Two Variables in Business and Economics
Marginal costs from joint cost functions
Marginal productivity
Marginal demand

14.4 Maxima and Minima
Maxima or minima
Saddle points

14.5 Constrained Optimization and Lagrange Multipliers
Maxima subject to constraint
Minima subject to constraint

Go to login.cengage.com to access additional resources.

These cards contain discussions, examples, and problems regarding the use of TI-83/84 graphing calculators and Excel spreadsheets. For additional specific instructions for using these technologies, see the Calculator Guide and Excel Guide on CourseMate for MATH APPS. Access at login.cengagebrain.com

1.2 GRAPHING EQUATIONS WITH GRAPHING CALCULATORS

Setting Windows
The window defines the highest and lowest values of x and y on the graph of the function that will be shown on the screen. To set the window manually, press the WINDOW key and enter the values that you want.

The values that define the viewing window can also be set by using ZOOM keys. Frequently the standard window (ZOOM 6) is appropriate. The standard window gives x- and y-values between −10 and 10. The window should be set to show the important parts of the graph and suggest the unseen parts. Such a graph is called complete.

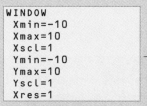
```
WINDOW
 Xmin=-10
 Xmax=10
 Xscl=1
 Ymin=-10
 Ymax=10
 Yscl=1
 Xres=1
```

Graphing Equations
To graph an equation in the variables x and y, first solve the equation for y in terms of x. If the equation has variables other than x and y, solve for the dependent variable and replace the independent variable with x. Press the Y= key to access the function entry screen.

To erase an equation, press CLEAR. To return to the homescreen, press 2nd MODE (QUIT). Determine an appropriate viewing window. Using the displayed coordinates from TRACE helps to determine an appropriate window. Pressing GRAPH or a ZOOM key will activate the graph.

The function $10y = x^4 - 200x^2 - 120$ is graphed by entering $0.1x^4 - 20x^2 - 12$ in Y1. Using a standard window gives an empty window, and using [−20, 20] by [−1200, 500] gives a better graph.

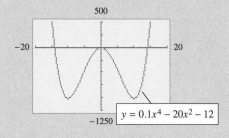
$$y = 0.1x^4 - 20x^2 - 12$$

1.2 GRAPHING A FUNCTION WITH EXCEL

1. Put headings on the two columns (x and $f(x)$, for example).
2. Fill the inputs (x-values) in Column A by hand or with a formula for them.
3. Enter the function formula for the function in B2.

   Enter = 6*A2 − 3 to represent $f(x) = 6x - 3$.
4. Select the cell containing the formula for the function (B2, for example).
5. Move the mouse to the lower right corner until there is a thin "+" sign.
6. Drag the mouse down to the last cell where formula is required, and press ENTER.

	A	B
1	x	$f(x) = 6x - 3$
2	−2	−15
3	−1	−9
4	0	−3
5	1	3
6	3	15
7	5	27
8	10	57

7. Highlight the two columns containing the values of x and $f(x)$.
8. Click the Chart Wizard icon and then select the XY (Scatter) chart type with the smooth curve option.
9. Click the Next button to get the Chart Source Data box. Then click Next to get the Chart Options box, and enter your chart title and labels for the x and y axes.
10. Click Next, select whether the graph should be within the current worksheet or on another, and click Finish.

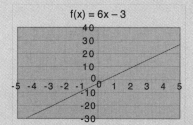

1.2 FINDING FUNCTION VALUES WITH GRAPHING CALCULATORS

Using TRACE on the Graph
Enter the function to be evaluated in Y1. Choose a window so that it contains the x-value whose y-value we seek.

(continued on reverse)

Press TRACE and then enter the selected x-value followed by ENTER. The cursor will move to the selected value and give the resulting y-value if the selected x-value is in the window. If the selected x-value is not in the window, Err: INVALID occurs. If the x-value is in the window, the y-value will occur even if it is not visible in the window.

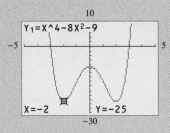

Using the TABLE ASK Feature

Enter the function with the Y= key. {Note: The = sign must be highlighted.} Press 2nd WINDOW (TBLSET), move the cursor to Ask opposite Indpnt:, and press ENTER. This allows you to input specific values for x. Pressing DEL will clear entries in the table. Then press 2nd TABLE and enter the specific values. This table evaluates $f(x) = x^4 - 8x^2 - 9$ at -2 and at 3.

X	Y₁
-2	-25
3	0
X =	

Making a Table of Values

If the Indpnt variable is on Auto, enter an initial x-value for the table in (TblStart), and enter the desired change (Δ Tbl) in the x-value in the table.

X	Y₁
-3	0
-2	-25
-1	-16
0	-9
1	-16
2	-25
3	0
X = 3	

Enter 2nd GRAPH TABLE to get a list of x-values and the corresponding y-values. The value of the function at the given value of x can be read from the table. Use the up or down arrows to find the x-values where the function is to be evaluated.

1.2 EVALUATING FUNCTIONS WITH EXCEL

1. Put headings on the two columns.
2. Fill the inputs in Column A by hand or with a formula for them. The formula =A2+1 gives 2 in A3 when ENTER is pressed.
3. Moving the mouse to the lower right corner of A3 until there is a thin "+" sign and dragging the mouse down "fills down" all required entries in column A.
4. Enter the function formula for the function in B2.

Enter $= 1000*(1.1)\wedge(A2)$ to represent $S = 1000(1.1^1)$. Pressing ENTER gives the value when $t = 1$.

5. Using Fill Down gives the output for all inputs.

	A	B
1	Year	Future Value
2	1	= 1000*(1.1) ˆ (A2)
3	= A2 + 1	

	A	B
1	Year	Future Value
2	1	1100
3	2	1210
4	3	1331

1.4 SOLVING SYSTEMS OF EQUATIONS IN TWO VARIABLES WITH GRAPHING CALCULATORS

To solve the system $\begin{cases} 3x + 2y = 12 \\ 4x - 3y = -1 \end{cases}$ graphically, we must solve the equations for y.

We graph $Y1 = 6 - (3/2)x$ and $Y2 = (1/3) + (4/3)x$ and then use Intersect under the CALC menu to find the point of intersection. If the two lines intersect in one point, the coordinates give the x- and y-values of the solution. The solution of the system above is $x = 2, y = 3$.

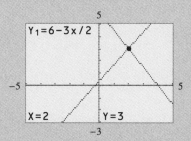

1.4 SOLVING SYSTEMS OF TWO EQUATIONS IN TWO VARIABLES WITH EXCEL

1. Write the two equations as linear functions in the form $y = mx + b$. The solution to $\begin{cases} 3x + 2y = 12 \\ 4x - 3y = -1 \end{cases}$ is found to be $x = 2, y = 3$, using Goal Seek as follows.

2. Enter the input variable x in cell A2 and the formula for each of the two equations in cells B2 and C2, respectively.
3. Enter = B2 − C2 in cell D2.

	A	B	C	D
1	x	= 6 − 1.5x	= 1/3 + 4x/3	= y1 − y2
2	1	= 6 − 1.5*A2	= 1/3 + 4*A2/3	= B2 − C2

4. Use Tools< Goal Seek after entering 0 in cell D2.
5. In the dialog box:
   a. Click the Set Cell box and click on the D2 cell.
   b. Enter 0 in the To Value box.
   c. Click the By Changing Cell box and click on the A2 cell.
6. Click OK in the Goal Seek dialog box, getting 0.
7. The x-value of the solution is in cell A2, and y-value is in both B2 and C2.

	A	B	C	D
1	x	= 6 − 1.5x	= 1/3 + 4x/3	= y1 − y2
2	2	3	3	0

Calculator and Excel guides are available on CourseMate for MATH APPS. Access at login.cengagebrain.com.

2.1 SOLVING EQUATIONS WITH A GRAPHING CALCULATOR

Solving Equations by the *x*-Intercept Method

To find the solutions to $f(x) = 0$ (the *x*-values where the graph crosses the *x*-axis):

1. Set one side of the equation to 0 and enter the other side as Y1 in the Y= menu.
2. Set the window so that the *x*-intercepts to be located can be seen.
3. Press 2nd TRACE to access the CALC menu and select 2:zero.
4. Answer the question "*left bound?*" with ENTER after moving the cursor close to and to the left of an *x*-intercept.
5. Answer the question "*right bound?*" with ENTER after moving the cursor close to and to the right of this *x*-intercept.
6. To the question "*guess?*" press ENTER. The coordinates of the *x*-intercept are displayed. The *x*-value is a solution.
7. Repeat to get all *x*-intercepts. The graph of a linear equation will cross the *x*-axis at most 1 time; the graph of quadratic equation will cross the *x*-axis at most 2 times, etc.

To solve $7x^2 = 16x - 4$, graph $y = -7x^2 + 16x - 4$, on a window with center at $x = 1$. The solutions are shown below.

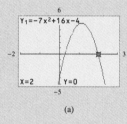

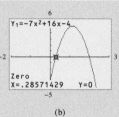

(a)   (b)

Solving Equations by the Intersection Method

1. Graph the left side of the equation as Y1 and the right side as Y2.
2. Find a point of intersection of the graphs as shown on the techcard for Section 1.4.
3. To find another point of intersection, repeat while keeping the cursor near the second point.

2.1 SOLVING QUADRATIC EQUATIONS WITH EXCEL

To solve a quadratic equation:

1. Enter *x*-values centered around the *x*-coordinate of the graph's vertex in column A and use the function formula to find the values of $f(x)$ in column B.
2. Graph the function, $f(x) = 2x^2 - 9x + 4$ in this case, and observe where the graph crosses the *x*-axis ($f(x)$ near 0).

	A	B	C	D	E	F
1	x	f(x)=2x^2-9x+4				
2	-1	15				
3	0	4				
4	1	-3				
5	2	-6				
6	3	-5				
7	4	0				
8	5	9				
9						

3. Use *Tools< Goal Seek*, entering a cell address with a function value in column B at or near 0, enter the set cell to the value 0, and enter the changing cell.
4. Click *OK* to find the *x*-value of the solution in cell A2. The solution may be approximate. The spreadsheet shows $x = 0.5001$, which is an approximation of the exact solution $x = 0.5$.

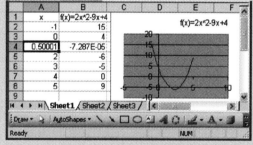

5. After finding the first solution, repeat the process using a second function value at or near 0. The second solution is $x = 4$ in this case.

2.2 GRAPHING QUADRATIC FUNCTIONS WITH A GRAPHING CALCULATOR

To graph a quadratic function:

1. Solve for *y* in terms of *x* and enter it in the Y= menu.
2. Find the coordinates of its vertex.
3. Set the window so the *x*-coordinate of its vertex is near its center and the *y*-coordinate is visible.

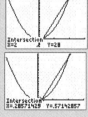

3. Press Graph. The graph of $P(x) = -0.1x^2 + 300x - 1200$ is shown on a window with the center at $x = 1500$.

2.4 GRAPHING POLYNOMIAL FUNCTIONS WITH A GRAPHING CALCULATOR

To graph a polynomial function:
1. See Table 2.1 in the text to determine possible shapes for the graph.
2. Graph the function, $y = x^4 - 2x^2$ in this case, in a window large enough to see the shape of the complete graph.
3. If necessary, adjust the window for a better view of the graph.

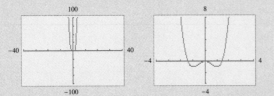

2.4 GRAPHING POLYNOMIAL FUNCTIONS WITH EXCEL

To graph a polynomial function ($f(x) = 2x^2 - 9x + 4$, in this case):
1. Use the function to create a table containing values for x and $f(x)$.
2. Highlight the two columns containing the values of x and $f(x)$.
3. Click the Chart Wizard icon and then select XY (Scatter) chart type with the smooth curve option.
4. Click the Next button to get the Chart Source Data box. Then click Next to get the Chart Options box, and enter your chart title and labels for the x and y axes.
5. Click Next, select whether the graph should be within the current worksheet or on another, and click Finish.

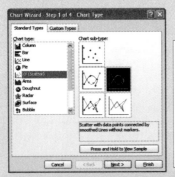

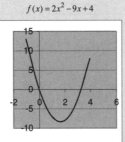

$$f(x) = 2x^2 - 9x + 4$$

2.4 GRAPHING RATIONAL FUNCTIONS WITH A GRAPHING CALCULATOR

To graph a rational function:

$$y = \frac{12x + 8}{3x - 9}$$

1. Determine the vertical and horizontal asymptotes.
2. Set the window so that the asymptotes are near its center.
3. Graph the function in a window large enough to see the shape of the complete graph.
4. If necessary, adjust the window for a better view of the graph.

2.4 GRAPHING DISCONTINUOUS FUNCTIONS WITH EXCEL

An Excel graph will connect all points corresponding to values in the table, so if the function you are graphing is discontinuous for some x-value a, enter x-values near this value and leave (or make) the corresponding $f(a)$ cell blank.

2.4 GRAPHING PIECEWISE-DEFINED FUNCTIONS WITH A GRAPHING CALCULATOR

A piecewise-defined function is defined differently over two or more intervals.
To graph a piecewise-defined function

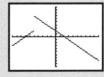

$$y = \begin{cases} f(x) & \text{if } x \leq a \\ g(x) & \text{if } x > a \end{cases}$$

1. Go to the Y= key and enter $Y_1 = (f(x))/(x \leq a)$ and $Y_2 = (g(x))/(x > a)$.

(The inequality symbols are found under the TEST menu.)

2. Graph the function using an appropriate window.

Evaluating a piecewise-defined function at a given value of x requires that the correct equation ("piece") be selected.

The following screens are used to graph

$$y = \begin{cases} x + 7 & \text{if } x \leq -5 \\ -x + 2 & \text{if } x > -5 \end{cases}$$

Details for modeling are on the next Card.

2.5 MODELING WITH GRAPHING CALCULATORS

A. Create a Scatter Plot

1. Press STAT and under EDIT press 1:Edit. This brings you to the screen where you enter data into lists.
2. Enter the *x*-values (input) in the column headed L1 and the corresponding *y*-values (output) in the column headed L2.
3. Go to the Y= menu and turn off or clear any functions entered there. To turn off a function, move the cursor over the = sign and press ENTER.
4. Press 2nd STAT PLOT, 1:Plot 1. Highlight ON, and then highlight the first graph type (Scatter Plot), Enter Xlist:L1, Ylist:L2, and pick the point plot mark you want.
5. Choose an appropriate WINDOW for the graph and press GRAPH, or press ZOOM, 9:ZoomStat to plot the data points.

Finding a Model for Life Expectancy

  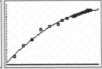

B. Find an Equation That Models a Set of Data Points

1. Observe the scatter plot to determine what type function would best model the data. Press STAT, move to CALC, and select the function type to be used to model the data.
2. Press the VARS key, move to Y-VARS, and select 1:Function and 1:Y1. Press ENTER.

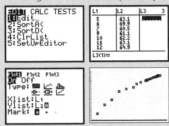

The coefficients of the equation will appear on the screen and the regression equation will appear as Y1 on the Y = screen.

Pressing ZOOM 9 shows how well the model fits the data. The model is $y = -0.00213x^2 + 530x + 45.2$.

2.5 MODELING WITH EXCEL

To create a scatter plot of data:
1. Enter the inputs (*x*-values) in Column A and the outputs (*y*-values) in Column B.
2. Highlight the two columns and use Chart Wizard, *XY*(Scatter) and click Next.
3. In Step 3 of the Chart Wizard, enter the title and the *x*- and *y*-axes labels.

To find the equations of lines or curves that best fit a given set of data points:

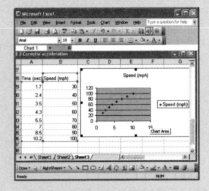

1. Place the scatter plot of the data in the worksheet.
2. Single click on the scatter plot in the workbook.
3. From the *Chart* menu choose *Add Trendline*.
4. Click the *regression type* that appears to be the best function fit for the scatter plot. If *Polynomial* is selected, choose the appropriate Order (degree).
5. Click the *Options* tab and check the *Display equation on chart* box.
6. Click *OK* and you will see the graph of the selected function that is the best fit, along with its equation.

The equation that models Corvette acceleration is shown here.

The speed of the Corvette is given by $y = 21.875x^{0.6663}$ where *x* is the time in seconds.

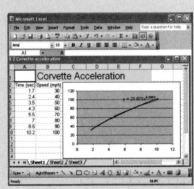

Calculator and Excel guides are available on CourseMate for MATH APPS. Access at login.cengagebrain.com.

3.1 ENTERING DATA INTO MATRICES WITH CALCULATORS

To enter data into matrices, press the MATRX key. Move the cursor to EDIT. Enter the number of the matrix into which the data is to be entered. Enter the dimensions of the matrix, and enter the value for each entry of the matrix. Press ENTER after each entry.

For example, we enter the matrix below as [A].

$$\begin{bmatrix} 1 & 2 & 3 \\ 2 & -2 & 1 \\ 3 & 1 & -2 \end{bmatrix}$$

1. Enter 3's to set the dimension, and enter the numbers.
2. To perform operations with the matrix or leave the editor, first press 2nd QUIT.
3. To view the matrix, press MATRX, the number of the matrix, and ENTER.

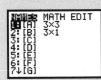 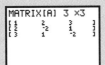 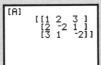

3.1 OPERATIONS WITH MATRICES USING CALCULATORS

To find the sum of two matrices, [A] and [D], enter [A] + [D], and press ENTER.

For example, the sum $\begin{bmatrix} 1 & 2 & 3 \\ 2 & -2 & 1 \\ 3 & 1 & -2 \end{bmatrix} + \begin{bmatrix} 7 & -3 & 2 \\ 4 & -5 & 3 \\ 0 & 2 & 1 \end{bmatrix}$ is shown below.

To find the difference, enter [A] − [D] and press ENTER.
We can multiply a matrix [D] by a real number (scalar) k by entering k [D].

  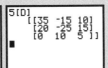

3.1 OPERATIONS WITH MATRICES USING EXCEL

1. Type a name A in A1 to identify the first matrix.
2. Enter the matrix elements of matrix A in the cells B1:D3.
3. Type a name B in A5 to identify the second matrix.
4. Enter the matrix elements of matrix B in the cells B5:D7.
5. Type a name A+B in A9 to indicate the matrix sum.
6. Type the formula =B1+B5 in B9 and press ENTER.
7. Use Fill Across to copy this formula across the row to C9 and D9.
8. Use Fill Down to copy the row B9:D9 to B11:D11, which gives the sum.
9. To subtract the matrices, change the formula in B9 to "=B1−B5" and proceed as with addition.

	A	B	C	D
1	A	1	2	3
2		4	5	6
3		7	8	9
4				
5	B	-2	-4	3
6		1	4	-5
7		3	6	-1
8				
9	A+B	-1	-2	6
10		5	9	1
11		10	14	8

3.2 MULTIPLYING TWO MATRICES USING CALCULATORS

To find the product of two matrices [C] and [A], enter [C][A] and press ENTER.

For example, we compute the product $\begin{bmatrix} 1 & 2 & 4 \\ 3 & 2 & -1 \end{bmatrix}\begin{bmatrix} 1 & 2 & 3 \\ 2 & -2 & 1 \\ 3 & 1 & -2 \end{bmatrix}$

[C][A] below. Note that [A][C] cannot be computed because their dimensions do not match. See the second screen below.

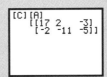 

3.2 MULTIPLYING TWO MATRICES USING EXCEL

1. Enter the names and elements of the matrices.
2. Enter the name A×B in A9 to indicate the matrix product.
3. Select a range of cells that is the correct size to contain the product (B9:D11 in this case).
4. Type "=mmult"(in the formula bar, and then select the cells containing the elements of matrix A (B1:D3).
5. Stay in the formula bar, type a comma and select the matrix B elements (B5:D7), and close the parentheses.
6. Press the CTRL, SHIFT, and ENTER keys *all at the same time*, giving the product.

3.3 SOLUTION OF SYSTEMS – REDUCED ECHELON FORM USING CALCULATORS

To solve a system of 3 equations in 3 variables:
1. Enter the augmented matrix as [A], then press 2nd QUIT.
2. In MATRIX, select MATH and 8:rref(.
3. Enter [A]) to get rref ([A]) and press ENTER.

For example, the system $\begin{cases} 2x - y + z = 6 \\ x + 2y - 3z = 9 \\ 3x \quad - 3z = 15 \end{cases}$ is solved at below.

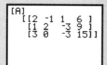 

If each row in the coefficient matrix (first 3 columns) contains a 1 with the other elements 0's, the solution is unique and the number in column 4 of a row is the value of the variable corresponding to a 1 in that row. The solution to the system above is unique: $x = 4$, $y = 1$, and $z = -1$.

3.4 FINDING THE INVERSE OF A MATRIX USING CALCULATORS

To find the inverse of a matrix, enter the elements of the matrix using MATRX and EDIT. Press 2nd QUIT.

Press MATRX, the number of the matrix, and ENTER, then press the x^{-1} key and ENTER.

For example, the inverse of $E = \begin{bmatrix} 2 & 0 & 2 \\ 0 & 0 & 1 \\ 0 & 2 & 0 \end{bmatrix}$ is shown below.

To see the entries as fractions, press MATH and press 1:▸Frac, and press ENTER.

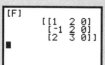 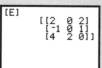 

Not all matrices have inverses. Matrices that do not have inverses are called singular matrices.

 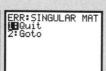

3.4 SOLVING SYSTEMS OF LINEAR EQUATIONS WITH MATRIX INVERSES USING CALCULATORS

The matrix equation $AX = B$ can be solved by computing $X = A^{-1}B$ if a unique solution exists.

The solution to $\begin{cases} 25x + 20y + 50z = 15{,}000 \\ 25x + 50y + 100z = 27{,}500 \\ 250x + 50y + 500x = 92{,}250 \end{cases}$

is found below to be $x = 254$, $y = 385$, $z = 19$.

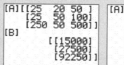

3.4 FINDING THE INVERSE OF A MATRIX WITH EXCEL

1. Enter the name A in A1 and the elements of the matrix in B1:D3 as in Section 3.2 of this techcard.
2. Enter the name "Inverse(A)" in A5 and select a range of cells that is the correct size to contain the inverse (B5:D7) in this case.
3. Enter "=minverse(", select matrix A (B1:D3), and close the parentheses.
4. Press the CTRL, SHIFT, and ENTER keys *all at the same time*, getting the inverse.

3.4 SOLVING SYSTEMS OF LINEAR EQUATIONS WITH MATRIX INVERSES USING EXCEL

A system of linear equations can be solved by multiplying the matrix containing the augment by the inverse of the coefficient matrix. The steps

used to solve $\begin{cases} 2x + y + z = 8 \\ x + 2y \quad = 6 \\ 2x + \quad z = 5 \end{cases}$ follow.

1. Enter the coefficient matrix A in B1:D3.
2. Compute the inverse of A in B5:D7. (See above.)
3. Enter B in cell A9 and enter the augment matrix in B9:B11.
4. Enter X in A13 and select the cells B13:B15.
5. In the formula bar, type "=mmult(", then select matrix inverse(A) in B5:D7, type a comma, select matrix B in B9:B11, and close the parentheses.
6. Press the CTRL, SHIFT, and ENTER keys *all at the same time*, getting the solution.
7. Matrix X gives the solution $x = 0$, $y = 3$, $z = 5$.

Calculator and Excel guides are available on CourseMate for MATH APPS. Access at login.cengagebrain.com.

4.1 GRAPHING SOLUTION REGIONS OF LINEAR INEQUALITIES WITH GRAPHING CALCULATORS

To graph the solution of a linear inequality in two variables, first solve the inequality for the independent variable and enter the other side of the inequality in Y1, so that $Y1 = f(x)$.

For example, to solve $4x - 2y \leq 6$, which is equivalent to $y \geq 2x - 3$, we graph $Y1 = 2x - 3$.

If the inequality has the form $y \leq f(x)$, shade the region below the graphed line and if the inequality has the form $y \geq f(x)$, shade the region above the line.

In this example, we shade the region above the line with SHADE under the DRAW menu and enter Shade (Y1, 10) on the homescreen.

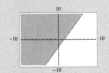

4.1 GRAPHING SOLUTIONS OF SYSTEMS OF LINEAR INEQUALITIES WITH GRAPHING CALCULATORS

To graph the solution region for a system of linear inequalities in two variables, write the inequalities as equations solved for y, and graph the equations.

For example, to find the region defined by the inequalities

$$\begin{cases} 5x + 2y \leq 54 \\ 2x + 4y \leq 60 \\ x \geq 0, y \geq 0 \end{cases}$$

choose a window with $x\text{min} = 0$ and $y\text{min} = 0$ because the inequalities $x \geq 0, y \geq 0$ limit the graph to Quadrant I. Write $y = 27 - 5x/2$ and $y = 15 - x/2$ and graph.

Testing points determines the region that satisfies the inequalities.

Using TRACE or INTERSECT with the pair of equations and finding the intercepts give the corners of the solution region, where the borders intersect. These corners of the region are (0, 0) (0, 15), (6, 12), and (10.8, 0).

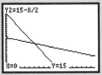 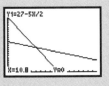

Use SHADE to shade the region determined by the inequalities. Shade under the border from $x = 0$ to a corner and shade under the second border from the corner to the x-intercept.

 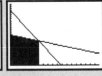 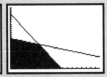

4.2 LINEAR PROGRAMMING WITH GRAPHING CALCULATORS

To solve a linear programming problem involving constraints in two variables:

Graph the constraint inequalities as equations, solved for y.

Test points to determine the region and use TRACE or INTERSECT to find each of the corners of the region, where the borders intersect.

Then evaluate the objective function at each of the corners.

For example, to maximize $f = 5x + 11y$ subject to the constraints

$$\begin{cases} 3x + 2y \leq 54 \\ 2x + 4y \leq 60 \\ x \geq 0, y \geq 0 \end{cases}$$

we graphically find the constraint region, and evaluate the objective function at the coordinates of each of the corners of the region. Evaluating $f = 5x + 11y$ at each of the corners determines where this objective function is maximized or minimized.

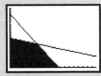

The corners of the region determined by the inequalities are (0, 0) (0, 15), (6, 12), and (10.8, 0).

At (0, 0), $f = 0$ At (0, 15), $f = 165$
At (6,12), $f = 162$ At (10.8, 0), $f = 54$

The maximum value of f is 165 at $x = 0, y = 15$.

4.3 LINEAR PROGRAMMING WITH EXCEL

To maximize $f = 6x + 13y + 20z$ subject to the constraints

$$\begin{cases} 3x + 7y + 4z \le 90{,}000 \\ x + 3y + 4z \le 30{,}000 \\ x + y + z \le 9{,}000 \end{cases}$$

1. On a blank spreadsheet, type a heading in cell A1, followed by the variable descriptions in cells A3–A5 and the initial values (zeros) in cells B3–B5.
2. a. Enter the heading Objectives in cell A7.
   b. Enter a description of the objective in cell A9 and the formula for the objective function in B7. The formula is =6*B3+13*B4+20*B5.
3. a. Type in the heading Constraints in A11 and descriptive labels in A13–A15.
   b. Enter left side of the constraint inequalities in B13–B15 and the maximums from the right side in C13–C15.

	A	B	C
1	Variables		
2			
3	# scientific calculators (x)	0	
4	# business calculators (y)	0	
5	# graphing calculators (z)	0	
6			
7	Objectives		
8			
9	Maximize profit	= 6*B3 + 13*B4 + 20*B5	
10			
11	Constraints		
12		Amount used	Maximum
13	Circuit components	= 5*B3 + 7*B4 + 10*B5	90000
14	Labor hours	= B3 + 3*B4 + 4*B5	30000
15	Cases	= B3 + B4 + B5	9000

4. Select Solver under the Tools menu. A dialog box will appear.
5. a. Click the Set Target Cell box and C9 (containing the formula for the objective function).
   b. Check the button Max for maximization.
   c. Click the By Changing Cells box and select cells B3 –B5.

6. a. Click the Subject to Constraints entry box. Press the Add button to add the first constraint.
   b. Click the left entry box and click cell B13 (containing the formula for the first constraint.
   c. Set the middle entry box to <=.
   d. Click the right entry box and C13 to enter the constraint.
   e. Click Add and repeat the steps 6b–6d for the remaining constraints.
   f. Click in the left entry box for the constraints and select the variables in B3– B5. Set the middle entry to >=, and type 0 in the right entry box.

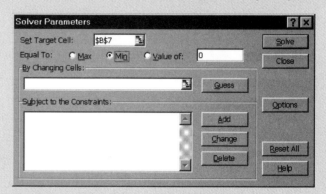

7. Click Solve in the Solver dialog box. A dialog box states that Solver found a solution. To see the Solver results, click Keep Solver Solution and also select Answer.
8. Go back to the Spreadsheet. The new values in B3–B5 are the values of the variables that give the maximum, and the value in B9 is the maximum value of the objective function.

	A	B	C
1	Variables		
2			
3	# scientific calculators (x)	2000	
4	# business calculators (y)	0	
5	# graphing calculators (z)	7000	
6			
7	Objectives		
8			
9	Maximize profit	152000	
10			
11	Constraints		
12		Amount used	Maximum
13	Circuit components	80000	90000
14	Labor hours	30000	30000
15	Cases	9000	9000

In Solver, minimization of an objective function is handled exactly the same as maximization, except min is checked and the inequality signs are $\ge$.

When mixed constraints are used, simply enter them with "mixed" inequalities.

Calculator and Excel guides are available on CourseMate for MATH APPS. Access at login.cengagebrain.com.

5.1 GRAPHING EXPONENTIALS WITH GRAPHING CALCULATORS

Graphing Exponentials

1. Enter the function as Y1 in the Y= menu
2. Set the x-range centered at $x = 0$
3. Set the y-range to reflect the function's range of $y > 0$. (See the graph of $y = 4^x$).

Note that some graphs appear to eventually merge with the negative x-axis, adjusting the windows can show that these graphs never touch the x-axis.

It may also be helpful to use TABLE to find values of y for selected values of x. (See the graph of $y = 1000e^{(.12x)}$).

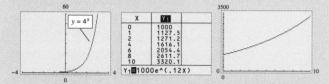

5.1 GRAPHING EXPONENTIALS WITH EXCEL

Exponential functions are entered into Excel differently for base e and for bases other than e.

1. Graphing $y = a^x$ uses the formula =a^x
2. Graphing $y = e^x$ uses the formula =exp(x)

To graph $y = 1.5x$ and $y = e^x$ on the same axes:

1. Type x in cell A1 and numbers centered at 0 in Column A.
2. Type $f(x)$ in cell B1, enter the formula =1.5^A2 in cell B2 and fill down.
3. Type $g(x)$ in cell C1, enter the formula =exp(A2) in cell C2 and fill down.
4. Select the entire table and use Chart Wizard to graph, as described on the Section 1.2 Tech Card.

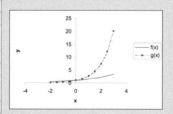

	A	B	C
1	x	f(x)	g(x)
2	-2	=1.5^A2	=EXP(A2)
3	-1.5	=1.5^A3	=EXP(A3)
4	-1	=1.5^A4	=EXP(A4)
5	-0.5	=1.5^A5	=EXP(A5)
6	0	=1.5^A6	=EXP(A6)
7	0.5	=1.5^A7	=EXP(A7)
8	1	=1.5^A8	=EXP(A8)
9	1.5	=1.5^A9	=EXP(A9)
10	2	=1.5^A10	=EXP(A10)
11	2.5	=1.5^A11	=EXP(A11)
12	3	=1.5^A12	=EXP(A12)

5.1 MODELING EXPONENTIAL FUNCTIONS WITH GRAPHING CALCULATORS

Find the exponential model for the following data

x	1	2	3	4	5	6	7	8	9	10
y	43	38	33	29	25	22	19	15	14	12

1. Create a Scatter Plot for the data, as described on the Section 2.5 Tech Card.
2. Choose STAT, then CALC. Scroll down to 0:ExpReg and press ENTER, then VARS, Y-VARS, FUNCTION, Y1, and ENTER. Recall that this both calculates the requested exponential model and enters its equation as Y1 in the Y= menu.

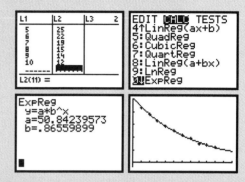

5.1 MODELING EXPONENTIAL FUNCTIONS WITH EXCEL

To find the exponential model for the following data

x	0	5	10	15	16	17	18
y	170	325	750	1900	2000	2200	2600

1. Create a Scatter Plot for the data (as described on the Section 2.5 Tech Card).
2. From the *Chart* menu choose *Add Trendline*. (See the Section 2.5 Tech Card.)
3. Click *exponential* regression type when that function type appears to be the best fit for the scatter plot.
4. Click the options tab on this box and select display equation.
5. Click Next and Finish to see equation and graph.

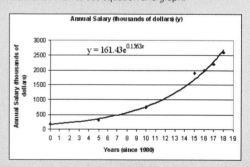

5.2 GRAPHING BASE *e* AND BASE 10 LOGARITHMIC FUNCTIONS WITH GRAPHING CALCULATORS

Enter the function as Y1 in the Y= menu.

1. For $y = \ln(x)$ use the LN key.
2. For $y = \log(x)$ use the LOG key.
3. Set the window x-range to reflect that the function's domain is $x > 0$.
4. Center the window y-range at $y = 0$.

(continued on reverse)

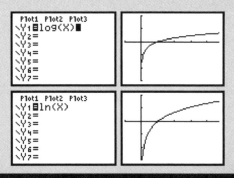

5.2 GRAPHING BASE *e* AND BASE 10 LOGARITHMIC FUNCTIONS WITH EXCEL

1. Create a table of values for *x*-values to reflect the function's domain is *x* > 0 (as described in the Section1.2 Tech Card).
2. For *y* = ln(*x*) use the formula =ln(*x*). For *y* = log(*x*) use the formula =log10(*x*).
3. Select the entire table and use Chart Wizard to graph. (See the Section 1.2 Tech Card.)

x	*f*(*x*) = ln(*x*)
0.9	−0.1053605
0.5	−0.6931472
1	0
2	0.69314718
3	1.09861229
4	1.38629436
5	1.60943791
6	1.79175947
7	1.94591015
8	2.07944154
9	2.19722458
10	2.30258509
11	2.39789527

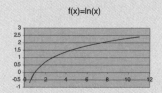

f(x)=ln(x)

x	*f*(*x*) = log(*x*)
0.9	−0.0457575
0.5	−0.30103
1	0
2	0.30103
3	0.47712125
4	0.60205999
5	0.69897
6	0.77815125
7	0.84509804
8	0.90308999
9	0.95424251
10	1
11	1.04139269

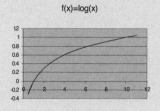

f(x)=log(x)

5.2 MODELING LOGARITHMIC FUNCTIONS WITH GRAPHING CALCULATORS

Find the logarithmic model for the following data

x	10	20	30	38
y	2.21	3.79	4.92	5.77

1. Create a scatter Plot for the data (as described on the Section 2.5 Tech Card.
2. Choose STAT, then CALC. Scroll down to 9:LnReg (see the Figure).

5.2 MODELING LOGARITHMIC FUNCTIONS WITH EXCEL

1. Create a Scatter Plot for the data (as described on the Section 2.5 Tech Card).
2. In the Chart menu, choose *Add Trendline* and click *logarithmic regression.*

x	*y*
10	2.31
20	3.79
30	4.92
38	5.77

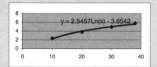

y = 2.5457Ln(x) - 3.6542

5.2 GRAPHING LOGARITHMIC FUNCTIONS WITH OTHER BASES

With a Graphing Calculator
1. Use the Change of Base Formula to rewrite the logarithmic function with base *e*,

$$\log x = \frac{\log x}{\log e} = \frac{\ln x}{\ln e}$$

2. Proceed as described above for graphing base *e* and base 10 logarithms

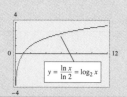

$$y = \frac{\ln x}{\ln 2} = \log_2 x$$

With Excel
1. The graph of a logarithm with any base *b* can be created by entering the formula =log(*x*, *b*). (Also, see the Section 1.2 Tech Card.)
2. Proceed as described above for graphing base e and base 10 logarithms.

The graph of *f*(*x*) = log₇ *x* is found below.

x	*f*(*x*) = log(*x*, 7)
0.9	−0.0541446
0.5	−0.3562072
1	0
2	0.35620719
3	0.56457503
4	0.71241437
5	0.82708748
6	0.92078222
7	1
8	1.06862156
9	1.2915007
10	1.18329466
11	1.23227441

Calculator and Excel guides are available on CourseMate for MATH APPS. Access at login.cengagebrain.com.

6.2 FUTURE VALUE OF A LUMP SUM WITH A GRAPHING CALCULATOR

To find the future value of a lump-sum investment:
1. Press the APPS key and select Finance, press ENTER.
2. Select TMV Solver, press ENTER.
3. Set N = the total number of periods, set I% = the annual percentage rate.
4. Set the PV = the lump sum preceded by a "-" to indicate to lump sum is leaving your possession.
5. Set PMT = 0 and set both P/Y and C/Y = the number of compounding periods per year.
6. Put the cursor on FV and press ALPHA ENTER to get the future value. The future value for $10,000 for 17 years at 9.8% compounded quarterly is $51,857.73

6.2 FINDING THE FUTURE VALUE OF A LUMP SUM WITH EXCEL

To find the future value of a lump-sum investment:
1. Type the headings in row A, and enter their values in row B (with the interest rate as a decimal).
2. Enter the formula =D2/E2 to compute the rate per period.
3. Enter Future Value in A4.
4. In cell B4, type the formula =fv(F2,B2,C2,A2,0) to compute the future value.

This spreadsheet gives the future value of an investment of $10,000 for 17 years at 9.8%, compounded quarterly.

Principal	Number of Periods	Payment	Annual Rate	Periods per year	Periodic Rate
10000	68	0	0.098	4	0.0245
Future Value	($51,857.73)				

6.2 FUTURE VALUE OF AN ANNUITY WITH A GRAPHING CALCULATOR

To find the future value of an ordinary annuity:
1. Press the APPS key and select Finance, press ENTER.
2. Select TMV Solver, press ENTER.
3. Set N = the total number of periods, set I% = the annual percentage rate.
4. Set the PV = 0 and set both P/Y and C/Y = the number of compounding periods per year. END should be highlighted.
5. Set PMT = the periodic payment preceded by a "-" to indicate to lump sum is leaving your possession.
6. Put the cursor on FV and press ALPHA ENTER to get the future value.

The future value of an ordinary annuity of $200 deposited at the end of each quarter for 2¼ years, with interest at 4% compounded quarterly, is shown.

For annuities due, all steps are the same except that BEGIN is highlighted.

Ordinary annuity

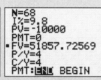
Annuity due

6.3 FINDING THE FUTURE VALUE OF AN ANNUITY WITH EXCEL

To find the future value of an ordinary annuity:
1. Type the headings in row A, and enter their values in row B (with the interest rate as a decimal).
2. Enter the formula =D2/E2 to compute the rate per period.
3. Enter Future Value in A4.
4. In cell B4, type the formula =fv(F2,B2,C2,A2,0) to compute the future value. (The 0 indicates that the deposits are made at the end of the periods.) The future value is in B4.

For annuities due, use =fv(F2,B2,C2,A2,1).
The payments are at the period beginning. See B5.

This spreadsheet gives the future value of an ordinary annuity of $200 deposited at the end of each quarter for 2 ¼ years, with interest at 4%, compounded quarterly.

Principal	Number of Periods	Payment	Annual Rate	Periods per year	Periodic Rate
0	9	200	0.04	4	0.01
Ordinary	annuity				
Future Value	($1873.71)				
Future Value	(1,892.44)				

6.4 PRESENT VALUE OF AN ANNUITY WITH A GRAPHING CALCULATOR

To find the present value of an ordinary annuity:
1. Press the APPS key and select Finance, press ENTER.
2. Select TMV Solver, press ENTER.
3. Set N = the total number of periods, set I% = the annual percentage rate.
4. Set the FV = 0 and set both P/Y and C/Y = the number of compounding periods per year. END should be highlighted.
5. Set PMT = the periodic payment preceded by a "-" to indicate the deposit is leaving your possession.
6. Put the cursor on PV and press ALPHA ENTER to get the present value.

(continued on reverse)

The lump sum needed to deposit to receive $1000 at the end of each month for 16 years if the annuity pays 9%, compounded monthly is shown below.

For annuities due, BEGIN is highlighted.

Ordinary annuity   Annuity due

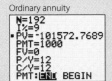

6.4 FINDING THE PRESENT VALUE OF AN ANNUITY WITH EXCEL

To find the present value of an ordinary annuity:
1. Type the headings in row A, and enter their values in row B (with the interest rate as a decimal).
2. Enter the formula =D2/E2 to compute the rate per period.
3. Enter Present Value in A4.
4. In cell B4, type the formula =fv(F2,B2,C2,A2,0) to compute the future value. (The 0 indicates that the deposits are made at the end of the periods.)

For annuities due, use =fv(F2,B2,C2,A2,1).
The payments are at the period beginning.

This spreadsheet gives the present value of an ordinary annuity paying $1000 at the end of each month for 16 years, with interest at 9%, compounded monthly.

Future Value	Number of Periods	Payment	Annual Rate	Periods per year	Periodic Rate
0	192	1000	0.09	12	0.0075
Ordinary	annuity				
Present Value	$101,572.77				

Annuity due					
Present Value	$102,334.56				

6.5 FINDING PAYMENTS TO AMORTIZE A LOAN WITH A GRAPHING CALCULATOR

To find the size of periodic payments to amortize a loan.
1. Press the APPS key and select Finance, press ENTER.
2. Select TMV Solver, press ENTER.
3. Set N = the total number of periods, set I% = the APR.
4. Set the PV = loan value and set both P/Y and C/Y = the number of periods per year
5. Set FV = 0
6. Put the cursor on PMT and press ALPHA ENTER to get the payment.

To repay a loan of $10,000 in 5 annual payments with annual interest at 10%, each payment must be $2637.97.

To repay a loan of $10,000 in 5 annual payments.

6.5 FINDING PAYMENTS TO AMORTIZE A LOAN WITH EXCEL

To find the periodic payment to pay off a loan:
1. Type the headings in row A, and enter their values in row B (with the interest rate as a decimal).
2. Enter the formula =D2/E2 in F2 to compute the rate per period.
3. Enter Payment in A4.
4. In cell B4, type the formula =Pmt(F2,B2,A2, C2,0) to compute the payment.

This spreadsheet gives the annual payment of a loan of $10,000 over 5 years when interest is 10% per year.

Loan Amount	Number of Periods	Future Value	Annual Rate	Periods per year	Periodic Rate
10,000	5	0	0.1	1	0.1
Payment	($2,637.97)				

6.5 FINDING THE NUMBER OF PAYMENTS TO AMORTIZE A LOAN WITH A GRAPHING CALCULATOR

To find the number of payments needed to amortize a loan.
1. Press the APPS key and select Finance, press ENTER.
2. Select TMV Solver, press ENTER.
3. Set the PV = loan value, set both P/Y and C/Y = the number of periods per year, set I% = the APR.
4. Set PMT = required payment and set FV = 0
5. Put the cursor on N and press ALPHA ENTER to get the payment.

The number of monthly payments to pay a $2500 credit card loan with $50 payments and 18% interest is 77.

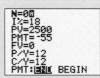

The number of monthly payments needed.

Calculator and Excel guides are available on CourseMate for MATH APPS. Access at login.cengagebrain.com.

7.5 EVALUATING FACTORIALS WITH A GRAPHING CALCULATOR

To evaluate factorials:
1. Enter the number whose factorial is to be calculated.
2. Choose MATH, then PRB. Scroll to 4: ! and press ENTER.
3. Press ENTER again to find the factorial. 7! is shown.

7.5 EVALUATING PERMUTATIONS WITH A GRAPHING CALCULATOR

To evaluate permutations:
1. For a "permutation of n objects taken r at a time" (such as $_{20}P_4$), first enter the value of n (such as $n = 20$).
2. Choose MATH, then PRB. Scroll to 2:nPr and press ENTER.

3. Enter the value of r (such as $r = 4$), and press ENTER to find the value of nPr. (See the Figure) $_{20}P_4$ is shown below.

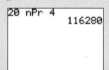

7.5 EVALUATING COMBINATIONS WITH A GRAPHING CALCULATOR

To evaluate combinations:
1. For a "combination of n objects taken r at a time" (such as $_{20}C_4$), first enter the value of n (such as $n = 20$).
2. Choose MATH, then PRB. Scroll to 3:nCr and press ENTER.
3. Enter the value of r (such as $r = 4$), and press ENTER to find the value of nCr.

Note that nCr = nC(n − r), as the Figure on the lower right shows for $_{20}C_4$ and $_{20}C_{16}$.

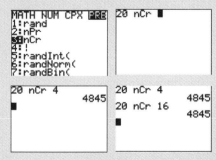

7.6 FINDING PROBABILITIES USING PERMUTATIONS AND COMBINATIONS WITH A CALCULATOR

To solve a probability problem that involves permutations or combinations:
1. Determine if permutations or combinations should be used.
2. Enter the ratios of permutations or combinations to find the probability.
3. If desired, use MATH, then 1:Frac to get the probability as a fraction.

If there are 5 defective computer chips in a box of 10, the probability that 2 chips drawn from the box will both be defective is

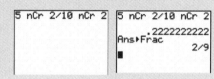

Calculator and Excel guides are available on CourseMate for MATH APPS. Access at login.cengagebrain.com.

8.1 BINOMIAL PROBABILITIES WITH A GRAPHING CALCULATOR

2nd DISTR A:binompdf(n,p,x) computes the probability of x successes in n trials of a binomial experiment with probability of success p. Using MATH 1:Frac gives the probabilities as fractions.

The probability of 3 heads in 6 tosses of a fair coin is found using 2nd DISTR, binompdf(6,.5,3)

The probabilities can be computed for more than one number in one command, using

2nd DISTR, binompdf(n,p,{$x_1,x_2,...$}).

The probabilities of 4, 5, or 6 heads in 6 tosses of a fair coin are found using 2nd DISTR, binompdf(6,.5,{4,5,6})

2nd DISTR, binomcdf(n,p,x) computes the probability that the number of successes is less than or equal to x for the binomial distribution with n trials and probability of success p.

The probability of 4 or fewer heads in 6 tosses of a fair coin is found using 2nd DISTR, binomcdf(6,.5,4).

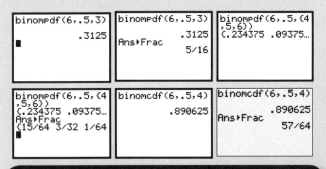

8.1 BINOMIAL PROBABILITIES WITH EXCEL

1. Type headings in cells A1:A3 and their respective values in cells B1:B3.
2. Use the function =binomdist(B1,B2,B3,cumulative) where
   - B1 is the number of successes
   - B2 is the number of independent trials
   - B3 is the probability of success in each trial
   - True replaces cumulative if a cumulative probability is sought; it is replaced by false otherwise
   - The probability of exactly 3 heads in 6 tosses is found by evaluating =binomdist(B1,B2,B3,false) in B3
   - The probability of 3 or fewer heads in 6 tosses is found by evaluating =binomdist(B1,B2,B3,true) in B5

This spreadsheet gives the probability of 3 heads in 6 tosses of a fair coin and the probability of 3 or fewer heads in 6 tosses.

	A	B
1	Number of successes	3
2	Number of trials	6
3	Probability of success	0.5
4	Prob of 3 successes	.3125
5	Probability of 3 or fewer successes	.65625

8.2 HISTOGRAMS WITH A GRAPHING CALCULATOR

To find a frequency histogram for a set of data:
1. Press STAT, EDIT, 1:edit to enter each number in a column headed by L1 and the corresponding frequency of each number in L2.
2. Press 2nd STAT PLOT, 1:Plot 1. Highlight ON, and then press ENTER on the histogram icon.

Enter L1 in xlist and L2 in Feq.

3. Press ZOOM, 9:ZoomStat or press Graph with an appropriate window.

The frequency histogram for the scores 38, 37, 36, 40, 35, 40, 38, 37, 36, 37, 39, 38 is shown.

4. If the data is given in interval form, a histogram can be created by using the steps above with the class marks used to represent the intervals.

8.2 BAR GRAPHS WITH EXCEL

To construct a bar graph for the given table of test scores:
1. Copy the entries of the table to cells A2:B6.
2. Select the range A2:B6.
3. Click on the Chart Wizard icon.
4. Select the graph option with the first sub-type.
5. Click Next.
6. Click Next through Steps 2–4. You can add a title in Step 3 if you desire.

Grade Range	Frequency
90-100	2
80-89	5
70-79	7
60-69	3
0-59	2

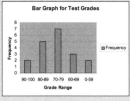

8.2 FINDING THE MEAN AND STANDARD DEVIATION OF RAW DATA WITH CALCULATORS

To find descriptive statistics for a set of data:
1. Enter the data in list L1 and the frequencies in L2.
2. To find the mean and standard deviation of the data in L1, press STAT, move to CALC, and press 1:1-Var Stats, and ENTER.

The mean and sample standard deviation of the data 1, 2, 3, 3, 4, 4, 4, 4, 5, 5 6, 7 are 4 and 1.65, respectively.

8.2 FINDING THE MEAN AND STANDARD DEVIATION OF GROUPED DATA WITH CALCULATORS

To find descriptive statistics for a set of data:
1. Enter the data in list L1 and the frequencies in L2.
2. To find the mean and standard deviation of the data in L1, press STAT, move to CALC, and press 1:1-Var Stats L1, L2 and ENTER.

For the data in the table

Salary	Number	Salary	Number
$59,000	1	$31,000	1
30,000	2	75,000	1
26,000	7	35,000	1
34,000	2		

The mean is $34,000 and the sample standard deviation is $14,132.84.

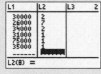

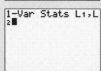

8.2 FINDING THE MEAN AND STANDARD DEVIATION OF RAW DATA WITH EXCEL

To find the mean, standard deviation and median of a raw data set:
1. Enter the data in row 1 (cells A1:L1).
2. Type the heading Mean in cell A3.
3. Type the formula =average(A1:L1) in cell B4.
4. Type the heading Standard Deviation in cell A4.
5. In cell B4, type the formula =stdev(A1:L1).
6. In cell A5, type the heading Median.
7. In cell B5, type the formula =median(A1:L1).

The mean, Standard Deviation, and mean for the data
1, 1, 1, 3, 3, 4, 4, 5, 6, 6, 7, 7 is found below.

	A	B	C	D	E	F	G	H	I	J	K	L
1	1	1	1	3	3	4	4	5	6	6	7	7
2												
3	Mean	4										
4	Standard Deviation	2.2563										
5	Median	4										

8.2 FINDING THE MEAN AND STANDARD DEVIATION OF GROUPED DATA WITH EXCEL

To Find the Mean
1. Enter the data in the cells A1:C6.
2. In D1, type the heading Class mark*frequency.
3. In D2, type the formula =B2*C2.
4. Copy the formula in D2 to D3:D6.
5. In B7, type the heading Total.
6. In cell C7, type the formula for the total frequencies, =sum(C2:C6).
7. In cell D7, type the formula for the total, =sum(D2:D6).
8. In cell A8, type the heading Mean.
9. In cell A9, type in the formula =D7/C7. See the calculation below.

To Find the Standard Deviation
10. In cell E1, type in the heading freq *(x − mean)^2.
11. In cell E2, type the formula =C2*(B2-A9)^2. (The A9 gives the value in A9; the reference doesn't change as we copy down.)
12. Copy the formula in E2 to E3:E6.
13. In cell C7, type the formula =sum(E2:E6).
14. In cell A10 type the heading Standard Deviation.
15. In cell A11, type the formula =sqrt(E7/(C7-1)). See the calculation below.

Grade Range	Class Marks	Frequency
90-100	95	3
80-89	84.5	4
70-79	74.5	7
60-69	64.5	0
50-59	54.5	2

	A	B	C	D	E
1	Grade Range	Class marks	frequency	class mark*frequency	freq*(x-x_mean)^2
2	90-100	95	3	285	832.2919922
3	80-89	84.5	4	338	151.5976563
4	70-79	74.5	7	521.5	103.4208984
5	60-69	64.5	0	0	0
6	50-59	54.5	2	109	1137.048828
7		Total	16	1253.5	2224.359375
8	Mean				
9	78.344				
10	Standard Deviation				
11	12.177				

8.2 CALCULATING NORMAL PROBABILITIES WITH CALCULATORS

To calculate normal probabilities:

The command 2ⁿᵈ DISTR, 2:normalcdf(lowerbound,upperbound, μ, σ) gives the probability that x lies between the lowerbound and the upperbound when the mean is μ and the standard deviation is σ.

To graph the normal distribution, press Y= and enter 2ⁿᵈ DIST 1:normalpdf (x, μ, σ) into y₁.

Then set the window values xmin and xmax so the mean μ falls between them and press ZOOM0:Zoomfit.

The probability that a score lies between 33 and 37 when the mean is 35 and the standard deviation is 2 is found below.

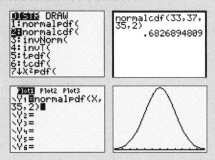

8.2 CALCULATING NORMAL PROBABILITIES WITH EXCEL

To calculate normal probabilities:
1. Type headings in A1:A4 and their respective values in cells B1:B4.
2. To find the probability that a score X is less than the x1 value in B3, enter the formula =normdist(B3, B1,B2,true) in cell B5.
3. To find the probability that X is less than the x2 value in B4, enter the formula =normdist(B4, B1,B2,true) in cell B6.
4. To find the probability that a score X is more than the value in B3 and less than the x2 value in B4, enter the formula =B6-B5 in cell B7.

Entries in B5, B6, and B7 give the probabilities of a score X being less than 100, less than 115, and between 100 and 115, respectively, when the mean is 100 and the standard deviation is 15.

	A	B
1	Mean	100
2	Standard deviation	15
3	x1	100
4	x2	115
5	Pr(X<x1)	0.5
6	Pr((X<x2)	0.841345
7	Pr(x1<X<x2)	0.341345

Calculator and Excel guides are available on CourseMate for MATH APPS. Access at login.cengagebrain.com.

9.1 EVALUATING LIMITS WITH A GRAPHING CALCULATOR

To evaluate $\lim_{x \to c} f(x)$:

1. Enter the function as Y1 in the Y= menu
2. Set the x-range so it contains $x = c$
3. Evaluate $f(x)$ for several x-values near $x = c$ and on each side of c by using one of the following methods.

   (a) Graphical Evaluation

   TRACE and ZOOM near $x = c$. If the values of y approach the same number L as x approaches c from the left and the right, there is evidence that the limit is L.

   (b) Numerical Evaluation

   Use TBLSET with Indpnt set to *Ask*. Enter values very close to and on both sides of c. The y-values will approach the same limit as above.

Evaluate $\lim_{x \to 3} \dfrac{x^2 - 9}{x - 3}$

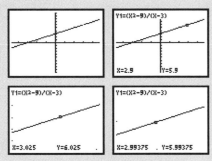

The y-values seem to approach 6.

The limit as x approaches 3 of $f(x)$ appears to be 6.

9.1–9.2 MORE LIMITS WITH A GRAPHING CALCULATOR: Piecewise-Defined Functions and Limits as $x \to \infty$

Limits of Piecewise Functions:

Enter the function as described on the Section 1.4 Tech Card, then use one of the Methods for Evaluating Limits with a Graphing Calculator (see Section 9.1 above).

Find $\lim_{x \to -5} f(x)$ where $f(x) = \begin{cases} x + 7 & \text{if } x \le -5 \\ -x + 2 & \text{if } x > -5 \end{cases}$

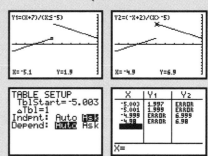

First enter Y1 $= (x + 7)/(x \le -5)$ and Y2 $= (-x + 2)/(x > -5)$. Both methods indicate that the limit does not exist (DNE).

Limits as $x \to \infty$

Enter the function as Y1, then use large values of x with one of the Methods for Evaluating Limits with a Graphing Calculator (see Section 9.1 above). Note: Limits as $x - \infty$ are done similarly.

Evaluate $\lim_{x \to \infty} \dfrac{3x - 2}{1 - 5x}$

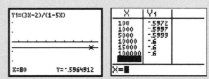

Both methods suggest that the limit is -0.6.

9.1–9.2 EVALUATING LIMITS WITH EXCEL

To evaluate $\lim_{x \to c} f(x)$:

1. Make a table of values for $f(x)$ near $x = c$. Include values on both sides of $x = c$.
2. Use the table of values to predict the limit (or that the limit does not exist).

Note: All limit evaluations with Excel use appropriate tables of values of $f(x)$. This is true when $f(x)$ is piecewise-defined and for limits as $x \to \infty$.

Find $\lim_{x \to 2} \dfrac{x^2 - 4}{x - 2}$

	A	B		C	D
1	x	f(x)		x	f(x)
2	2.1	4.1		1.9	3.9
3	2.05	4.05		1.95	3.95
4	2.01	4.01		1.99	3.99
5	2.001	4.001			

The tables suggest that $\lim_{x \to 2} \dfrac{x^2 - 4}{x - 2} = 4$.

9.3–9.7 APPROXIMATING DERIVATIVES WITH A GRAPHING CALCULATOR

To find the numerical derivative (approximate derivative) of $f(x)$ at $x = c$.

Method 1:

1. Choose MATH, then 8:nDeriv(and press ENTER.
2. Enter the function, x, and the value c, so the display shows nDeriv($f(x)$, x, c) then press ENTER. The approximate derivative at the specified value will be displayed.

 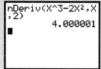

Find the numerical derivative of $f(x) = x^3 - 2x^2$ at $x = 2$.
The numerical derivative is approximately 4.

Method 2:

1. Enter the function as Y1 in the Y= menu, and graph in a window that contains both c and $f(c)$.
2. Choose CALC by using 2nd TRACE, then 6:dy/dx, enter the x-value, c, and press ENTER. The approximate derivative at the specified value will be displayed.

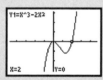 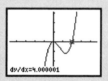

The value of the derivative is approximately 4.

Warning: Both approximation Methods above require that the derivative exists at $x = c$ and will give incorrect information when $f'(c)$ does not exist.

9.3–9.7 CHECKING DERIVATIVES WITH A GRAPHING CALCULATOR

To check the correctness of the derivative of a function $f(x)$:

1. In the Y= menu, enter as Y1 the derivative $f'(x)$ that you found, and graph it in a convenient window.
2. Enter the following as Y2: nDeriv($f(x)$, x, x).
3. If the second graph lies on top of the first, the derivative is correct.

Verify that the derivative of $f(x) = x^3 - 2x^2$ is $f(x) = 3x^2 - 4x$.

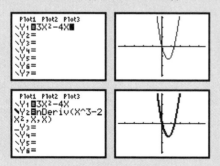

9.3–9.7 APPROXIMATING DERIVATIVES WITH EXCEL

To approximate $f'(c)$.

1. Numerically investigate the limit in the definition of derivative:

$$f(x) = \lim_{h \to 0} \frac{f(x + h) - f(x)}{h}$$

2. Use the given $f(x)$ and $x = c$ to create a table of values for h near 0 (and on both sides of $h = 0$).

Investigate $f'(1)$ for $f(x) = x^3$.

	A	B	C	D	E
1	h	1+h	f(1)	f(1+h)	(f(1+h)-f(1))/h
2	0.1	1.1	1	1.331	3.31
3	0.01	1.01	1	1.030301	3.0301
4	0.001	1.001	1	1.003003	3.003001
5	0.0001	1.0001	1	1.0003	3.00030001
6	-0.0001	0.9999	1	0.9997	2.99970001
7	-0.001	0.999	1	0.997003	2.997001
8	-0.01	0.99	1	0.970299	2.9701
9	-0.1	0.9	1	0.729	2.71

The table suggests that $f'(1) = 3$, which is the actual value.

Note: Excel has no built-in derivative approximation tool.

9.8 APPROXIMATING THE SECOND DERIVATIVE WITH A GRAPHING CALCULATOR

To approximate $f''(c)$:

1. Enter $f(x)$ as Y1 in the Y= menu.
2. Enter nDeriv(Y1, x, x) as Y2.
3. Estimate $f''(c)$ by using nDeriv(Y2, x, c).

Find the second derivative of $f(x) = x^3 - 2x^2$ at $x = 2$.

 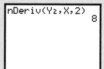

Thus $f''(2) = 8$.

Calculator and Excel guides are available on CourseMate for MATH APPS. Access at login.cengagebrain.com.

10.1 FINDING CRITICAL VALUES WITH A GRAPHING CALCULATOR

Methods I and II show how to find the critical values of $f(x) = \frac{1}{3}x^3 - 4x$. Note that the derivative is $f'(x) = x^2 - 4$.

To find or approximate critical values of $f(x)$, that is, x-values that make the derivative equal to 0 or undefined:
I. Find the derivative of $f(x)$.
II. Use Method 1 or Method 2 to find the critical values:

Method 1

1. Enter the derivative in the Y= menu as Y1 and graph it in a convenient window.
2. Find where Y1 = 0 by one of the following:
   (a) Using TRACE to find the x-intercepts of Y1.
   (b) Using 2nd CALC then 2:ZERO (or ROOT).
   (c) Using TBLSET then TABLE to find the values of x that give Y1 = 0.
3. Use the graph of Y1 (and TRACE or TABLE) to find the values of x that make the derivative undefined.

The only critical values are $x = 2$ and $x = -2$.

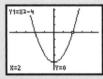

Method 2

1. Use MATH then 0:SOLVER to solve Y1 = 0. This gives the critical values for which the derivative equals zero.
2. If appropriate, use 0:SOLVER to solve (Denominator of Y1) = 0 to find the critical values for which the derivative is undefined.

The only critical values are $x = 2$ and $x = -2$.

10.1 RELATIVE MAXIMA AND MINIMA WITH A GRAPHING CALCULATOR OR EXCEL

WITH A GRAPHING CALCULATOR:

1. In the Y= menu enter the function as Y1 and the derivative as Y2.
2. Use TBLSET and TABLE to evaluate the derivative to the left and to the right of each critical value.
3. Use the signs of the values of the derivative to determine whether f is increasing or decreasing around the critical values, and thus to classify the critical values as relative maxima, relative minima, or horizontal points of inflection.
4. Graph the function to confirm your conclusions.

Find the relative maxima and minima of $f(x) = \frac{1}{3}x^3 - 4x$ with a Graphing Calculator. Note that the derivative is $f'(x) = x^2 - 4$.

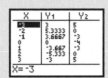 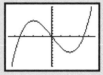

Relative max @ $(-2, 16/3)$ and min @ $(2, -16/3)$

WITH EXCEL

1. Make a table with columns for x-values, the function, and the derivative.
2. Extend the table to include x-values to the left and to the right of all critical values.
3. Use the signs of the values of the derivative to determine whether f is increasing or decreasing around the critical values, and thus to classify the critical values as relative maxima, relative minima, or horizontal points of inflection. You may want to graph the function to confirm your conclusions.

Use Excel to find the relative maxima and minima of $f(x) = x^2$. Note that the derivative is $f'(x) = 2x$.

	A	B	C
	x	f(x)	f'(x)
1			
2	-2	4	-4
3	-1.5	2.25	-3
4	-1	1	-2
5	-0.5	0.25	-1
6	0	0	0
7	0.5	0.25	1
8	1	1	2
9	1.5	2.25	3
10	2	4	4

10.1 CRITICAL VALUES AND VIEWING WINDOWS ON A GRAPHING CALCULATOR

To use critical values to set a viewing window that shows a complete graph.
1. Once the critical values for a function have been found
   (a) Enter the function as Y1 and the derivative as Y2. Use TABLE to determine where the function is increasing and where it is decreasing
   (b) In WINDOW menu set x-min so that it is smaller than the smallest critical value and set x-max so that it is larger than the largest critical value.
2. Use TABLE or VALUE to determine the y-coordinates of the critical values. Set y-min and y-max to contain the y-coordinates of the critical points.
3. Graph the function.

Let $f(x) = 0.0001x^3 + 0.003x^2 - 3.6x + 5$. Given that the critical values for $f(x)$ are $x = -120$ and $x = 100$, set a window that shows a complete graph and graph the function.

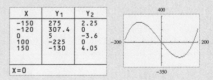

10.2 EXPLORING f, f', AND f'' RELATIONSHIPS WITH A GRAPHING CALCULATOR OR EXCEL

To explore relationships among the graphs of a function and its derivatives.
1. Find the functions for f' and f''.
2. Graph all three functions in the same window.
   - Notice that f increases when f' is above the x-axis (+) and decreases when f' is below the x-axis (–).
   - Notice that f is concave up when f'' is above the x-axis (+) and is concave down when f'' is below the x-axis (–).

Note: These graphs can be done with either a graphing calculator or Excel.

Let $f(x) = x^3 - 9x^2 + 24x$. Graph f', f'', and f'' on the interval [0, 5] to explore the relationships among these functions. [Note: the following screens show this done with Excel. A graphing calculator can also be used with the three functions entered as Y1, Y2, and Y3.]

10.3–10.4 FINDING OPTIMAL VALUES WITH A GRAPHING CALCULATOR

To find the optimal values of a function when the goal is not to produce a graph.
1. Enter the function as Y1 in the Y= menu.
2. Select a window that includes the x-values of interest and graph the function.
3. While looking at the graph of the function, choose the CALC menu, scroll to 3:minimum or 4:maximum depending on which one is to be found, and press ENTER. This will result in a "Left Bound?" prompt.
   (a) Move the cursor to a point to the left of the point of interest. Press ENTER to select the left bound.
   (b) Move the cursor to the right of the point of interest. Press ENTER to select the right bound.
   (c) Press ENTER at the "Guess?" prompt. The resulting point is an approximation of the desired optimum value.

Let $f(x) = 100\sqrt{2304 + x^2} + 50(100 - x)$. Find the minimum value of $f(x)$ on the interval [0, 100] for x. Note: the following screens use the window x: [–10, 110] and y: [–2500, 12500]

The minimum value is $y \approx 9157$ and occurs when $x \approx 27.7$.

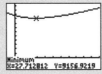

Note: Finding a maximum value works similarly.

10.3–10.4 FINDING OPTIMAL VALUES WITH EXCEL

To find the optimal value of a function when the goal is not to produce a graph.
1. Set up a spreadsheet that identifies the variable and the function whose optimal value is sought.
2. Choose Tools > Solver. Then, in the Dialog Box
   (a) Set the Target Cell as that of the objective function.
   (b) Check Max or Min according to your goal.
   (c) Set the Changing Cells to reference the variable.
3. Click on the Options box. Make sure "Assume Linear Model" is NOT checked. Then click OK.
4. Click Solve in the Solver dialog box. You will get a dialog box stating that Solver found a solution. Save the solution if desired, then click OK.
5. The cells containing the variable and the function should now contain the optimal values.

Minimize area $A = x^2 + \dfrac{160}{x}$ for $x > 0$.

The function is minimized for $x = 4.3089$ and the minimum value is $A = 55.699$.

10.5 ASYMPTOTES AND WINDOW SETTING WITH A GRAPHING CALCULATOR

To use asymptotes and critical values to set a viewing window that shows a complete graph.
1. Once the asymptotes and critical values for a function have been found
   (a) Determine where the function is increasing and where it is decreasing (by using TABLE and with Y2 as the derivative).
   (b) In WINDOW menu set x-min so that it is smaller than the smallest x-value that is either a vertical asymptote or a critical value and set x-max so that it is larger than the largest of these important x-values.
2. Use TABLE or VALUE to determine the y-coordinates of the critical values. Set y-min and y-max so they contain the y-coordinates of any horizontal asymptotes and critical points.
3. Graph the function.

Let $f(x) = \dfrac{x + 10}{x^2 + 300}$. Given that $f(x)$ has no vertical asymptotes, the line $y = 0$ is a horizontal asymptote, and the critical values are $x = -30$ and $x = 10$. Set the window and graph $y = f(x)$.

The critical points for $f(x)$ are $(-30, -1/60)$ and $(10, 1/20)$. An x-range that contains –30 and 10 is needed. Because $y = 0$ is a horizontal asymptote, the relative extrema are absolute, and the y-range must be quite small to see the shape of the graph.

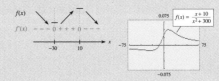

11.1 DERIVATIVES OF LOGARITHMIC FUNCTIONS WITH GRAPHING CALCULATORS

To check that the derivative of $y = \ln(2x^6 - 3x + 2)$ is Y1 $= \dfrac{2x^5 - 3}{2x^6 - 3x + 2}$, we show that the graph of Y2 $=$ nDeriv(ln($2x^6 - 3x + 2$), x, x) lies on the graph of $y_1 = \dfrac{2x^5 - 3}{2x^6 - 3x + 2}$ (see the Sections 9.3–9.7 Tech Card).

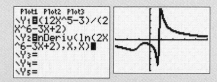

11.1 FINDING CRITICAL VALUES WITH GRAPHING CALCULATORS

To find the critical values of a function $y = f(x)$:
1. Enter the function as Y1 in the Y= menu, and the derivative as Y2.
2. Press MATH 0 (Solver). Press the up arrow revealing EQUATION SOLVER equ:0=, and enter Y2 (the derivative).
3. Press the down arrow or ENTER and the variable appears with a value (not the solution). Place the cursor on the variable whose value is sought.
4. Press ALPHA SOLVER (ENTER). The value of the variable changes to the solution of the equation that is closest to that value.
5. To find additional solutions (if they exist), change the value of the variable and press ALPHA SOLVE (ENTER). The value of the variable gives the solution of Y2 = 0 that is closest to that value.

To find the critical values of $y = x^2 - 8\ln x$, we solve $0 = 2x - \dfrac{8}{x}$.

With

The two critical values are $x = 2$ and $x = -2$.

11.1 FINDING OPTIMAL VALUES WITH GRAPHING CALCULATORS

To find the optimal values of a function $y = f(x)$:
1. Find the critical values of a function $y = f(x)$.
   (Find the steps for finding critical values above.)
2. Graph $y = f(x)$ on a window containing the critical values.
3. The y-values at the critical values (if they exist) are the optimal values of the function.

To find the optimal values of $y = x^2 - 8 \ln x$, we solve $0 = 2x - \dfrac{8}{x}$ and evaluate $y = x^2 - 8 \ln x$ at the solutions.

With

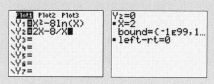

The two critical values are $x = 2$ and $x = -2$.

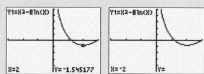

The minimum of $y = x^2 - 8 \ln x$ is -1.545 at $x = 2$.
The function $y = x^2 - 8 \ln x$ is undefined for all negative values because $\ln x$ is undefined for all negative values. Thus there is no optimum value of the function at $x = -2$.

11.2 DERIVATIVES OF EXPONENTIAL FUNCTIONS WITH GRAPHING CALCULATORS

To check that the derivative of $y = 5^{x^2 + x}$ is $y = 5^{x^2 + x}(2x + 1)\ln 5$ we show that the graph of Y2 $= nDeriv(5^{x^2 + x}, x, x)$ lies on the graph of Y1 $= 5^{x^2 + x}(2x + 1)\ln 5$ (see the Sections 9.3–9.7 Tech Card).

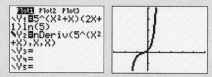

11.2 FINDING CRITICAL VALUES WITH GRAPHING CALCULATORS

To find the critical values of a function $y = f(x)$:
1. Enter the function as Y1 in the Y= menu, and the derivative as Y2.
2. Press MATH 0 (Solver). Press the up arrow revealing EQUATION SOLVER equ:0=, and enter Y2 (the derivative).
3. Press the down arrow or ENTER and the variable appears with a value (not the solution). Place the cursor on the variable whose value is sought.
4. Press ALPHA SOLVER (ENTER). The value of the variable changes to the solution of the equation that is closest to that value.
5. To find additional solutions (if they exist), change the value of the variable and press ALPHA SOLVE (ENTER). The value of the variable gives the solution of Y2 = 0 that is closest to that value.

To find the critical values of $y = e^x - 3x^2$, we solve $0 = e^x - 6x$.

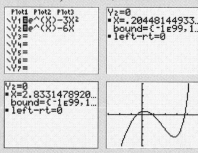

A relative maximum of $y = e^x - 3x^2$ occurs at $x = 0.204$ and a relative minimum occurs at $x = 2.833$.

12.1–12.2 CHECKING INDEFINITE INTEGRALS WITH GRAPHING CALCULATORS

To check a computed indefinite integral with the command fnInt:

1. Enter the integral of $f(x)$ (without the $+ C$) as Y1 in Y= menu.
2. Move the cursor in Y2, press MATH, 9:fnInt(and enter $f(x), x, 0, x$) so the equation is Y2 = fnInt $(f(x), x, 0, x)$
3. Press GRAPH with an appropriate window. If the second graph lies on top of the first, the graphs agree and the computed integral checks.
4. Pressing ENTER with the cursor to the left of Y2 changes the thickness of the second graph, making it more evident that the second lies on top of the first.

Checking that the integral of $f(x) = x^2$ is $\int x^2 \, dx = \dfrac{x^3}{3} + C$:

12.1–12.2 FAMILIES OF FUNCTIONS WITH GRAPHING CALCULATORS

To graph some functions in the family of indefinite integrals of $f(x)$:

1. Integrate $f(x)$.

2. Enter equations of the form $y = \int f(x) \, dx + C$ for different values of C.

3. Press GRAPH with an appropriate window. The graphs will be shifted up or down, depending on C.

The graphs of members of the family $y = \int (2x - 4) \, dx + C$ with $C = 0, 1, -2,$ and 3.

 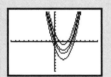

12.5 DIFFERENTIAL EQUATIONS WITH GRAPHING CALCULATORS

To solve initial value problems in differential equations:

1. Integrate $f(x)$, getting $y = F(x) + C$. If a value of x and a corresponding value of y in the integral $y = F(x) + C$ is known, this initial value can be used to find one function that satisfies the given conditions.
2. Press MATH, 0: Solver and press the up arrow to see EQUATION SOLVER.
3. Set 0 equal the integral minus y, getting $0 = F(x) + C - y$, and press the down arrow.
4. Enter the given values of x and y, place the cursor on C, and press ALPHA, SOLVE AND ENTER. Replace C with this value to find the function satisfying the conditions.

To solve $\dfrac{dy}{dx} = 2x - 4$, we note that the integral of both sides is $y = x^2 - 4x + C$. If $y = 8$ when $x = -1$ in this integral, we can find C, and thus a unique solution, shown below.

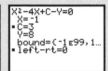 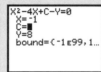 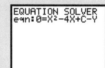

The unique solution is $y = x^2 - 4x + 3$.

Calculator and Excel guides are available on CourseMate for MATH APPS. Access at login.cengagebrain.com.

13.1 APPROXIMATING DEFINITE INTEGRALS USING RECTANGLES WITH EXCEL

To approximate the area under the graph of $y = f(x)$ and above the x-axis from a to b, using left-hand endpoints:

1. Divide $b - a$ by the number of rectangles to get the width of each rectangle.
2. Enter the x-values in Column A and the function values in column B
3. Add a third column with the heading Δx in C1, and the rectangle width in cells c2:c6.
4. Add a fourth column with the heading Rectangle Area in D1 and use the formula =B2*C2 to get the area of the first rectangle.
5. Copy this formula down to cells D3:D6.
6. In cell C7, type the heading Total Area, and in D7 enter the formula =sum(D2:D6) and press ENTER.

The approximate area under the graph of $f(x) = x^2$ from $x = 1$ to $x = 2$, with 5 rectangles, is found below. In the example, the width is 0.2 and the left-hand endpoints are 1.0, 1.2, 1.4, 1.6, 1.8.

x	$f(x)$
1	1
1.2	1.44
1.4	1.96
1.6	2.56
1.8	3.24

x	$f(x)$	Δx	Rectangle Area
1	1	0.2	0.2
1.2	1.44	0.2	0.288
1.4	1.96	0.2	0.392
1.6	2.56	0.2	0.512
1.8	3.24	0.2	0.648
		Total Area	2.04

13.1 APPROXIMATING DEFINITE INTEGRALS WITH GRAPHING CALCULATORS–AREAS UNDER CURVES

To approximate the area under the graph of $y = f(x)$ and above the x-axis:

1. Enter $f(x)$ under the Y= menu and graph the function with an appropriate window.
2. Press 2nd CALC and 7: $\int f(x)\,dx$.
3. Press ENTER. Move the cursor to, or enter, the lower limit (the left x-value).
4. Press ENTER. Move the cursor to, or enter, the upper limit (the right x-value).
5. Press ENTER. The area will be displayed.

The approximate area under the graph of $f(x) = x^2$ from $x = 0$ to $x = 3$ is found below.

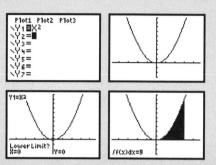

13.1 APPROXIMATING DEFINITE INTEGRALS WITH GRAPHING CALCULATORS–ALTERNATE METHOD

To approximate the definite integral of $f(x)$ from $x = a$ to $x = b$.

1. Press MATH, 9: fnInt(Enter $f(x)$, x, a, b so the display shows fnInt($f(x), x, a, b$).
2. Press ENTER to find the approximation of the integral.
3. The approximation may be made closer than that in step 3 by adding a fifth argument with a number (tolerance) smaller than 0.00001.

The approximation of $\int_1^3 (4x^2 - 2x)\,dx$ is found below.

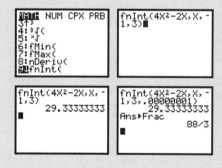

13.2 APPROXIMATING THE AREA BETWEEN TWO CURVES

To approximate the area between the graphs of two functions:

1. Enter one function as Y1 and the second as Y2. Press GRAPH using a window that shows all points of intersection of the graphs.
2. Find the x-coordinates of the points of intersection of the graphs, using 2nd CALC: intersect.
3. Determine visually which graph is above the other over the interval between the points of intersection.
4. Press MATH, 9: fnInt(. Enter $f(x)$, x, a, b) so the display shows fnInt($f(x)$, x, a, b) where $f(x)$ is Y2 − Y1 if the graph of Y2 is above the graph of y_1 between a and b, or Y2 − Y1 if Y1 is above Y2.

The area enclosed by the graphs of $y = 4x^2$ and $y = 8x$ is found as follows.

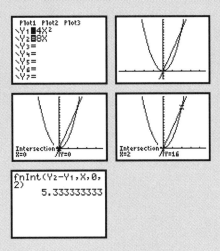

The area between the curves is 16/3.

13.2 APPROXIMATING THE AREA BETWEEN TWO CURVES— ALTERNATE METHOD

The area between the graphs can also be found by using 2nd CALC, $\int f(x)\, dx$.

1. Enter Y3 = Y2 − Y1 where Y2 is above Y1.
2. Turn off the graphs of Y1 and Y2 and graph Y3 with a window showing where Y3 > 0.
3. Press 2nd CALC and $\int f(x)\, dx$.
4. Press ENTER. Move the cursor to, or enter, the lower limit (the left x-value).
5. Press ENTER. Move the cursor to, or enter, the upper limit (the right x-value).
6. Press ENTER. The area will be displayed.

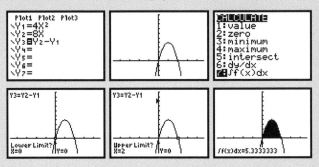

14.1 GRAPHS OF FUNCTIONS OF TWO VARIABLES WITH EXCEL

To create a surface plot for a function of two variables:

1. (a) Generate appropriate x-values beginning in B1 and continuing *across*.
   (b) Generate appropriate y-values beginning in A2 and continuing *down*.
2. Generate values for the function that correspond to the points (x, y) from step 1 as follows:

   In cell B2, enter the function formula with B$1 used to represent x and $A2 to represent y. See the online Excel Guide for additional information about the role and use of the $ in this step.
3. Select the entire table of values. Click Chartwizard and choose Surface in the Chart type menu.
4. Annotate the graph and click Finish to create the surface plot. By clicking into the resulting graph, it can be moved and viewed from a different perspective.

Let $f(x, y) = 10 - x^2 - y^2$. Plot the graph of this function for both x and y in the interval [−2, 2].

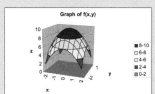

14.5 CONSTRAINED OPTIMIZATION AND LAGRANGE MULTIPLIERS WITH EXCEL

To solve a constrained optimization problem:

1. Set up the problem in Excel.
2. Choose Tools > Solver and do the following:
   - Choose the objective function as the Target Cell.
   - Check Max or Min depending on the problem.
   - Choose the cells representing the variables for the By Changing Cells box.
   - Click on the Constraints box and press Add. Then enter the constraint equations.
3. Click on the Options box and make sure that "Assume Linear Model" is NOT checked. Then click OK.
4. Click Solve in the Solver Dialog box. Then click OK to solve.

Maximize $p = 600\, l^{\frac{2}{3}} K^{\frac{1}{3}}$ subject to $40l + 100k = 3000$

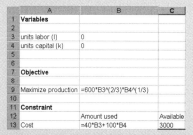

From the results above, we see that when $l = 50$ and $k = 10$, the maximum value for P is approximately 17,544.

Calculator and Excel guides are available on CourseMate for MATH APPS. Access at login.cengagebrain.com.